一流本科专业一流本科课程建设系列教材

Heat Transfer Bilingual Intensive Teaching

传热学双语精讲

主　编　吴金星
副主编　吴学红　王为术　郑慧凡　梁坤峰

机械工业出版社

本书采用双栏排版、英汉对照，以适应教育国际化发展的趋势，培养具有国际视野的专业人才。本书在内容上力求简明扼要，突出重点，精讲基本概念和基础理论，并融入了课程思政元素，同时注重理论联系实际，通过实例培养学生的工程实践能力和创新思维能力。

本书围绕热传导、热对流和热辐射三种基本传热方式，系统地阐述了稳态热传导和非稳态热传导的理论与数值解法、对流传热的理论基础与实验关联式、热辐射的基本定律和辐射传热计算，以及传热过程分析与换热器热计算方法。章节后的思考题、自测题和习题可以帮助读者加深对专业知识的理解，掌握专业理论的工程应用方法。

本书可作为能源动力类、建筑环境类、化工类、机械类、材料类、航空航天类及电气类等专业的专业基础课或专业英语课教材，也可供相关技术人员参考。

图书在版编目（CIP）数据

传热学双语精讲 ：英汉对照 / 吴金星主编. 北京 ：机械工业出版社，2024. 8. -- (一流本科专业一流本科课程建设系列教材). -- ISBN 978-7-111-76633-9

Ⅰ. TK124

中国国家版本馆 CIP 数据核字第 2024YQ9497 号

机械工业出版社（北京市百万庄大街22号　邮政编码100037）

策划编辑：尹法欣　　责任编辑：尹法欣　段晓雅

责任校对：宋学敏　　封面设计：王　旭

责任印制：张　博

北京建宏印刷有限公司印刷

2025年1月第1版第1次印刷

184mm×260mm・26.5印张・654千字

标准书号：ISBN 978-7-111-76633-9

定价：79.80 元

电话服务

客服电话：010-88361066

010-88379833

010-68326294

网络服务

机　工　官　网：www.cmpbook.com

机　工　官　博：weibo.com/cmp1952

金　　书　　网：www.golden-book.com

机工教育服务网：www.cmpedu.com

前 言

PREFACE

传热现象无处不在，无时不有。从自然界的春、夏、秋、冬四季更替，到日常生活中的衣、食、住、行各种活动，都与传热现象密切相关。工业生产和高新技术领域更是传热学应用的主场。在化工和炼油等流程性工业中，物料的加热和冷却都是热能的传递过程，其传热效率决定了生产效率和能源利用效率。在能源动力、制冷空调和建筑环境等领域的设备系统中，各种形式的换热器都是系统的关键设备。在航空航天和电子电气等高新技术领域，航天器舱内需要保持恒定温度，在大气层飞行时其外壳需要绝热良好；电子器件在运行时，需要高效散热，这些都与传热息息相关。

传热学是研究温差作用下热能传递规律的科学，是能源与动力工程、过程装备与控制工程、建筑环境与能源应用工程、新能源科学与工程、储能科学与工程等专业的重要专业基础课，也是专业核心课程。传热学课程注重理论与实践相结合，在学生能够熟练掌握热量传递的基本规律、计算方法和应用技术的基础上，培养学生分析与解决工程传热问题的能力，为后续专业课程的学习及将来从事相关专业技术工作，奠定坚实的理论与技术基础。因此，学好传热学对于学生未来的事业发展至关重要。

本书在内容上突出了以下特点：

（1）英汉对照　目前，国内外传热学教材已有许多版本，但还没有英汉对照的双语版本。为适应教育国际化发展的趋势，培养既懂专业理论，又有国际交流能力的高素质人才，提高学生的英语综合应用能力，增强学生将来的国际竞争力，教育部出台了一系列倡导高校开展双语教学的政策。为了实现上述目标，由多所高校教学经验丰富的任课老师合作编写了本书。为了便于英汉对照，让学生既能方便学习专业知识，又能准确理解相关专业英语的含义，本书采用双栏排版方式，并按照知识内容的专业层次，尽量将其划分成小段落。

（2）重点精讲　由于传热学的概念和理论较多，国内外学者在传热领域的新理论和新技术成果不断涌现，所以目前传热学教材内容普遍较多。这虽然给学生提供了大量有价值的知识和资料，但也容易令初学者难以把握学习重点，产生畏难情绪。为此本书力求简化理论推导过程和工程中应用较少的内容，精讲基本概念和基础理论，先打牢基础，再结合实际工程需求，举例说明传热学解决实际问题的一般方法和步骤，从而培养学生思考、分析和解决问题的能力，引导学生自主扩展学习。针对工程中常见的各类问题，编者对课后思考题、自测题和习题进行了精选，给学生留出思考的空间，让学生通过精准练习来掌握方法和提升能力，避免题海战术。

（3）课程思政　党的二十大报告明确指出要“全面贯彻党的教育方针，落实立德树人根本任务，培养德智体美劳全面发展的社会主义建设者和接班人。”传热学内容与生活及生产息息相关，可挖掘的思政元素很多，在专业知识讲授中开展课程思政，不仅可增强专业知识的趣味性和工程性，提升学习效果，而且有助于培养学生的社会主义核心价值观，使学生成为理想远大、信念坚定、拼搏进取的优秀专门人才，为我国的可持续发展和民族复兴做出贡献。

（4）微课视频　本书为每一小节配套了重点讲解视频，视频附有双语字幕，以二维码形式附在小节标题处，读者扫描二维码即可观看或收听本节内容讲解。读者也可登录中国大学 MOOC（https://www.icourse163.org/course/ZZU-1461084170）开展免费线上学习，并参与回答问题和线上讨论等互动活动，提高学习效果。

本书不仅适用于能源动力类、建筑环境类专业，而且同样适用于化工、机械、材料、航空航天和电气等专业。书中内容可以满足 56~64 学时的教学需求，若根据专业需求选择相关章节内容，也可满足 32~48 学时的教学需求。

本书由郑州大学吴金星任主编，并负责制订教材编写大纲和全书定稿；郑州轻工业大学吴学红、华北水利水电大学王为术、中原工学院郑慧凡、河南科技大学梁坤峰任副主编；郑州大学李松歌、肖嘉邦、徐耀、马宇翔参与了翻译、校对工作。此外，郑州大学赵金辉、王丹、张东伟、李哲、陈雅博、王永庆等为本书编写提供了帮助，王定标、曹海亮、王军雷、涂维峰审阅了书稿并提出了中肯的修改意见，在此对他们表示感谢。

本书的编写得到了郑州大学本科重点教材建设项目的资助，也是国家级一流本科专业、河南省一流本科专业、河南省一流本科课程、河南省本科高等学校精品在线开放课程、高等学校能源动力类教学研究与实践项目、郑州大学一流本科重点课程、郑州大学课程思政教育教学改革示范课程等项目的建设成果。

在本书编写过程中，编者参考了很多专家学者的相关教材、著作和论文，在此一并致谢！

限于编者水平，书中难免有不妥之处，敬请广大读者批评指正。

编　者

目　录
CONTENTS

Chapter 1 The Basic Modes of Heat Transfer

第 1 章 热量传递的基本方式

授课视频

1.1 Importance of Heat Transfer

Heat transfer is an important professional basic course for energy and power engineering, process equipment and control engineering, building environment and energy application engineering, etc. It is also an important elementary theoretical course for chemical engineering, machinery, materials, metallurgy, electrical and other majors, which plays an important supporting role in learning the subsequent professional courses.

First, let's learn about the importance of heat transfer. Heat is the energy of a large number of irregularly moving microscopic particles inside matter, and heat is also called thermal energy or heat energy. Therefore, heat or thermal energy is related to the motion state of microscopic particles. The more intense the motion of microscopic particles, the more heat the matter has, and the higher the temperature in the macro view.

According to statistics, thermal energy is the most used form of energy, and 85% to 90% of other forms of energy must be converted into thermal energy and then be used. Therefore, in the nature, daily life, industrial production process and high-tech fields, heat transfer exists everywhere.

In nature, with the alternation of spring, summer, autumn, and winter, the temperature has changed. People will feel the warm spring, the scorching summer, the refreshing

1.1 传热学的重要性

传热学是能源与动力工程、过程装备与控制工程、建筑环境与能源应用工程等专业的重要专业基础课。它也是化工、机械、材料、冶金、电气等专业的重要理论基础课，对学习后续的专业课程具有重要的支撑作用。

首先认识一下传热学的重要性。热是物质内部大量微观粒子做无规则运动所具有的能量，也称为热能/热量。由此可见，热或者热能与微观粒子的运动状态有关。微观粒子运动越剧烈，则物质所具有的热量就越多，宏观上表现为温度越高。

据统计，热能是人们利用最多的能量形式，其他形式的能量 85%~90%都要转换成热能，然后再加以利用。因此，在自然界、日常生活、工业生产过程中，以及高新技术领域，到处都存在传热现象。

在自然界中，伴随着春夏秋冬四季更替，气温发生了变化。人们会感受到春意融融、夏日炎炎、秋高气爽

autumn and the cold winter. As day and night change within a day, the ambient temperature is constantly increasing or decreasing. Heat transfer occurs all the time between objects and the environment. Therefore, for the common thermal equipment (e. g. boilers, heat exchangers) and steam pipelines, in order to reduce heat loss, a layer of thermal insulation must be added to their outer surface, as shown in Fig. 1. 1.

和寒风凛冽。在一天之内随着昼夜变换，环境温度也在不停地升高或降低。可见，物体与环境之间时时刻刻都在发生传热现象。因此，常用的热设备（如锅炉、换热器）及蒸汽管道，为了减少热量损失，其外表面都要加一层保温层，如图 1. 1 所示。

Fig. 1. 1 External insulation layer of boilers and steam pipelines
图 1. 1 锅炉及蒸汽管道外加保温层

In our daily life, we need to heat food when we cook, and wear cotton-padded clothes when it's cold, and use radiators or air conditioners to keep the house warm. These phenomena are closely related to heat transfer. For example, when boiling water, aluminum kettles will burn easily once the water boils dry, while iron ones will not. Because aluminum has a lower melting point. But why do we often use aluminum kettles? This is mainly because the thermal conductivity of aluminum is greater than that of iron.

在日常生活中，做饭时要加热饭菜，天冷了要穿棉衣，室内要通过暖气片或空调来取暖。这些现象都与热量传递密切相关。再比如烧水时，一旦水烧干了，铝壶就很容易被烧毁，而铁壶不容易被烧毁。因为铝的熔点较低。但是为什么我们还常用铝壶？这主要因为铝的导热系数大于铁的导热系数。

The thermal conductivity is a very important concept in heat transfer, which reflects the heat transfer rate of material. Therefore, it is faster to boil water in aluminum kettles, which is conducive to energy saving. In addition, the handle of the kettle is always covered with a layer of rubber, which has a good thermal insulation effect. Because the thermal conductivity of rubber is relatively small.

导热系数这个概念在传热学中非常重要，它反映了材料传递热量的快慢。所以用铝壶烧水比较快，有利于节约能源。另外，水壶的提手上都会包上一层橡胶，这一层橡胶具有很好的隔热作用。因为橡胶的导热系数比较小。

There are also thermos cups, thermos bottles and the double glazing widely used in northern buildings in daily

在生活中还有保温杯、暖水瓶，以及北方建筑上广泛采用的双层玻璃，

life, all of which are the double vacuum structure. This structure increases the resistance of heat transfer process and is more conducive to heat insulation.

都属于双层真空结构。这种结构增大了传热过程的阻力，更有利于隔热保温。

Question In winter, a wooden rod and an iron rod are placed outdoors for a long time, does it feel the same when you touch them with your hands? Why? We should all have the experiences of feeling that the iron rod is colder, which means its temperature is lower. But in fact, this is an illusion. The wooden rod and iron rod must have reached the thermal equilibrium with the environment after being placed outdoors for a long time, that is to say, their temperature are same, but why do you feel their temperatures are different?

思考题 在冬天室外长时间放置一根木棒和一根铁棒，用手摸上去感觉会一样吗？为什么？人们应该都有这样的经验，感觉铁棒会更凉些，即温度更低。但事实上这是一个错觉。木棒和铁棒在室外长时间放置，肯定与环境达到了热平衡，即温度相同，但为什么感觉温度不同呢？

Because the thermal conductivity of the iron rod is much greater than that of the wooden rod, the iron rod transfers heat faster and takes away more heat in the same time. The human body's perception of cold and warmth is determined by the heat dissipating capacity, not by the temperature. The human body feels cold when it dissipates too much heat, on the contrary, the body feels hot when dissipates less heat.

这是因为铁棒的导热系数远远大于木棒，所以铁棒传递热量更快，同样时间内带走的热量更多。人体对冷暖感觉的衡量指标是散热量多少，而不是温度的高低。当人体散热量多时就会感觉冷，反之散热量少时就会感觉热。

Energy power, refrigeration and air conditioning, building environment and other systems are the home courts of heat transfer application. In the energy power systems, boiler, economizer, air preheater, steam condenser, cold water tower and other key equipment are all heat transfer equipment. In refrigeration air conditioning and heat pump systems, the key equipment, evaporator and condenser (generally known as two devices), are heat transfer equipment. The heat transfer efficiency of these heat transfer equipment determines the total energy efficiency of the system.

能源动力、制冷空调、建筑环境等系统更是传热学应用的主场。在能源动力系统中，锅炉、省煤器、空气预热器、凝汽器、凉水塔等关键设备都是热量传递设备。在制冷空调及热泵系统中，其关键设备蒸发器、冷凝器（俗称“两器”）都属于传热设备。这些传热设备的传热效率决定了系统能源利用的总效率。

In chemical engineering, oil refining, pharmaceutical and other industrial production systems, materials usually need to be heated or cooled, which are heat transfer processes. The investment of heat transfer equipment accounts

在化工、炼油、制药等工业生产系统中，物料常常需要加热或冷却，这些都是传热过程。这些生产系统中传热设备的投资占系统总投资的

for 30%-40% of the total investment in these production systems and 50%-60% by weight. Therefore, heat transfer equipment is very important.

30%~40%，按重量计算占 50%~60%。因此，传热设备的重要性可见一斑。

In the high-tech fields such as aerospace, electronics and electricity, the internal environment of space shuttle needs to maintain a constant temperature, and the surface material requires good thermal insulation. The electronic devices often need to strengthen heat dissipation in operation. Therefore, heat transfer technology is very important to ensure the stable performance of these devices.

在航空航天和电子电气等高新技术领域，航天飞机的内部环境需要保持恒定温度，表面材料又要求绝热良好。对于电子器件，在运行中常常需要加强散热。所以传热技术对于保证这些设备的性能稳定至关重要。

In history there was once a major space accident caused by a heat transfer problem. On February 1, 2003, the American space shuttle Columbia crashed on its return trip. An investigation later found that the cause of the accident was damage to a heat shield tile on the left wing of the plane, which caused high temperature gas of thousands of degrees Celsius to enter the cabin, resulting in the destruction of the aircraft and human death.

在历史上曾经发生过一次由传热问题导致的重大航天事故。2003 年 2 月 1 日，美国哥伦比亚号航天飞机在返回途中坠毁了。事后经过调查发现，事故的原因是飞机左翼上一片隔热瓦损伤了，导致上千摄氏度的高温气体进入舱体，从而造成机毁人亡。

The aerospace accident warns us that as future engineers, we must improve our sense of social responsibility, be meticulous and serious in our work, and let the concept of "life first, safety first" be internalized in our heart and externalized in our practice.

航天事故警示我们：作为未来的工程师，我们必须要提高自己的社会责任感，工作中要做到一丝不苟，严肃认真，让"生命至上、安全第一"的理念内化于心、外化于行。

All the above examples illustrate a fact: heat transfer is significantly important in all walks of life. Therefore, heat transfer, engineering thermodynamics and fluid mechanics are the three core courses for energy and power majors, which lay a solid theoretical foundation for the development of professional technology. Therefore, learning heat transfer well can not only promote the rapid development of professional technology and equipment, but also certainly make greater contributions for improving China's energy utilization efficiency, and energy conservation and emission reduction.

上述种种事例都说明了一个事实：传热学在各行各业中都非常重要。因此，传热学、工程热力学、流体力学并称能源动力类专业的三大核心课程，为专业技术发展奠定了坚实的理论基础。因此，学好传热学不仅能够促进专业技术和设备的快速发展，而且对提高我国的能源利用效率和节能减排必将会做出更大的贡献。

Summary This section illustrates that the phenomenon

小结 本节阐述了传热现象无处不

of heat transfer is everywhere and at all times, and has great influence on our work and life.

在、无时不有，对我们的工作和生活影响巨大。

Question Which heat transfer phenomena illustrate the importance of heat transfer in nature, daily life, industrial production and high-tech fields?

思考题 在自然界、日常生活、工业生产及高新技术领域中，哪些传热现象说明了传热学的重要性？

Self-tests

自测题

1. Why should a layer of thermal insulation be added to the outer surface of boilers, heat exchangers and steam pipelines?

2. What are the key heat transfer devices in the energy power system?

1. 为什么锅炉、换热器及蒸汽管道等外表面都要加一层保温层？

2. 在能源动力系统中有哪些关键的热量传递设备？

授课视频

1.2 Research Content and Methods of Heat Transfer

1.2 传热学的研究内容及方法

Heat transfer is a subject which studies the laws of heat transfer, that is, the relationship between the heat transfer quantity and the factors such as temperature difference, time, and physical properties. The specific research contents include principle or law of heat transfer, heat transfer mechanism, calculation methods and test techniques, and so on.

传热学是研究热能（热量）传递规律的学科，即热量传递的多少与温差、时间、物性等因素之间的关系。具体研究内容包括热量传递原理或规律、传热机理、计算方法和测试技术等。

The principle or law of heat transfer is summed up by observation and test, so it is a macroscopic problem. The heat transfer mechanism is a reasonable speculation based on the heat transfer phenomena and laws. It is the physical mechanism inside of matter, so it is a microscopic problem. Calculation methods and test techniques are carried out according to principles or formulas and focus on practical applications.

传热原理或规律是通过观察和测试而总结出来的，所以是一个宏观的问题。传热机理是基于传热现象、规律而进行的合理推测，是物质内部的物理机制，因而是一个微观的问题。计算方法和测试技术是根据原理或公式而进行的，侧重于实际应用。

Why does heat transfer occur? This is determined by the second law of thermodynamics. The second law of thermodynamics tells us: heat can be spontaneously transferred from high temperature heat source to low temperature heat source. As long as a high temperature heat source and a low temperature heat source exist, heat transfer will occur. It can also be said that temperature difference is the driving force of heat transfer.

为什么会发生热量传递呢？这是由热力学第二定律所决定的。热力学第二定律告诉我们：热量可以自发地由高温热源传给低温热源。只要有一个高温热源，一个低温热源，就会发生热量传递，也可以说，温差是热量传递的推动力。

In the field of engineering and technology, there are two main types of heat transfer problems: ① Pay attention to the heat transfer rate and its control. The rate of heat transfer is how fast or slow heat transfer. In engineering practice, some heat transfer processes need to be enhanced, that is, to increase the heat transfer rate; while some heat transfer processes need to be weakened such as heat preservation of equipment to reduce heat loss. ② Pay attention to the temperature distribution and its control in the research object. Because the temperature distribution in the object affects not only the heat transfer rate, but also the heat transfer effect.

在工程技术领域中，传热问题主要有两大类：① 关注传热速率大小及其控制，传热速率就是传热快慢。在工程实际中，有的传热过程需要强化传热，即提高传热速率；而有的传热过程需要弱化传热，比如对设备进行保温，减少热量散失。② 关注研究对象内温度分布及其控制。因为物体内的温度分布既会影响传热速率，也会影响传热效果。

In the process of learning, only when we have mastered the basic law of heat transfer, have the ability to analyze heat transfer problems, and be able to skillfully apply heat transfer calculation methods and testing technology, can we solve these two kinds of engineering heat transfer problems.

在学习过程中，只有掌握了热量传递的基本规律，具有分析传热问题的能力，并能够熟练应用传热计算方法和测试技术，才能很好地解决这两类工程传热问题。

In engineering, according to the relationship between temperature and time, the heat transfer process can be divided into two categories: unsteady-state heat transfer process and steady-state heat transfer process. The characteristic of unsteady-state heat transfer process is that the temperature of each point changes continuously with time. That is, the temperature distribution is related to time. For example, thermal equipment like boilers, heat exchangers, the temperature of each point increases continuously when starting, and decreases continuously when stopping; Therefore, the processes of start-up and shutdown belong to

在工程中，按照温度随时间的变化关系，传热过程可分为非稳态传热过程和稳态传热过程两类。非稳态传热过程的特征是：各点的温度随着时间不断地变化，即温度分布与时间有关。比如，热力设备像锅炉、换热器等，起动时各点的温度不断升高；停机时温度不断降低；因此，起动和停机的过程都属于非稳态传热过程。

unsteady-state heat transfer process.

The characteristic of steady-state heat transfer process is that temperature at each point does not change with time. That is, temperature distribution is independent of time. For example, in the continuous normal operation of thermal equipment, although the temperature varies from point to point. For example, a fluid with a high temperature at the inlet and a low temperature at the outlet, but the temperature at each point does not change with time. Most of the heat transfer problems we will encounter in the future, in addition to the obvious start-up, shutdown and other unstable heat transfer processes, are generally treated as steady-state heat transfer process.

稳态传热过程的特征是：各点的温度不随时间而变化，即温度分布与时间无关。比如，尽管热力设备在连续正常运行时各点温度有高有低，一种流体的进口处温度高、出口处温度低，但是每个点的温度都不随时间变化。在今后遇到的大多数传热问题，除了明显的起动、停机等非稳态传热过程以外，一般都按照“稳态传热过程”来处理。

Heat transfer is an applied basic subject closely combined with engineering practice. There are four commonly used research methods: theoretical analysis method, numerical method, experimental method and analogy method.

传热学是一门与工程实际结合紧密的应用性基础学科。常用的研究方法有四种：理论分析法、数值解法、实验法和比拟法。

1. Theoretical analysis method

1. 理论分析法

The method generally includes the following three steps:

理论分析法一般包括以下三个步骤：

① Make reasonable simplifications and assumptions on actual physical problems to obtain a geometric model and a physical model; ② Establish a mathematical description of the physical problem to obtain a mathematical model; ③ Use the method of theoretical deduction (i. e. Advanced Mathematics) to obtain the theoretical solution/analytical solution of the problem. Overall, the theoretical analysis method is more rigorous, and the results and conclusions are more accurate.

① 对实际物理问题进行合理的简化与假设，得到几何模型和物理模型；② 对所研究的物理问题进行数学描述，从而得到数学模型；③ 用理论推导（即高等数学）的方法来获得问题的理论解/分析解。总体来看，理论分析法思路比较严谨，得到的结果及结论比较精确。

Here, the geometric model refers to the geometry and size of the research object, that is, what the research object looks like; The physical model includes the physical parameters of the material, whether there is an internal heat source, and whether it is a steady-state process, etc. The physical characteristics are relatively abstract. Mathematical

这里，几何模型是指研究对象的几何形状和尺寸，即研究对象是什么模样；物理模型的内容包括材料的物性参数、是否有内热源、是否为稳态过程等，物理特性相对来说比较抽象。数学模型是指所研究物理

model refers to the governing equation (group) of the physical problem.

问题的控制方程（组）。

2. Numerical solution method (also called semi theoretical method)

2. 数值解法（也叫作半理论法）

Why is it called a semi-theoretical method? Because the first two steps are exactly the same as the theoretical method.

为什么称为半理论法呢？因为其前两步与理论分析法完全相同。

① Make reasonable simplifications and assumptions on actual physical problems to obtain a geometric model and a physical model. ② Establish a mathematical description of the physical problem to obtain a mathematical model; Because the geometric or mathematical models of actual physical problems are more complicated and cannot be solved by theoretical methods. So, we can only find a new way, using computers and software to solve them by numerical methods.

① 对实际物理问题进行合理的简化与假设，得到几何模型和物理模型；② 对所研究的物理问题进行数学描述，从而得到数学模型。因为实际物理问题的几何模型或数学模型比较复杂，理论方法解不了，所以只能另辟蹊径，利用计算机和软件采用数值方法求解。

③ Discretization of calculation area and equation; The so-called computational domain discretization is to artificially divide the continuum into many discrete small regions. Equation discretization is to convert differential equation into algebraic equation for easy solution. ④ Conduct computer programming or use commercial software to obtain numerical results through calculation. Generally speaking, there are calculation errors in numerical results, which need to be verified or corrected by experimental and theoretical methods.

③ 进行计算区域的离散和方程的离散。所谓计算区域离散，就是人为地把连续体化分成许多离散的小区域；方程离散就是把微分方程转换成代数方程，以便于求解。④ 进行计算机编程或应用商业软件，通过计算获得数值结果。一般来讲，数值结果都存在计算误差，需要用实验法和理论分析法来验证或修正。

Academician Tao Wenquan of Xi'an Jiaotong University is a well-known expert on numerical heat transfer in China. He has long been engaged in the research and engineering applications of numerical simulation methods of heat transfer, trained a large number of numerical calculation experts, and made outstanding contributions to the formation and development of computational heat transfer disciplines in China. Nowadays, he is over the age of rare, but he is still guiding postgraduate students and giving classes to undergraduates.

西安交通大学教授陶文铨院士是我国在数值传热学方面的知名专家。他长期从事传热学数值模拟方法及其工程应用的研究，培养了大批数值计算专家，为我国计算传热学科的形成与发展做出了杰出贡献。如今他已年过古稀，但依然在指导研究生、给本科生上课。

Academician Tao said that the "Westward Relocation Spirit" with the overall situation in mind, selfless dedication, carrying forward traditions, and arduous pioneering is the spiritual driving force that leads him to engage in academic research. This "Westward Relocation Spirit" has also been inspiring every one of us to enhance our sense of national pride and determination to serve the country through science and technology, establish lofty ideals and goals in life, and actively devote ourselves to the construction of the motherland.

陶院士说过，胸怀大局、无私奉献、弘扬传统、艰苦创业的“西迁精神”是引领他做学术研究的精神动力。这种“西迁精神”同样也一直激励着我们每一个人，增强民族自豪感和科技报国的决心，树立远大的人生理想和目标，积极投身于祖国建设。

3. Experimental method (this is an empirical method)

3. 实验法（是一种具有经验性质的方法）

Under the guidance of the similarity principle, an experimental platform is built, and the empirical correlation is fitted out through the experimental test. The results and conclusions obtained from the experiments can be widely used in engineering practice. Disadvantage of the method is that it takes a lot of manpower, material and financial resources to carry out experiments.

在相似原理的指导下，搭建实验台，通过实验测试拟合出经验关联式，从实验中得出的结果和结论可普遍应用于工程实际中。这种方法的缺点是实验需付出大量的人力、物力和财力。

4. Analogy method

4. 比拟法

The two physical phenomena are similar in mechanism, so they should be similar in mathematical description, so the solution of unknown phenomenon can be deduced from the solution of known phenomenon. For example, the mechanism of heat conduction and electric conduction is similar (both by the movement of free electrons), so heat conduction and electric conduction can be analogized.

两种物理现象在机理上具有相似性，因而在数学描述上也应该有相似，从而可以由已知现象的解推出未知现象的解。比如，导热和导电的机理相似（靠自由电子运动），所以导热与导电可以比拟。

Heat transfer and engineering thermodynamics are the core courses of energy and power majors. They are both different and closely related.

传热学与工程热力学这两门课都是能源动力类专业的核心课程。它们既有区别，又有密切的联系。

(1) Distinction. Thermodynamics deals with how much heat is transferred from one equilibrium to another. Therefore, the related parameters of thermodynamics such as heat transfer quantity and temperature have nothing to do with time;

（1）区别。热力学研究物体从一个“平衡态”到另一个“平衡态”传递热量多少。所以，热力学的相关参数（如传热量、温度等）都与

Heat transfer focus on the process of heat transfer, that is the rate of heat transfcr. In thc process of heat transfer, the relevant physical quantities such as heat transfer rate and temperature are functions of time. From the research contents, the two courses are complementary.

时间无关。传热学关注的是热量传递的过程，即热量传递的速率。在传热过程中，相关物理量（如热流量、温度等）都是时间的函数。从研究内容上来说，两门课程是互补的。

(2) Connection. Heat transfer is based on the first and second laws of thermodynamics, that is, heat is always conserved during the transfer process, and satisfies the first law of thermodynamics—the law of conservation of energy. And the heat is always transferred from the hot reservoir to the cold reservoir, that's in accordance with the second law of thermodynamics.

(2) 联系。传热学以热力学第一定律和第二定律为基础，即热量在传递过程中始终是守恒的，满足热力学第一定律——能量守恒定律；并且，热量始终从高温热源向低温热源传递，这符合热力学第二定律。

Summary　This section illustrates the research contents of heat transfer, classification of heat transfer problem and heat transfer process, and four research methods of heat transfer. Through the deeds of Academician Tao Wenquan, we can feel the heart of the older generation of educators to repay the country, and through the "Westward Relocation Spirit", we have also greatly stimulated our patriotic feelings.

小结　本节阐述了传热学的研究内容、传热问题和传热过程的分类，以及传热学的四种研究方法。通过陶文铨院士的事迹，我们可以感受到老一辈教育工作者的报国之心，通过“西迁精神”也极大地激发了我们的爱国情怀。

Question　Is steady-state heat transfer equivalent to thermal equilibrium?

思考题　稳态传热是否等价于热平衡态？

Self-tests

自测题

1. What are the research contents of heat transfer, two kinds of heat transfer problems and two heat transfer processes?

1. 传热学的研究内容、两类传热问题、两个传热过程是什么？

2. What are the four research methods and learning methods of heat transfer?

2. 传热学有哪四种研究方法？以及哪些学习方法？

1.3 Heat Conduction and Its Laws

1.3 热传导及其规律

The heat transfer phenomenon in nature is very complex. According to the heat transfer mechanism, heat transfer includes three basic modes: heat conduction, heat convection and heat radiation, as shown in Fig. 1.2.

在自然界中，传热现象非常复杂。按照传热机理，热能传递有三种基本方式：热传导（导热）、热对流和热辐射，如图 1.2 所示。

Fig. 1.2 Three basic modes of heat transfer

图 1.2 热能传递的三种基本方式

The definition of heat conduction: the process of heat transfer within an object by the thermal motion of microscopic particles, when two objects at different temperatures come into contact with each other, or the parts of an object have different temperatures and there is no relative displacement.

热传导的定义：温度不同的两个物体相互接触，或者一个物体的各部分之间温度不同，并且不发生相对位移时，在物体内部依靠微观粒子热运动而进行的热能传递过程。

The definition of heat conduction includes two working conditions: ① Two objects at different temperatures have to come into contact and become one; ② The parts of an object have different temperatures and cannot move. In this case, the object can only transfer heat by the internal microscopic particle motion. The microscopic particles mentioned here include molecules, atoms and free electrons.

热传导的定义中包含了两种工况：① 两个温度不同的物体要接触，合二为一；② 一个物体的各部分之间温度不同，并且不能移动。在这种情况下，物体只能依靠内部的微观粒子运动进行热能传递。这里所说的微观粒子包括分子、原子和自由电子。

Heat conduction has the following five characteristics: ① There must be a temperature difference. Temperature difference is the driving force of heat transfer, so this is a

热传导具有以下五个特点：① 必须有温差。温差是热能传递的推动力，所以这是个基本条件。② 两个

basic condition. ② The two objects have to contact directly. ③ There is no relative macroscopic movement between objects. The above two conditions belong to external conditions. Objects must touch to each other without movement. ④ Heat transfers by the thermal motion of microscopic particles. This is the mechanism of heat conduction, and the most fundamental difference from other basic heat transfer modes. ⑤ Pure heat conduction in gravitational field occurs in dense solids and fluids with higher upper temperatures. Such as metal, plastic, glass and so on, are dense solids.

物体之间必须直接接触。③ 物体间不发生相对宏观运动。以上两个条件属于外部条件，物体间既要接触，又不能移动。④ 靠微观粒子热运动而传递热量。这是导热的机理，也是与其他基本传热方式最根本的区别。⑤ 在引力场下单纯的导热发生在密实固体和上部温度较高的流体中。如金属、塑料、玻璃等，都属于密实固体。

If the medium in a vessel is not solid but fluid in gravity field, when the lower part of the fluid is heated and expanded, the density goes down, the hot fluid goes up, and the cold fluid drops at the same time, the macroscopic motion will occur between the mediums. This heat transfer process is not heat conduction, but another basic heat transfer mode—heat convection. When the no-gravitational field or the upper part of the fluid is heated, only pure heat conduction occurs inside the fluid.

如果在重力场的一个容器中，里面的介质不是固体而是流体，当流体的下部被加热时，流体受热膨胀，密度就会减小，热流体上升，同时冷流体下降，介质之间就产生了宏观运动。这个传热过程就不属于导热，而是另外一种基本传热方式——热对流。当无引力场或流体的上部被加热时，流体内只会发生纯导热。

Next let's go to learn about heat conduction mechanism of matter. Different kinds of matter are made up of different microscopic particles and their heat conduction mechanisms are different.

接下来认识一下物质的导热机理。构成物质的微观粒子不同，其导热机理也不一样。

For nonmetallic bodies, they are made up of atoms, chains of atoms (also called a lattice) are formed in a physical way between adjacent atoms. When some of the atoms are heated, the vibration of atoms is accelerated and the vibration of atomic chains is also accelerated, resulting in elastic waves, so heat is transferred to the adjacent atoms. This is the heat conduction mechanism of nonmetallic bodies.

对于非金属体，它们是由原子组成的，相邻原子之间以物理方式形成原子链（也称为晶格）。当一部分原子受热时，原子的振动加速，原子链的振动也随之加速，从而产生弹性波，这样就把热量传递到相邻原子。这就是非金属体的导热机理。

For metal bodies, they are also made up of atoms and there are a large number of free electrons outside the nucleus. When a metal body is heated, there is not only the vibration of atoms, but also the irregular motion of a large number of free electrons. Moreover, the contribution of free electron motion to heat transfer is greater. This is the heat conduction mechanism of metal bodies. Therefore, all metals are good heat conductors and good electric conductors.

对于金属体，它们也是由原子组成的，同时原子核外存在大量自由电子。当金属体受热时，不仅有原子的振动，还有大量自由电子的无规则运动，并且自由电子运动对传热的贡献更大。这就是金属体的导热机理。因此，所有金属都是良好的导热体，同时也是良好的导电体。

For gases, they are composed of molecules. With the increase of temperature, the motion of molecules is accelerated, and heat is transferred by the collision between molecules. This is the heat conduction mechanism of gases.

对于气体，它们是由分子组成的。随着温度升高，分子的运动速度加快，依靠分子间的相互碰撞来传递热量。这就是气体的导热机理。

Fig. 1.3 shows a schematic diagram of the one-dimensional heat conduction of a flat plate, where δ denotes the thickness of the flat plate, the surface area of both sides is A, and their temperature is t_{w1} and t_{w2} respectively. In steady state, how to calculate the heat transferred through the flat plate?

图 1.3 所示为平板的一维导热示意图，图中平板的厚度为 δ，两侧面的面积均为 A，温度分别为 t_{w1} 和 t_{w2}。在稳态情况下，平板传递的热量怎么计算呢？

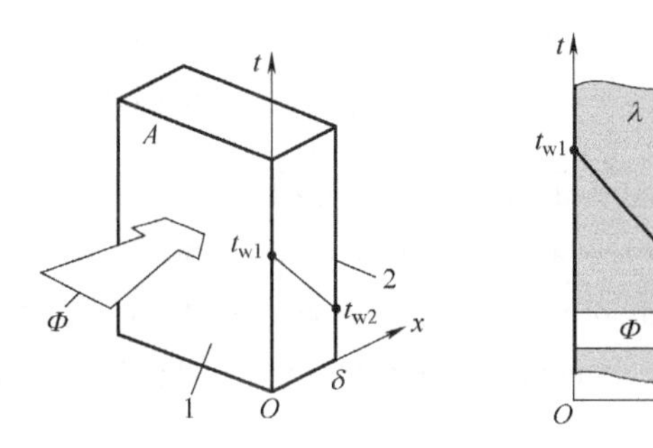

Fig. 1.3 One-dimensional heat conduction of a flat plate

图 1.3 平板的一维导热示意图

As early as 1822 French physicist Fourier discovered the basic law of heat conduction, so we call it Fourier's law: heat conduction quantity through the flat plate per unit time is proportional to the local temperature change rate and the flat plate area A. Calculation formula is:

早在 1822 年，法国物理学家傅里叶就发现了导热的基本规律，称为傅里叶定律：单位时间内通过平板的导热量，与当地的温度变化率及平板面积 A 成正比，计算公式为

$$\Phi=-\lambda A\frac{\mathrm{d}t}{\mathrm{d}x} \tag{1.1}$$

Both sides of the equation are divided by area A to get:

等式的两边同除以面积 A，从而得到

$$q=\frac{\Phi}{A}=-\lambda\frac{\mathrm{d}t}{\mathrm{d}x} \tag{1.2}$$

These are the two expressions of Fourier's law, which are completely identical. The negative sign indicates that the direction of heat flux is opposite to the direction of temperature change rate. Heat flux always transfers from high temperature region to low temperature region, but the direction

这是傅里叶定律的两种表达形式，完全等价。式中负号表示热流方向与温度变化率方向相反。热流总是从高温区传向低温区，而温度变化率的方向是从低温区指向高温区。

of temperature change rate is from low temperature region to high temperature region. They are in opposite directions, so a negative sign should be added.

两者方向相反，所以式中要加负号。

The physical meaning of other symbols is as follows: Φ is heat transfer rate: the heat transfer quantity per unit time (W=J/s); q is heat flux: the heat transfer quantity per unit area per unit time (W/m^2); A is heat conduction area: surface/cross-sectional area perpendicular to the heat flux direction (m^2); λ is thermal conductivity [W/(m · K) or W/(m · ℃)]. The temperature in unit can be absolute temperature or degrees Celsius.

式中，其他符号的物理意义：Φ 是热流量，指单位时间传递的热量（W=J/s）；q 是热流密度，指单位时间通过单位面积传递的热量（W/m^2）；A 是导热面积，指垂直于热流方向的表面/截面面积（m^2）；λ 是导热系数 [W/(m · K) 或 W/(m · ℃)]。单位中的温度可以用热力学温度，也可以用摄氏度。

The following focuses on thermal conductivity. Thermal conductivity characterizes the heat transmission capacity of material and is an inherent property of material. It is also a kind of thermophysical parameter, mainly related to the type of material and temperature.

下面重点讨论一下导热系数（热导率）。导热系数表征材料导热能力大小，是材料的固有属性，也是一种热物性参数，主要与材料种类和温度有关。

The thermal conductivity of different materials varies greatly, as shown in Fig. 1. 4. From the figure, we can clearly see the relative magnitude of thermal conductivity of several materials. To sum up, the order of thermal conductivity of materials is as follows:

不同材料的导热系数差别很大，如图 1.4 所示。从图中可以明显看出几种材料导热系数的相对大小。归纳起来，材料的导热系数大小顺序是：

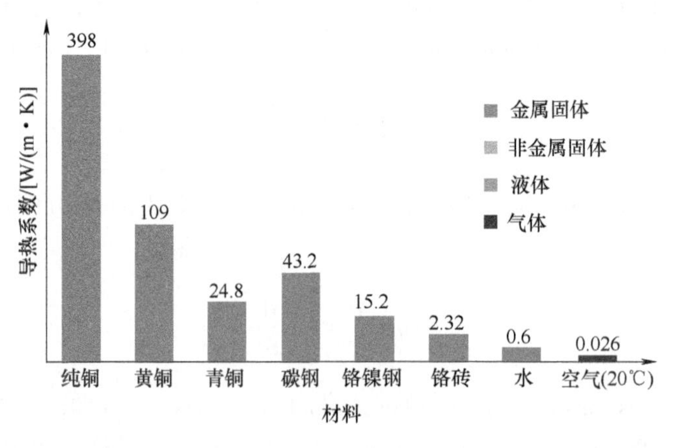

Fig. 1. 4 The thermal conductivity of different materials

图 1.4 不同材料的导热系数

$$\lambda_{\text{金属(metallic material)}} > \lambda_{\text{非金属(non metallic material)}} > \lambda_{\text{液体(liquid)}} > \lambda_{\text{气体(gas)}}$$

Fourier's law reveals the general law of heat conduction. For a specific problem of one-dimensional steady-state heat conduction, how to calculate?

傅里叶定律揭示了导热的一般规律。对于一个具体的一维稳态导热问题，如何计算？

Fig. 1.3 shows a flat plate with thermal conductivity of λ and a thickness of δ, the area of both sides is A, and the temperature is t_{w1} and t_{w2} respectively, which means that the temperature on each side is uniform, so the temperature difference between the two sides keeps constant. In steady state, calculate the heat conduction quantity of the flat plate. Obviously, this is a one-dimensional heat conduction problem, and the heat is transferred only along the direction of plate thickness.

图 1.3 所示为一块导热系数为 λ 的平板，厚度为 δ，两侧面的面积均为 A，温度分别为 t_{w1} 和 t_{w2}，即每个侧面上温度是均匀的，因此两侧面之间的温差也保持不变。在稳态情况下，计算平板的导热量。显然这是一维导热问题，热量只沿着板厚方向传递。

When the thermal conductivity, heat transfer area and temperature difference of the material are constant, according to Fourier's law, the heat transfer quantity between the two sides is a constant. Thus, the heat transfer quantity per unit area, i. e. heat flux q, is also a constant. Based on the above analyses and Fourier formula:

在材料的导热系数、传热面积和温差都不变的情况下，根据傅里叶定律，两侧面之间传递的热量是一个常数。那么，单位面积上传递的热量即热流密度 q 也是一个常数。根据上述分析和傅里叶公式：

$$q = -\lambda \frac{\mathrm{d}t}{\mathrm{d}x}$$

We separate the variables of this formula, then integrate the thickness coordinates x and temperature coordinates t respectively:

对这个式子进行分离变量，然后分别对厚度坐标 x 和温度坐标 t 积分：

$$q \int_0^{\delta} \mathrm{d}x = -\lambda \int_{t_{w1}}^{t_{w2}} \mathrm{d}t$$

The algebraic equations can be obtained:

可得到代数方程：

$$q = \lambda \frac{t_{w1} - t_{w2}}{\delta} \tag{1.3}$$

Fourier formula is a differential equation, which is suitable for various heat conduction processes. For one-dimensional steady-state heat conduction, the formula of heat flux becomes an algebraic equation. It is more convenient to calculate the steady-state heat conduction by using this formula.

傅里叶公式是一个微分方程式，适用于各种导热过程。而对于一维稳态导热问题，热流密度的计算公式变成了一个代数方程式。用这个公式计算稳态导热问题会更加简便。

In fact, this formula was not derived from Fourier formula, but was first put forward by French physicist Biot in 1804, so it can be called Biot formula. Next, we will make a further analysis of Biot formula. The mechanism of electrical conduction and heat conduction is similar, so they can be analogized, but how? The law of conducting electricity is Ohm's law:

事实上，这个计算式并不是由傅里叶公式导出的，而是法国物理学家毕渥在 1804 年首先提出的，因此可称为毕渥公式。接下来对毕渥公式做进一步的分析。导电与导热机理相似，因而它们之间可以比拟。怎么比拟呢？导电遵循的规律是欧姆定律：

$$I = U/R \tag{1.4}$$

Let's change the mathematical form of Biot formula. The numerator and denominator on the right side of the equation divided by the thermal conductivity λ and Eq. (1.3) becomes:

把毕渥公式做一个数学形式上的变化。等式右侧的分子和分母同除以导热系数 λ，式（1.3）变为：

$$q = \frac{t_{w1} - t_{w2}}{\delta/\lambda} = \frac{\Delta t}{r_\lambda} \tag{1.5}$$

Expressed by heat transfer rate:

用热流量表示：

$$\Phi = \frac{t_{w1} - t_{w2}}{\delta/(A\lambda)} = \frac{\Delta t}{R_\lambda} \tag{1.6}$$

In these two formulas, what is the physical meaning of r_λ and R_λ? The analogy with Ohm's law shows that r_λ and R_λ are equivalent to the resistance of the thermal conduction process, so both r_λ and R_λ can be called thermal conductivity resistance. There is no area in r_λ which is also called area thermal resistance. The thermal circuit diagram of the heat conduction process is shown in Fig. 1. 5.

这两个式子中，r_λ、R_λ 有什么物理意义呢？通过与欧姆定律表达式的对比发现，r_λ、R_λ 相当于导热过程的阻力，因此可以把 r_λ、R_λ 都称为导热热阻。r_λ 中不含面积，又称为面积热阻。导热过程的热路图如图 1. 5 所示。

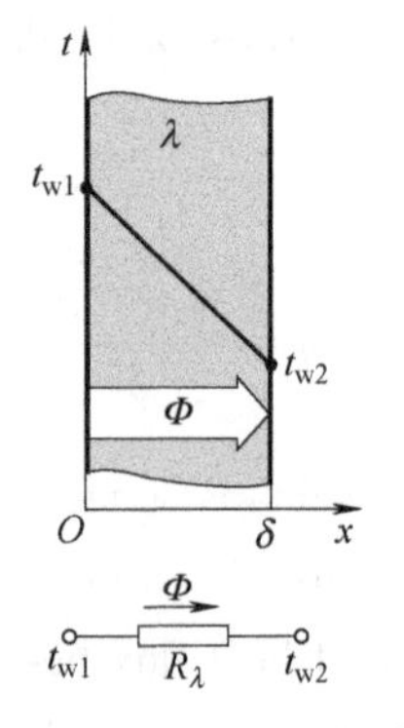

Fig. 1. 5 The thermal circuit diagram of the heat conduction process

图 1. 5 导热过程的热路图

The above is a simple mathematical deformation of Biot formula, which makes it easier to understand its physical meaning. The heat flux or heat transfer rate is proportional to the temperature difference, but inversely proportional to the thermal conductivity resistance. This analysis method of heat transfer process is called thermal resistance analysis method. In addition, it should be noted that the thermal resistance analysis method is only applicable for steady-state heat transfer process, and there is no internal heat source.

上面对毕渥公式做了一个简单的数学变形，毕渥公式的物理意义就更容易理解了。热流密度或热流量都与温差成正比，而与导热热阻成反比。这种传热过程的分析方法，称为热阻分析法。另外需要注意，热阻分析法仅适用于稳态传热过程，且没有内热源。

Summary　This section illustrates the definition, characteristics and mechanism of heat conduction, and the basic law of heat conduction, the calculation method of one-dimensional steady-state heat conduction, and the thermal resistance analysis method.

小结　本节阐述了导热的定义、特点和机理，以及导热基本定律、一维稳态导热的计算方法和热阻分析法。

Question　Put spoons of different materials (e. g. metal, plastic) into hot water, which spoon of butter melts faster? Why?

思考题　不同材质的汤勺（如金属、塑料）放入热水中，哪个汤勺中的黄油融化更快？为什么？

Self-tests

自测题

1. Is thermal conductivity an inherent property of materials? What are the main factors involved?

1. 导热系数是材料的固有属性吗？其主要与哪些因素有关？

2. Who first proposed the relationship of heat conduction to temperature difference and wall thickness?

2. 最早提出导热热量和温差及壁厚的关系的人是谁？

授课视频

1.4　Heat Convection and Its Laws

1.4　热对流及其规律

Heat convection refers to the heat transfer process caused by relative displacement (i. e. macroscopic motion) between different parts of the fluids with different temperatures, and the mixing of cold and hot fluids. According to the definition, heat convection has the following three characteristics:

热对流，是指温度不同的各部分流体之间发生相对位移（即宏观运动），冷热流体之间相互掺混而导致的热量传递过程。根据定义，热对流有以下三个特点：

① There must be temperature difference between the parts of the fluid. This is the basic condition of heat transfer. ② The relative macroscopic movement occurs between the parts of the fluid. This is the necessary condition and the mechanism of heat convection. ③ Thermal convection occurs only inside the fluid and has nothing to do with the external wall. As the macroscopic motion of the fluid causes heat convection, heat conduction caused by the thermal movement of microparticles is ignored.

① 流体内部各部分之间必须有温差，这是传热的基本条件。② 流体各部分之间发生相对的宏观运动。这是热对流的必要条件，也是热对流的机理。③ 热对流仅发生在流体内部，与外部壁面无关。由于流体的宏观运动引起了热对流，此时微观粒子的热运动所引起的导热忽略不计。

Definition of convection heat transfer: when the fluid flows over the surface of an object, the heat transfer process between the fluid and the solid surface.

对流传热的定义：流体流过一个物体表面时，流体与固体表面间的热量传递过程。

Convection heat transfer has the following five characteristics: ① Convection heat transfer is a complex heat transfer process in which both heat convection and heat conduction exist simultaneously; Convection heat transfer is not a basic heat transfer mode, including both heat convection and heat conduction. Inside the fluid, the heat transfer mode belongs to heat convection, and the heat transfer mode between the fluid and the wall surface belongs to heat conduction. ② Convection heat transfer is the heat transfer process when a fluid flows over the surface of an object. ③ The fluid must be in direct contact with the wall and have macro motion. This is the necessary condition of convection heat transfer. ④ There must be a temperature difference between the fluid and the wall. ⑤ A boundary layer with large velocity gradient and temperature gradient will be formed on the wall surface. The boundary layer is a thin layer of fluid near the wall, with a lot of velocity changes and temperature changes.

对流传热具有以下五个特点：① 对流传热是热对流与热传导同时存在的复杂传热过程。对流传热不是一种基本传热方式，既包含热对流，又包含热传导。在流体内部，传热方式属于热对流；在流体与壁面之间的传热方式属于热传导。② 对流传热是流体流过物体表面时的传热过程。③ 流体与壁面必须直接接触并发生宏观运动，这是对流传热的必要条件。④ 流体与壁面之间必须有温差。⑤ 壁面处会形成速度梯度和温度梯度很大的边界层。边界层就是在壁面附近有一薄层流体，其内部的速度变化和温度变化都很大。

According to the cause of fluid flow, convection heat transfer can be divided into forced convection heat transfer and natural convection heat transfer.

按照引起流体流动的原因，对流传热可分为强制对流传热和自然对流传热。

Forced convection heat transfer refers to the fluid flow and heat transfer caused by water pumps, fans or other differential pressure. In industrial production, due to the existence

强制对流传热：是指由水泵、风机或其他压差作用所造成的流体流动和传热。在工业生产中，由于流体

of various resistances in the process of fluid flow, pumps, fans, compressors and other equipment are usually needed to transport the fluid. Therefore, most of the heat transfer between the fluid and the wall surface is forced convection heat transfer.

流动过程中存在各种阻力，通常需要泵、风机、压缩机等设备对流体进行输送，因此，流体与壁面的传热多数都是强制对流传热。

Natural convection heat transfer refers to the fluid flow and heat transfer caused by the density difference between cold and hot fluids inside the fluid. Such as around the radiator, natural convection heat transfer occurs between the air and the radiator; When boiling water in the pot, the flow of water is also natural convection. In the state of weightlessness, although there is a density difference in the fluid, there is no natural convection. So the existence of gravitational field is a necessary condition for natural convection.

自然对流传热：是由于流体内部冷、热流体之间的密度差而引起的流体流动和传热。如在暖气片周围，空气与暖气片之间会发生自然对流传热；在锅中烧水时，水流也是自然对流。在失重状态下，尽管流体中存在密度差，也不会产生自然对流。因此，重力场的存在是自然对流的必要条件。

The law of convection heat transfer: Newton ' s law of cooling. It is usually expressed directly by Newton cooling formula:

对流传热遵循的规律：牛顿冷却定律。通常直接用牛顿冷却公式来表示：

$$\Phi = hA(t_w - t_f) \tag{1.7}$$

Both sides of the equation are divided by area A to obtain the formula of heat flux:

等式两边同除以面积 A，得到热流密度的计算式：

$$q = \Phi/A = h(t_w - t_f) \tag{1.8}$$

In the formula, the physical meaning of each symbol is: Φ is heat transfer rate (W); q is heat flux (W/m^2); A is the contact area between fluid and wall surface (m^2), which can be a plane or a curved surface; t_w is the temperature of wall surface (℃); t_f is main stream temperature of fluid (℃); h is surface heat transfer coefficient [$W/(m^2 \cdot K)$], also often called as convective heat transfer coefficient.

式中，各个符号的物理意义：Φ 是热流量（W）；q 是热流密度（W/m^2）；A 是流体与壁面的接触面积（m^2），壁面可以是平面，也可以是曲面；t_w 是壁面温度（℃）；t_f 是流体主流温度（℃）；h 是表面传热系数 [$W/(m^2 \cdot K)$]，也常称为对流换热系数。

By rearranging the form of Newton cooling formula, h can be expressed as:

把牛顿冷却公式变换一下，即可把 h 表示出来：

$$h = \Phi/[A(t_w - t_f)] \tag{1.9}$$

Its physical meaning is: when the temperature difference between the fluid and the wall is 1 degree Celsius, the

其物理意义是：当流体与壁面的温度相差1℃时，单位壁面面积上、

heat transfer quantity per unit wall surface area and per unit time. That is, how much heat is transferred per unit temperature difference, per unit area and per unit time, then what is the convective heat transfer coefficient. This formula is only used to calculate convective heat transfer coefficient h in the experiment, which cannot explain the influencing factors of h.

单位时间内所传递的热量，即单位温差、单位面积、单位时间内传递的热量是多少，那么表面传热系数就是多少。这个公式仅用于实验时计算表面传热系数 h，并不能说明 h 的影响因素。

The main influencing factors of h include fluid flow velocity and fluid physical properties (such as density, viscosity, thermal conductivity, specific heat capacity), the shape, size and arrangement of heat transfer wall surface, etc. As can be seen, the convective heat transfer coefficient h is not a constant or a physical parameter, which is related to the heat transfer process. In heat transfer, improving convective heat transfer coefficient h has always been a key research content.

h 的影响因素主要有流体流速，流体物性（密度、黏度、导热系数、比热容），传热壁面的形状、大小及布置等。可见，表面传热系数 h 不是一个常数，也不是物性参数，它与传热过程有关。在传热学中，提高对表面传热系数 h 一直都是重点的研究内容。

Fig. 1.6 shows that the variation range of the convective heat transfer coefficient h during several typical convective heat transfer processes. The following conclusions can be drawn from the figure:

图 1.6 中给出了几种典型对流换热过程中表面传热系数 h 的变化范围。从图中可以得出以下结论：

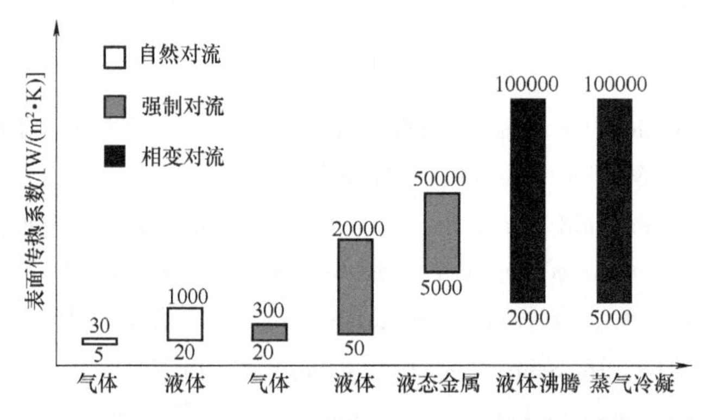

Fig. 1.6 The convective heat transfer coefficient during convective heat transfer processes

图 1.6 对流传热过程中的表面传热系数

phase-change (boiling, condensing) convection h≫forced convection h≫natural convection h

相变（沸腾、冷凝）对流 h≫强制对流 h≫自然对流 h

liquid metal h≫liquid h≫gas h

液态金属 h≫液体 h≫气体 h

Next, imitate the thermal resistance analysis method of the heat conduction process to conduct the thermal resistance

下面仿照导热过程的热阻分析法，进行对流传热的热阻分析。对牛顿

analysis of convection heat transfer. Transform the Newton cooling formula in mathematical form, divide the numerator and denominator on the right-side of the equal sign by hA or h, the following two expressions can be obtained:

冷却公式进行数学形式的变换，等号右侧分子、分母同除以 hA 或 h，即可得到下面两个表达式：

$$\Phi=\frac{\Delta t}{1/(hA)}=\frac{\Delta t}{R_h},\quad q=\frac{\Delta t}{1/h}=\frac{\Delta t}{r_h} \tag{1.10}$$

In the two formulas, the molecules are temperature difference, and the denominators are respectively:

在这两个公式中，分子都是温差，分母分别是：

$$R_h=1/(hA),\quad r_h=1/h \tag{1.11}$$

Both of these expressions are called convective heat transfer resistance.

这两个表达式都称为对流传热热阻。

Example 1.1 The heat transfer area of an indoor radiator is 3m^2, the surface temperature $t_w=50℃$, and it conducts natural convection heat transfer with indoor air at a temperature of 20℃, surface heat transfer coefficient $h=4\text{W}/(\text{m}^2\cdot℃)$. How much power is the radiator equivalent to an electric heater?

例 1.1 一室内暖气片的散热面积为 3m^2，表面温度 $t_w=50℃$，与温度为20℃的室内空气进行自然对流传热，表面传热系数 $h=4\text{W}/(\text{m}^2\cdot℃)$，试问该暖气片相当于多大功率的电暖器？

Solution: Let's analyze the problem firstly. Actually, this question has nothing to do with the electric heater, but to calculate the heat dissipation power of the radiator. Assuming that there is steady-state natural convection heat transfer between the radiator and indoor air. Whether natural convection or forced convection, both need to be calculated according to the Newton cooling formula.

解： 首先做一个分析。此题实际上与电暖器没有任何关系，就是要计算暖气片的散热功率。假设暖气片与室内空气之间是稳态的自然对流传热。不管是自然对流，还是强制对流，都需要根据牛顿冷却公式计算。

$$\Phi=hA(t_w-t_f)=4\text{W}/(\text{m}^2\cdot\text{K})\times3\text{m}^2\times(50-20)\text{K}=360\text{W}$$

Substituting the known data, the heat transfer rate of the radiator can be obtained, that is power. Conclusion: The radiator is equivalent to 360W electric heater.

代入已知数据，就可以得到暖气片的热流量，即功率。结论：该暖气片相当于功率为360W的电暖器。

Summary This section illustrates heat convection and convection heat transfer and their characteristics, and the convective heat transfer laws and calculation methods.

小结 本节阐述了热对流和对流换热及其特点，以及对流传热的规律和计算方法。

Question Why does the daytime breeze come from the ocean, and the night breeze come from the mainland?

思考题 为什么白天的微风来自海洋，而夜晚的微风来自大陆？

Self-tests

1. What are the types of convection heat transfer? What does it have to do with heat convection?

2. What are the factors that affect the surface heat transfer coefficient?

自测题

1. 对流传热有哪些类型？与热对流有什么关系？

2. 表面传热系数的影响因素有哪些？

授课视频

1.5 Thermal Radiation and Its Laws

Thermal radiation is one of the three basic ways of heat transfer. The definition of thermal radiation is given. The way in which an object transmits energy through electromagnetic waves is called radiation. The radiation caused by heat is called thermal radiation. Thermal radiation has the following four characteristics:

(1) As long as the temperature is above absolute zero, that is 0K, any object will continuously emit thermal radiation. The temperature of any object we come into contact with is above absolute zero, so they are constantly emitting thermal radiation.

(2) Thermal radiation can propagate in vacuum with higher efficiency than in medium. Because when thermal radiation propagates in a medium, part of it will be absorbed and reflected by the medium, so the propagation efficiency is reduced.

(3) Thermal radiation is accompanied by a conversion of energy forms. When an object emits thermal radiation, the energy is converted from thermal energy to radiant energy.

(4) The energy of thermal radiation depends on the fourth power of the thermodynamic temperature of the object.

1.5 热辐射及其规律

热辐射是热量传递的三种基本方式之一。首先给出热辐射的定义。物体通过电磁波传递能量的方式，称为辐射；由于热的原因而发生的辐射，称为热辐射。热辐射具有以下四个特点：

(1) 只要温度高于绝对零度，即0K，任何物体都会不停地发出热辐射。我们周围所接触到的任何物体，温度都高于绝对零度，因此都在不停地发出热辐射。

(2) 热辐射可以在真空中传播，并且传播效率比在介质中更高。因为热辐射在介质中传播时，会被介质吸收和反射一部分，所以传播效率降低。

(3) 热辐射伴随能量形式的转换。物体发出热辐射时，能量由热能转换成辐射能。

(4) 热辐射的能量多少取决于物体热力学温度的4次方。

In daily life, there are many examples of the thermal radiation. Solar energy is the thermal radiation emitted by the sun, which is the most common and typical radiant energy. People like to bask in the sun in winter. Is it right to say that you will feel warmer if you bask in the sun through the glass in an airtight room?

在日常生活中，热辐射的例子很多。太阳能就是由太阳所发出的热辐射，是最常见、最典型的辐射能。在冬天人们都喜欢晒太阳。如果在密闭的房间里隔着玻璃晒太阳，会感觉更暖和，这个说法对吗？

The answer is yes. Because the solar radiant energy belongs to short wave, and can pass through the glass into the room, and after being absorbed by the object, the radiant energy emitted by the object belongs to the long wave, and can't pass through the glass. This keeps the indoor temperature rising, so it will be warmer. There is a technical name for this phenomenon, called "greenhouse effect". It is often said that only one side is warm when warming yourself by a fire in the field. Because only the side facing the fire can receive thermal radiation.

答案是肯定的。因为太阳的辐射能属于短波，能够穿过玻璃进入室内，而被物体吸收以后，物体发出的辐射能属于长波，无法穿过玻璃。这样使室内温度不断升高，所以会更暖和。这种现象有一个专业名称，称为“温室效应”。人们常说：野地里烤火一面热。因为烤火时只有面对火的一面才能接受到热辐射。

The calculation of thermal radiation requires the assumption of an ideal model, i. e. (absolute) black body: It is an object that absorbs all thermal radiation projected onto its surface. Thermal radiation in all directions and at all wavelengths can be fully absorbed. Therefore, at the same temperature, the black body is the most absorbent, and also the most radiating.

要对热辐射进行计算，需要先假设一个“理想模型”，即（绝对）黑体：能吸收投射到其表面的所有热辐射的物体。各个方向和所有波长的热辐射都能被黑体所吸收。因此，在相同温度下黑体的吸收能力最强，同时其辐射能力也最强。

The law of black body radiation: the Stefan-Boltzmann law. Its expression is as follows:

黑体辐射的规律：斯特藩-玻尔兹曼定律。其表达式如下：

$$\Phi = \sigma A T^4 \tag{1.12}$$

$$q = \sigma T^4 \tag{1.13}$$

These are two formulas for calculating black body thermal radiation. Note that T in the formula is the thermodynamic temperature of the black body, and $\sigma = 5.67 \times 10^{-8}$ W/(m^2 · K^4).

这是黑体热辐射的两个计算公式。注意，式中 T 是黑体的热力学温度；$\sigma = 5.67\times10^{-8}$W/(m^2 · K^4)。

When calculating the thermal radiation of an actual object, the modified coefficient ε shall be obtained by experiments and comparisons based on black body. ε is called the emissivity or blackness of an actual object, and it must be less

计算实际物体的热辐射时，要以黑体为基准，通过实验和比较从而获得修正系数 ε。ε 称为实际物体的发射率或黑度，必定小于1。实际物体

than 1. The thermal radiation calculation of actual objects is based on the calculation formula of black body thermal radiation multiplied by the modified coefficient ε.

的热辐射计算，是在黑体热辐射计算公式的基础上，乘以修正系数 ε。

$$\Phi=\varepsilon\sigma AT^4,\quad \varepsilon<1 \tag{1.14}$$

Thermal radiation is the outward emission of energy from an object, and is unidirectional. Thermal radiation is always absorbed by another object, and as it absorbs, it also emits thermal radiation, which is absorbed by the previous object. This results in radiation heat transfer, that is the heat transfer process between objects by thermal radiation. It is a bidirectional radiation and absorption process between two or more objects. That is, every object is constantly emitting thermal radiation and also absorbing thermal radiation at the same time. Radiation heat transfer has three characteristics:

热辐射是物体向外发出能量，是单向的。热辐射总要被另一个物体所吸收，而此物体在吸收的同时也在发出热辐射，从而被前一物体吸收。这就产生了辐射传热，即物体之间通过热辐射进行的热量传递过程。它是两个或多个物体之间双向的辐射和吸收过程，即每个物体都不停地发出热辐射，同时也在吸收热辐射。辐射传热具有以下三个特点：

(1) It does not require direct contact between hot and cold objects, and energy can be transferred in vacuum.

(1) 不需要冷热物体直接接触，在真空中即可传递能量。

(2) During the radiation heat transfer process, it is accompanied by the conversion of energy forms. When an object emits thermal radiation, heat energy is converted into radiant energy. When an object absorbs thermal radiation, radiant energy is converted into heat energy as well.

(2) 在辐射传热过程中，伴有能量形式的转换。当物体发出热辐射时，热能转换成了辐射能；当物体吸收热辐射时，辐射能又转换成了热能。

(3) No matter the temperature is higher or lower, objects are constantly emitting and absorbing thermal radiation from one another. The heat radiated from a high-temperature object to a low-temperature object is always greater than the heat radiated from a low-temperature object to a high-temperature object. The final result is still heat transfer from high temperature to low temperature.

(3) 无论温度高低，物体间都在不停地发出及吸收热辐射。高温物体辐射给低温物体的热量，总是大于低温物体辐射给高温物体的热量，最终结果仍然是热量由高温传向低温。

Two simple radiation heat transfer cases. Case 1: A finite object is contained by an infinite cavity. The surface area of the object is A_1, the surface temperature is T_1, the emissivity is ε_1, contained by a large cavity, just like a small object in a large house. The surface temperature of the cavity is T_2, and $T_1>T_2$, the radiation heat transfer

两种简单辐射传热情形。情形 1：一个有限大物体被无限大空腔包容。物体的表面面积为 A_1，表面温度为 T_1，发射率为 ε_1，被很大的空腔包容，就像大房子里放一个小物体。空腔的表面温度为 T_2，并且

between the object and the surface of the cavity is calculated as follows:

$T_1>T_2$，该物体与空腔表面之间的辐射传热量按下式计算：

$$\Phi=\varepsilon_1 A_1 \sigma(T_1^4-T_2^4) \tag{1.15}$$

Case 2: Between two very close parallel black body walls. The physical meaning of "very close" is that the heat loss through the gap between the two walls is not considered. Calculation of the radiation heat transfer:

情形2：两块非常接近的互相平行的黑体壁面之间。"非常接近"的物理意义是，不考虑通过两壁面之间缝隙的热量损失。辐射传热量计算：

$$\Phi=\sigma A(T_1^4-T_2^4) \tag{1.16}$$

$$q_{1\to 2}=\sigma(T_1^4-T_2^4) \tag{1.17}$$

Example 1.2 The outer diameter of the insulation layer of a horizontal steam pipe $d=583\text{mm}$, the measured average temperature of the outer surface is 48℃, the air temperature is 23℃. Natural convection heat transfer takes place between the air and the outer surface of the pipe, and the surface heat transfer coefficient $h=3.42\text{W}/(\text{m}^2\cdot℃)$. The emissivity of the outer surface of the insulation layer $\varepsilon=0.9$.

例1.2 一根水平放置的蒸汽管道，其保温层外径 $d=583\text{mm}$，外表面实测平均温度为48℃，空气温度为23℃。空气与管道外表面间进行自然对流传热，表面传热系数 $h=3.42\text{W}/(\text{m}^2\cdot℃)$，保温层外表面的发射率 $\varepsilon=0.9$。

Questions (1) What heat transfer modes must be considered for heat dissipation of this pipe? (2) Calculate the total heat dissipating capacity per unit length pipe.

问：(1) 此管道的散热必须考虑哪些热量传递方式？(2) 计算单位长度管道的总散热量。

Before solving the problem, we need to make three assumptions: ① The given parameters such as temperature and physical properties keep constant along the length of the pipe; ② The steady-state heat transfer process; ③ The surface temperature of the factory building around the pipe is equal to the air temperature.

在求解问题之前，需要先做出三点假设：① 沿管长方向给定参数（如温度、物性等）保持不变；② 稳态传热过程；③ 管道周围厂房表面温度等于空气温度。

Solution: (1) How to consider the heat dissipation of the pipe? Pipe surface 48℃, air 23℃, there is a temperature difference between them, therefore, there must be natural convection heat transfer on the pipe surface.

解：(1) 管道散热该怎么考虑？管道表面温度为48℃，空气温度为23℃，两者之间有温差，因此，管道表面必然存在自然对流传热。

Notes: ① Air neither absorbs nor emits thermal radiation. ② The steam pipe is generally placed in the factory building, radiation heat transfer will occur between the pipe

注意：① 空气既不吸收也不发出热辐射。② 一般蒸汽管道放置在厂房里，管道与厂房壁面之间会产

and the wall of the factory building. Therefore, the answer to the first question is: natural convection heat transfer and radiation heat transfer should be considered for the heat dissipation of the pipe.

生辐射传热。因此，第 1 个问题的答案是：管道散热应考虑自然对流传热和辐射传热两种方式。

(2) Since there are two ways of heat dissipation of the pipe, the heat dissipating capacity of the pipe should also be calculated separately.

(2) 既然管道散热有两种方式，管道的散热量也要分别进行计算。

1) When only natural convection is considered, natural convection heat dissipating capacity per unit length pipe. Using Newton cooling formula and substituting the known parameters, the natural convection heat dissipating capacity can be calculated to be 156. 5W/m.

1) 只考虑自然对流时，单位长度管道的自然对流散热量。采用牛顿冷却公式，代入已知参数，即可求出自然对流散热量为 156. 5W/m。

$$q_{l,c} = \pi dh(t_w - t_f) = 3.14 \times 0.583 \times 3.42 \times (48-23)\,\mathrm{W/m} = 156.5\,\mathrm{W/m}$$

2) Radiation heat dissipating capacity per unit length pipe can be calculated according to the simple radiation heat transfer case one, that is, the finite object is contained by the infinite cavity. When substituting the known parameters, we should pay attention to: The centigrade temperature given in the topic needs to be added 273. 15 to the thermodynamic temperature and then the radiation heat dissipating capacity can be calculated to be 275. 1W/m.

2) 单位长度管道的辐射散热量，可按照简单辐射传热情形 1 来计算，即有限大物体被无限大空腔包容。代入已知参数时，需要注意：题目中给的摄氏温度，需要加上 273. 15 变为热力学温度，然后就可求出辐射散热量为 275. 1W/m。

$$q_{l,r} = \pi d\sigma\varepsilon(T_1^4 - T_2^4) = 3.14 \times 0.583 \times 5.67 \times 10^{-8} \times 0.9 \times [(48+273.15)^4 - (23+273.15)^4]\,\mathrm{W/m} = 275.1\,\mathrm{W/m}$$

Through the above calculations, we can draw the following conclusion: For a kind of surface heat dissipation problem with surface temperature of only tens of centigrade temperature, the natural convection heat dissipating capacity has the same order of magnitude as the radiation heat dissipating capacity which must be considered at the same time.

通过上述计算，可得出以下结论：对于表面温度只有几十摄氏度的这类表面的散热问题，自然对流散热量与辐射散热量具有相同的数量级，必须同时予以考虑。

Summary This section illustrates radiation, radiation heat transfer and its characteristics, as well as radiation heat transfer laws and calculation methods.

小结 本节阐述了热辐射、辐射传热及其特点，以及辐射传热的规律和计算方法。

Question What are the heat transfer modes of indoor heating package in winter application? Which heat transfer mode is the main one?

思考题 冬天室内暖气片在应用中存在哪些传热方式？以哪种传热方式为主？

Self-tests

自测题

1. What are the heat transfer mechanisms of the three basic modes of heat transfer? What are their characteristics? What laws do they follow?

1. 热能传递的三种基本方式的传热机理是什么？各有什么特点？各遵循什么规律？

2. What are the connections and differences between heat conduction, convection heat transfer and radiation heat transfer?

2. 热传导、热对流、热辐射三种传热方式的联系与区别是什么？

授课视频

1.6 Heat Transfer Process and Calculation of Its Heat Transfer Coefficient

1.6 传热过程及其传热系数计算

Heat transfer process is a specific concept in heat transfer, which refers to the heat exchange process between two kinds of fluids through the solid wall. Therefore, the heat transfer process includes three heat transfer links as shown in Fig. 1.7.

"传热过程"在传热学中是一个特指的概念，是指两种流体通过固体壁面进行的热量交换过程。因此，"传热过程"包含了三个传热环节，如图 1.7 所示。

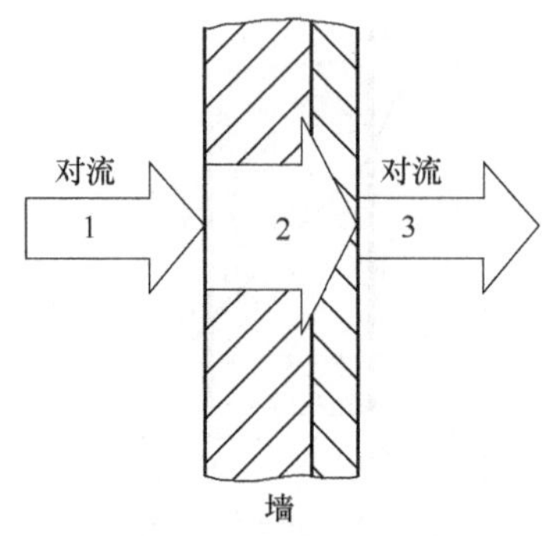

Fig. 1.7 The heat transfer process of two fluids through the wall

图 1.7 两种流体通过壁面的传热过程

First, a fluid transfers heat to the left wall through convection heat transfer; Second, the wall transfers heat from

第 1 个环节，一种流体通过"对流传热"把热量传给左侧壁面；第 2

the left to the right through heat conduction; Third, another fluid takes heat away from thc right wall by convection heat transfer. This heat transfer process is obviously a one-dimensional heat transfer process in series, and we consider it as steady-state heat transfer, that is, the heat transfer rate in each heat transfer link is equal.

个环节，壁面通过“导热”把热量从左侧传给右侧；第3个环节，另一种流体与右侧壁面通过“对流传热”把热量带走。这个“传热过程”显然是一个串联的一维传热过程，并且认为是稳态传热，即每个传热环节的热流量都相等。

Next, the heat transfer rate of each heat transfer link in the one-dimensional steady-state heat transfer process is calculated separately. According to Newton cooling formula, the convective heat transfer quantity of the left fluid:

接下来，分别计算出一维稳态传热过程中每个传热环节的热流量。按照牛顿冷却公式，左侧流体的对流传热量：

$$\Phi=h_1A(t_{f1}-t_{w1})$$

According to the Fourier heat conduction formula, the heat conduction quantity on the middle wall:

按照傅里叶导热公式，中间壁面的导热量：

$$\Phi=\frac{\lambda A}{\delta}(t_{w1}-t_{w2})$$

According to Newton cooling formula, the convective heat transfer quantity of the right fluid:

按照牛顿冷却公式，右侧流体的对流传热量：

$$\Phi=h_2A(t_{w2}-t_{f2})$$

According to the thermal resistance analysis method, the thermal resistance of the three heat transfer links can be obtained respectively. The thermal resistance analysis of the heat transfer process is shown in Fig. 1. 8.

按照热阻分析法，可求出三个传热环节的热阻。传热过程的热阻分析如图1.8所示。

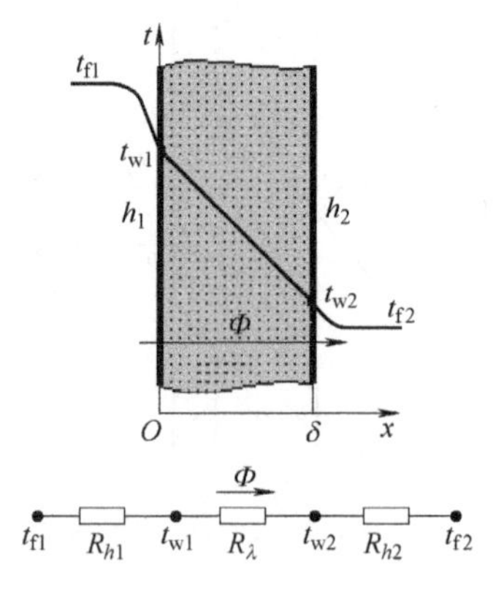

Fig. 1. 8 The thermal resistance analysis of the heat transfer process

图1.8 传热过程的热阻分析

Convective heat transfer resistance on the left:

$$1/h_1A$$

左侧的对流传热热阻：

Thermal conduction resistance of the wall in the middle:

$$\delta/\lambda A$$

中间的壁面导热热阻：

Convection heat transfer resistance on the right:

$$1/h_2A$$

右侧的对流传热热阻：

In the expressions of the three heat transfer links, the thermal resistance moves to the left side of the equal sign, and the right side is written in the form of temperature difference:

在三个传热环节的表达式中，热阻移到等号的左侧，右侧写为温差形式：

$$\frac{1}{h_1A}\Phi = t_{f1} - t_{w1}, \quad \frac{\delta}{\lambda A}\Phi = t_{w1} - t_{w2}, \quad \frac{1}{h_2A}\Phi = t_{w2} - t_{f2}$$

Then, add the left and right sides of the equal signs of the three expressions respectively, the temperature on both sides of the wall cancels out in the three expressions, and the result is as follows:

然后，把三个表达式的等号左侧和右侧分别相加，三个式子中壁面两侧的温度正好约去，整理后得到：

$$\Phi = \frac{t_{f1} - t_{f2}}{\frac{1}{h_1A} + \frac{\delta}{\lambda A} + \frac{1}{h_2A}} \tag{1.18}$$

The following conclusions can be drawn from the calculation of the heat transfer rate of heat transfer process and heat transfer link: The heat transfer rate in the heat transfer process is equal to the total temperature difference in the heat transfer process divided by its total heat resistance. The heat transfer rate of a single heat transfer link is equal to the temperature difference of a single heat transfer link divided by its heat resistance.

通过传热过程和传热环节的热流量计算，可以得出以下结论：传热过程的热流量等于传热过程的总温差除以传热过程总热阻。单个传热环节的热流量等于单个传热环节的温差除以单个传热环节的热阻。

In the steady-state heat transfer process, the heat transfer rate of the heat transfer process should be equal to that of a single heat transfer link. So we obtain two ways to calculate the heat transfer rate. In the calculation formula of the heat transfer rate in the heat transfer process, the numerator and denominator are multiplied by the area A, and this formula can be obtained:

在稳态传热过程中，传热过程的热流量应等于单个传热环节的热流量，由此得到了计算热流量的两种方法。在传热过程的热流量计算式中，分子、分母同乘以面积 A，可得到以下表达式：

$$\Phi=\frac{t_{f1}-t_{f2}}{\frac{1}{h_1A}+\frac{\delta}{\lambda A}+\frac{1}{h_2A}}=\frac{A(t_{f1}-t_{f2})}{\frac{1}{h_1}+\frac{\delta}{\lambda}+\frac{1}{h_2}} \tag{1.19}$$

The denominator of the formula is just the sum of the (area) thermal resistance of the three series heat transfer links, which is called the total thermal resistance and recorded as $1/k$.

该计算式的分母正好是三个串联传热环节的（面积）热阻之和，称为总热阻，记作 $1/k$。

$$\frac{1}{h_1}+\frac{\delta}{\lambda}+\frac{1}{h_2}=\frac{1}{k} \tag{1.20}$$

An important principle is also obtained, that is the superposition principle of series thermal resistance: In a series heat transfer process, if the heat transfer rate through each link is the same (i. e., steady-state heat transfer without internal heat source), then the total thermal resistance of the series links is equal to the sum of the thermal resistance of each link.

由此也得到了一个重要的原理，即串联热阻叠加原理：在一个串联的热量传递过程中，如果通过各个环节的热流量都相同（即稳态传热且无内热源），则串联环节的总热阻等于各串联环节热阻之和。

With the total thermal resistance, the reciprocal of the total thermal resistance is a new important parameter, which is called the overall heat transfer coefficient k.

有了总热阻之后，总热阻的倒数是一个新的重要参数，称为总传热系数 k。

$$k=\frac{1}{\frac{1}{h_1}+\frac{\delta}{\lambda}+\frac{1}{h_2}} \tag{1.21}$$

The unit of the overall heat transfer coefficient is the same as that of convective heat transfer coefficient. This coefficient characterizes the intensity of heat transfer process, and is determined by the three links of heat transfer process. With the overall heat transfer coefficient, the total heat transfer equation can be abbreviated as the following form:

总传热系数的单位与表面传热系数的单位相同 [$W/(m^2 \cdot K)$]，该系数表征了传热过程的强烈程度，由传热过程的三个环节所决定。有了总传热系数以后，总传热方程式可简写为以下形式：

$$\Phi=kA\Delta t=kA(t_{f1}-t_{f2}) \tag{1.22}$$

There are three points to be explained about the total heat transfer equation: ① In the process of heat transfer, the greater the overall heat transfer coefficient k is, the better the heat transfer effect will be. ② Increasing the overall heat transfer coefficient k, can be achieved either by increasing the h_1, h_2 and λ, or by reducing the wall thickness δ.

关于总传热方程式，有三点需要说明：① 在传热过程中，总传热系数 k 越大，传热效果越好。② 要增大总传热系数 k，可通过增大 h_1、h_2、λ 或减小壁厚 δ 来实现。③ 当为非稳态传热过程或有内热源

③ Thermal resistance analysis method cannot be applied for the unsteady-state heat transfer process or the presence of internal heat source.

时，不能用热阻分析法。

Example 1.3 The heat transfer process of a freon condenser is preliminarily calculated, and the following data are obtained: The convective heat transfer coefficient of water in the tube $h_1 = 8700\text{W}/(\text{m}^2 \cdot \text{K})$, the condensation heat transfer coefficient of freon steam outside tube $h_2 = 1800\text{W}/(\text{m}^2 \cdot \text{K})$, and the wall thickness of heat transfer tube $\delta = 1.5$ mm. The thermal conductivity of copper (the material of the tube) is $\lambda = 383\text{W}/(\text{m} \cdot \text{K})$.

例 1.3 对一台氟利昂冷凝器的传热过程进行初步测算，得到以下数据：管内水对流换热的表面传热系数 $h_1 = 8700\text{W}/(\text{m}^2 \cdot \text{K})$，管外氟利昂蒸气的凝结传热的表面传热系数 $h_2 = 1800\text{W}/(\text{m}^2 \cdot \text{K})$，换热管的壁厚 $\delta = 1.5\text{mm}$。管子的材料铜的导热系数 $\lambda = 383\text{W}/(\text{m} \cdot \text{K})$。

(1) Calculate the thermal resistance of the three links and the overall heat transfer coefficient of the condenser. (2) To enhance heat transfer, which link should we start from?

(1) 计算三个环节的热阻及冷凝器的总传热系数。(2) 想要增强传热，应从哪个环节入手?

Assumptions: ① Steady-state heat transfer process; ② The tube is treated as a plate, which is related to the calculation of thermal resistance.

假设：① 稳态传热过程；② 把圆管当作平板处理，关系到热阻计算。

Solution: Area thermal resistance of convection heat transfer on water side:

解: 水侧的对流传热“面积热阻”：

$$\frac{1}{h_1} = \frac{1}{8700\text{W}/(\text{m}^2 \cdot \text{K})} = 1.15 \times 10^{-4}\text{m}^2 \cdot \text{K/W}$$

Area thermal resistance of thermal conduction of the tube wall:

管壁的导热“面积热阻”：

$$\frac{\delta}{\lambda} = \frac{1.5 \times 10^{-3}\text{m}}{383\text{W}/(\text{m} \cdot \text{K})} = 3.92 \times 10^{-6}\text{m}^2 \cdot \text{K/W}$$

Area thermal resistance of steam condensation heat transfer:

蒸气的凝结传热“面积热阻”：

$$\frac{1}{h_2} = \frac{1}{1800\text{W}/(\text{m}^2 \cdot \text{K})} = 5.56 \times 10^{-4}\text{m}^2 \cdot \text{K/W}$$

The three thermal resistances can be obtained by substituting known parameters, and calculate the reciprocal of the sum of the three thermal resistances, then obtain the overall heat transfer coefficient of the condenser:

代入已知参数即可求出三个热阻，对三个热阻之和求倒数，即可得到冷凝器的总传热系数：

$$k=\frac{1}{\frac{1}{h_1}+\frac{\delta}{\lambda}+\frac{1}{h_2}}=\frac{1}{1.15\times10^{-4}+3.92\times10^{-6}+5.56\times10^{-4}}\text{W/(m}^2\cdot\text{K)}=1482\text{W/(m}^2\cdot\text{K)}$$

The calculation results show that the area thermal resistance of the water side accounts for 17.0% of the total thermal resistance; The area thermal resistance of the tube wall accounts for 0.6% of the total thermal resistance, which is almost negligible; The area thermal resistance of the freon steam side accounts for 82.4% of the total thermal resistance, this thermal resistance is the biggest. Therefore, in order to enhance heat transfer, the thermal resistance on the freon steam side should be reduced.

计算结果表明，水侧的面积热阻占总热阻 17.0%；管壁的面积热阻占总热阻 0.6%，几乎可以忽略；氟利昂蒸气侧的面积热阻占总热阻 82.4%，热阻最大。因此，要增强传热，应降低氟利昂蒸气侧的热阻。

Summary This section illustrates the heat transfer process and the calculation method of overall heat transfer coefficient, and the superposition principle of series thermal resistance.

小结 本节阐述了传热过程和总传热系数的计算方法，以及串联热阻叠加原理。

Question If the temperature in the room remains at 25℃ in both summer and winter, can people wear the same clothes in the room? Why?

思考题 如果房间里的温度在夏天和冬天都保持 25℃，人在房间里所穿的衣服能否一样？为什么？

Self-tests

自测题

1. What is the "heat transfer process", which heat transfer links does it include?

1. 什么是“传热过程”？它包括哪些传热环节？

2. What is the overall heat transfer coefficient? What is the relationship between it and the convective heat transfer coefficient?

2. 什么是总传热系数？与表面传热系数的关系是什么？

3. What is the superposition principle of series thermal resistance? What are its applicable conditions?

3. 什么串联热阻叠加原理？其适用条件是什么？

Exercises

习题

1.1 The water supply capacity of the shower nozzle is generally 100cm^3 per minute when it works normally. Cold water is heated from 15℃ to 43℃ by an electric heater. What is the heating power of the electric heater? To

1.1 淋浴器的喷头正常工作时的供水量一般为 100cm^3/min。冷水通过电热器从 15℃ 被加热到 43℃。试问电热器的加热功率是多少？为

save energy, it has been suggested that the used hot water (at 38℃) could be fed into a heat exchanger to heat the cold water coming into the shower. If the heat exchanger can heat cold water to 27℃ , try to calculate how much energy can be saved by using a waste heat recovery heat exchanger after taking a bath for 15min?

了节省能源，有人提出可以将用过后的热水（温度为 38℃）送入一个换热器去加热进入淋浴器的冷水。如果该换热器能将冷水加热到 27℃，试计算采用余热回收换热器后洗澡 15min 可以节省多少能源?

1.2 For the two horizontal interlayers shown in Fig. 1.9, try to analyze the differences of heat exchange between cold and hot surfaces. If the thermal conductivity of a fluid in a interlayers is experimentally determined, which arrangement should be used?

1.2 对于图 1.9 所示的两种水平夹层，试分析冷、热表面间热量交换的方式有何不同? 如果要通过实验来测定夹层中流体的导热系数，应采用哪一种布置?

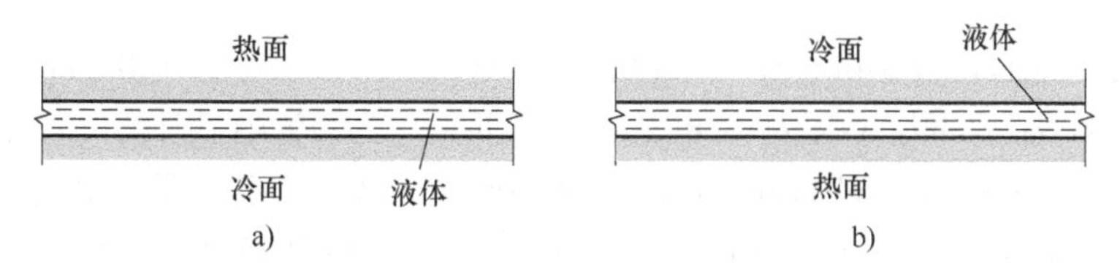

Fig. 1.9 Figure for the exercise 1.2

图 1.9 习题 1.2 图

1.3 In summer, the sun shines on the outer surface of a wooden door made of laminate with a thickness of 40mm. The heat flux of the inner surface of the wooden door measured by a heat flow meter is 15W/m^2. The outer surface temperature is 40℃ , and the inner surface temperature is 30℃. Try to estimate the thermal conductivity of the wooden door in the direction of thickness.

1.3 夏天，阳光照在一厚度为 40mm 的用层压板制成的木门外表面，用热流计测得木门内表面热流密度为 15W/m^2，外表面温度为 40℃，内表面温度为 30℃。试估算此木门在厚度方向上的导热系数。

1.4 In a convective heat transfer experiment of air transversely flowing through a single circular tube, the following data were obtained: The average temperature of the tube wall t_w = 69℃ , the air temperature t_f = 20℃ , the outer diameter of the tube d = 14mm, the length of the heating section is 80mm, and the power of the input heating section is 8.5W. If all the heat is transferred to the air through convection heat transfer, what is the convective heat transfer coefficient at this time?

1.4 在一次测定空气横向流过单根圆管的对流传热实验中，得到下列数据：管壁平均温度 t_w = 69℃，空气温度 t_f = 20℃，管子外径 d = 14mm，加热段长 80mm，输入加热段的功率为 8.5W，如果全部热量通过对流传热传给空气，试问此时的对流传热表面传热系数多大?

1.5 The experiment of boiling heat transfer in saturated water at pressure of 1.01×10^5Pa was carried out by means

1.5 对置于水中的不锈钢管采用电加热的方法进行压力为 1.01×10^5Pa

of electric heating for the stainless steel tube placed in water. The measured heating power is 50W, the outer diameter of the stainless steel tube is 4mm, the length of the heating section is 10mm, and the average surface temperature is 109℃. Try to calculate the convective heat transfer coefficient of boiling heat transfer at this time.

的饱和水沸腾传热实验。测得加热功率为 50W，不锈钢管外径为 4mm，加热段长 10mm，表面平均温度为 109℃。试计算此时沸腾传热的表面传热系数。

1.6 Cosmic space can be seen as 0K vacuum space approximately. A spacecraft is flying in space, the average temperature of the outer surface is 250℃ and surface emissivity is 0.7, try to calculate the heat transfer quantity per unit surface of the spacecraft.

1.6 宇宙空间可近似地看成为 0K 的真空空间。一航天器在太空中飞行，其外表面平均温度为 250℃，表面发射率为 0.7，试计算航天器单位表面上的传热量。

1.7 A spherical spacecraft with a radius of 0.5m flies in space with the surface emissivity of 0.8. The total heat dissipation of the spacecraft's electronics is 175W. Try to estimate the average surface temperature of the spacecraft, assuming that it receives no radiant energy from space.

1.7 半径为 0.5m 的球状航天器在太空中飞行，其表面发射率为 0.8。航天器内电子元件的散热总共为 175W。假设航天器没有从宇宙空间接受任何辐射能量，试估算其表面的平均温度。

1.8 There is a gas cooler, the convective heat transfer coefficient of the gas side $h_1 = 95\text{W}/(\text{m}^2 \cdot \text{K})$, wall thickness $\delta = 2.5\text{mm}$, $\lambda = 46.5\text{W}/(\text{m} \cdot \text{K})$, the convective heat transfer coefficient of the water side $h_2 = 5800\text{W}/(\text{m}^2 \cdot \text{K})$. Suppose that the heat transfer wall can be regarded as a flat wall, and try to calculate the thermal resistance per unit area of each link and the overall heat transfer coefficient from gas to water. In order to enhance this heat transfer process, which link should be started from first? Give your reasons.

1.8 有一台气体冷却器，气侧的表面传热系数 $h_1 = 95\text{W}/(\text{m}^2 \cdot \text{K})$，壁面厚 $\delta = 2.5\text{mm}$，$\lambda = 46.5\text{W}/(\text{m} \cdot \text{K})$，水侧的表面传热系数 $h_2 = 5800\text{W}/(\text{m}^2 \cdot \text{K})$。设传热壁可以看成平壁，试计算各个环节单位面积的热阻及从气到水的总传热系数。为了强化这一传热过程，应首先从哪一环节着手？并说明理由。

1.9 It is assumed that the two sides of the wall surface shown in Fig. 1.10 are maintained at 20℃ and 0℃ respectively, and the high-temperature side is heated by the fluid, $\delta = 0.8\text{m}$, $t_{f1} = 100$℃, $h_1 = 200\text{W}/(\text{m}^2 \cdot \text{K})$, and the process is steady-state. Try to determine the thermal conductivity of the wall material.

1.9 设图 1.10 所示壁面两侧分别维持在 20℃ 及 0℃，且高温侧受到流体的加热，$\delta = 0.8\text{m}$，$t_{f1} = 100$℃，$h_1 = 200\text{W}/(\text{m}^2 \cdot \text{K})$，过程是稳态的，试确定壁面材料的导热系数。

1.10 The cavity shown in Fig. 1.11 is composed of two parallel black body surfaces. The cavity is vacuumed, and

1.10 图 1.11 所示的空腔由两个平行黑体表面组成，空腔内抽成真空，

the thickness of the cavity is much smaller than its height and width. Other known conditions are shown in Fig. 1. 11. Surface 2 is one side of a flat plate with a thickness of $\delta = 0.1\text{mm}$, the other surface 3 is heated by a high temperature fluid, the thermal conductivity of the flat plate $\lambda = 17.5\text{W/(m·K)}$. What is the temperature of surface 3 under steady-state conditions?

且空腔的厚度远小于其高度与宽度。其余已知条件如图 1. 11 所示。表面 2 是厚为 $\delta = 0.1\text{mm}$ 的平板的一侧面，其另一侧表面 3 被高温流体加热，平板的导热系数 $\lambda = 17.5\text{W/(m·K)}$。试问在稳态工况下表面 3 的温度 t_{w3} 为多少？

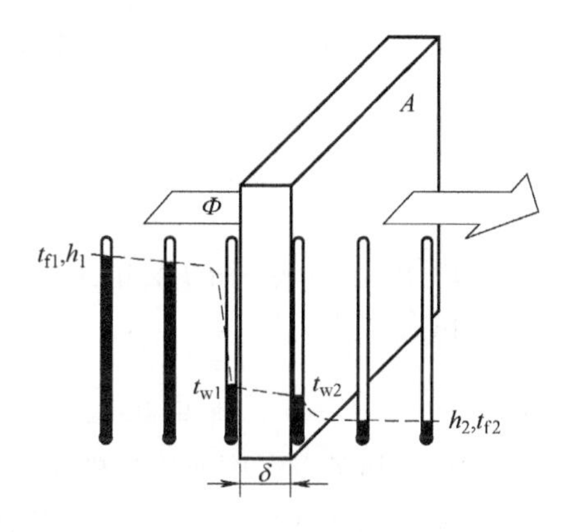

Fig. 1. 10　Figure for the exercise 1. 9

图 1. 10　习题 1. 9 图

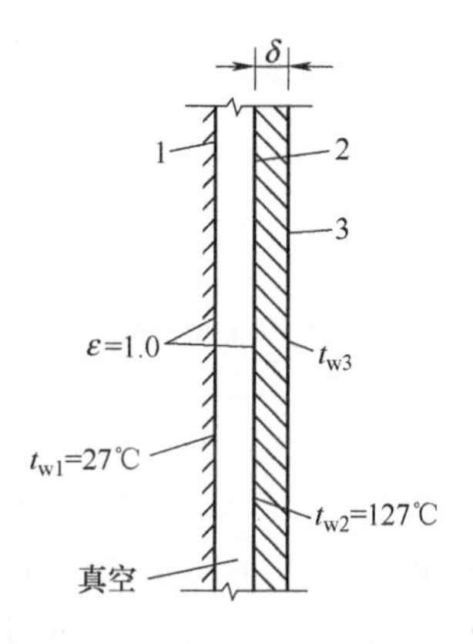

Fig. 1. 11　Figure for the exercise 1. 10

图 1. 11　习题 1. 10 图

Chapter 2 Steady-state Heat Conduction

第2章 稳态热传导

2.1 The Basic Law of Heat Conduction

2.1 导热基本定律

The following are several basic concepts of heat conduction problem.

以下是热传导问题的几个基本概念。

(1) Temperature field: The collection of temperature of each space point at each moment in a thermal conductor, also known as temperature distribution. It's a function of space and time, its expression is:

(1) 温度场：在导热体中各个时刻空间各点温度所组成的集合，也称为温度分布。它是空间和时间的函数，表达式为：

$$t=f(x,y,z,\tau) \tag{2.1}$$

Where the symbol t denotes temperature, and τ denotes time. The mathematical meaning of this expression is that temperature t is a function of coordinate x, y, z and time τ. Its physical meaning is that the temperature of the point with coordinates x, y, z is t in the moment τ.

式中，符号 t 表示温度，而 τ 表示时间。这个表达式的数学含义是：温度 t 是坐标 (x,y,z) 和时间 τ 的函数。它的物理意义是：坐标为 (x,y,z) 的点在 τ 时刻的温度是 t。

According to whether the temperature varies with time at each point, the temperature field can be divided into steady-state temperature field and unsteady-state temperature field. In steady-state temperature field, temperature t is only a function of coordinates x, y, z, and independent of time τ. Temperature t does not vary with time τ, as can be expressed as:

按照各点温度是否随时间变化，温度场可分为稳态温度场和非稳态温度场。对于稳态温度场，温度 t 只是坐标 (x,y,z) 的函数，与时间 τ 无关。温度 t 不随时间 τ 变化，可表示为：

$$\frac{\partial t}{\partial \tau}=0 \tag{2.2}$$

In unsteady-state temperature field, temperature t is a function of both coordinates x, y, z and time τ, also called transient temperature field. A three-dimensional body

对于非稳态温度场，温度 t 既是坐标 (x,y,z) 的函数，又是时间 τ 的函数，又称为瞬态温度场。图2.1中

as depicted in Fig. 2. 1, heating from one side, the temperature at each point varies with time, its temperature field $t=f(x,y,z,\tau)$, where t_1, t_2 are instantaneous isothermal surfaces.

所示的三维物体，从其一侧加热，各点温度会随着时间而变化，其温度场为 $t=f(x,y,z,\tau)$，其中，t_1、t_2 为瞬时等温表面。

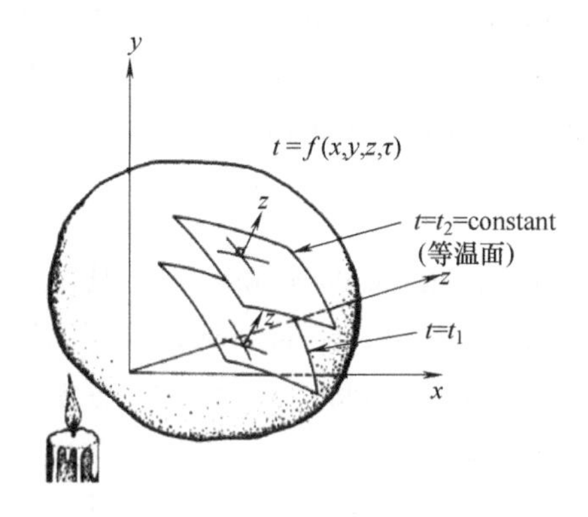

Fig. 2. 1　A three-dimensional transient temperature field

图 2. 1　三维瞬态温度场

For the temperature field of an object surface, we cannot observe it with the naked eye. However, with the help of a thermal infrared imager, the temperature level of each point can be qualitatively distinguished.

对于物体表面的温度场，当然无法通过肉眼观察到，但是借助红外热像仪就可以定性地分辨出各点温度的相对高低。

(2) **Isothermal surface**: A surface formed by points of the same temperature at the same time in a temperature field. The isothermal surface may be either plane or curved surface.

(2) **等温面**：温度场中同一时刻温度相同的点连成的面。等温面可能是平面，也可能是曲面。

(3) **Isotherm**: Cut the isothermal surfaces by a plane, you can get a cluster of isotherms on the plane.

(3) **等温线**：用一个平面去截等温面，在这个平面上就会得到一簇等温线。

Both isothermal surfaces and isotherms have the following characteristics: ① The isothermal surfaces/isotherms with different temperature do not intersect with each other. ② In a continuous temperature field, the isothermal surfaces or isotherms will not be interrupted. They are either completely enclosed surfaces/curves in the object, or they terminate at the boundary of the object.

等温面与等温线都具有以下特点：① 温度不同的等温面/等温线彼此不会相交。② 在连续的温度场中，等温面或等温线不会中断。它们要么是物体中完全封闭的曲面/曲线，要么终止于物体的边界上。

③ There is no temperature difference and heat transfer on the isothermal surfaces/isotherms. There is a temperature

③ 等温面/等温线上没有温差，不会有热量传递。不同的等温面/等温

difference and heat transfer between different isothermal surfaces/isotherms. ④ The temperature difference between two adjacent isotherms is equal, but the geometric distances are not necessarily equal on isotherm diagram. It shows that the temperature gradient varies everywhere. Therefore, the density of isotherm distribution also intuitively reflects the relative magnitude of heat flux in different regions, as shown in Fig. 2. 2.

线之间有温差，有热量传递。④ 等温线图上相邻两条等温线间的温差相等，但几何距离未必相等，说明各处的温度梯度是变化的。因此，等温线分布的疏密也直观反映出不同区域的热流密度相对大小，如图 2. 2所示。

From Fig. 2. 2, we can clearly see that the isotherm distribution is sparse in the central region, while the isotherm distribution is relatively dense in the boundary region. Therefore, We can get the central region has a lower heat flux than the boundary region.

从图 2. 2 中可明显看出，中心区的等温线分布比较稀疏，而边界区的等温线分布比较稠密。因此可以判断出，中心区比边界区的热流密度小。

(4) **Heat flux line**: In the normal direction of the isotherm in a two-dimensional temperature field, the line between the intersection points of the normal line and the adjacent isotherms is called the heat flux line. It reflects the direction of the heat flux, as shown in Fig. 2. 3. The temperature difference Δt between adjacent isotherms is equal, but the distance Δs between any two points is generally different, in which the normal distance Δn is the shortest. So, we give the definition of temperature gradient.

(4) **热流线**：在二维温度场中等温线法线方向上，法线与相邻等温线的交点的连线就是热流线，它反映了热流的方向，如图 2. 3 所示。虽然相邻两条等温线之间的温差 Δt 是相等的，但任意两点间的距离 Δs 一般不等，其中，法向距离 Δn 最短。据此给出温度梯度的定义。

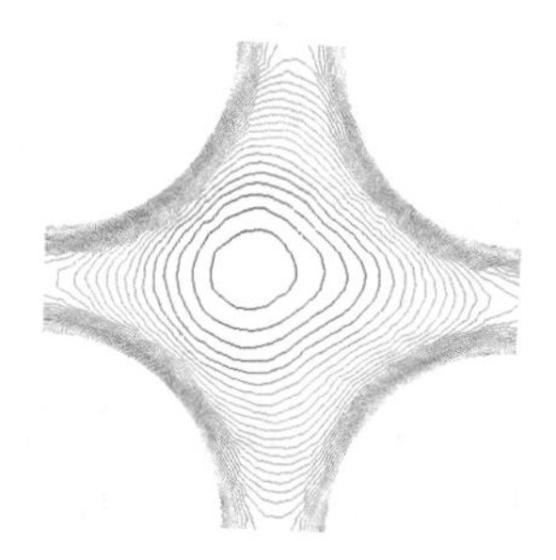

Fig. 2. 2 The isotherm distribution

图 2. 2 等温线分布

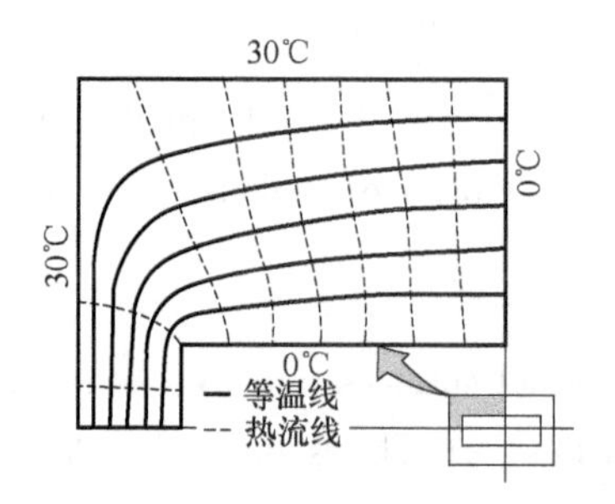

Fig. 2. 3 The heat flux line distribution

图 2. 3 热流线分布

(5) **Temperature gradient**: The limit of the ratio of temperature increment to normal distance along the normal direction of isothermal surface is temperature gradient. Its expression is:

(5) **温度梯度**：沿等温面法线方向温度增量 Δt 与法向距离 Δn 比值，取极限就是温度梯度。表达式为：

$$\mathrm{grad}t=\lim_{\Delta n\to 0}\frac{\Delta t}{\Delta n}\boldsymbol{n}=\frac{\partial t}{\partial n}\boldsymbol{n} \tag{2.3}$$

In the rectangular coordinate system, the temperature gradient can be written as:

在直角坐标系中，温度梯度可写为：

$$\mathrm{grad}\ t=\frac{\partial t}{\partial x}\boldsymbol{i}+\frac{\partial t}{\partial y}\boldsymbol{j}+\frac{\partial t}{\partial z}\boldsymbol{k} \tag{2.4}$$

Note that the temperature gradient is a vector, its positive direction points in the direction of temperature rise.

注意：温度梯度是矢量，正向指向温度升高的方向。

(6) Heat flux: The heat transfer quantity per unit area per unit time. If the heat conduction surface is not a plane, the magnitude of the heat flux in different directions is different, so the heat flux is a vector. The direction of the heat flux vector is the direction of the maximum heat flux at a point of the isothermal surface. That is the reverse direction of the temperature gradient. In the rectangular coordinate system, the heat flux vector can be decomposed along three coordinate axis:

(6) 热流密度： 单位时间、单位面积上所传递的热量。如果导热面不是平面，则不同方向上热流密度的大小不同，所以热流密度是一个矢量。热流密度矢量的方向是等温面上某点的热流密度最大的方向，即温度梯度方向的反向。在直角坐标系中，热流密度矢量可沿三个坐标轴进行分解：

$$\boldsymbol{q}=q_x\boldsymbol{i}+q_y\boldsymbol{j}+q_z\boldsymbol{k} \tag{2.5}$$

On the basis of the above basic concepts, the basic law of heat conduction is given below.

在上述基本概念的基础上，下面给出导热基本定律。

Fourier's law: The heat flux perpendicular to the isothermal surface is proportional to the temperature gradient at that point, and its direction is opposite to the direction of temperature gradient. Its expression is:

傅里叶定律： 垂直导过等温面的热流密度正比于该处的温度梯度，方向与温度梯度方向相反。表达式为：

$$\boldsymbol{q}=-\lambda\,\mathrm{grad}\ t \tag{2.6}$$

Where grad t denotes the temperature gradient, corresponding to the temperature change rate of one-dimensional heat conduction problem. The heat flux vector $\boldsymbol{q}$ can be decomposed along the coordinate axes x, y and z:

式中，grad t 表示温度梯度，对应于一维导热问题的温度变化率。热流密度矢量 $\boldsymbol{q}$ 可以沿坐标轴 (x,y,z) 进行分解：

$$q_x=-\lambda\frac{\partial t}{\partial x},\quad q_y=-\lambda\frac{\partial t}{\partial y},\quad q_z=-\lambda\frac{\partial t}{\partial z} \tag{2.7}$$

In the rectangular coordinate system, the expression of the basic law of heat conduction is:

在直角坐标系中，导热基本定律的表达式为：

$$\boldsymbol{q}=-\lambda\frac{\partial t}{\partial x}\boldsymbol{i}-\lambda\frac{\partial t}{\partial y}\boldsymbol{j}-\lambda\frac{\partial t}{\partial z}\boldsymbol{k} \tag{2.8}$$

Note: Fourier's law only applies to isotropic materials, that is, thermal conductivity λ is equal in all directions.

注意：傅里叶定律仅适用于各向同性材料，即导热系数λ沿各个方向相等。

Summary This section illustrates the concepts of temperature field, temperature gradient, isothermal surfaces, heat flux line and heat flux etc. Please focus on the Fourier's law of heat conduction.

小结 本节阐述了温度场、温度梯度、等温面、热流线和热流密度等概念，要重点掌握傅里叶导热定律。

Question Please look at Fig. 2.4, This is a steady-state temperature field, and each point temperature keeps constant, that is, the left side temperature is lower and the temperature of the other three sides is higher. What kind of external heating/cooling conditions can maintain the steady-state temperature field?

思考题 请看图2.4，这是一个稳态温度场，各点温度均保持不变，即左侧面温度较低，其他三个侧面温度较高。在什么样的外部加热/冷却条件下才能保持这种稳态温度场呢？

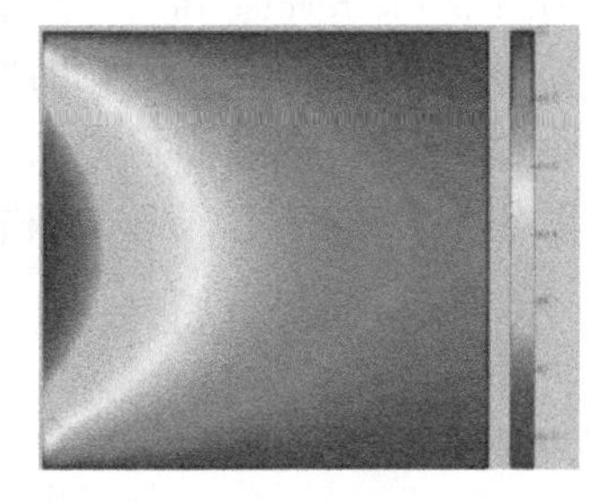

Fig. 2.4 The steady-state temperature field

图2.4 稳态导热温度场

Self-tests

自测题

1. What are the main physical quantities to solve the heat conduction problem?

1. 求解导热问题主要求哪些物理量？

2. Why does the minus sign appear in the Fourier heat conduction formula?

2. 为什么傅里叶导热公式中出现负号？

授课视频

2.2 Thermal Conductivity

2.2 导热系数/热导率

When studying heat conduction problem, thermal conductivity,

研究导热问题时，必然用到热导率，

i. e. thermal conduction coefficient, is necessary. The thermal conductivity denotes the thermal conduction capacity of a substance, and is the inherent thermophysical parameter of a substance. However, thermal conductivity can only be measured by experiments. The thermal conductivity of commonly used materials can be found in the appendix at the back of the textbook.

即导热系数。导热系数表示物质导热能力的大小，是物质固有的热物性参数。但是导热系数只能通过实验来测定，常用材料的导热系数可查教材后附录。

According to the expression of Fourier's law, the definition formula of thermal conductivity can be derived:

根据傅里叶定律的表达式，可以导出导热系数的定义式：

$$\lambda = \frac{|\boldsymbol{q}|}{|\mathbf{grad}\ t|} \tag{2.9}$$

From the formula, the thermal conductivity is equal to the thermal conduction quantity of an object, per unit temperature gradient, per unit area and per unit time.

由表达式可以看出，导热系数等于物体中单位温度梯度、单位时间通过单位面积的导热量。

The main influencing factors of thermal conductivity are the sort and temperature of substance. Different substances have different thermal conductivity, which varies with the temperature. In addition, the composition and density of material, humidity, pressure and other factors can also affect the thermal conductivity of different substances. But their effects are so small that they can't be considered normally. For substances of different materials and phases, the variation trend of thermal conductivity is as follows:

导热系数的主要影响因素是物质的种类和温度。不同的物质，导热系数不同，并随着温度的变化而变化。另外，材料成分及密度、湿度、压力等因素，也会对不同物质的导热系数产生影响。但这些因素影响很小，一般情况下可不考虑。对于不同材质、不同相态的物质，导热系数的变化趋势是：

$$\lambda_{\text{金属(metallic material)}} > \lambda_{\text{非金属(nonmetallic material)}}$$

$$\lambda_{\text{固体(solid)}} > \lambda_{\text{液体(liquid)}} > \lambda_{\text{气体(gas)}}$$

Different substances have different thermal conductivity. It is mainly because their microstructures and heat conduction mechanisms are different.

不同物质的导热系数不同，主要是因为它们的微观结构及导热机理不同。

The heat conduction mechanism of gases is the thermal motion and collisions of molecules. The range of thermal conductivity of gases is:

气体的导热机理：是分子的热运动及相互碰撞。气体的导热系数范围：

$$\lambda_{gas} \approx 0.006 \sim 0.6\ \text{W/(m} \cdot {}^\circ\text{C)}$$

For example:

如：

$$0℃\ \lambda_{air}=0.0244W/(m\cdot℃),\quad 20℃\ \lambda_{air}=0.026W/(m\cdot℃)$$

It can be seen that as the temperature increases, the thermal conductivity of air increases slightly. The variation of the thermal conductivity of gases with temperature is shown in Fig. 2. 5. From the figure, the thermal conductivity of various gases increases with the increase of temperature.

可见，随着温度升高，空气的导热系数略有增大。气体的导热系数随温度的变化关系如图 2. 5 所示。从图中可以看出，各种气体的导热系数都随温度升高而增大。

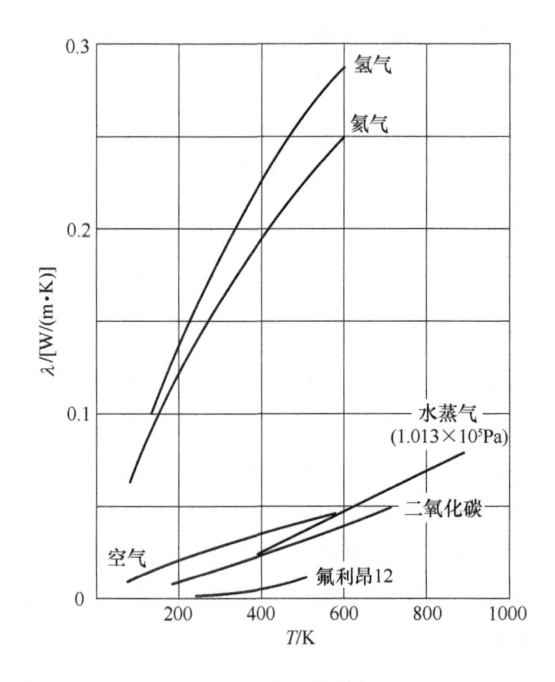

Fig. 2. 5 The variation of the thermal conductivity of gases with temperature

图 2. 5 气体的导热系数随温度的变化关系

The heat conduction mechanism of liquid is more complicated. There is not only lattice vibration but also molecular motion. The range of thermal conductivity of liquids is:

液体的导热机理：比较复杂，不但有晶格振动，还有分子运动。液体的导热系数范围：

$$\lambda_{liquid}\approx 0.07\sim 0.7W/(m\cdot℃)$$

For example:

例如：

$$20℃\ \lambda_{water}=0.6W/(m\cdot℃)$$

The variation of the thermal conductivity of liquids with temperature is shown in Fig. 2. 6. From the figure, the variation of the thermal conductivity of liquids is complicated, and lacks of overall regularity. With the increase of temperature, the thermal conductivity of ammonia and Freon

液体的导热系数随温度的变化关系如图 2. 6 所示。从图中可以看出，液体的导热系数变化比较复杂，缺乏整体的规律性。随着温度升高，氨水和氟利昂的导热系数都减小，

decreases, especially the variation of the thermal conductivity of ammonia is very obvious; While the thermal conductivity of glycerol increases slowly and the thermal conductivity of engine oil is basically unchanged; The thermal conductivity of water is first increasing and then decreasing. When the absolute temperature is about 420K, i. e. 150℃, the thermal conductivity of water reaches its maximum value.

尤其是氨水的导热系数变化很明显；而甘油的导热系数缓慢增大，机油的导热系数基本不变；水的导热系数先增大而后减小，在热力学温度约 420K 即 150℃时，水的导热系数为最大值。

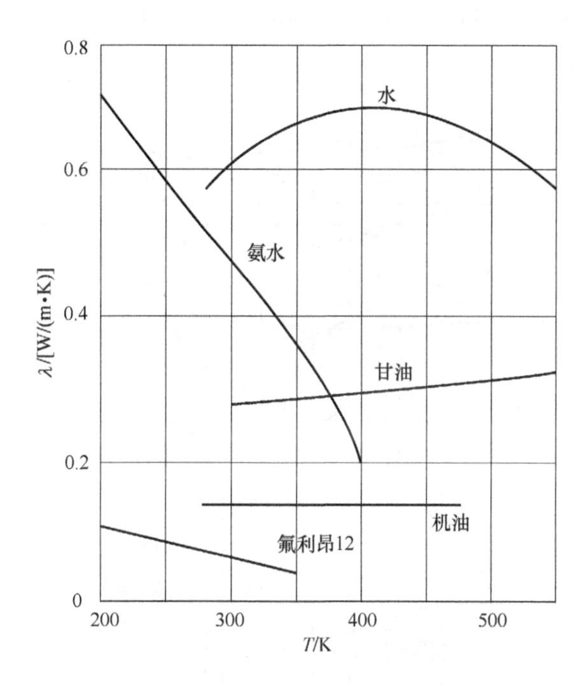

Fig. 2. 6　The variation of the thermal conductivity of liquids with temperature

图 2. 6　液体的导热系数随温度的变化关系

The heat conduction mechanism of pure metals is free electron motion and lattice vibration, mainly the former. Thermal conductivity of metals ranges from:

纯金属的导热机理：是自由电子运动及晶格振动，以前者为主。金属的导热系数范围：

$$\lambda_{\text{mental}} \approx 12 \sim 418\text{W/(m} \cdot ℃)$$

The variation of the thermal conductivity of metals with temperature is shown in Fig. 2. 7. From the figure, the variation trend of thermal conductivity of common metals is:

金属的导热系数随温度的变化关系如图 2. 7 所示。从图中可以看出，常见金属的导热系数变化趋势：

$$\lambda_{银(\text{silver})} > \lambda_{铜(\text{copper})} > \lambda_{金(\text{gold})} > \lambda_{铝(\text{aluminum})} > \lambda_{铁(\text{iron})} > \lambda_{不锈钢(\text{stainless steel})}$$

The metal with the highest thermal conductivity is silver, but it belongs to precious metal, which is generally used

导热系数最大的金属是银，但银属于贵金属，一般用在精密的电子器

in precision electronic devices. Copper is often used to enhance thermal conduction in industry. Iron and stainless steel are commonly used in heat exchangers under the general working conditions. As temperature rises, the thermal conductivity of most metals decreases, for example:

件上。工业上为了增强导热，常采用铜。一般工况下换热器常用铁和不锈钢。随着温度升高，大多数金属的导热系数会减小，如：

$$10\text{K}\ \lambda_{Cu}=12000\text{W}/(\text{m}\cdot℃),\quad 15\text{K}\ \lambda_{Cu}=7000\text{W}/(\text{m}\cdot℃)$$

The reason is that the lattice vibration is strengthened with the increase of temperature, which interferes with the movement of free electrons.

其原因是：随着温度升高，晶格振动加强，干扰了自由电子运动。

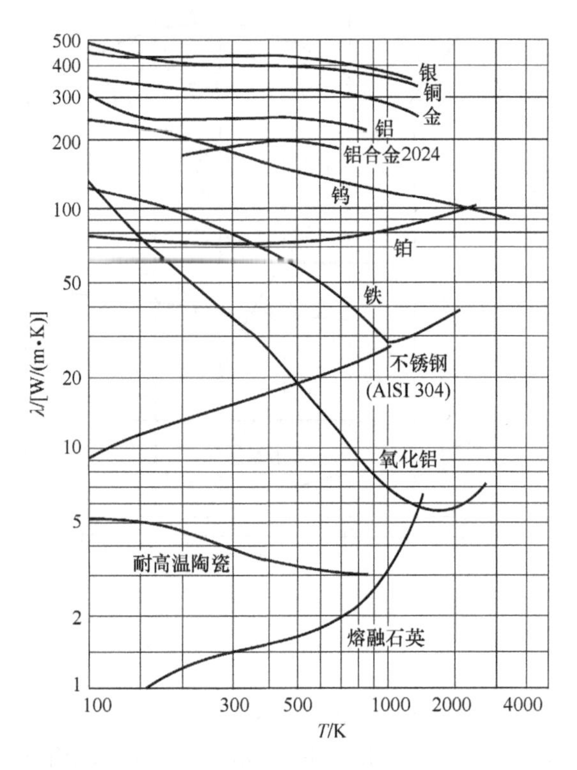

Fig. 2. 7 The variation of the thermal conductivity of metals with temperature

图 2. 7 金属的导热系数随温度的变化关系

The thermal conductivity of alloys is generally lower than that of pure metals. Because the doping of impurities into the metal destroys the integrity of the lattice and interferes with the movement of free electrons, resulting in a decrease in the thermal conductivity. For example, at room temperature, the thermal conductivity of pure copper is 398W/(m · K), after adding zinc, pure copper becomes brass and its thermal conductivity drops to 109W/(m · K). In addition,

合金的导热系数一般都小于纯金属。因为金属中掺入杂质，破坏了晶格的完整性，干扰了自由电子运动，导致了导热系数减小。如常温下，纯铜的导热系数是 398W/(m · K)，加入锌元素之后，变成了黄铜，其导热系数降到了 109W/(m · K)。另外，金属的加工过程（如锻压）也会造

metal processing such as forging can also cause lattice defects, resulting in a decrease in thermal conductivity.

成晶格的缺陷，导致导热系数减小。

The heat conduction mechanism of alloys: It is still the movement of free electron and lattice vibration, but mainly lattice vibration. With the increase of temperature, their thermal conductivity increases. This is contrary to the change of the thermal conductivity of pure metal with temperature. The reason is that the lattice vibration is strengthened with the increase of temperature, so the thermal conductivity is enhanced.

合金的导热机理：仍然是自由电子运动及晶格振动，但以晶格振动为主。随着温度升高，合金的导热系数增大，这与纯金属导热系数随温度的变化关系正好相反。其原因是：随着温度升高，晶格振动加强，所以导热性能增强。

The heat conduction mechanism of non-metallic solids: It is just lattice vibration, so the thermal conductivity is relatively small. With the increase of temperature, the thermal conductivity increases. The difference of thermal conductivity makes the materials have different applications. Materials with small thermal conductivity can be used for thermal insulation. Such as the thermal insulation materials on buildings. The range of its thermal conductivity is:

非金属固体的导热机理：仅仅是依靠晶格振动，所以导热系数都比较小。随着温度升高，其导热系数增大。导热系数大有大的用途，小有小的用途。导热系数小的材料可用于隔热保温，如建筑上的隔热保温材料，其导热系数的范围是：

$$\lambda \approx 0.025 \sim 3\mathrm{W/(m \cdot ℃)}$$

In order to further reduce the thermal conductivity, the building materials are mostly porous or fiber structure such as hollow brick. The thermal conductivity of porous materials is also related to density and humidity. The thermal conductivity increases with the increase of density and humidity. According to the national standard, when the temperature is not higher than 350℃, the materials with thermal conductivity less than 0.12W/(m · K) can be used as thermal insulation materials.

为了进一步减小导热系数，建筑材料多采用多孔或纤维结构，如空心砖。多孔材料的导热系数还与密度和湿度有关。随着密度和湿度增加，材料导热系数增大。国家标准规定，温度不高于350℃时，导热系数小于0.12W/(m · K) 的材料，都可以作为保温材料（或称为绝热材料）。

Summary This section illustrates the definition formula and influencing factors of the thermal conductivity, and the variation laws of thermal conductivity of gas, liquid and solid.

小结 本节阐述了导热系数的定义式和影响因素，以及气体、液体和固体的导热系数变化规律。

Question Why is the thermal conductivity of hydrogen and helium far greater than that of other gases?

思考题 为什么氢气和氦气的导热系数远远大于其他气体呢？

Self-tests

1. What are the variation laws of the thermal conductivity of different materials?

2. What is the principle of using porous materials such as hollow bricks as insulation materials?

自测题

1. 不同材料导热系数的变化规律是什么?

2. 常把多孔材料（如空心砖）作为保温材料的原理是什么?

授课视频

2.3 The Differential Equation of Heat Conduction

2.3 导热微分方程式

The differential equation of heat conduction is the mathematical description of the heat conduction problem. In the final analysis, solving the heat conduction problem is to solve two physical quantities, namely the heat flux and temperature distribution in the heat conductor.

导热微分方程式也就是导热问题的数学描写。求解导热问题归根到底是求解两个物理量，也就是导热体内的热流密度和温度分布。

According to Fourier's law, first figure out the temperature distribution in conductor, then figure out the temperature gradient, finally by substituting temperature gradient into Fourier's law, the heat flux can be calculated. Therefore, the first task in solving the heat conduction problem is to figure out the temperature distribution in heat conductor.

由傅里叶定律可知，只要求出物体内的温度分布，就可以求出温度梯度，代入傅里叶定律，即可求出热流密度。因此，解决导热问题的首要任务是求出导热体内的温度分布。

Then we deduce the differential equation of heat conduction. Our purpose is to establish a functional relationship among temperature t, coordinate (x, y, z) and time τ, and to obtain the temperature distribution in the heat conductor. Two theories are applied in the derivation: the first law of thermodynamics and Fourier's law. This is a theoretical analysis process. According to the general steps of theoretical analysis, first make assumptions as follows:

下面推导导热微分方程式，目的是建立温度 t 与坐标 (x,y,z) 及时间 τ 之间的函数关系，得到导热体内的温度分布。推导时要用到两个理论：热力学第一定律和傅里叶定律。这是一个理论分析过程。按照理论分析的一般步骤，首先要做出如下假设：

(1) The object under study is an isotropic continuum. This condition is naturally satisfied in most engineering situations.

(1) 所研究的物体是各向同性的连续介质。工程中多数情况都自然满足这一条。

(2) Thermophysical parameters such as thermal conductivity, specific heat capacity and density are known. Thermophysical parameters of commonly used materials can be found easily.

(2) 导热系数、比热容和密度等热物性参数均为已知。常用材料的热物性参数均可查到。

(3) There is an internal heat source inside the object which is uniformly distributed, and the heat source intensity is $\dot{\Phi}$. $\dot{\Phi}$ represents heat released per unit volume in the heat conductor per unit time (W/m^3). What is the internal heat source? The nuclear energy in nuclear fuel is automatically converted into heat energy, which is equivalent to an internal heat source in an object.

(3) 物体内具有内热源，且均匀分布，热源强度为 $\dot{\Phi}$。$\dot{\Phi}$ 表示单位体积的导热体，在单位时间内放出的热量，单位为 W/m^3。什么是内热源呢？核燃料中核能自动转换为热能，相当于物体中有内热源。

Based on the above assumptions, the physical model of the heat conduction problem is determined. Then we build the geometric model and mathematical model, as shown in Fig. 2. 8.

通过以上假设确定了导热问题的物理模型。下面建立几何模型和数学模型，如图 2. 8 所示。

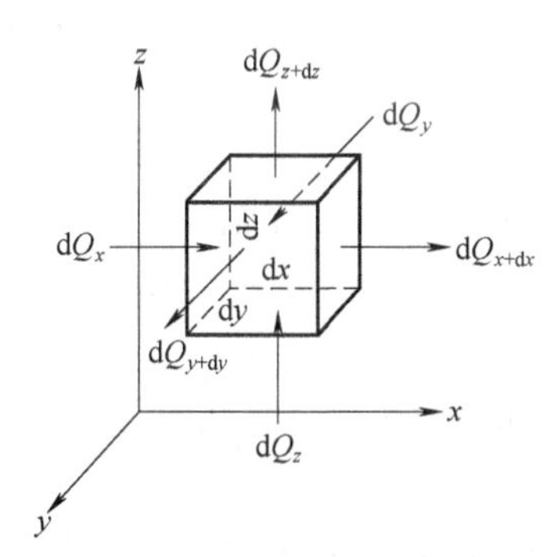

Fig. 2. 8 Thermal conductivity analysis of microelement

图 2. 8 微元体的导热过程分析

Take any micro-element body, i. e. hexahedron, in the thermal conductor: the side lengths are dx, dy and dz respectively. According to the first law of thermodynamics: [The heat variation of the micro-element body Q] = [The variation of thermodynamic internal energy ΔU] + [Work done to the outside W]. When the work done by the micro-element body to the outside is 0, that is $W=0$, during $d\tau$ time, in the micro-element body: [The net heat imported into and exported from the micro-element body 1] + [The heat generated by an internal heat source 2] = [The variation of thermodynamic internal energy 3].

在导热体中任取一微元体，即六面体，边长分别为 dx、dy、dz。根据热力学第一定律：[微元体的热量变化量 Q] = [热力学内能的变化量 ΔU] + [对外界做功 W]。当微元体对外界做功为 0 时，即 $W=0$，在 $d\tau$ 时间内微元体中：[导入与导出微元体的净热量 1] + [内热源发热量 2] = [热力学内能的变化量 3]。

Next solve [The net heat imported into and exported from thc micro element body 1]. The import heat of the micro-element body along the x-axis and through the surface of x during $\mathrm{d}\tau$ time:

下面求解［导入与导出微元体的净热量1］。在 $\mathrm{d}\tau$ 时间内，沿 x 轴方向经 x 表面导入微元体的热量：

$$\mathrm{d}Q_x = q_x \mathrm{d}y\mathrm{d}z\mathrm{d}\tau$$

The export heat of the micro-element body along the x-axis and through the surface of $x+\mathrm{d}x$ during $\mathrm{d}\tau$ time:

在 $\mathrm{d}\tau$ 时间内，沿 x 轴方向经 $x+\mathrm{d}x$ 表面导出微元体的热量：

$$\mathrm{d}Q_{x+\mathrm{d}x} = q_{x+\mathrm{d}x} \mathrm{d}y\mathrm{d}z\mathrm{d}\tau$$

Where

其中，

$$q_{x+\mathrm{d}x} = q_x + \frac{\partial q_x}{\partial x}\mathrm{d}x$$

The right side of this formula can be regarded as the first two terms of Taylor series expansion; It can also be viewed as the difference of physical quantity between two adjacent points, which is the change rate of physical quantity between two points times the distance between them.

这个式子右侧可看作泰勒级数展开式的前两项，也可看作相邻两点的物理量之差，为两点间物理量的变化率乘以两点间的距离。

The net heat imported into and exported from the micro-element body along the x-axis during $\mathrm{d}\tau$ time can be written as:

在 $\mathrm{d}\tau$ 时间内，沿 x 轴方向导入与导出微元体的净热量可写为：

$$\mathrm{d}Q_x - \mathrm{d}Q_{x+\mathrm{d}x} = -\frac{\partial q_x}{\partial x}\mathrm{d}x\mathrm{d}y\mathrm{d}z\mathrm{d}\tau$$

The same method can be used to obtain the net heat imported into and exported from the micro-element body along the y-axis and z-axis respectively during $\mathrm{d}\tau$ time:

同样方法可求出：在 $\mathrm{d}\tau$ 时间内分别沿 y 轴和 z 轴方向导入与导出微元体的净热量：

$$\mathrm{d}Q_y - \mathrm{d}Q_{y+\mathrm{d}y} = -\frac{\partial q_y}{\partial y}\mathrm{d}x\mathrm{d}y\mathrm{d}z\mathrm{d}\tau$$

$$\mathrm{d}Q_z - \mathrm{d}Q_{z+\mathrm{d}z} = -\frac{\partial q_z}{\partial z}\mathrm{d}x\mathrm{d}y\mathrm{d}z\mathrm{d}\tau$$

The sum of the net heat imported into and exported from micro-element body along the three axes = [1]. According to Fourier's law, the expressions of heat flux are as follows:

沿三个坐标轴方向导入与导出微元体的净热量相加 = [1]。按照傅里叶定律，热流密度的表达式为：

$$q_x = -\lambda\,\frac{\partial t}{\partial x},\quad q_y = -\lambda\,\frac{\partial t}{\partial y},\quad q_z = -\lambda\,\frac{\partial t}{\partial z}$$

Substitute them into the expression of [1] to get:

把它们代入 [1] 的表达式得：

$$[1]=\left[\frac{\partial}{\partial x}\left(\lambda\frac{\partial t}{\partial x}\right)+\frac{\partial}{\partial y}\left(\lambda\frac{\partial t}{\partial y}\right)+\frac{\partial}{\partial z}\left(\lambda\frac{\partial t}{\partial z}\right)\right]\mathrm{d}x\mathrm{d}y\mathrm{d}z\mathrm{d}\tau$$

[The heat generated by internal heat source in micro-element body during dτ time 2]:

[在 dτ 时间内微元体中内热源的发热量 2]：

$$[2]=\dot{\Phi}\mathrm{d}x\mathrm{d}y\mathrm{d}z\mathrm{d}\tau$$

[The variation in thermodynamic internal energy of micro-element body during dτ time 3]:

[在 dτ 时间内微元体热力学能的变化量 3]：

$$[3]=\rho c\frac{\partial t}{\partial \tau}\mathrm{d}x\mathrm{d}y\mathrm{d}z\mathrm{d}\tau$$

The calculation method of internal energy variation is based on the middle school physics knowledge: When the variation in temperature of an object is Δt, the variation of internal energy is equal to:

内能变化量的计算方法是根据中学物理知识：一个物体温度变化为 Δt 时，内能的变化量等于：

$$cm\Delta t=c\rho\mathrm{d}x\mathrm{d}y\mathrm{d}z\frac{\partial t}{\partial \tau}\mathrm{d}\tau \tag{2.10}$$

In this way, the variation of internal energy is obtained. By the first law of thermodynamics, [1]+[2]=[3], after arranging, the differential equation of heat conduction is obtained, which is also known as the governing equation of heat conduction process:

这样就得到了内能的变化量。由热力学第一定律：[1]+[2]=[3]，整理后得到导热微分方程式，也称为导热过程的控制方程：

$$\rho c\frac{\partial t}{\partial \tau}=\frac{\partial}{\partial x}\left(\lambda\frac{\partial t}{\partial x}\right)+\frac{\partial}{\partial y}\left(\lambda\frac{\partial t}{\partial y}\right)+\frac{\partial}{\partial z}\left(\lambda\frac{\partial t}{\partial z}\right)+\dot{\Phi} \tag{2.11}$$

Unsteady-state term　Thermal diffusion terms　Heat source term

非稳态项　热扩散项 x　y　z　热源项

The differential equation of heat conduction is complicated, and can be simplified according to specific working conditions. Assuming that the physical parameters ρ, c and λ are constant and there is no internal heat source, the differential equation of heat conduction becomes:

导热微分方程式比较复杂，针对具体工况可进行简化。如果物性参数 ρ、c、λ 均为常数，且无内热源，导热微分方程式变为：

$$\frac{\partial t}{\partial \tau}=a\left(\frac{\partial^2 t}{\partial x^2}+\frac{\partial^2 t}{\partial y^2}+\frac{\partial^2 t}{\partial z^2}\right) \tag{2.12}$$

Where

式中，

$$a=\frac{\lambda}{\rho c} \tag{2.13}$$

The thermal diffusivity: reflects the relationship between the thermal conductivity and the heat storage capacity of materials in heat conduction. The greater the value of a, then the greater the value of λ or the smaller the value of ρc, which means that once a certain part of the object gets heat, and the heat can quickly diffuse (transfer) in the entire object. Therefore, the thermal diffusivity a reflects the ability of parts of an object to achieve uniform temperature when the object is heated or cooled.

热扩散率（导温系数）a：反映了导热过程中材料的导热能力（λ）与沿途物质储热能力（ρc）间的关系。a 值大，即 λ 值大或 ρc 值小，说明物体的某一部分一旦获得热量，该热量在整个物体中能很快扩散（传递）。因此，热扩散率 a 反映了物体被加热或冷却时物体内各部分温度趋向于均匀一致的能力。

Under the same heating conditions, the greater the thermal diffusivity of an object, the smaller the temperature difference inside the object. Heat diffusivity a reflects the dynamic characteristics of heat conduction process, and is an important physical quantity to study unsteady-state heat conduction.

在同样加热条件下，物体的热扩散率越大，物体内部各处的温差越小。热扩散率 a 反映了导热过程的动态特性，是研究非稳态导热的重要物理量。

Summary This section illustrates a basic method for theoretical analysis of heat transfer problems, and the differential equation of heat conduction is obtained by theoretical derivation.

小结 本节阐述了对传热问题进行理论分析的一种基本方法，通过理论推导得到了导热微分方程式。

Question Which physical parameters are independent of the steady-state heat conduction of solid?

思考题 固体的稳态导热过程与哪些物性参数无关？

Self-tests

自测题

1. What laws should be adopted to solve the differential equation of heat conduction?

1. 求解导热微分方程式要采用哪些定律？

2. Write the differential equation of heat conduction in rectangular coordinates.

2. 写出直角坐标系下的导热微分方程式。

3. Assuming that the physical parameters are constant, and no internal heat source and two-dimensional steady-state heat conduction, how to simplify the differential equation of heat conduction?

3. 如果物性参数均为常数、无内热源、二维稳态导热，导热微分方程式如何简化？

2.4 The Definite Solution Conditions of a Heat Conduction Process

2.4 导热过程的定解条件

In the previous section, the physical model and the geometric model are established by simplifying the actual heat transfer problem, and the differential equation of thermal conduction is obtained through theoretical derivation, which is the mathematical formulation of the heat conduction problem.

前面通过对实际导热问题的简化，建立了物理模型和几何模型，经过理论推导得到了导热微分方程式，也就是导热问题的数学描写。

The differential equation of thermal conduction describes the relationship between temperature t of an object with time τ and spatial coordinates x, y and z. It is a general equation suitable for all heat conduction problems. However, the equation neither involves the specific boundary conditions of the heat conduction process, nor shows the explicit functional relationship between physical quantities. So, for a specific heat conduction process, in order to obtain the unique solution that satisfies the equation, it is necessary to supply the definite solution conditions, which are the supplementary conditions to determine the unique solution of the equation.

导热微分方程式描述了物体的温度 t 随时间 τ 和空间坐标 x、y、z 的变化关系，是适合于所有导热问题的通用方程式。但是该方程式没有考虑导热过程的具体边界条件，也无法看出物理量之间明确的函数关系。因此，对于一个具体的导热过程，要得到满足该方程式的唯一解，需要补充定解条件，即确定该方程唯一解的补充说明条件。

For a specific heat conduction problem, a complete mathematical formulation includes the differential equation of thermal conduction and the definite solution conditions. The differential equation of thermal conduction is the governing equation or mathematical model of heat conduction problem. The definite solution conditions are also known as the unique value conditions to determine solution. That is to say, the solution of the equation is unique with these conditions.

对于一个具体的导热问题，完整的数学描写包括导热微分方程式和定解条件。导热微分方程式即导热问题的控制方程或数学模型，定解条件又称为单值性条件。就是说有了这些条件之后方程式的解是唯一的。

The definite solution conditions include four categories: geometric conditions, physical conditions, time conditions and boundary conditions.

定解条件包括四类：几何条件、物理条件、时间条件和边界条件。

(1) Geometric conditions, i. e. geometric model, illustrate the geometry and size of a heat conductor. For example, is it a plane wall or a cylindrical wall, the thickness of the plane wall and the diameter of the cylindrical wall are all geometric conditions.

(1) 几何条件：即几何模型，说明了导热体的几何形状和大小。比如是平壁还是圆筒壁，平壁的厚度、圆筒壁的直径等都属于几何条件。

(2) Physical conditions, i. e. physical model, explain the physical characteristics of a heat conductor. For example, whether the physical parameters λ, c, ρ and etc. vary with temperature, is there an internal heat source? If so, how about its size and distribution? Is the material isotropic and so on?

(2) 物理条件：即物理模型，说明了导热体的物理特征。比如物性参数 λ、c 和 ρ 等数值是否随温度变化，是否有内热源？如果有，其大小和分布怎样，材料是否为各向同性等。

(3) Time conditions, show the characteristics of heat conduction process in time, or whether is heat conduction process related to time. Time is independent for the steady-state heat conduction process, so there's no time condition. It is related to time for the unsteady-state heat conduction process, then the temperature distribution in a heat conductor should be given at the beginning of the process. The time condition is also called the initial condition.

(3) 时间条件：说明了在时间上导热过程进行的特点，或者说导热过程是否与时间有关。对于稳态导热过程，与时间无关，所以没有时间条件。对于非稳态导热过程，与时间有关，那么应给出过程开始时刻导热体内的温度分布。时间条件又称为初始条件。

(4) Boundary conditions, show the characteristics of the heat conduction process at the boundary of a thermal conductor, and reflect the interaction conditions between the heat conduction process and the surrounding environment. Generally, there are three types of boundary conditions:

(4) 边界条件：说明了导热体边界上导热过程进行的特点，反映了导热过程与周围环境之间相互作用的条件。边界条件一般有三类：

The first type of boundary condition: gives the temperature value at the boundary of the thermal conductor. For steady-state heat conduction, the boundary temperature t_w is constant. For the unsteady-state heat conduction, the boundary temperature t_w is a function of time when time $\tau>0$. For example, the boundary conditions on both sides of the plane plate in Fig. 2. 9:

第一类边界条件：给出了导热体边界上的温度值。对于稳态导热，边界温度 t_w 为常数。对于非稳态导热，当时间 $\tau>0$ 时，边界温度 t_w 为时间 τ 的函数。例如，图 2.9 中平板两侧的边界条件：

$$x=0, \quad t=t_{w1}$$

$$x=\delta, \quad t=t_{w2}$$

The second type of boundary condition: gives the heat flux value at the boundary of the object, as shown in Fig. 2. 10. When time $\tau>0$, the heat flux at the wall is:

第二类边界条件：给出了物体边界上的热流密度值，如图 2. 10 所示。当时间 $\tau>0$ 时，壁面处的热流密度为：

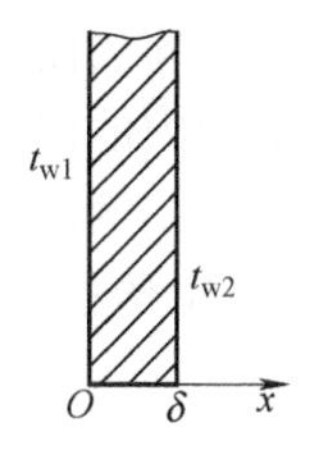

Fig. 2. 9 The first type of boundary condition

图 2. 9 第一类边界条件

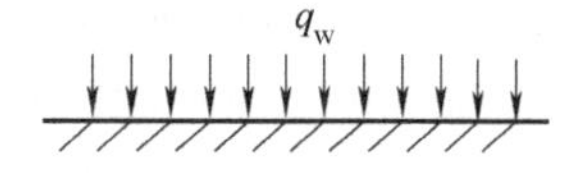

Fig. 2. 10 The second type of boundary condition

图 2. 10 第二类边界条件

$$q_w = -\lambda \left(\frac{\partial t}{\partial n} \right)_w \tag{2.14}$$

Dividing $-\lambda$ on both sides of the equation to obtain:

等式两边同除以 $-\lambda$，得到：

$$\left(\frac{\partial t}{\partial n} \right)_w = -\frac{q_w}{\lambda} \tag{2.15}$$

Where λ is the thermal conductivity of the wall. For the second type of boundary condition, the known heat flux is equivalent to the known temperature gradient in the normal direction of the wall. For steady-state heat conduction, the heat flux value is constant. For unsteady-state heat conduction, the heat flux value is a function of time, and the expression is:

式中，λ 是壁面的导热系数。对于第二类边界条件，已知热流密度相当于已知壁面法向的温度梯度值。对于稳态导热，热流密度值为常数。对于非稳态导热，热流密度值是时间的函数。表达式为：

$$q_w = f(\tau) \tag{2.16}$$

Another special case is an adiabatic boundary, namely the heat flux value is 0, which is equivalent to the temperature gradient of 0.

还有一个特例：绝热边界，即热流密度值为 0，相当于温度梯度为 0。

The third type of boundary condition: gives the surface heat transfer coefficient between the fluid and the wall on the object boundary and the fluid temperature when the convective heat transfer occurs between the object wall and the fluid, as shown in Fig. 2. 11.

第三类边界条件：当物体壁面与流体进行对流传热时，给出了物体边界上流体与壁面之间的表面传热系数和流体的温度，如图 2. 11 所示。

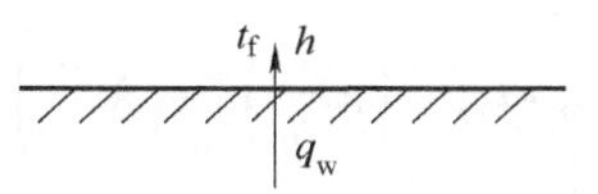

Fig. 2. 11 The third type of boundary condition

图 2. 11 第三类边界条件

According to Newton's law of cooling, the heat flux between the fluid and the wall is equal to the surface heat transfer coefficient h multiplied by the temperature difference between the wall and the fluid. According to Fourier's law, the heat flux of the wall is equal to the thermal conductivity multiplied by the temperature gradient in the wall. According to the law of conservation of energy, the heat conduction quantity of the wall should be equal to the convective heat transfer quantity of the fluid. So the expression is as follows:

根据牛顿冷却定律，流体与壁面之间的热流密度等于表面传热系数 h 乘以壁面与流体的温差。根据傅里叶定律，壁面的热流密度等于导热系数乘以壁面内的温度梯度。根据能量守恒定律，壁面导热量应该等于流体的对流传热量。所以有表达式：

$$-\lambda(\partial t/\partial n)_{w}=h(t_{w}-t_{f}) \tag{2.17}$$

Where the wall temperature and the temperature gradient are unknowns, λ is the thermal conductivity of the wall.

式中，壁温、温度梯度都是未知数；λ 是壁面的导热系数。

There are two complex boundary conditions in engineering:

在工程中还会遇到两种复杂的边界条件：

1) Radiation boundary condition. When only radiation heat transfer occurs between the surface of the heat conductor and the external environment with a temperature of T_e, such as the heating elements on the spacecraft to dissipate heat into space, then the expression of the boundary condition is as follows:

1）辐射边界条件。当导热体表面与温度为 T_e 的外部环境只发生辐射传热时，如航天器上的发热元件向太空散热，则边界条件的表达式如下：

$$-\lambda\frac{\partial T}{\partial n}=\varepsilon\sigma(T_{w}^{4}-T_{e}^{4}) \tag{2.18}$$

Namely, the heat conduction quantity of the object is equal to the radiation heat transfer quantity between them.

即物体的导热量等于两者之间传辐射传热量。

2) Interface continuous condition. When two kinds of materials are in good contact, the physical meaning of good contact is that the temperature and heat flux should be continuous at the interface, then the expressions of boundary conditions are as follows:

2）界面连续条件。当两种材料接触良好时，接触良好的物理意义是：其界面上应满足温度与热流密度连续的条件，则边界条件的表达式如下：

$$t_1=t_2,\quad\left(\lambda\frac{\partial t}{\partial n}\right)_1=\left(\lambda\frac{\partial t}{\partial n}\right)_2 \tag{2.19}$$

Namely, the temperature at the contact surface of two materials is equal, so is the heat flux.

即两种材料的接触面处温度相等，热流密度也相等。

After the definite solution conditions are determined, together with the differential equation of heat conduction, a complete mathematical formulation of heat conduction problem is formed, which is the governing equation group. Then figuring out the temperature field, and the temperature gradient can be obtained from the temperature field. Substituting the temperature gradient into the Fourier's law of heat conduction, and the heat flux can be obtained.

定解条件确定之后，与导热微分方程式一起，构成了导热问题完整的数学描写，即控制方程组。然后可以求出温度场，由温度场可以求出温度梯度，把温度梯度代入傅里叶导热定律即可求出热流密度。

Summary This section illustrates the definite solution conditions of the heat conduction process, including geometric conditions, physical conditions, time conditions and boundary conditions. We should focus on mastering the first, second and third type of boundary conditions.

小结 本节阐述了导热过程的定解条件，包括几何条件、物理条件、时间条件和边界条件。要重点掌握第一、二、三类边界条件。

Question Under what conditions can the third type of boundary condition of the heat conduction problem be transformed into the first type of boundary condition?

思考题 导热问题的第三类边界在什么条件下可转化第一类边界条件？

Self-test What are the definite solution conditions for solving the heat conduction problem? What types of boundary conditions are there?

自测题 求解导热问题的定解条件包括哪些？边界条件有哪几类？

授课视频

2.5 The Steady-state Heat Conduction of a Single-layer Plane Wall

2.5 单层平壁的稳态导热

First of all, a typical one-dimensional steady-state heat conduction problem is described: one-dimensional, steady-state, constant physical properties and no internal heat source. Under these conditions, investigate the heat conduction situation of the plane plate or the cylinder wall, that is to solve two physical quantities: temperature distribution and heat flux. The general solving steps of heat conduction problems are as follows:

首先给出典型一维稳态导热问题：一维、稳态、常物性、无内热源。在这样的条件下，考察平板/圆筒壁的导热状况，即求解两个物理量：温度分布和热流密度。导热问题的一般求解步骤如下：

(1) List the differential equation of heat conduction in the corresponding coordinate system; Rectangular coordinate system is used to solve the heat conduction of the plane wall; Cylindrical coordinate system is used to solve the heat conduction of cylinder wall or tube wall.

(1) 列出相应坐标系下导热微分方程式。求解平壁导热时，用直角坐标系；求解圆筒壁/管壁导热时，用圆柱坐标系。

(2) Give the definite solution conditions, and simplify the differential equation according to the geometric conditions, physical conditions and time conditions.

(2) 给出定解条件，并根据几何条件、物理条件和时间条件，简化微分方程式。

(3) Solve the differential equation to obtain the temperature distribution combined with the boundary conditions.

(3) 结合边界条件，求解微分方程，从而得出温度分布。

(4) Figure out the temperature gradient, and substitute it into Fourier's law to obtain the heat flux.

(4) 求出温度梯度，并代入傅里叶定律，即可求出热流密度。

The definite solution conditions for the thermal conduction of a single layer plane wall are given as follows. **Geometry conditions**: single layer plane plate with the thickness of δ; **Physical conditions**: known thermal physical parameters ρ, c, λ, and no internal heat source; **Time condition**: the steady-state heat conduction; **Boundary conditions**: the surface temperatures on both sides of the plane plate is uniform and constant, which are t_1 and t_2 respectively.

下面给出单层平壁导热的定解条件。**几何条件**：单层平壁，厚度为 δ；**物理条件**：热物性参数 ρ、c、λ 均已知，并且无内热源；**时间条件**：稳态导热；**边界条件**：平板两侧表面温度均匀且恒定，分别为 t_1 和 t_2。

According to the general steps to solve the heat conduction problem, first list the differential equations of heat conduction in rectangular coordinate system:

根据求解导热问题的一般步骤，先列出直角坐标系下导热微分方程式：

$$\rho c \frac{\partial t}{\partial \tau} = \frac{\partial}{\partial x}\left(\lambda \frac{\partial t}{\partial x}\right) + \frac{\partial}{\partial y}\left(\lambda \frac{\partial t}{\partial y}\right) + \frac{\partial}{\partial z}\left(\lambda \frac{\partial t}{\partial z}\right) + \dot{\Phi}$$

Then according to the geometric conditions, physical conditions and time conditions, the differential equation is simplified to obtain the governing equation:

然后，根据几何条件、物理条件和时间条件，对微分方程式进行简化，可得到控制方程：

$$\frac{\mathrm{d}^2 t}{\mathrm{d}x^2} = 0$$

Boundary conditions:

边界条件：

left boundary (左边界):
right boundary (右边界):

$$\begin{cases} x = 0, t = t_1 \\ x = \delta, t = t_2 \end{cases}$$

Directly integrate the differential equation twice to obtain the temperature distribution expression:

对微分方程直接积分两次，得温度分布表达式：

$$t=c_1x+c_2$$

Substitute the boundary conditions into the expression to obtain the undetermined coefficients c_1 and c_2:

把边界条件代入表达式，即可求出待定系数 c_1、c_2：

$$\Rightarrow \begin{cases} c_1=\dfrac{t_2-t_1}{\delta} \\ c_2=t_1 \end{cases}$$

Substitute c_1 and c_2 into the temperature distribution expression to obtain the temperature distribution in the plane plate:

把 c_1、c_2 代入温度分布表达式，可得平板内的温度分布：

$$t=\frac{t_2-t_1}{\delta}x+t_1$$

This is a linear function. Then derivation of the temperature t to coordinate x obtains the temperature change rate. Substitute it into Fourier's law of heat conduction to obtain the calculation formula of heat flux and heat transfer rate:

这是一个线性函数。然后，温度 t 对坐标 x 求导，得到温度变化率。把它代入傅里叶导热定律，即可得到热流密度和热流量的计算式：

$$q=-\lambda\frac{t_2-t_1}{\delta}=\frac{\Delta t}{\delta/\lambda} \tag{2.20}$$

$$\Phi=\frac{\Delta t}{\delta/(A\lambda)} \tag{2.21}$$

We find that the calculation formula of heat flux is the same as the Biot formula.

可以看到，热流密度的计算式与毕渥公式是一致的。

Example 2.1 The wall of a boiler is made of cement perlite with a density of 300kg/m^3 and wall thickness δ is 120mm, known the inner wall temperature $t_1=500$℃ and the outer wall temperature $t_2=50$℃. Try to figure out the heat loss per square meter of the furnace wall per hour.

例 2.1 一台锅炉，炉墙采用密度为 300kg/m^3 的水泥珍珠岩，壁厚 $\delta=120$mm，已知内壁温度 $t_1=500$℃，外壁温度 $t_2=50$℃。试求每平方米炉墙每小时的热损失。

Solution: For the actual heat transfer problem, the first thing is to simplify the problem and make assumptions as follows: ① One-dimensional problem; ② Steady-state heat conduction.

解：对实际传热问题，首先要进行简化和假设：① 一维问题；② 稳态导热。

For the cement perlite with a density of 300kg/m^3, the average thermal conductivity is a function of average temperature. So, it is necessary to calculate the heat transfer rate according to the thermal conductivity under the average temperature of the furnace wall. To figure out the average thermal conductivity, the average temperature of the material needs to be calculated first:

密度为 300kg/m^3 的水泥珍珠岩，平均导热系数是平均温度的函数。因此，需按炉墙平均温度下的导热系数来计算热流量。为求出平均导热系数，需要先算出材料的平均温度：

$$\bar{t}=\frac{500℃+50℃}{2}=275℃$$

Substitute it into the calculation formula of the average thermal conductivity to obtain the average thermal conductivity.

代入平均导热系数的计算式，可得平均导热系数。

$$\begin{aligned}\bar{\lambda}&=(0.0651+0.000105\times275)\mathrm{W/(m\cdot℃)}\\&=(0.0651+0.0289)\mathrm{W/(m\cdot℃)}\\&=0.094\mathrm{W/(m\cdot℃)}\end{aligned}$$

Substitute the average thermal conductivity into Biot formula, the heat loss per square meter of furnace wall can be obtained:

把平均导热系数代入毕渥公式，可求出每平方米炉墙的热损失：

$$q=\frac{\lambda}{\delta}(t_1-t_2)=\frac{0.094\mathrm{W/(m\cdot℃)}}{0.12\mathrm{m}}\times(500℃-50℃)=353\mathrm{W/m^2}$$

Summary　This section illustrates the solution method of the steady-state heat conduction of a single layer plane wall. We should focus on the analysis method of heat conduction problem and the simplification method of differential equation of heat conduction.

小结　本节阐述了单层平壁稳态导热问题的求解方法。要重点掌握导热问题的分析方法，以及导热微分方程式的简化方法。

Question　If the thermal conductivity of the plane plate is not constant, is the temperature distribution in the plane plate still linear?

思考题　如果平板的导热系数不是常数，平板内温度分布还是线性吗？

Self-test　What are the main steps to solve the heat conduction problem?

自测题　求解导热问题主要步骤有哪些？

2.6 The Steady-state Heat Conduction of a Multilayer Plane Wall

2.6 多层平壁的稳态导热

A multilayer plane wall is composed of several layers of plane wall with different materials. For example, the walls of a house are composed of a lime inner layer, a cement mortar layer, a red brick main layer and so on, as shown in Fig. 2. 12.

多层平壁是由几层不同材料的平壁组合而成的。例如，房屋的墙壁由白灰内层、水泥砂浆层、红砖主体层等组成，如图 2. 12 所示。

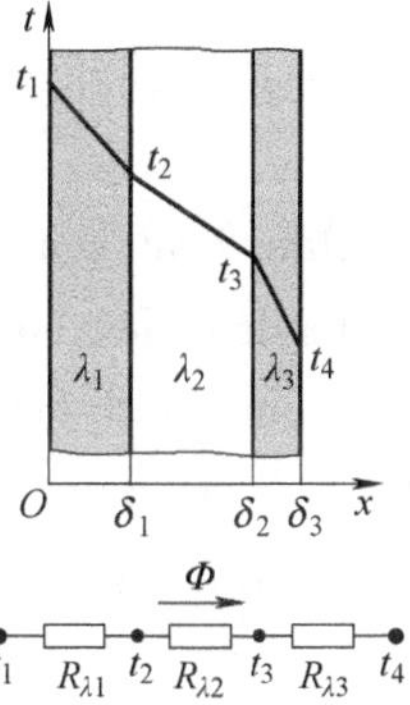

Fig. 2. 12 The thermal resistance of the walls of house

图 2. 12 房屋墙壁热阻图

Assuming good contact between layers, the temperature on the contact surface can be considered equal. Boundary conditions:

假设各层之间接触良好，可认为接触面上各处温度相等。边界条件：

$$\text{left boundary(左边界)}: \quad x=0, \quad t=t_1$$

$$\text{right boundary(右边界)}: \quad x=\sum_{i=1}^{3}\delta_i, \quad t=t_4$$

The thermal resistance of each layer is its thickness divided by its thermal conductivity:

各层热阻分别是其厚度除以其导热系数：

$$r_i=\frac{\delta_i}{\lambda_i}$$

According to the superposition principle of series thermal resistance, the heat flux of the whole heat conduction process can be obtained.

由串联热阻叠加原理，可求出整个导热过程的热流密度。

$$q = \frac{\Delta t}{\sum_{i=1}^{3} r_i} = \frac{t_1 - t_4}{\sum_{i=1}^{3} \frac{\delta_i}{\lambda_i}} \tag{2.22}$$

After calculating the heat flux q, for each layer of plane wall, the calculation formula of unknown temperature can be derived by using Biot formula respectively. For example, the first layer:

热流密度 q 求出后，对每一层平壁分别采用毕渥公式，可导出未知温度的计算式。比如第一层：

$$q = \frac{\lambda_1}{\delta_1}(t_1 - t_2) \quad \Rightarrow \quad t_2 = t_1 - q\frac{\delta_1}{\lambda_1} \tag{2.23}$$

In the same method, t_3 can be obtained. The second layer:

同样方法，可求出 t_3。第二层：

$$q = \frac{\lambda_2}{\delta_2}(t_2 - t_3) \quad \Rightarrow \quad t_3 = t_2 - q\frac{\delta_2}{\lambda_2}$$

Next, the boundary conditions of steady-state heat conduction of a multilayer plane wall are extended. How to calculate the heat flux q of a multilayer plane wall for steady-state heat transfer under the third type of boundary condition? Similarly, adopt the superposition principle of series thermal resistance:

下面对多层平壁稳态导热问题的边界条件进行拓展。多层平壁在第三类边界条件下，稳态传热的热流密度 q 该如何计算呢？同样，采用串联热阻叠加原理：

$$q = \frac{t_{f1} - t_{f2}}{\frac{1}{h_1} + \sum_{i=1}^{n} \frac{\delta_i}{\lambda_i} + \frac{1}{h_2}} \tag{2.24}$$

To sum up, no matter what type of boundary conditions, as long as it is steady-state heat transfer, then, the heat flux of the heat transfer process is equal to the total temperature difference of the heat transfer process divided by the corresponding total thermal resistance.

总结：不论是第几类边界条件，只要是稳态传热，那么，传热过程的热流密度就等于传热过程的总温差除以对应的总热阻。

Example 2.2 The furnace wall of a boiler is made up of three layers of materials, the inner layer is a chamotte brick with a thickness of 115mm, the middle layer is a grade B diatomite brick with a thickness of 125mm, and the outer layer is asbestos board with a thickness of 70mm. It is known that the temperatures of inner and outer surface of the furnace wall are $t_1 = 495$℃、$t_4 = 60$℃

例 2.2 一台锅炉的炉墙由三层材料叠合而成，最里面是耐火黏土砖，厚为 115mm，中间是 B 级硅藻土砖，厚为 125mm，最外层为石棉板，厚为 70mm。已知炉墙内、外表面温度分别为 $t_1 = 495$℃、$t_4 = 60$℃，试求每平方米炉墙每小时的

respectively, try to calculate hourly heat loss per square meter of furnace wall and the temperature t_2 at the interface between the chamotte brick and the diatomite brick.

热损失以及耐火黏土砖与硅藻土砖分界面上的温度 t_2。

Solution: For the actual heat transfer problem, first simplify and assume: one-dimensional, steady-state, and no thermal contact resistance. The thermal conductivity of chamotte bricks and diatomite bricks is a function of temperature, and it can be calculated according to the average temperature, but the interface temperature t_2 and t_3 are unknown quantities, we need to use the iterative method, i. e. trial and error method.

解: 对实际传热问题，首先要进行简化和假设：一维、稳态、无接触热阻。耐火黏土砖与硅藻土砖的 λ 都是温度的函数，可按照平均温度计算，但界面温度 t_2 和 t_3 都是未知量，需用迭代法（试错法）。

The basic solution steps of the iterative method are as follows: First assume t_2 and t_3 according to the outer wall temperature, and then calculate the average temperature:

迭代法的基本求解步骤是：先根据外壁面温度，假设 t_2 和 t_3，然后求出平均温度：

$$\bar{t}_1=\frac{t_1+t_2}{2}\quad \bar{t}_2=\frac{t_2+t_3}{2}$$

Substitute the average temperature into the calculation formula of the average thermal conductivity to obtain the average thermal conductivity of chamotte brick is:

把平均温度代入平均导热系数的计算式，可求出耐火黏土砖的平均导热系数：

$$\{\bar{\lambda}_1\}_{W/(m\cdot ℃)}=0.0477+0.0002\{\bar{t}_1\}_{℃}$$

The average thermal conductivity of diatomite brick is:

硅藻土砖的平均导热系数：

$$\{\bar{\lambda}_2\}_{W/(m\cdot ℃)}=0.7+0.00058\{\bar{t}_2\}_{℃}$$

Based on the superposition principle of series thermal resistance:

由串联热阻叠加原理：

$$q=\frac{t_1-t_4}{\dfrac{\delta_1}{\bar{\lambda}_1}+\dfrac{\delta_2}{\bar{\lambda}_2}+\dfrac{\delta_3}{\lambda_3}}$$

After the heat flux q is known, t_2 and t_3 can be calculated:

求出热流密度 q 之后，再反算出 t_2 和 t_3：

$$t_2=t_1-q\frac{\delta_1}{\bar{\lambda}_1},\quad t_3=t_2-q\frac{\delta_2}{\bar{\lambda}_2}$$

The calculated values of t_2 and t_3 are compared with the hypothetical values, if the error of both is within 10%, then the hypothetical value is considered reasonable, which is also known as iterative computation convergence. If not, we need to re-assume t_2 and t_3, and recalculate until the calculation convergence.

t_2 和 t_3 的反算值与假设值进行比较，如果两者误差在 10%以内，认为假设值合理，称为迭代计算收敛。如果超过 10%，需要重新假设 t_2 和 t_3，重新计算，直到计算收敛。

Summary　This section illustrates the calculation methods of heat flux and intermediate wall temperature during steady-state heat conduction of a multilayer plane wall.

小结　本节阐述了多层平壁的稳态导热过程中热流密度和中间壁温的计算方法。

Question　In the steady-state heat conduction process of a multi-layer plane wall, what parameters does the slope of temperature distribution in each layer depend on?

思考题　多层平壁的稳态导热过程中，各层壁面内温度分布的斜率取决于什么参数？

Self-test　What principle is used to solve the steady-state heat conduction problem of a multilayer plane wall?

自测题　求解多层平壁稳态导热问题采用什么原理？

授课视频

2.7 The Steady-state Heat Conduction of a Single-layer Cylinder Wall

2.7 单层圆筒壁的稳态导热

It is inconvenient to use a rectangular coordinate system to solve the steady-state heat conduction problem of cylindrical wall, so a cylindrical coordinate system is needed. In a cylindrical coordinate system, as shown in Fig. 2.13, the differential equation of heat conduction of a single-layer cylinder wall, i. e. circular tube, is as follows:

求解圆筒壁的稳态导热问题，用直角坐标系求解不便，需使用圆柱坐标系。在圆柱坐标系下，如图 2.13 所示，单层圆筒壁（即圆管）的导热微分方程式如下：

$$\rho c \frac{\partial t}{\partial \tau} = \frac{1}{r}\frac{\partial}{\partial r}\left(\lambda r \frac{\partial t}{\partial r}\right) + \frac{1}{r^2}\frac{\partial}{\partial \varphi}\left(\lambda \frac{\partial t}{\partial \varphi}\right) + \frac{\partial}{\partial z}\left(\lambda \frac{\partial t}{\partial z}\right) + \dot{\Phi} \tag{2.25}$$

A brief description of this formula is as follows: Just like the differential equation of heat conduction in rectangular coordinate system, the left side of the equal sign is the

对该式加以简单说明：与直角坐标系下的导热微分方程一样，等号左侧为非稳态项，即温度与时间有关，

unsteady-state term, which indicates that temperature is related to time; The first three terms on the right side of the equal sign are respectively the net heat imported into and exported from the micro element body along the coordinate r, φ, z direction, and the last term is the internal heat source.

等号右侧前三项分别是沿坐标 r、φ、z 方向导入、导出微元体的净热量，最后一项为内热源项。

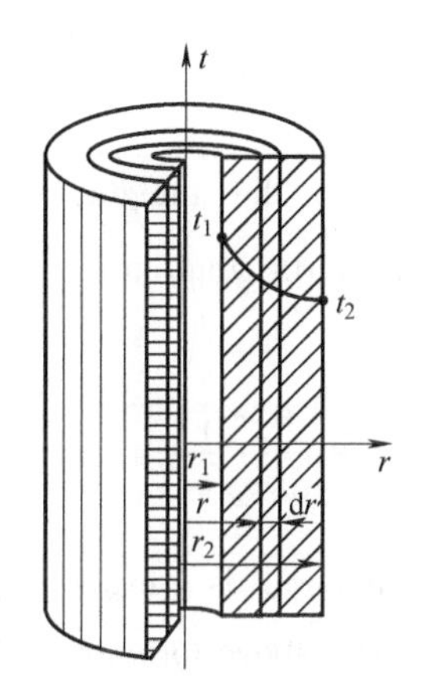

Fig. 2. 13 The temperature distribution of the cylinder wall

图 2. 13 圆筒壁的温度分布

Assuming that the tube length is l, the outer radius of the cylinder wall is less than 1/10 of the tube length and the inner/outer wall temperature is known. The definite solution conditions are one-dimensional, steady-state, no internal heat source and constant physical property. Simplify the differential equation according to the definite solution conditions, to obtain the simplified equation:

假设管长为 l，圆筒壁的外半径小于管长的 1/10，并已知内/外壁面温度。定解条件是一维、稳态、无内热源、常物性。根据定解条件对微分方程式进行简化，可得到简化方程：

$$\frac{d}{dr}\left(r\frac{dt}{dr}\right)=0 \tag{2.26}$$

The boundary conditions are the first type:

边界条件为第一类：

$$\begin{aligned}&\text{inner boundary(内边界):}\\&\text{outer boundary(外边界):}\end{aligned}\quad\begin{cases}r=r_1, & t=t_1\\ r=r_2, & t=t_2\end{cases}$$

Similarly, integrate the simplified equation twice to get the expression of temperature distribution:

同样，对简化方程积分两次，可得到温度分布表达式：

$$r\frac{dt}{dr}=c_1 \quad\Rightarrow\quad t=c_1\ln r+c_2$$

Apply the boundary conditions to get the binary linear equation group:

应用边界条件得到二元一次方程组：

$$\begin{cases} t_1 = c_1 \ln r_1 + c_2 \\ t_2 = c_1 \ln r_2 + c_2 \end{cases}$$

Solve the equation group to obtain the undetermined coefficients c_1 and c_2:

解方程组，即可得到待定系数 c_1、c_2：

$$c_1 = \frac{t_2 - t_1}{\ln(r_2/r_1)}, \quad c_2 = t_1 - \frac{t_2 - t_1}{\ln(r_2/r_1)} \ln r_1$$

Substitute the coefficients c_1 and c_2 into the integral formula to obtain the formula of temperature distribution in the cylinder wall:

把系数 c_1、c_2 代入积分表达式，得到圆筒壁内的温度分布表达式：

$$t = t_1 + \frac{t_2 - t_1}{\ln(r_2/r_1)} \ln(r/r_1) \tag{2.27}$$

The temperature distribution in the cylinder wall is a logarithmic curve along the radial direction. The above formula of temperature distribution in the cylinder wall is an explicit form, and therc is no regularity between physical quantities, so it is difficult to grasp. After simple mathematical deformation, it can be written in an implicit form:

圆筒壁内温度分布沿径向呈对数曲线。上面的圆筒壁内温度分布表达式为显式的形式，物理量之间缺乏规律性，因此不容易掌握。经过简单的数学变形，可写成隐式的形式：

$$\frac{t - t_1}{t_2 - t_1} = \frac{\ln \dfrac{r}{r_1}}{\ln \dfrac{r_2}{r_1}} \tag{2.28}$$

t and r correspond to each other in the formula.

式中 t 与 r 前后是对应的。

Next, solve for the heat flux and heat transfer rate on the arbitrary circular cross-section of a single-layer cylinder wall. According to the above formula of temperature distribution in the cylinder wall, the derivative of temperature t with respect to coordinate r gives the temperature change rate:

接下来求解单层圆筒壁任意圆截面上的热流密度和热流量。根据上述的圆筒壁内温度分布表达式，温度 t 对坐标 r 求导，得到温度变化率：

$$\frac{\mathrm{d}t}{\mathrm{d}r} = \frac{1}{r} \frac{t_2 - t_1}{\ln(r_2/r_1)}$$

Substitute it into Fourier's law, we can obtain the heat flux:

代入傅里叶定律即可求出热流密度：

$$q = -\lambda \frac{\mathrm{d}t}{\mathrm{d}r} = \frac{\lambda}{r} \frac{t_1 - t_2}{\ln(r_2/r_1)} \tag{2.29}$$

It is steady-state heat conduction in the single-layer cylinder wall, but the heat flux decreases with the increase of radius, so the heat flux at each cross-section of the cylindrical wall is not a constant. Then calculate the heat transfer rate of the cylinder wall:

虽然单层圆筒壁内是稳态导热，但由于热流密度 q 随半径 r 的增大而减小，所以，圆筒壁内各个截面上的热流密度 q 并不是常数。然后计算圆筒壁的热流量：

$$\Phi=2\pi rlq=\frac{t_1-t_2}{\dfrac{\ln(r_2/r_1)}{2\pi\lambda l}}=\frac{t_1-t_2}{R_\lambda} \tag{2.30}$$

The formula (2.30) shows that the heat transfer rate of the cylinder wall is still a constant in steady state. Where R_λ denotes the thermal conduction resistance of the cylinder wall with length l.

式（2.30）表明，在稳态情况下，圆筒壁的热流量仍然是常数。其中，R_λ 表示长度为 l 的圆筒壁的导热热阻。

$$R_\lambda=\frac{1}{2\pi\lambda l}\ln\left(\frac{r_2}{r_1}\right) \tag{2.31}$$

The heat transfer rate of the cylinder wall can be denoted by the heat transfer rate per unit tube length:

圆筒壁的热流量，可用单位管长热流量表示：

$$q_l=\frac{t_1-t_2}{\dfrac{1}{2\pi\lambda}\ln\left(\dfrac{d_2}{d_1}\right)} \tag{2.32}$$

Example 2.3 There is a circular tube with an inner diameter of 30mm and an outer diameter of 50mm, its thermal conductivity is 25W/(m·℃), the temperature of the inner wall is 40℃, and the temperature of the outer wall is 20℃. Try to figure out the heat transfer rate in the wall per unit tube length and the formula of the temperature distribution in the tube wall.

例 2.3 有一圆管内径为 30mm，外径为 50mm，其导热系数为 25W/(m·℃)，内壁面温度为 40℃，外壁面温度为 20℃。试求通过壁面的单位管长热流量和管壁内温度分布的表达式。

Solution: By the heat transfer rate calculation formula of the cylinder wall, the heat transfer rate per unit tube length can be calculated:

解：由圆筒壁的热流量计算式，可求出单位管长热流量：

$$q_l=\frac{t_1-t_2}{\dfrac{1}{2\pi\lambda}\ln\dfrac{d_2}{d_1}}=\frac{40-20}{\dfrac{1}{2\times25\pi}\ln\dfrac{50}{30}}\text{W/m}=6150.0295\text{W/m}$$

Then according to the temperature distribution formula of the cylinder wall (implicit):

再由圆筒壁的温度分布表达式（隐式）：

$$\frac{t-t_1}{t_2-t_1}=\frac{\ln\frac{r}{r_1}}{\ln\frac{r_2}{r_1}}$$

Substitute into the known data, the formula of temperature distribution inside the tube wall can be obtained:

代入已知数据，即可得到管壁内温度分布的表达式：

$$\frac{t-40}{20-40}=\frac{\ln r-\ln 0.015}{\ln\frac{25}{15}}$$

$$t=-39.1520\ln r-124.428$$

Summary This section illustrates the solution method of the steady-state heat conduction of a single-layer cylinder wall with the first type of boundary condition, the temperature distribution and the calculation formula of heat transfer rate in the tube wall are obtained.

小结 本节阐述了单层圆筒壁的稳态导热问题在第一类边界条件下的求解方法，得到了管壁内温度分布及热流量计算式。

Question In the case of steady-state heat conduction, the temperature distribution in the single-layer plane wall is a straight line, why is the temperature distribution in the single-layer cylinder wall a logarithmic curve?

思考题 在稳态导热情况下，单层平壁内的温度分布为直线，为什么单层圆筒壁内温度分布呈对数曲线？

Self-tests

自测题

1. What coordinate system is suitable for solving a single-layer cylinder wall?

1. 求解单层圆筒壁用什么坐标系比较适合？

2. Write the expression of thermal conductivity resistance of a single-layer cylinder wall.

2. 写出单层圆筒壁的导热热阻表达式。

授课视频

2.8 The Steady-state Heat Conduction of a Multilayer Cylindrical Wall

2.8 多层圆筒壁的稳态导热

In the case of steady-state heat conduction, the calculation method of heat conduction quantity of a multilayer cylinder wall composed of different materials is the same as that of

在稳态导热情况下，由不同材料所构成的多层圆筒壁，其导热量计算方法与多层平壁是一样的。总导热

the multilayer plane wall. The total heat conduction quantity is equal to the total temperature difference divided by the total heat conduction resistance. So under the first type of boundary condition, namely, the surface temperatures of the inner/outer walls are known, the calculation formula of the heat conduction quantity of the three-layer cylindrical wall can be written as:

量等于总温差除以总导热热阻。因此，在第一类边界条件下，即已知内/外壁面的表面温度，三层圆筒壁的导热量计算公式可写为：

$$\Phi=\frac{t_1-t_4}{\sum_{i=1}^{3}\frac{1}{2\pi\lambda_i L}\ln\frac{r_{i+1}}{r_i}} \tag{2.33}$$

The denominator is the sum of heat conduction resistance of the three-layer cylindrical wall. Divide the both sides of the formula by the length L of the cylinder wall to obtain the heat transfer rate per unit length of the three-layer cylindrical wall:

分母为三层圆筒壁的热阻之和。对该式两边同除以圆筒壁长度 L，可得到单位长度三层圆筒壁的热流量。

$$q_l=\frac{t_1-t_4}{\sum_{i=1}^{3}\frac{1}{2\pi\lambda_i}\ln\frac{r_{i+1}}{r_i}} \tag{2.34}$$

The denominator is the sum of heat conduction resistance per unit length of the three-layer cylindrical wall. For a single-layer cylindrical wall of unit tube length, the steady-state heat transfer process under the third type of boundary condition, we can figure out the convective heat transfer quantity inside the tube:

分母是三层圆筒壁的单位长度热阻之和。对于单位管长的单层圆筒壁，在第三类边界条件下的稳态传热过程，可计算出管内对流传热量：

$$q_l\,|_{h1}=2\pi r_1 h_1(t_{f1}-t_{w1}) \tag{2.35}$$

The heat conduction quantity of the tube wall:

管壁的导热量：

$$q_l=\frac{t_{w1}-t_{w2}}{\frac{1}{2\pi\lambda}\ln\frac{r_2}{r_1}} \tag{2.36}$$

The convective heat transfer quantity outside the tube:

管外对流传热量：

$$q_l\,|_{h2}=2\pi r_2 h_2(t_{w2}-t_{f2}) \tag{2.37}$$

For the total heat transfer process, the calculation formula of heat transfer rate per unit tube length is:

对于总传热过程，单位管长的热流量计算式：

$$q_l = \frac{t_{f1} - t_{f2}}{\frac{1}{2\pi r_1 h_1} + \frac{1}{2\pi\lambda}\ln\frac{r_2}{r_1} + \frac{1}{2\pi r_2 h_2}} = \frac{\Delta t_f}{R_l} \tag{2.38}$$

Where R_l is the total heat resistance of the heat transfer process per unit length of the cylindrical wall.

式中，R_l 是单位长度圆筒壁的传热过程的总热阻。

Example 2.4 The outer diameter of a tube is $2r$ and the temperature of the outer wall is t_1. If it is wrapped with two layers of thermal insulation material with a thickness of r(i.e. $\delta_2 = \delta_3 = r$) outside the tube, the thermal conductivity is λ_2 and λ_3 respectively, and $\lambda_2 = 2\lambda_3$, the surface temperature of the outer layer is t_2. If the positions of the two layers of thermal insulation material are reversed and other conditions remain unchanged, how about the thermal insulation situation? What conclusions can be drawn from it?

例 2.4 某管道外径为 $2r$，外壁温为 t_1。如果外包两层厚度均为 r 的保温材料（即 $\delta_2 = \delta_3 = r$），导热系数分别为 λ_2 和 λ_3，且 $\lambda_2 = 2\lambda_3$，外层材料的外表面温度为 t_2。如将两层保温材料的位置对调，其他条件不变，保温情况如何变化？由此能得出什么结论？

Analysis: The research object is two thermal insulation layers, which have nothing to do with the tube. How to reflect the change of insulation situation? It should be the heat dissipation.

分析：研究对象是两层保温层，与管道无关。“保温情况”变化怎么体现呢？应该是散热量。

Solution: Suppose that the inner and outer radii of the two insulation layers are r_2 and r_3, r_3 and r_4 respectively, then:

解：设两层保温层的内、外半径分别为 r_2 和 r_3、r_3 和 r_4，则：

$$r_3/r_2 = 2, \quad r_4/r_3 = 3/2$$

When the thermal insulation material with high thermal conductivity is placed inside, the heat dissipating capacity per unit length q_L can be calculated:

当导热系数大的保温材料放在里面时，可求出单位长度的散热量 q_L：

$$q_L = \frac{t_1 - t_2}{\frac{1}{2\pi\lambda_2}\ln\frac{r_3}{r_2} + \frac{1}{2\pi\lambda_3}\ln\frac{r_4}{r_3}} = \frac{\Delta t}{\frac{1}{2\pi \cdot 2\lambda_3}\ln 2 + \frac{1}{2\pi\lambda_3}\ln\frac{3}{2}} = \frac{\lambda_3 \Delta t}{0.11969}$$

When the thermal insulation material with high thermal conductivity is placed outside, the heat dissipating capacity per unit length q'_L can be calculated:

当导热系数大的保温材料放在外面时，可求出单位长度的散热量 q'_L：

$$q'_L = \frac{t_1 - t_2}{\frac{1}{2\pi\lambda_3}\ln 2 + \frac{1}{2\pi \cdot 2\lambda_3}\ln\frac{3}{2}} = \frac{\lambda_3 \Delta t}{0.1426}$$

Compare the heat dissipating capacity of the two cases and eliminate the same terms:

两种情况的散热量相比，消去相同的量：

$$\frac{q_L}{q'_L}=\frac{0.1426}{0.11969}=1.19$$

It can be seen that the first case, that is, when the thermal insulation material with high thermal conductivity is placed inside, the heat dissipating capacity is greater.

可见，第一种情况，即导热系数大的保温材料放在里面时，散热量更大。

Conclusion: When the thermal insulation material with high thermal conductivity is placed outside, and thermal insulation material with low thermal conductivity is placed inside, it is better for the overall insulation, namely, the heat dissipating capacity is smaller.

结论：导热系数大的保温材料放在外面，导热系数小的保温材料放在里面，对整体保温更有利，即散热量更小。

Summary This section illustrates the calculation method of heat transfer rate for steady-state heat conduction of a multilayer cylindrical wall.

小结 本节阐述了多层圆筒壁稳态导热的热流量计算方法。

Question What principle is used to solve the steady-state heat conduction problem of a multilayer cylindrical wall?

思考题 求解多层圆筒壁的稳态导热问题采用什么原理？

Self-tests

自测题

1. Write the expression of the thermal conductivity resistance of a multilayer cylindrical wall.

1. 写出多层圆筒壁的导热热阻表达式。

2. In steady-state heat conduction, why is the heat transfer rate through any circular section of a cylinder wall constant, but the heat flux q through any circular section of the cylinder wall not constant?

2. 在稳态导热情况下，为什么通过圆筒壁任意圆截面的热流量是常数，而通过圆筒壁任意圆截面的热流密度 q 不是常数？

授课视频

2.9 The Steady-state Heat Conduction with Complex Boundary Conditions

2.9 复杂边界条件的稳态导热

The complex boundary condition refer to the second and

复杂边界条件是指第二类和第三类

third types of boundary condition. Next, learn the solution method through an cxamplc, as shown in Fig. 2. 14.

边界条件。下面结合例题来学习求解方法，如图 2. 14 所示。

Example 2. 5 An electric iron with electric power of 1200W, its bottom is placed vertically in a room at ambient temperature of 25℃. The thickness of the metal bottom plate is 5mm and the thermal conductivity is $\lambda = 15\text{W}/(\text{m} \cdot ℃)$, the area of the bottom plate A is 300cm^2, the convective heat transfer coefficient including radiation effect $h = 80\text{W}/(\text{m}^2 \cdot ℃)$. Try to figure out the surface temperature on both sides of the bottom plate under the steady-state condition.

例 2. 5 一个电熨斗电功率为 1200W，底面竖直放置于环境温度为 25℃的房间中。金属底板厚度为 5mm，导热系数 $\lambda = 15\text{W}/(\text{m} \cdot ℃)$，底板面积 $A = 300\text{cm}^2$，考虑辐射作用在内的表面传热系数 $h = 80\text{W}/(\text{m}^2 \cdot ℃)$。试确定稳态条件下底板两侧表面的温度。

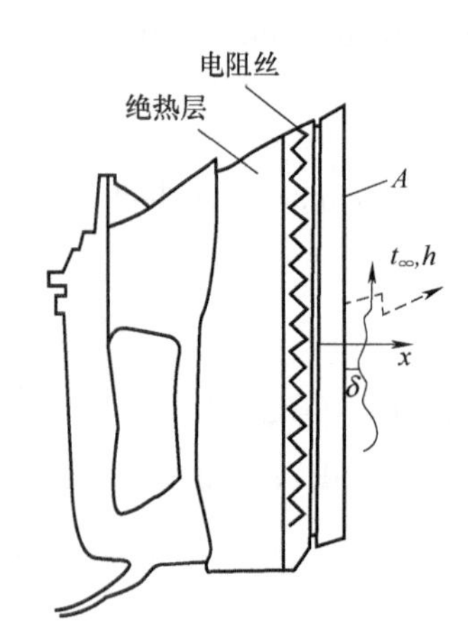

Fig. 2. 14 Heat dissipation diagram of electric iron underside

图 2. 14 电熨斗底面散热图

Solution: For the actual heat transfer problem, focus on how to analyze the problem. First of all, determine the research object. Obviously, the research object is not the entire electric iron, but a part of it, namely the metal bottom plate. The geometric model of the problem can be simplified to a plane plate.

解： 对于实际传热问题，要重点掌握分析问题的方法。首先确定研究对象。显然研究对象不是整个电熨斗，而是其中一部分：金属底板。问题的几何模型可简化为一个平板。

Next, simplify and assume the physical model of the actual problem. ① The thermal insulation layer has good performance; all heat is lost to the environment through the bottom plate. ② Heat is transferred only along the thickness of the plate, so this is a one-dimensional steady-state heat conduction problem.

接着对实际问题的物理模型进行简化和假设。① 绝热层性能良好，全部热量都通过底板散失到环境中。② 热量只沿着板厚方向传递，因此，属于一维稳态导热问题。

Known conditions: heat flux on the left side of the bottom plate, belongs to the second type boundary condition; convection heat transfer on the right side of the bottom plate, belongs to the third type boundary condition. According to the electric power of the electric iron, the heat flux q_0 at the left boundary of the bottom plate can be calculated.

已知条件：底板左侧的热流密度，属于第二类边界条件；底板右侧为对流传热，属于第三类边界条件。根据电熨斗的电功率，可计算出底板左侧边界的热流密度 q_0。

The mathematical formulation of the temperature field of the bottom plate includes: ① Governing equation, namely, the second derivative of t with respect to x is equal to 0:

底板温度场的数学描写包括：① 控制方程，即温度 t 对横坐标 x 的二阶导数等于 0：

$$\frac{\mathrm{d}^2 t}{\mathrm{d}x^2}=0$$

② Boundary conditions. On the left side: the heat conduction quantity of the plane plate is equal to the heat flux q_0;

② 边界条件。左侧面：平板的导热量等于热流密度 q_0：

$$x=0,-\lambda\frac{\mathrm{d}t}{\mathrm{d}x}=q_0$$

On the right side: the heat conduction quantity of the plane plate is equal to the convective heat transfer quantity.

右侧面：平板的导热量等于对流传热量。

$$x=\delta,-\lambda\frac{\mathrm{d}t}{\mathrm{d}x}=h(t-t_\infty)$$

The general solution of the above differential equation is:

上述微分方程的通解：

$$t=c_1x+c_2$$

Where c_1 and c_2 are undetermined constants. Substitute them into the boundary conditions on the left to get:

式中，c_1、c_2 为待定常数。代入左侧边界条件，得：

$$-\lambda c_1=q_0$$

Substitute them into the boundary conditions on the right to get:

代入右侧边界条件，得：

$$-\lambda c_1=h[(c_1\delta+c_2)-t_\infty]$$

Combine the two equations to get the expressions of c_1 and c_2:

两式联立可求得 c_1、c_2 的表达式：

$$c_1=-\frac{q_0}{\lambda},\quad c_2=t_\infty+\frac{q_0}{h}+\frac{q_0}{\lambda}\delta$$

Substitute c_1 and c_2 into the general solution to get the expression of temperature distribution along the thickness direction in the metal bottom plate:

把 c_1、c_2 代入通解，可得到金属底板内沿厚度方向的温度分布表达式：

$$t=t_{\infty}+q_0\left(\frac{\delta-x}{\lambda}-\frac{1}{h}\right)$$

Obviously, the temperature t is a linear function of the coordinate x. Next, substitute the known parameters and the coordinates of both sides of the wall to get the temperature of both sides of the wall:

显然，温度 t 是坐标 x 的线性函数。下面代入已知参数和壁面两侧的坐标，可得到两侧壁面的温度：

Left side $x=0$:

左侧面 $x=0$：

$$t_{x=0}=t_{\infty}+q_0\left(\frac{\delta}{\lambda}+\frac{1}{h}\right)=538℃$$

Right side $x=\delta$:

右侧面 $x=\delta$：

$$t_{x=\delta}=t_{\infty}+q_0\left(0+\frac{1}{h}\right)=525℃$$

Summarize of solution ideas: The solution of the one-dimensional steady-state heat conduction problem starts from solving the differential equation of the heat conduction. As long as the expression of temperature distribution in the plane plate is obtained, the temperature of the boundary surface on both sides of the plane plate can be obtained.

求解思路总结：该一维稳态导热问题的求解，首先求解导热微分方程式，只要求出平板内的温度分布表达式，即可求出平板两侧边界面上的温度。

In the previous solution of the heat conduction problem, the area of heat conduction and the thermal conductivity are considered to be constants. For steady state, no internal heat source and the first type of boundary condition, how to solve the one-dimensional heat conduction problem with variable cross-section and variable thermal conductivity? Also according to Fourier's law:

在前面的导热问题求解中，认为导热面积和导热系数都是常数。对于稳态、无内热源、第一类边界条件下，变截面、变导热系数一维导热问题，如何求解？同样要根据傅里叶定律：

$$\Phi=-\lambda A\frac{\mathrm{d}t}{\mathrm{d}x}$$

When $\lambda=\lambda(t)$ and $A=A(x)$, namely, both parameters are variables, Fourier formula becomes:

当 $\lambda=\lambda(t)$，$A=A(x)$ 时，即两个参数都是变量，傅里叶公式变成：

$$\Phi=-\lambda(t)A(x)\frac{\mathrm{d}t}{\mathrm{d}x} \tag{2.39}$$

Separate variables for temperature t and coordinate x and integrate them separately, and notice that the heat transfer rate Φ has nothing to do with the coordinate x during the steady-state heat conduction, so Φ can be moved outside the integral symbol:

对温度 t 和坐标 x 分离变量，分别进行积分，并注意到稳态导热时，热流量 Φ 与坐标 x 无关，所以 Φ 可提到积分号外边：

$$\Phi \int_{x_1}^{x_2} \frac{1}{A(x)} \mathrm{d}x = -\int_{t_1}^{t_2} \lambda(t) \mathrm{d}t$$

$$\Phi \int_{x_1}^{x_2} \frac{\mathrm{d}x}{A(x)} = -\int_{t_1}^{t_2} \lambda(t) \mathrm{d}t \frac{t_2-t_1}{t_2-t_1} = -\frac{\int_{t_1}^{t_2} \lambda(t) \mathrm{d}t}{t_2-t_1}(t_2-t_1)$$

$$\frac{\int_{t_1}^{t_2} \lambda(t) \mathrm{d}t}{t_2-t_1} = \bar{\lambda} \tag{2.40}$$

When λ changes linearly with temperature t, the average thermal conductivity at the average temperature can be directly obtained. Finally, the calculation formula of heat transfer rate is as follows:

当 λ 随温度 t 呈线性变化时，可直接求出平均温度下的平均导热系数。最后，热流量的计算式为：

$$\Phi = \frac{\bar{\lambda}(t_1-t_2)}{\int_{x_1}^{x_2} \frac{\mathrm{d}x}{A(x)}} \tag{2.41}$$

The important conclusions can be drawn from the above analysis: In the actual heat conduction problem, no matter how the thermal conductivity λ changes, as long as we figure out the average thermal conductivity, we can use the formula of constant thermal conductivity to calculate.

通过上述分析，可以得到一个重要结论：在实际导热问题中，无论导热系数 λ 如何变化，只要计算出平均导热系数，就可利用所有的“定导热系数”公式，进行计算。

Summary This section illustrates the solution methods for the one-dimensional steady-state heat conduction problem under the second and third types of boundary conditions, and the calculation methods of one-dimensional steady-state heat conduction with variable cross-section or variable thermal conductivity.

小结 本节阐述了第二、三类边界条件下一维稳态导热问题的求解方法，以及变截面或变导热系数一维稳态导热问题的计算方法。

Question For the steady-state heat conduction of a plane wall, if boundary conditions of both sides are the second or third type of boundary condition, can the temperature field be determined?

思考题 对于平壁的稳态导热问题，如果两个边界条件均为第二类或者第三类，温度场能否得出确定的解？

Self-tests

1. If the thermal conductivity is a variable, how to deal with it in the actual thermal conductivity calculation?

2. Under what boundary conditions must there be a solution to the steady-state heat conduction problem?

3. When the temperature on both sides of the plane plate is different, the thermal conductivity is also different at the cross-sections at different thickness even if it is a steady-state heat transfer, why is the thermal conductivity always taken as a constant in the calculation?

自测题

1. 若导热系数为变量，实际导热计算时如何处理？

2. 在什么边界条件下稳态导热问题肯定有解？

3. 当平板两侧的温度不同时，即使是稳态导热，不同厚度截面处的导热系数也不同，为什么计算时导热系数常取常数？

授课视频

2.10 The Enhanced Heat Transfer Methods for a Heat Transfer Process

2.10 传热过程的强化传热方法

Under the third type of boundary condition, of one-dimensional steady-state heat conduction in plane wall is shown in Fig. 2.15, and the heat transfer rate is calculated as follows:

在第三类边界条件下，平壁的一维稳态导热如图 2.15 所示，热流量计算如下：

$$\Phi = \frac{t_{f1} - t_{f2}}{\dfrac{1}{h_1 A} + \dfrac{\delta}{\lambda A} + \dfrac{1}{h_2 A}} \tag{2.42}$$

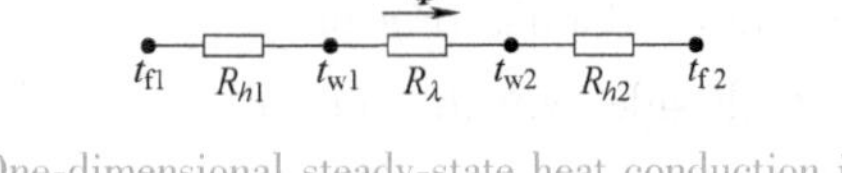

Fig. 2.15 One-dimensional steady-state heat conduction in plane walls

图 2.15 平壁的一维稳态导热

Where the numerator is the total temperature difference, which is the temperature difference between the two fluids on both sides of the wall; The denominator is the total thermal resistance, which includes the thermal conduction resistance of the wall $\frac{\delta}{\lambda A}$ and the convection thermal resistance of the fluid on both sides $\frac{1}{h_1 A}$, $\frac{1}{h_2 A}$.

式中，分子是总温差，即壁面两侧两种流体的温差；分母是总热阻，包括壁面的导热热阻$\frac{\delta}{\lambda A}$以及两侧流体的对流传热热阻$\frac{1}{h_1 A}$、$\frac{1}{h_2 A}$。

According to the calculation formula of heat transfer rate, in order to increase the heat transfer rate, the so-called enhanced heat transfer, the following measures can be taken:

根据热流量计算式，要想增大热流量，即所谓的强化传热，可采取以下措施：

(1) Increase the numerator, that is to increase the total temperature difference. The temperature difference can be increased by increasing the temperature of hot fluid or decreasing the temperature of cold fluid. However, it is sometimes difficult to achieve these requirements due to the restriction of the process conditions.

(1) 增大分子，即增大总温差。通过提高热流体的温度，或者降低冷流体的温度，都可以实现增大温差。但常常受到工艺条件限制，有时难以实现。

(2) Reduce the denominator, that is to reduce the total thermal resistance. The total thermal resistance includes the thermal conduction resistance of the wall and the convection thermal resistance of two fluids on both sides of the wall.

(2) 减小分母，即减小总热阻。总热阻包括壁面的导热热阻以及壁面两侧两种流体的对流传热热阻。

Next, the feasibility of the second type of methods is analyzed specifically as follows:

1) The metal wall is generally very thin, while the thermal conductivity λ is large, so the thermal conduction resistance is very small. It can almost be ignored, so it is difficult to enhance heat conduction.

下面具体分析一下第二类方法的可行性：

1) 金属壁一般很薄，而导热系数λ较大，所以导热热阻很小。它几乎可以被忽略，所以强化导热的空间很小。

2) In order to reduce the convection thermal resistance, we can try to improve the convective heat transfer coefficient of fluids on both sides. These methods are very effective and very commonly used, but it is difficult to improve the convective heat transfer coefficient, so it is always a hot topic in the heat transfer research.

2) 要减小对流传热热阻，我们可以想办法增大两侧流体的表面传热系数。这些方法很有效，也很常用，但提高表面传热系数难度较大，所以，一直是传热学研究的热点。

3) Increasing the heat transfer area A is a very commonly used method and easy to achicvc. It nceds to be explained that increasing the heat transfer area as mentioned here is not to simply increase the number of heat transfer tubes, instead increase the heat transfer area in unit volume by using extended surface in the case of constant volume, which is commonly used especially in gas heat transfer.

3）增大传热面积A，这是一种很常用的方法，并且容易实现。需要说明一下，这里所说的“增大传热面积”，并不是简单地增加传热管子数，而是在体积不变的情况下，通过扩展表面来增大单位体积内的传热面积，尤其在气体传热时很常用。

Next, focus on increasing the heat transfer area per unit volume. The effective method to increase the heat transfer area is to install ribs or fins. The so-called fins or ribs are the extended surfaces attached to the base surface. It is generally considered that the ribs and fins are the other names of the extended surface and they have no obvious difference. This textbook believes that when the height of the flake is small, it is called rib, and when the height is large, it is called fin.

接下来重点讨论增大单位体积内的传热面积。增大传热面积的有效方法主要是加装肋片或翅片，所谓肋片或翅片，就是依附于基础表面上的扩展表面。一般认为肋片、翅片是扩展表面的两个名称，没有明显区别。本教材认为，片的高度较小时称为肋片，片的高度较大时称为翅片。

In engineering and nature, we often see some objects with extended surfaces, such as ribs or fins installed outside the tube. When the weather is hot, the puppy may instinctively stick out its tongue or raise its ears. The biologists believe that dinosaurs are warm-blooded animals, their fins can enhance the heat dissipation during movement.

在工程上和自然界中，常常会看到一些带有凸出表面的物体，比如管外加装的肋片或翅片。小狗在天热时，也会本能地伸舌头或者翘耳朵。生物学家认为：恐龙是一种温血动物，其身上的鳍片可加强运动时的热量散失。

On the heat exchange surface of some heat exchange equipment, adding ribs or fins is an important measure to increase heat transfer quantity. The circular finned tubes and the H-shaped finned tubes are shown in Fig. 2.16 respectively.

在一些换热设备的换热面上，加装肋片或翅片是增大传热量的重要措施。图2.16所示分别是圆形翅片管和H形翅片管。

Fig. 2.16　Circular finned tubes and H-shaped finned tubes

图2.16　圆形翅片管和H形翅片管

There are many forms of ribs or fins: The longitudinal finned tube and the transverse finned tube are two main forms of finned tube, and their difference is the angle between the fin plane and the tube axis, as shown in Fig. 2. 17. Fig. 2. 18 shows various finned tubes shown in a foreign textbook, including outer finned tube and inner finned tube.

肋片或翅片的形式有很多种：纵向翅片管和横向翅片管是翅片管的两种主要形式，其区别是片平面与管轴线的夹角不同，如图 2. 17 所示。图 2. 18 所示是国外教材上展示的各种翅片管，有外翅片管和内翅片管两大类。

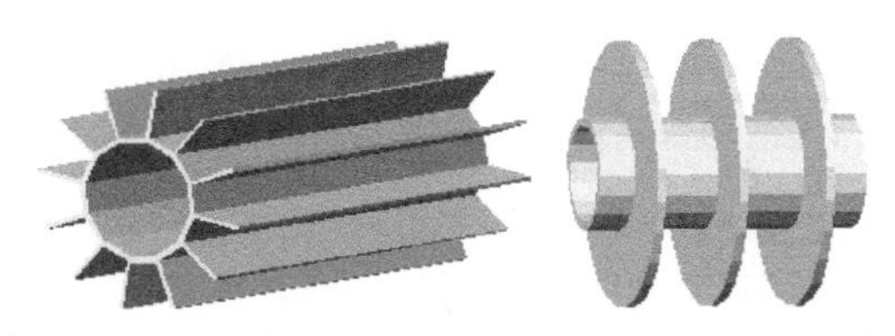

Fig. 2. 17 The longitudinal finned tube and the transverse finned tube

图 2. 17 纵向翅片管和横向翅片管

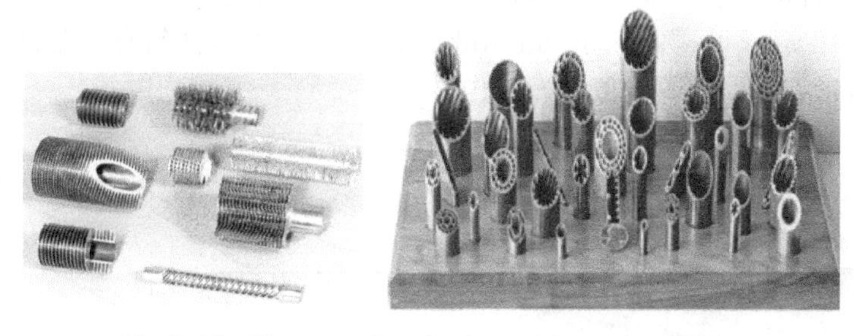

Fig. 2. 18 The outer finned tube and inner finned tube

图 2. 18 外翅片管和内翅片管

(From *A Heat Transfer Textbook* (4th edition) by John H. Lienhard V and John H. Lienhard IV)

From these finned/ribbed tubes above, we can know that the shapes of fin wall include straight fin, ring fin; there are uniform cross-sections and variable cross-sections. The typical structures of fins include needle fin, straight fin, ring fin, large plate fin and so on. No matter which kind of ribs or fins, the airflow direction is always parallel to the wall direction of the fins at runtime.

从以上这些翅片管/肋片管可以看出，肋壁的形状有直肋和环肋；有等截面，也有变截面。肋片的典型结构有针肋、直肋、环肋和大套片等形式。不管是哪一种肋片或翅片，运行时气流方向总是平行于肋片的壁面方向。

Summary This section illustrates the heat transfer enhancement methods of heat transfer process, and recognizes the structural characteristics of various fins.

小结 本节阐述了传热过程的强化传热方法，认识了各种翅片的结构特点。

Question What is the basic idea of heat transfer enhancement of a heat transfer process?

思考题 传热过程强化传热的基本思路是什么？

Self-tests

自测题

1. What are the enhanced heat transfer methods for one-dimensional steady-state heat conduction of a flat wall?

1. 平壁一维稳态导热的强化传热方法有哪些？

2. What is the main role of the fin? What are the types and structural forms?

2. 肋片的主要作用是什么？有哪些类型和结构形式？

授课视频

2.11 The Heat Conduction Analysis of the Straight Fin with Constant Cross-section

2.11 等截面直肋的导热分析

The straight fin with constant cross-section is a most common type of fin in engineering. Let's analyze the heat conduction characteristics of the straight fin with constant cross-section as follows, as shown in Fig. 2.19.

等截面直肋，是工程中最常见的一种肋片形式。下面分析等截面直肋的导热特点，如图 2.19 所示。

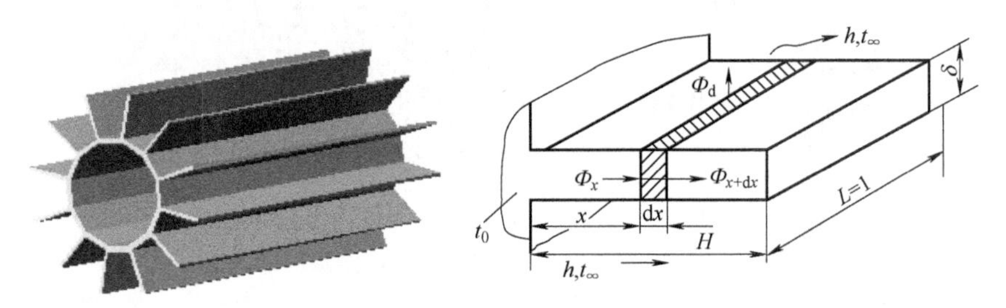

Fig. 2.19 The heat conduction model of the straight fin with constant cross-section

图 2.19 等截面直肋的导热模型

Known conditions: ① Thickness δ, length L and height H of a rectangular straight fin; ② The temperature of the fin root is t_0, and the temperature of the surrounding fluid/gas is t_∞; ③ The convective heat transfer coefficient between the fin and the fluid is h; ④ The thermal conductivity λ of the material, the convective heat transfer coefficient h and the cross-sectional area A_c of the fin are constant. Try to figure out the temperature field and heat transfer rate of the fin.

已知条件：① 矩形直肋的厚度 δ、长度 L、高度 H；② 肋根温度为 t_0，周围流体/气体的温度为 t_∞；③ 肋片与流体的表面传热系数为 h；④ 材料的导热系数 λ、表面传热系数 h 和肋片截面面积 A_c 均为常数。试求：肋片的温度场和热流量。

Heat conduction of the fin is a typical heat conduction problem, we should master the analysis ideas and methods of such problems. Strictly speaking, the temperature field in the actual fin is the heat conduction problem of three-dimensional, steady-state, no internal heat source, and

肋片导热，是一类典型的导热问题，要掌握分析这类问题的思路及方法。严格地说，实际肋片中的温度场是三维、稳态、无内热源、常物性，同时具有第一、第二、第三

constant physical properties, and with the first, second, and third type of boundary conditions. So the governing equation of the heat conduction process can be simplified as Laplace's equation:

类边界条件的导热问题。因此，导热过程的控制方程可简化为拉普拉斯方程：

$$\nabla^2 t=\frac{\partial^2 t}{\partial x^2}+\frac{\partial^2 t}{\partial y^2}+\frac{\partial^2 t}{\partial z^2}=0 \tag{2.43}$$

The governing equations of three-dimensional problems are more complicated and difficult to solve directly, so we need to ignore the secondary factors and simplify the problem, it is better to simplify the three-dimensional problem to one-dimensional problem. The basic ideas of simplification are as follows:

由于三维问题的控制方程比较复杂，直接求解很困难，所以需要忽略次要因素，对问题进行简化，最好将三维问题简化为一维问题。简化的基本思路如下：

(1) Fin length L is much greater than δ and H, so it can be considered that the temperature is uniform along the fin length direction; The unit length fin can be used for analysis, namely, $L=1$. The three-dimensional problem is reduced to two-dimension.

(1) 肋片长度 $L \gg \delta$、H，因此，可认为肋片长度方向温度均匀；可取单位长度来分析，即 $L=1$m。三维问题降了一维。

(2) The convection thermal resistance $1/h$ on the surface of the fin is much larger than the thermal conduction resistance of the fin δ/λ, so the thermal conduction resistance can be ignored, and the temperature is considered uniform along the thickness direction. In this way, one dimension is reduced again, and now only heat conduction along the height of the fin is left.

(2) 肋片表面的对流传热热阻 $1/h$ 远大于肋片的导热热阻 δ/λ，因此，导热热阻可忽略，认为温度沿厚度方向均匀。这样又降了一维，目前只剩下沿肋片高度方向的导热。

(3) Thermal conductivity of material λ, convective heat transfer coefficient h and cross-sectional area A_c of the fin are constant.

(3) 材料导热系数 λ、表面传热系数 h 和肋片截面面积 A_c 均为常数。

(4) The top surface area of the fin is small and the temperature is low, which can be regarded as adiabatic, namely, the heat flux is 0.

(4) 肋片顶端表面面积很小，温度较低，可视为绝热，即热流密度为 0。

The actual problem is ultimately simplified as one-dimensional steady-state heat conduction along the fin height, i.e., the x-axis direction. The simplified heat conduction model of straight fin with constant section is shown in Fig. 2. 20.

实际问题最终简化为沿肋片高度即 x 轴方向的一维稳态导热。等截面直肋的简化导热模型如图 2. 20 所示。注意：简化的只是几何模型，

Note that only the geometric model is simplified, the geometric shape and parameters of the actual object are unchanged.

实际物体的几何形状及参数都不变。

Boundary conditions of the model: The left boundary, i. e. the temperature t_0 of the fin root is given; the right boundary, i. e. the fin end is adiabatic, namely $q = 0$, the temperature change rate $dt/dx = 0$. The difficulty of the problem is how to deal with the upper and lower surfaces of the fin?

模型的边界条件：左边界——肋根给定温度 t_0；右边界——肋端绝热，$q = 0$，等价于温度变化率 $dt/dx = 0$。问题的难点：对肋片的上下表面该如何处理呢？

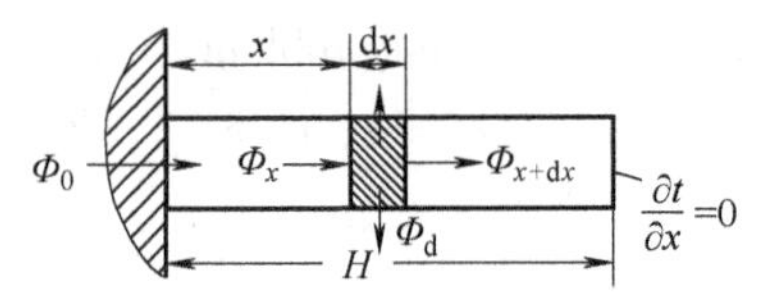

Fig. 2. 20 The simplified heat conduction model of straight fin with constant section

图 2. 20 等截面直肋的简化导热模型

Because the fin is simplified to a one-dimension heat conduction problem, namely, there is only heat conduction along the direction of the fin height. So there are only the left and right boundaries, and upper and lower surfaces of the fin are not computational boundary. In fact, the heat from the fin root transferred to the fin is mainly transferred to the fluid through convection heat transfer between the upper and lower surfaces and the fluid.

因为肋片已简化为一维导热问题，即只有沿肋片高度方向的导热。因而只有左边界和右边界，而肋片的上下表面不是计算边界。实际上从肋片根部导入肋片的热量主要通过上下表面与流体的对流传热而传给流体。

Through the above simplification and hypothesis of the actual problem, the geometric model, physical model, and boundary conditions of the actual problem are obtained. Next, there are two methods to establish the mathematical formulation of heat conduction of the fin: ① The law of conservation of energy and Fourier's law, which is the method of deducing differential equation of heat conduction; ② Solve the differential equation of heat conduction with the definite solution conditions, which is the general method for solving the heat conduction problem. As a result, the two methods lead to the same result, the governing equations for the heat conduction of the fin are the exactly same. Here we just learn the solution method ①.

上面通过对实际问题的简化和假设，得到了实际问题的几何模型、物理模型，以及边界条件。接下来，进行肋片导热问题的数学描写，这里有两种方法：① 能量守恒定律+傅里叶定律，即推导导热微分方程的方法；② 导热微分方程+定解条件进行求解，即求解导热问题的一般方法。从结果看，两种方法殊途同归，得到的肋片导热控制方程完全相同。这里只介绍求解方法①。

First take the micro-element body dx, and then according

首先取微元体 dx，然后根据能量守

to the law of conservation of energy: the heat imported into the micro-element body minus the heat exported from the micro-element body, and then plus the calorific value of the internal heat source, is equal to the increment of internal energy. Because it is the steady-state heat conduction, the increment of internal energy is 0.

恒定律：导入微元体的热量，减去导出微元体的热量，再加上内热源的发热量，等于内能增量。因为是稳态导热，内能增量为0。

$$\Phi_x - \Phi_{x+\mathrm{d}x} - \Phi_\mathrm{d} = 0 \tag{2.44}$$

Then according to Fourier's law: The heat imported into the micro-element body is written as equation (2.45):

再根据傅里叶定律：导入微元体的热量记作式（2.45）：

$$\Phi_x = -\lambda A_\mathrm{c} \frac{\mathrm{d}t}{\mathrm{d}x} \tag{2.45}$$

The heat exported from the micro-element body is written as equation (2.46):

导出微元体的热量记作式（2.46）：

$$\Phi_{x+\mathrm{d}x} = \Phi_x + \frac{\mathrm{d}\Phi_x}{\mathrm{d}x}\mathrm{d}x = \Phi_x - \lambda A_\mathrm{c} \frac{\mathrm{d}^2 t}{\mathrm{d}x^2}\mathrm{d}x \tag{2.46}$$

Here is the key step: the upper and lower surfaces of the fin are not computational boundaries, but there is convection heat transfer, then we can convert the convective heat transfer quantity into calorific value of the internal heat source. This is an innovative solution. According to Newton cooling formula: the convective heat transfer quantity between the upper and lower surfaces of the fin and the fluid is written as equation (2.47):

下面是关键一步：肋片的上下表面虽然不是计算边界，但有对流传热，不妨把“对流传热量”折算成“内热源发热量”。这是一种创新的解决方法。根据牛顿冷却公式：肋片的上下表面与流体的对流传热量记作式（2.47）：

$$\Phi_\mathrm{d} = h(P\mathrm{d}x)(t - t_\infty) \tag{2.47}$$

Where P is the perimeter of the cross-section of the fin involved in heat transfer. Since the fins dissipate heat to the environment, it is equivalent to a minus source term, so a minus must be added to the source term. Substitute the equations (2.45), (2.46) and (2.47) into the energy conservation equation, after arranging to get:

式中，P 为参与传热的肋片截面周长。由于肋片向环境散热，相当于负的源项，因而源项前要加负号。把式（2.45）~式（2.47）三个式子代入能量守恒方程，整理后得到：

$$\frac{\mathrm{d}^2 t}{\mathrm{d}x^2} - \frac{hP}{\lambda A_\mathrm{c}}(t - t_\infty) = 0 \tag{2.48}$$

This formula is the governing equation for the heat conduction of fins, which is a second-order non-homogeneous ordinary differential equation about temperature. Next, solve

这个表达式就是肋片导热的控制方程，是关于温度的二阶非齐次常微分方程。接下来，求解该微分方程。

the differential equation. In order to homogenize the differential equation, we need to introducc a ncw parameter, namely excess temperature:

为了使微分方程齐次化，需引入一个新的参数——过余温度：

$$t-t_{\infty}=\theta \tag{2.49}$$

The complete mathematical formulation of the heat conduction problem of fins is as follows:

肋片导热问题完整的数学描写如下：

$$\text{Governing equation:}\quad \frac{\mathrm{d}^2\theta}{\mathrm{d}x^2}=m^2\theta$$

$$\text{Boundary conditions:}\quad \begin{cases} x=0,\theta=\theta_0=t_0-t_{\infty} \\ x=H,\dfrac{\mathrm{d}\theta}{\mathrm{d}x}=0 \end{cases}$$

Where $m=\sqrt{\dfrac{hP}{\lambda A_{\mathrm{c}}}}$ is a constant.

式中，$m=\sqrt{\dfrac{hP}{\lambda A_{\mathrm{c}}}}$为一个常量。

The solution of the homogeneous equation is still complicated, the general solution of the equation is directly given as:

齐次方程的求解仍然比较复杂，直接给出方程的通解为：

$$\theta=c_1\mathrm{e}^{mx}+c_2\mathrm{e}^{-mx} \tag{2.50}$$

Substitute the boundary conditions into the general solution, we can figure out the undetermined constants c_1 and c_2. And substitute c_1 and c_2 into the general solution to get the temperature distribution expression in the fin:

把边界条件代入通解，可求出待定常数 c_1、c_2；把 c_1、c_2 代入通解，可得到肋片中的温度分布表达式：

$$\theta=\theta_0\frac{\mathrm{e}^{m(H-x)}+\mathrm{e}^{-m(H-x)}}{\mathrm{e}^{mH}+\mathrm{e}^{-mH}}=\theta_0\frac{\mathrm{ch}[m(H-x)]}{\mathrm{ch}(mH)} \tag{2.51}$$

Here are the formulas of hyperbolic sine function, hyperbolic cosine function and hyperbolic tangent function.

这里给出双曲正弦函数、双曲余弦函数、双曲正切函数的表达式。

$$\mathrm{sh}(x)=\frac{\mathrm{e}^x-\mathrm{e}^{-x}}{2},\quad \mathrm{ch}(x)=\frac{\mathrm{e}^x+\mathrm{e}^{-x}}{2},\quad \mathrm{th}(x)=\frac{\mathrm{e}^x-\mathrm{e}^{-x}}{\mathrm{e}^x+\mathrm{e}^{-x}} \tag{2.52}$$

With the temperature distribution formula in the fin, the excess temperature of the fin end can be figured out:

有了肋片中的温度分布表达式，就可以求出肋端的过余温度：

$$t_H-t_{\infty}=\frac{t_0-t_{\infty}}{\mathrm{ch}(mH)} \tag{2.53}$$

According to the temperature distribution formula, the temperature distribution figure in the fins with constant cross-section can be drawn, as shown in Fig. 2.21.

根据温度分布表达式，可画出等截面肋片内的温度分布图，如图 2.21 所示。

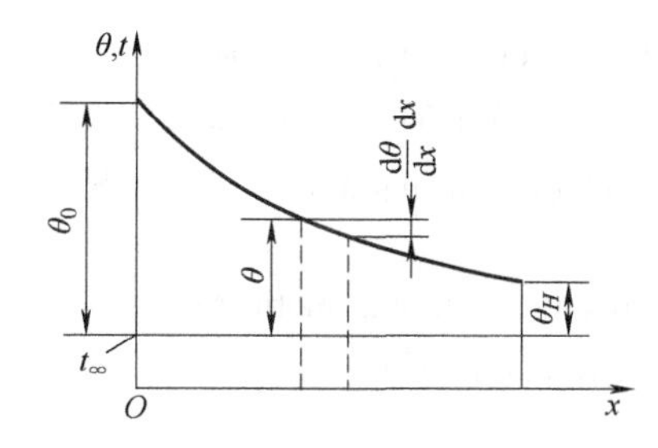

Fig. 2. 21 Temperature distribution within a fin of constant cross-section

图 2. 21 等截面肋片内的温度分布

From Fig. 2. 21, the temperature distribution in the fins under steady-state conditions can be seen, namely, temperature gradually decreases from the fin root to the fin end. It can also be calculated that the actual heat dissipating capacity on the surface of the fin equals to the heat imported into the fin through the fin root. Therefore, Fourier's law can be used to calculate.

从图 2. 21 中可以看出在稳态条件下肋片内的温度分布，即从肋根到肋端温度逐渐降低。还可计算出：肋片表面的实际散热量=通过肋根导入肋片的热量。因此，可用傅里叶定律来计算。

$$\Phi_{x=0} = -\lambda A_c \frac{d\theta}{dx}\bigg|_{x=0} = \lambda A_c \theta_0 m \frac{\mathrm{sh}(mH)}{\mathrm{ch}(mH)} = \frac{hP}{m}\theta_0 \mathrm{th}(mH) \tag{2.54}$$

Some explanations of the above solving process are as follows:

对上述求解过程说明如下：

(1) In the above derivation the heat dissipation of the fin end is ignored, that is to say that the fin end is considered to be adiabatic. For most fin heat conduction problems, especially the high and thin fins, the calculation is accurate enough. In order to improve the calculation accuracy, half of the fin thickness can be added to the fin height.

（1）上述推导中忽略了肋端的散热，即认为肋端绝热。对于多数肋片导热问题，尤其是高而薄的肋片，这样计算足够精确。如果要提高计算精度，可把肋片厚度的一半加到肋片高度上。

(2) The temperature distribution law in the fin is that the temperature gradually decreases from the fin root to the fin end, so the higher the fin is not the better.

（2）肋片中温度分布规律是：从肋根到肋端温度逐渐降低，所以肋片并非越高越好。

(3) The above analysis approximately assumes that the temperature field in the fin is one-dimensional, and when $\frac{\delta/\lambda}{1/h} \leqslant 0.05$, the calculation error is less than 1%, which is accurate enough. But for the short and thick fins, the temperature field is two-dimensional and the error of the above calculation is large.

（3）上述分析近似认为肋片温度场为一维，当$\frac{\delta/\lambda}{1/h} \leqslant 0.05$时，计算误差小于 1%，足够精确。但对于短而厚的肋片，温度场为二维，上述计算误差较大。

(4) In fact, the convective heat transfer coefficient h between the fins and the fluid is not uniform. This situation can only be solved by numerical calculation methods.

(4) 实际上，肋片与流体的表面传热系数 h 并非均匀一致。这种情况只能用数值计算方法求解。

Summary This section illustrates the solution methods of the steady-state heat conduction problem of the straight-fin with constant cross-section.

小结 本节阐述了等截面直肋稳态导热问题的求解方法。

Question From the perspective of efficient heat conduction and material saving, what is the optimal cross-sectional shape of the fin?

思考题 从高效导热又节约材料的角度考虑，肋片的最优截面形状是什么样?

Self-tests

自测题

1. How to simplify the analysis of heat conduction of straight fin with constant cross-section?

1. 分析等截面直肋导热问题时是怎么简化的?

2. The upper and lower interfaces of straight fin with constant cross-section are not computational boundaries, but there is heat transfer. How to build mathematical models?

2. 等截面直肋片的上下界面不是计算边界，但有热量传递。建立数学模型时如何处理?

3. When there are no fins on the heat exchange wall, the calculation formula of local heat dissipating capacity is $\Phi=A_c h(t_0-t_\infty)$, the heat dissipation area increases by N times after installing fins, does the heat dissipating capacity also increases by N times?

3. 换热壁面无肋片时，局部散热量的计算式为 $\Phi=A_c h(t_0-t_\infty)$，加装肋片后散热面积增大 N 倍，散热量也增加 N 倍吗?

授课视频

2.12 An Application Case of Heat Conduction of Fins—a Thermowell

2.12 肋片导热的应用案例——温度计套管

Next, Let's look at an application case of heat conduction of fins, that is a thermowell.

下面看一个肋片导热的应用案例——温度计套管。

Example 2.6 In the air reservoir of the compressor equipment, the air temperature is measured by a glass mercury thermometer, which is inserted into the oil-filled iron thermowell, as shown in Fig. 2.22. It is given that the

例 2.6 在压气机设备的储气筒里，空气温度用一支玻璃水银温度计来测量，温度计插入装油的铁套管中，如图 2.22 所示。已知温度计的读数

reading of the thermometer is 100℃, and the temperature at the junction between the air reservoir and the thermowell is $t_0 = 50℃$, the height H of the thermowell is 140mm, and the wall thickness $\delta = 1\text{mm}$, the thermal conductivity of the tube is 58.2W/(m·℃) and the convective heat transfer coefficient h of the outer surface of the thermowell is 29.1W/(m^2·℃). Try to discuss that: (1) Can the reading of thermometer accurately represent the air temperature of the measured point? (2) If not, analyze how much the thermometric error is.

为100℃，储气筒与温度计套管连接处的温度 $t_0 = 50℃$，套管的高度 $H = 140\text{mm}$，壁厚 $\delta = 1\text{mm}$，管材的导热系数 $\lambda = 58.2\text{W/(m·℃)}$，套管外表面的表面传热系数 $h = 29.1\text{W/(m}^2\text{·℃)}$。试分析：(1) 温度计的读数能否准确代表被测点的空气温度？(2) 如果不能，分析其误差有多大？

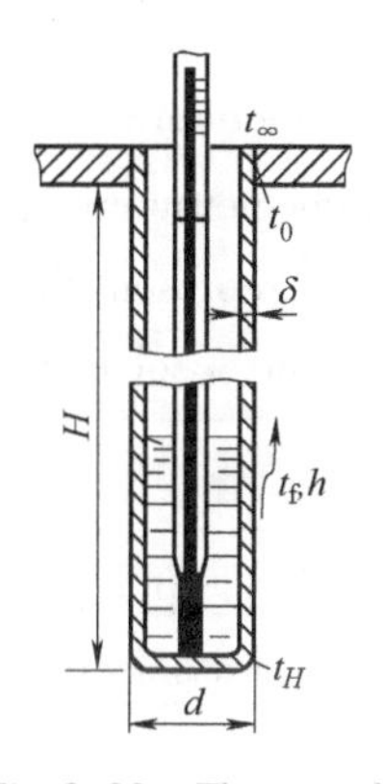

Fig. 2.22 Thermowell

图 2.22 温度计套管

First, determine the research object: thermowell.

首先确定研究对象：温度计套管。

The heat transfer process is analyzed as follows: The thermosensitive bubble of the thermometer is in contact with the thermowell top, $t_H = 100℃$, it can be considered that the reading t_H is the wall temperature of the thermowell top, which is 100℃. Whether the reading of the thermometer accurately represents the air temperature at the measured point, depends on whether there is a temperature difference between the wall of the thermowell top and the air in the air reservoir. That is the difference between t_H and t_f.

下面进行传热过程分析：温度计的感温泡与套管顶端接触，可认为其读数即是套管顶端的壁面温度，$t_H = 100℃$。温度计的读数能否准确代表被测点的空气温度，取决于套管顶端壁面与储气筒里空气之间是否存在温差，即 t_H 与 t_f 之差。

From Fig. 2.22, there are three heat transfer modes between the thermometer thermowell and the surrounding environment: ① The temperature of the thermowell top t_H is 100℃ and the temperature of the thermowell root is $t_0 = 50℃$. So there is heat conduction from the thermowell top

从图 2.22 中可以看出，温度计套管与四周环境之间有三种热量传递方式：① 套管顶端 $t_H = 100℃$，根部 $t_0 = 50℃$，因此，从顶端到根部存在导热；② 套管外表面（50~100℃）

to the thermowell root. ② There is radiation heat transfer between the outer surface of the thermowell (at 50~100℃) and the cylinder body of the air reservoir (at 50℃). ③ Convection heat transfer from the compressed air (at t_f) to the thermowell surface (at 50~100℃).

向储气筒筒身（50℃）存在辐射传热。③ 压缩空气（t_f）向套管表面（50~100℃）的对流传热。

In the steady-state the convective heat transfer quantity imported into the thermowell is exactly equal to the sum of the heat conduction quantity imported into the root and the radiation heat transfer quantity imported into the cylinder body from the thermowell. So, the temperature t_H at the thermowell top must be lower than the temperature t_f of compressed air. That is to say, there is thermometric error, the reading of thermometer cannot accurately represent the air temperature at the measured point. Now let's solve for the temperature of the compressed air.

稳态时，套管获得的对流传热量，正好等于套管向根部的导热量及向筒身的辐射传热量之和。因此，套管顶端的温度 t_H 必然低于压缩空气的温度 t_f，即存在测温误差，温度计的读数不能准确代表被测点的空气温度。下面来求解压缩空气的温度。

The key to solve such problems is to establish an appropriate geometric model. The temperature of each cross-section in the thermowell is equal, and the circular cross-section of the thermowell can be expanded into a plane. Then, the thermowell can be regarded as a straight fin with constant cross-sectional area of $\pi d\delta$. Where d is the thermowell diameter.

解决此类问题的关键是：建立适当的几何模型。套管中每一横截面上温度可认为相等，可把套管的圆截面展开为平面，那么，套管可看成截面面积为 $\pi d\delta$ 的等截面直肋，其中，d 为套管直径。

The thermometric error is the excess temperature of the thermowell top, $\theta_H = t_H - t_f$, t_f is the temperature of the compressed air, which is equivalent to the temperature t_∞ of the surrounding fluid in the heat conduction problem of the fin. In the literature of Heat Transfer, the temperature of the fluid in a finite space is usually denoted as t_f, while the temperature of the fluid in a large space is denoted as t_∞, the two are corresponding to each other. According to the thermal conduction analysis of the fin in the previous lecture, the calculation formula of the temperature of the fin end is as follows:

测温误差，即套管顶端的过余温度，$\theta_H = t_H - t_f$，t_f 就是压缩空气的温度，相当于肋片导热问题中周围流体温度 t_∞。在传热学文献中，一般把位于有限空间内流体温度记作 t_f，而把位于大空间内流体温度记作 t_∞，两者是对应的。根据上一节中肋片导热分析结果，肋端温度的计算式：

$$t_H - t_f = \frac{t_0 - t_f}{\mathrm{ch}(mH)} \tag{2.55}$$

After arranging, the formula of air temperature can be obtained:

整理后，可得出空气温度的表达式：

$$t_{\mathrm{f}}=\frac{t_{H}\mathrm{ch}(mH)-t_{0}}{\mathrm{ch}(mH)-1} \tag{2.56}$$

In this case, the perimeter of the heat transfer cross-section is $P=\pi d$, and the cross-section area of the thermowell is $A_{\mathrm{c}}=\pi d\delta$. Thus, substitute the known parameters into the formula and the value of mH in the equation is equal to 3.13.

本例中，传热截面周长 $P=\pi d$，套管截面面积 $A_{\mathrm{c}}=\pi d\delta$。于是，代入已知参数，可求出式中 mH 的值为 3.13。

$$mH=\sqrt{\frac{hP}{\lambda A_{\mathrm{c}}}}H=\sqrt{\frac{h}{\lambda\delta}}H=\sqrt{\frac{29.1\mathrm{W}/(\mathrm{m}^2\cdot\mathrm{K})}{58.2\mathrm{W}/(\mathrm{m}\cdot\mathrm{K})\times0.001\mathrm{m}}}\times0.14\mathrm{m}=3.13$$

From the math manual, we can get ch(3.13) = 11.5. Substitute it into the calculation formula to find that t_{f} is equal to 104.7℃. The reading of the thermometer is 100℃ and the calculated value is 104.7℃. So, the thermometric error is 4.7℃ and such a large error is often unacceptable. Then, how to reduce the thermometric error? From the one-dimensional heat conduction process of the thermowell, the thermal resistance analysis figure can be drawn, as shown in Fig. 2.23.

查数学手册可得 ch(3.13) = 11.5。代入 t_{f} 计算式可求出 t_{f} 等于 104.7℃。温度计读数 100℃，计算值为 104.7℃。所以，测温误差是 4.7℃。这么大的误差往往是不容许的。那么，怎样才能减小测温误差呢？从温度计套管的一维导热过程来看，可以画出热阻分析图，如图 2.23 所示。

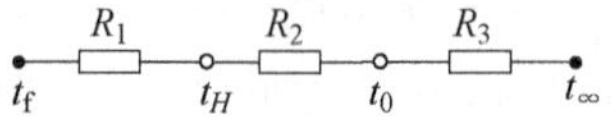

t_{f} —R_1— t_H —R_2— t_0 —R_3— t_∞

Fig. 2.23 The thermal resistance analysis figure

图 2.23 热阻分析图

In Fig. 2.23, t_∞ is the environment temperature outside the air reservoir, R_3 denotes the heat transfer resistance between the outside of the air reservoir and the environment, R_1 denotes the convective heat transfer resistance between the thermowell and the compressed air, R_2 denotes the thermal conduction resistance between the top and root of the thermowell.

图 2.23 中，t_∞ 为储气筒外的环境温度，R_3 代表储气筒外侧与环境间的传热热阻，R_1 代表套管与压缩空气间的对流传热热阻，R_2 代表套管顶端与根部间的导热热阻。

To reduce the thermometric error, namely t_H should be as close to t_{f} as possible, R_1 should be reduced, while R_2 and R_3 should be increased as far as possible. The following measures could be taken: ① Enhance convection heat

要减小测温误差，即 t_H 尽量接近 t_{f}，应尽量减小 R_1，而增大 R_2 及 R_3。可采取以下措施：① 强化套管与压缩空气间的对流传热，以减

transfer between the thermowell and the compressed air to reduce convective heat transfer resistance R_1; ② The thermowell should be made of materials with lower thermal conductivity or increase the height, or decrease the wall thickness of the thermowell, to increase the thermal conduction resistance R_2; ③ Add insulation material outside the gas reservoir to increase the heat transfer resistance R_3 between the outer side of the air reservoir and the environment.

小对流传热热阻 R_1；② 套管选用导热系数更小的材料，或增加高度，或减小壁厚，以增大导热热阻 R_2；③ 在储气筒外加装保温材料，以增大储气筒外侧与环境间的传热热阻 R_3。

Summary This section illustrates an application case of heat conduction of the fin—the heat conduction process of thermowell.

小结 本节阐述了肋片导热的一个应用案例——温度计套管的导热过程。

Question What is the basic idea to simplify the heat conduction problem of the thermowell?

思考题 套管导热问题简化的基本思路是什么？

Self-tests

自测题

1. What is the physical significance of "excess temperature" introduced in solving the nonhomogeneous differential equation of heat conduction problem?

2. What measures should be taken to reduce the temperature measurement error of the thermowell in the gas reservoir?

1. 求解导热问题非齐次微分方程式引入了"过余温度"，其物理意义是什么？

2. 要减小储气筒中温度计套管的测温误差，需采用哪些措施？

授课视频

2.13 Fin Efficiency and Contact Thermal Resistance

2.13 肋片效率与接触热阻

In order to evaluate the heat transfer effect after installing fins from the perspective of heat dissipation, the fin efficiency is introduced. The definition of fin efficiency is denoted as the following mathematical formula, which makes the physical meaning clearer:

为了从散热的角度来评价加装肋片后的传热效果，引入肋片效率。肋片效率的定义用下面这个数学表达式表示，物理意义更清晰：

$$\eta_f = \frac{\Phi}{\Phi_0} \tag{2.57}$$

The fin efficiency is equal to the actual heat dissipating capacity of the fin divided by the heat dissipating capacity of the whole fin surface at the temperature of the fin root. The temperature distribution in the fin is that the temperature drops gradually from the root to the end of the fin.

肋片效率等于肋片的实际散热量除以假设整个肋表面都处于肋根温度下的散热量。肋片内的温度分布：从肋根到肋端温度是逐渐降低的。

If assume that the entire fin surface is at the fin root temperature, at this time the heat dissipating capacity is much larger than the actual heat dissipating capacity and is called the maximum heat dissipating capacity under the ideal conditions, which can be calculated by Newton cooling formula.

如果假设整个肋表面都处于肋根温度下，此时，散热量会远大于实际散热量。称为理想状况下的最大散热量，可用牛顿冷却公式计算：

$$\Phi_0 = hPH\theta_0 \tag{2.58}$$

Where P denotes the perimeter of the fin cross-section, H denotes the height of the fin, and $\theta_0 = t_0 - t_\infty$. We have already obtained before: the actual heat dissipating capacity on the fin surface equals to the heat imported into the fin through the fin root, write down as equation (2.59).

式中，P 是肋片截面周长；H 是肋片高度；$\theta_0 = t_0 - t_\infty$。前面已经得到了：肋片表面的实际散热量=通过肋根导入肋片的热量，记作式（2.59）。

$$\Phi_{x=0} = \frac{hP}{m}\theta_0 \mathrm{th}(mH) \tag{2.59}$$

For the straight fin with constant cross-section, divide Eq. (2.59) by Eq. (2.58) to obtain the fin efficiency.

对于等截面直肋，用式（2.59）除以式（2.58）得肋片效率：

$$\eta_f = \frac{\frac{hP}{m}\theta_0 \mathrm{th}(mH)}{hPH\theta_0} = \frac{\mathrm{th}(mH)}{mH}$$

For circular fin, the longitudinal sectional area of the fin is δH, denoted by A_L, multiply the numerator and denominator on the right side of the equation $mH = \sqrt{\frac{hP}{\lambda A_c}}H = \sqrt{\frac{h2l}{\lambda l\delta}}H = \sqrt{\frac{2h}{\lambda\delta}}H$ by H to the power of 1/2 to get:

对于环肋，肋片的纵截面面积为 δH，记作 A_L，等式 $mH = \sqrt{\frac{hP}{\lambda A_c}}H = \sqrt{\frac{h2l}{\lambda l\delta}}H = \sqrt{\frac{2h}{\lambda\delta}}H$ 右侧的分子、分母同乘以 $H^{1/2}$ 得：

$$mH = \sqrt{\frac{2h}{\lambda\delta H}}H^{3/2} = \sqrt{2}\left(\frac{h}{\lambda A_L}\right)^{\frac{1}{2}}H^{\frac{3}{2}} \tag{2.60}$$

The fin efficiency is related to mH, so it is related to the parameters $\left(\frac{h}{\lambda A_L}\right)^{\frac{1}{2}} H^{\frac{3}{2}}$, its relationship curve is shown in Fig. 2. 24.

可见肋效率与 mH 有关，所以就与参量$\left(\frac{h}{\lambda A_L}\right)^{\frac{1}{2}} H^{\frac{3}{2}}$有关，其关系曲线如图 2. 24 所示。

After we have the graph, we can directly use Fig. 2. 24 to find out the fin efficiency and calculate the actual heat dissipating capacity, instead of using equation (2. 59). The factors affecting the fin efficiency include the thermal conductivity of the material λ, the convective heat transfer coefficient of the medium around the fin h and the geometry and size of the fin, such as perimeter P, area A, height H, etc.

有了曲线图以后，环肋的散热量，不再用式（2. 59）计算，而直接用图 2. 24 查出肋片效率，即可算出实际散热量。影响肋片效率的因素包括材料导热系数 λ、肋片周围介质的表面传热系数 h、肋片的几何形状和尺寸，如周长 P、面积 A、高度 H 等。

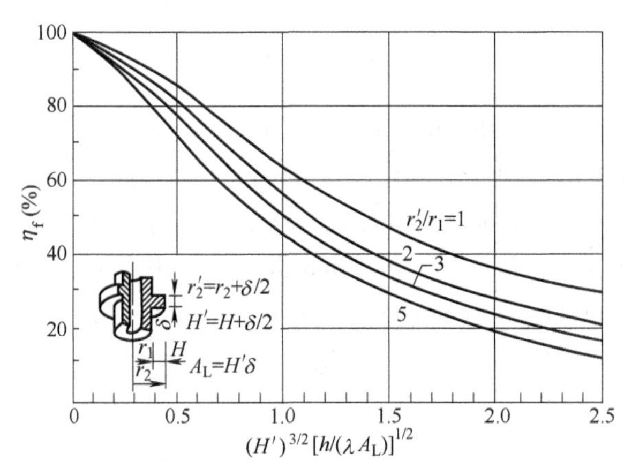

Fig. 2. 24 Efficiency curves of circular fins in rectangular section

图 2. 24 矩形剖面环肋的效率曲线

Next, calculate heat conduction quantity in the circular fin and the triangle cross-section straight fin. In order to reduce the weight of the fin and save materials and keep the heat dissipating capacity unchanged basically, it is necessary to use the fin with variable cross-section, such as the straight fin with triangular cross-section. For the fin with variable cross-section, the calculation formula of heat dissipating capacity of the fin derived from the heat conduction differential equation is quite complex, but it is convenient to calculate with the efficiency curve of the fin.

下面计算通过环肋及三角形截面直肋的导热量。为减轻肋片重量、节省材料，并保持散热量基本不变，需要采用变截面肋片，如三角形截面直肋。对于变截面肋片，从导热微分方程式导出的肋片散热量计算公式相当复杂，利用肋片效率曲线计算很方便。

Fig. 2. 24 and Fig. 2. 25 show the efficiency curves of the

图 2. 24 和图 2. 25 分别给出了矩形

circular fin with rectangular cross-section, and the straight fin with constant cross-section and triangular fin, respectively: take the parameters $\left(\frac{h}{\lambda A_L}\right)^{\frac{1}{2}} H^{\frac{3}{2}}$ as the x axis and take the fin efficiency η_f as y axis, find out the fin efficiency, then the heat dissipating capacity of the fin can be obtained.

剖面环肋、等截面直肋和三角形肋片的效率曲线：都以 $(H')^{\frac{3}{2}}\left(\frac{h}{\lambda A_L}\right)^{\frac{1}{2}}$ 为 x 轴，以肋片效率 η_f 为 y 轴，可查出肋片效率，即可求出肋片散热量。

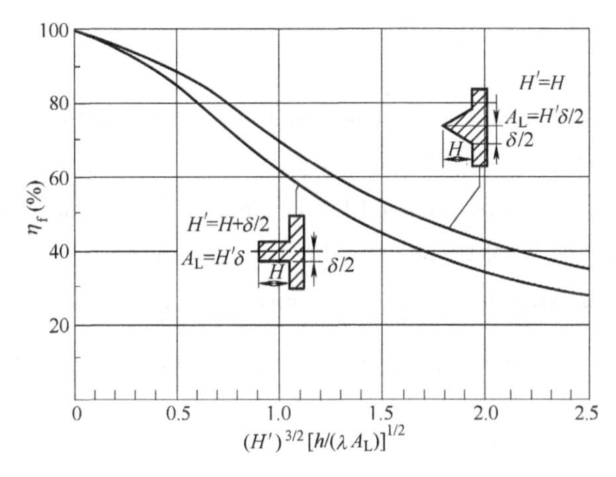

Fig. 2. 25 Efficiency curves of the straight fin with constant cross-section and triangular fin

图 2. 25 等截面直肋和三角形肋片的效率曲线

Sum up the calculation method of heat dissipating capacity of the fin: ① Find out the fin efficiency η_f from the curves; ② The maximum heat dissipating capacity under ideal condition can be calculated by Newton cooling formula; ③ The actual heat dissipating capacity can be calculated from the formula $\Phi=\eta_f\Phi_0$. When designing the fin, not only to select the shape, and calculate the heat dissipating capacity, but also to consider the mass, the difficulty in manufacturing and price, and the restriction of spatial position.

总结一下肋片散热量的计算方法：① 由图线查出肋片效率 η_f；② 用牛顿冷却公式可计算出理想状况下的最大散热量；③ 可计算出实际散热量 $\Phi=\eta_f\Phi_0$。设计肋片时，除了选择形状、计算散热量，还要考虑质量、制造难易程度，以及价格、空间位置的限制等因素。

The above formulas are used to calculate the fin efficiency and heat dissipating capacity of a single fin. In engineering practice, the fin is always used in groups, and the concrete parameters are shown in Fig. 2. 26. Therefore, it is of little engineering significance to calculate the efficiency of a single fin.

上述公式为单个肋片效率和散热量的计算公式，工程实际中总是成组采用肋片，具体参数如图 2. 26 所示。所以，计算单个肋片的效率，工程意义不大。

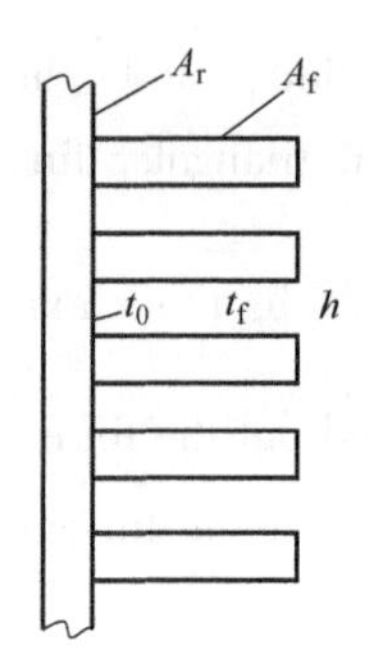

Fig. 2. 26 Schematic diagram of ribbed surface

图 2. 26 肋片表面示意图

The total efficiency of the fin surface is calculated below. Total heat transfer area: $A_0=A_r+A_f$, where A_r is the base surface area excluding the area occupied by the fins; A_f is the total area of the fins. Total temperature difference: $\Delta t=t_0-t_f$. So the total convective heat transfer quantity:

下面计算肋面总效率。总传热面积：$A_0=A_r+A_f$，其中，A_r 是除去肋片所占面积以外的基础表面面积；A_f 是肋片的总面积。总温差：$\Delta t=t_0-t_f$，那么，总对流传热量：

$$\begin{aligned}\Phi &= A_r h(t_0-t_f)+A_f\eta_f h(t_0-t_f)=h(t_0-t_f)(A_r+A_f\eta_f)\\ &= A_0 h\Delta t\,\frac{A_r+A_f\eta_f}{A_0}\end{aligned} \tag{2.61}$$

$$\frac{A_r+A_f\eta_f}{A_0}=\eta_0 \tag{2.62}$$

This is the total efficiency of the fin surface.

这就是肋面总效率。

Between the above fins and the base wall and between the multilayer plane wall or multilayer cylinder wall, it is assumed that the contact surface is in good contact. But in fact, on the interface of direct contact between two solid surfaces, point contact or local surface contact always appears, which brings additional thermal resistance to heat conduction. Due to the incomplete contact between the two contact surfaces, the additional thermal resistance is brought to the heat conduction, called contact thermal resistance.

在上述肋片与基础壁面之间，以及多层平壁或多层圆筒壁之间，均假设接触面“接触良好”。而实际上两固体表面直接接触的界面上，容易出现点接触或局部面接触，从而给导热带来额外的热阻。由于两接触面之间的不完全接触给导热带来额外的热阻，称为接触热阻。

When the interspace between the two interfaces is filled with gas with low thermal conductivity, the effect of contact thermal resistance on heat conduction is more obvious. When there is a temperature difference between the two solid walls, the heat transfer mechanism at the interface is the heat conduction of solid between the contact points and the

当两个界面的空隙中充满导热系数较小的气体时，接触热阻对导热的影响更突出。当两个固体壁面间具有温差时，界面处热量传递机理为接触点间的固体导热和空隙中的空气导热，对流和辐射对传热过程的

heat conduction of air in the interspace, the effect of convection and radiation on the heat transfer process is very small. With contact thermal resistance, the calculation formula of heat flux is:

影响很小。得到接触热阻之后，热流密度的计算式为：

$$q=\frac{t_1-t_3}{\frac{\delta_A}{\lambda_A}+r_c+\frac{\delta_B}{\lambda_B}} \tag{2.63}$$

In the total thermal resistance, besides the thermal conduction resistance of the two walls, there is also the contact thermal resistance r_c. The temperature difference can also be calculated by the above formula:

总热阻中，除了两层壁面的导热热阻以外，还增加了"接触热阻"r_c。通过式（2.63）也可计算出温差：

$$t_1-t_3=q\left(\frac{\delta_A}{\lambda_A}+r_c+\frac{\delta_B}{\lambda_{AB}}\right) \tag{2.64}$$

From the above two formulas, the following conclusions can be drawn: ① When the temperature difference is constant, as the contact thermal resistance increases the heat transfer rate will decrease. ② Under the condition of constant heat transfer rate, when the contact thermal resistance is large, a large temperature difference will be generated at the interface.

从上述两个表达式，可得出以下结论：① 当温差不变时，随着接触热阻增大，热流量必然减小。② 热流量不变，当接触热阻较大时，在界面上必然产生较大温差。

The analysis shows that the influencing factors of contact thermal resistance include: ① roughness of solid surface; ② hardness matching of two contact surfaces; ③ extrusion force on the contact surface; ④ properties of the gas medium in the interspace.

分析发现，接触热阻的影响因素包括：① 固体表面的粗糙度；② 两个接触表面的硬度匹配；③ 接触面上的挤压力；④ 空隙中气体介质的性质。

In experimental research and engineering applications, the main methods to eliminate contact thermal resistance include: ① Fill the interspace with high thermal conductivity materials (such as heat-conducting oil, silicone oil) and silver etc. ② Use advanced electronic packaging materials, the thermal conductivity is more than 400W/(m·K).

在实验研究与工程应用中，消除接触热阻的主要方法有：① 采用高导热材料（如导热油、硅油）、银等高导热材料填充空隙；② 采用先进的电子封装材料，导热系数达400W/(m·K) 以上。

Summary This section illustrates the calculation method of fin efficiency and the effect of contact thermal resistance on heat transfer process.

小结 本节阐述了肋片效率的计算方法，以及接触热阻对传热过程的影响。

Question What is the heat transfer mechanism between two solid walls when a contact thermal resistance is generated between them?

思考题 当两个固体壁面间产生接触热阻时，两壁面间的传热机理是什么？

Self-tests

自测题

1. What is fin efficiency? What is it used for?

2. What is contact thermal resistance? What effect does it have on the heat conduction process?

1. 什么是肋片效率？其有什么用途？

2. 什么是接触热阻？其对导热过程有什么影响？

授课视频

2.14 An Application Case of Heat Conduction of Fins—Plane-plate Type Solar Collector

2.14 肋片导热的应用案例——平板式太阳能集热器

There are two main types of solar collector: plane-plate type and vacuum tube type, as shown in Fig. 2.27. Next, focus on the plane-plate type. We learn the analysis method of heat conduction problem with examples.

太阳能集热器主要有平板式和真空管式两种形式，如图 2.27 所示。下面重点讨论平板式，结合例题来学习导热问题的分析方法。

a) 平板式

b) 真空管式

Fig. 2.27 Solar collectors

图 2.27 太阳能集热器

Example 2.7 Fig. 2.28 shows a simple endothermic plate structure of a plane-plate type solar collector. The side of the endothermic plate facing the sun is coated with a material with a high absorptivity of solar radiation, and a group of parallel tubes is set on the back of the endothermic plate. The tubes are filled with cooling water to absorb solar radiation, and the insulating material is full filled between the tubes. The front surface of the endothermic plate receives solar radiation, while is cooled by the

例 2.7 图 2.28 所示为平板式太阳能集热器的一种简单的吸热板结构。吸热板面向太阳的一面有一层对太阳辐射吸收比很高的材料，吸热板的背面设置了一组平行的管子，管内通以冷却水，以吸收太阳辐射，管子之间充满绝热材料。吸热板的正面在接受太阳辐射的同时受到环境冷却。设净吸收的太阳辐

environment. Let the net absorbed solar radiation be q_r, the convective heat transfer coefficient be h, the air temperature be t_∞ and the temperature at the junction between the tube and the endothermic plate be t_0. Try to write the mathematical formulation of determining the temperature distribution in the endothermic plate and solve it.

射为 q_r，表面传热系数为 h，空气温度为 t_∞，管子与吸热板结合处的温度为 t_0，试写出确定吸热板中温度分布的数学描写，并求解。

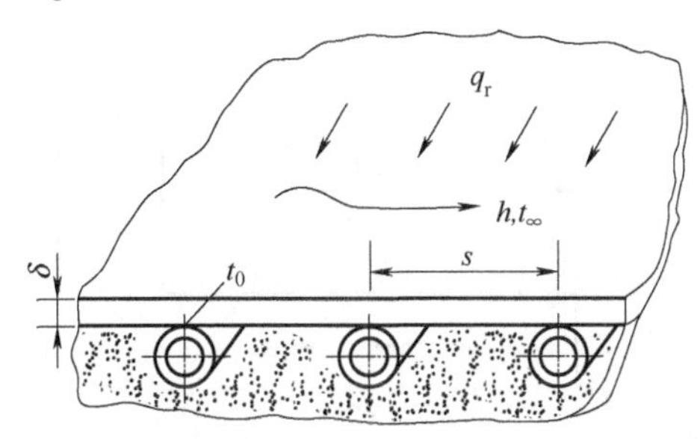

Fig. 2. 28 The endothermic plate of a plane-plate type solar collector

图 2. 28 平板式太阳能集热器的吸热板

Solution: The research object of this problem is the endothermic plate of a solar collector. The key problem is how to establish geometric models and physical models?

解： 该问题的研究对象是太阳能集热器的吸热板。关键问题是：怎样建立几何模型及物理模型？

The thermal conduction problem of the endothermic plate is analyzed and simplified as follow:

下面对吸热板导热问题进行分析和简化：

(1) The length of the endothermic plate is much greater than its thickness, so it can be considered that the temperature is uniform along the length, and a cross-section can be studied.

(1) 吸热板的长度远大于其厚度，可认为长度方向温度均匀，可取一个截面进行研究。

(2) The temperature distribution of the endothermic plate is symmetrical about the middle section between any two adjacent cooling water tubes, namely, the middle section is a symmetrical plane, which is also an adiabatic surface, and the mathematical formula is as follows:

(2) 任意两根相邻冷却水管之间，吸热板的温度分布关于中间截面对称，即中间截面是一个对称面（即绝热面），数学表达式为：

$$dt/dx = 0$$

(3) The back of the endothermic plate is insulated well, so the back is also an adiabatic surface, which is equivalent to a symmetrical plane. Since it is a symmetric plane, the geometric model can be supplemented by half about the symmetric plane, which is consistent with the geometric model and boundary conditions of the fin.

(3) 吸热板背面绝热良好，因而背面也是绝热面，相当于对称面；既然是对称面，可把几何模型关于对称面再补充一半，这样就与肋片的几何模型和边界条件一致了。

(4) Because the thermal conduction resistance is much less than the convective thermal resistance, then the temperature variation along the thickness direction at any section x can be ignored. The heat conduction problem is reduced to one dimension. A simplified model of an endothermic plate is shown in Fig. 2. 29.

(4) 由于导热热阻远远小于对流传热热阻（$\delta/\lambda \ll 1/h$），那么，任意截面 x 处沿厚度方向的温度变化可忽略不计。导热问题降到了一维。吸热板的简化模型如图 2. 29 所示。

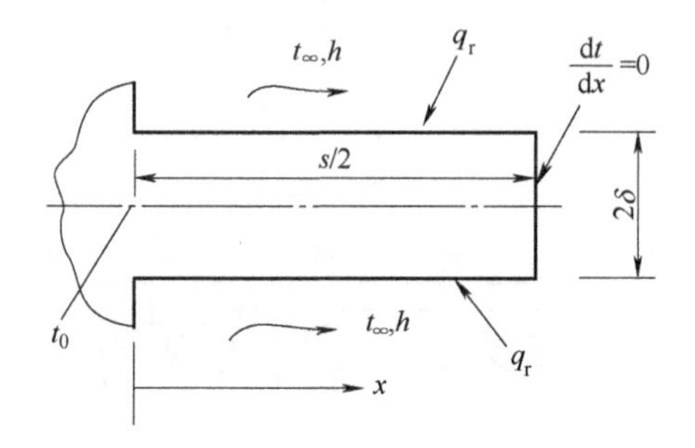

Fig. 2. 29 A simplified model of a endothermic plate

图 2. 29 吸热板的简化模型

The temperature distribution of the endothermic plate can be simplified as the heat conduction problem of the straight fin with constant cross-section. Its mathematical model is as follows:

吸热板的温度分布问题，可简化为等截面直肋的导热问题。其数学模型是：

$$\frac{d^2t}{dx^2}+\frac{\dot{\Phi}}{\lambda}=0$$

$$x=0, t=t_0; x=\frac{s}{2}, \frac{dt}{dx}=0$$

According to the analysis method of heat conduction of the fin, the formula of heat source term can be derived:

仿照肋片导热的分析方法，可进一步导出热源项 $\dot{\Phi}$ 的表达式：

$$\dot{\Phi}=-\frac{hP(t-t_\infty)-q_rP}{A_c}=-\frac{hP}{A_c}\left(t-t_\infty-\frac{q_r}{h}\right)$$

Substitute it into the governing equation to obtain:

代入控制方程中，得：

$$\frac{d^2t}{dx^2}=\frac{hP}{\lambda A_c}\left(t-t_\infty-\frac{q_r}{h}\right)$$

Only t in brackets is a variable. In order to solve the governing equation, it is necessary to homogenize it. Suppose $\theta=t-t_\infty-\frac{q_r}{h}$, then to obtain:

括号中只有 t 是变量。要求解控制方程，需要对它齐次化处理，定义 $\theta=t-t_\infty-\frac{q_r}{h}$，于是得：

$$\frac{d^2t}{dx^2}=m^2\theta$$

$$x=0,\theta=\theta_0;x=\frac{s}{2},\frac{d\theta}{dx}=0$$

Obviously, the solution of this problem is the same as the temperature distribution expression of the straight fin with constant cross-section, just need to replace H in the formula with $s/2$. The temperature distribution in the endothermic plate is as follows:

显然，这一问题的解与等截面直肋的温度分布表达式相同，只要将式中的 H 用 $s/2$ 来代替即可。吸热板内的温度分布是：

$$\theta=\theta_0\frac{\mathrm{ch}[m(H-x)]}{\mathrm{ch}(mH)}$$

Summary This section illustrates an application case of heat conduction of the fin, i. e. the heat conduction problem of the endothermic plate of the plane-plate type solar collector.

小结 本节阐述了肋片导热的一个应用案例——平板式太阳能集热器的吸热板导热问题。

Question What is the basic idea of solving the temperature distribution in the endothermic plate of the plane-plate type solar collector?

思考题 平板式太阳能集热器的吸热板内，求解温度分布的基本思路是什么？

授课视频

2.15 One-dimensional Steady-state Heat Conduction with an Internal Heat Source

2.15 有内热源的一维稳态导热

The working conditions 1: the first type of boundary condition.

工况 1：第一类边界条件。

Known: The plane wall with thickness of 2δ has a uniform internal heat source, the thermal conductivity coefficient λ is a constant, the temperature on both sides of the plane wall is t_w, and the heat conduction is steady-state.

已知：平壁厚度为 2δ，有均匀内热源，导热系数 λ 为常数，平壁两侧壁面的温度均为 t_w，稳态导热。

Analysis: The structure of the plane wall is symmetrical, and the heat transfer conditions are the same. Therefore,

分析：平壁的结构对称，传热条件也相同，因此，温度分布也应该对

the temperature distribution should also be symmetrical. Thus, take half a plane wall to study. The central section of the plane wall must be a symmetrical plane, which can also be regarded as an adiabatic plane.

称。由此可取半个平壁来研究。平壁的中心截面一定是对称面，也可看作是绝热面。

When building a geometric model, please remember the pithy formula: a plane of symmetry, take half of it. The origin of the coordinates is set on the center line of the wall.

建立几何模型时，记住口诀：对称面，取一半。坐标原点设在平壁的中心线上。

Mathematical model:

数学模型：

$$\begin{cases} \dfrac{d^2t}{dx^2}+\dfrac{\dot{\Phi}}{\lambda}=0 \\ x=0,\dfrac{dt}{dx}=0 \\ x=\delta,t=t_w \end{cases} \quad \text{第二类边界条件}$$

Integrate the differential equation twice, we can get:

对微分方程积分两次可得：

$$t=-\frac{\dot{\Phi}}{2\lambda}x^2+c_1x+c_2 \tag{2.65}$$

Substitute the boundary conditions into the equation to obtain the expression of temperature distribution:

把边界条件代入可得温度分布表达式：

$$t=t_w+\frac{\dot{\Phi}}{2\lambda}(\delta^2-x^2) \tag{2.66}$$

It can be seen that the temperature t is a quadratic function of coordinate x.

可见，温度 t 是坐标 x 的二次函数。

The working condition 2: the third type of boundary condition.

工况 2：第三类边界条件。

Known: Both sides of the plane wall with a uniform internal heat source are subjected to convection heat transfer with the fluid of temperature t_f, and the convective heat transfer coefficient is h, as shown in Fig. 2. 30. Try to find the temperature at any section x of the plane wall and the heat flux through this section.

已知： 具有均匀内热源 $\dot{\Phi}$ 的平壁，两侧同时与温度为 t_f 的流体进行对流传热，表面传热系数为 h，如图 2. 30 所示。试求：平壁任一截面 x 处的温度，以及通过该截面的热流密度。

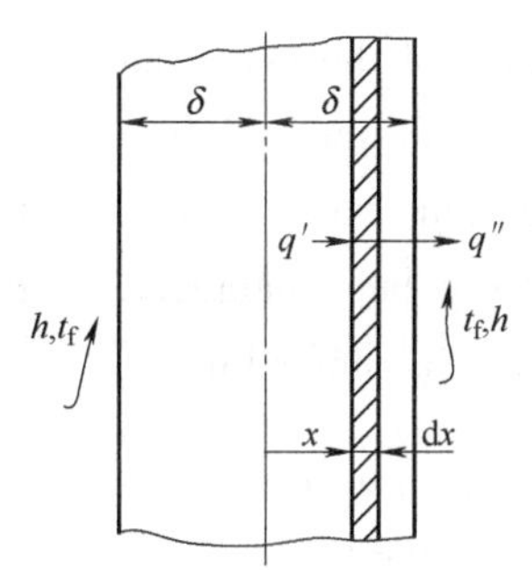

Fig. 2. 30　Schematic diagram of the plane wall

图 2. 30　平壁示意图

Solution: ① Geometric model: Due to the symmetry of plane wall structure and heat transfer conditions, just study half the thickness of the plane wall. Similarly, the coordinate origin is set on the center line of the wall. ② Mathematical model.

解：① 几何模型：由于平板的结构及传热条件的对称性，只需研究平壁厚的一半。同样，坐标原点设在平壁的中心线上。② 数学模型：

$$\begin{cases} \dfrac{d^2t}{dx^2}+\dfrac{\dot{\Phi}}{\lambda}=0 \\ x=0,\dfrac{dt}{dx}=0 & \text{第二类边界条件} \\ x=\delta,-\lambda\dfrac{dt}{dx}=h(t-t_f) & \text{第三类边界条件} \end{cases}$$

Integrate the governing equation twice, then we can get:

对控制方程积分两次就可以得到：

$$t=-\frac{\dot{\Phi}}{2\lambda}x^2+c_1x+c_2 \tag{2.67}$$

The undetermined constant c_1, c_2 is determined by the boundary conditions and the temperature distribution in the plane wall can be obtained as follows:

由边界条件确定待定常数 c_1、c_2，可得平壁中的温度分布：

$$t=\frac{\dot{\Phi}}{2\lambda}(\delta^2-x^2)+\frac{\dot{\Phi}\delta}{h}+t_f \tag{2.68}$$

It can be seen that temperature t is also a quadratic function of coordinate x. The heat flux at any section can be calculated by Fourier's law of heat conduction:

可见，温度 t 同样是坐标 x 的二次函数。通过任一截面处的热流密度可用傅里叶导热定律计算：

$$q=-\lambda\frac{dt}{dx}=\dot{\Phi}x \tag{2.69}$$

This is the relationship between heat flux and internal heat source.

这就是热流密度与内热源的关系式。

Discussion: Compared with the solution of the heat conduction problem of the plane wall without internal heat source, the heat flux is no longer a constant, the temperature distribution is no longer a straight line but a parabola, and the highest temperature point is on the center line

讨论： 与无内热源的平壁导热问题的解相比，热流密度不再是常数，温度分布也不再是直线，而是抛物线，温度最高点在中心线上。

Need to special note that the case of a given wall temperature t_w, i. e. the first type of boundary condition, it can be regarded as a special case where the fluid temperature t_f is equal to the wall temperature t_w when the convective heat transfer coefficient h tends to infinity.

特别说明： 给定壁面温度 t_w 的情形（即第一类边界条件），可看成当表面传热系数 h 趋于无穷大时，流体温度 t_f 等于壁面温度 t_w 的一个特例。

Example 2.8 Fig. 2.31 shows a simplified model of heat dissipation of fuel elements in a nuclear reactor. The model is a large plane wall composed of three layers of plates, and the middle layer is the nuclear fuel layer with $\delta_1 = 14\text{mm}$. Both sides are the aluminum plates with $\delta_2 = 6\text{mm}$, and the contact between layers is good. The fuel layer has an internal heat source $\dot{\Phi} = 1.5\times10^7\text{W/m}^3$, $\lambda_1 = 35\text{W/(m}\cdot\text{K)}$. The aluminum plates have no internal heat source, $\lambda_2 = 100\text{W/(m}\cdot\text{K)}$, its surface is cooled by high pressure water with temperature $t = 150℃$, and the convective heat transfer coefficient $h = 3500\text{W/(m}^2\cdot\text{K)}$. Regardless of the contact thermal resistance. Under steady state, try to determine the maximum temperature of the fuel layer, the interface temperature between the fuel layer and the aluminum plate, and surface temperature of the aluminum plate, and qualitatively draw the temperature distribution diagram of the simplified model.

例 2.8 图 2.31 所示为核反应堆中燃料元件散热的一个简化模型。该模型是一个三层平板组成的大平壁，中间是 $\delta_1 = 14\text{mm}$ 的核燃料层，两侧均为 $\delta_2 = 6\text{mm}$ 的铝板，层间接触良好。燃料层有 $\dot{\Phi} = 1.5\times10^7\text{W/m}^3$ 的内热源，$\lambda_1 = 35\text{W/(m}\cdot\text{K)}$；铝板中无内热源，$\lambda_2 = 100\text{W/(m}\cdot\text{K)}$，其表面受到温度为 $t = 150℃$ 的高压水冷却，表面传热系数 $h = 3500\text{W/(m}^2\cdot\text{K)}$。不计接触热阻，试确定稳态工况下燃料层的最高温度、燃料层与铝板的界面温度，以及铝板的表面温度，并定性画出简化模型中的温度分布图。

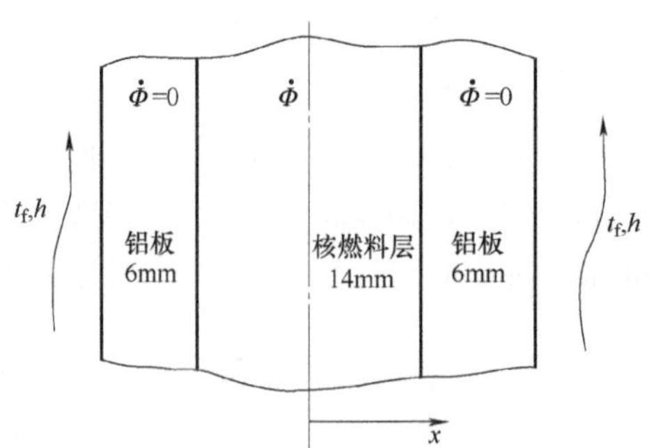

Fig. 2.31 A simplified model of heat dissipation of fuel elements in a nuclear reactor

图 2.31 核反应堆中燃料元件散热的简化模型

Solution: Firstly, the heat conduction problem is simplified and analyzed: ① Due to the symmetry of structure, just study half of the model. ② The intensity of the internal heat source is constant. Under steady-state conditions, the maximum temperature t_0 of the fuel layer must be on the center line $x=0$. ③ Regardless of contact thermal resistance. Under steady-state conditions, the heat released by nuclear fuel must be transferred to cooling water through the heat conduction of the aluminum plate finally, the heat flux from interface to cooling water is equal. ④ The interface temperature is denoted as t_1 and the surface temperature of the aluminum plate is denoted as t_2. Analysis shows that: $t_0>t_1>t_2$.

解：首先对该导热问题进行简化与分析：① 由于结构对称性，只需研究半个模型。② 内热源强度为常数，稳态工况下，燃料层最高温度 t_0 必然在其中心线上 $x=0$ 处。③ 不计接触热阻，稳态工况下，核燃料放出的热量必然通过铝板导热最终全部传给冷却水，且从界面到冷却水所传递的热流量均相等。④ 界面温度记为 t_1，铝板表面温度记为 t_2。分析可知：$t_0>t_1>t_2$。

Then, qualitatively draw the temperature distribution on the section and the thermal resistance from the interface to the cooling water, as shown in Fig. 2. 32. In the figure, R_{A1} is the thermal conduction resistance of the aluminum plate, R_{A2} is the convective heat transfer resistance, and q is the heat flux from the fuel layer into the aluminum plate. According to the thermal equilibrium, substitute the known parameters into the formula to get:

那么，可定性地画出截面上的温度分布，以及从界面到冷却水的热阻，如图 2. 32 所示。图中，R_{A1} 为铝板的导热热阻，R_{A2} 为对流传热热阻，q 为从燃料层进入铝板的热流密度。据热平衡关系，代入已知参数，可得

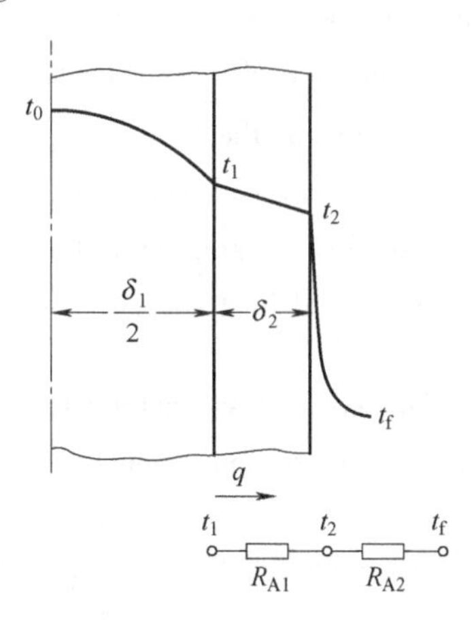

Fig. 2. 32 Temperature distribution on the section and the thermal resistance of the plane plate

图 2. 32 截面上的温度分布及平板热阻

$$q=\frac{\delta_1}{2}\dot{\Phi}=1.05\times10^5\,\text{W/m}^2$$

According to Newton cooling formula $q=h(t_2-t_f)$, the

按牛顿冷却公式 $q=h(t_2-t_f)$，可导

temperature can be derived $t_2 = t_f + \frac{q}{h}$. Substitute the known parameters into the formula to get $t_2 = 180℃$.

出温度 $t_2 = t_f + \frac{q}{h}$。代入已知参数，可求出 $t_2 = 180℃$。

For the outside aluminum plate, according to Fourier's law of heat conduction $q = \frac{\lambda_2 (t_1 - t_2)}{\delta_2}$, we can derive $t_1 = \frac{q\delta_2}{\lambda_2} + t_2$. Substitute the known parameters into the formula to get $t_1 = 186.3℃$.

对于外侧的铝板，按傅里叶导热定律 $q = \frac{\lambda_2 (t_1 - t_2)}{\delta_2}$，可导出 $t_1 = \frac{q\delta_2}{\lambda_2} + t_2$。代入已知参数，可求出 $t_1 = 186.3℃$。

According to the expression of temperature distribution of plane wall with an internal heat source under the first type of boundary condition:

根据有内热源的平板在第一类边界条件下的温度分布表达式：

$$t = \frac{\dot{\Phi}}{2\lambda}(\delta^2 - x^2) + t_w$$

Substitute the known parameters into the expression to get $t_0 = t_1 + \frac{\dot{\Phi}(\delta_1/2)^2}{2\lambda_1} = 196.8℃$.

代入已知参数，可求出 $t_0 = t_1 + \frac{\dot{\Phi}(\delta_1/2)^2}{2\lambda_1} = 196.8℃$。

Discussions:

(1) When analyzing the thermal resistance, why do we start with the interface temperature t_1 instead of t_0? This is because there is an internal heat source in the fuel layer, and the heat flux at different sections is not equal. Therefore, the thermal resistance analysis method is not applicable for the heat conduction problem with internal heat source.

(2) Can we use the thermal resistance analysis method to find the interface temperature t_1 first and then t_2? Certainly, we can find t_1 from the equation $q = \frac{t_1 - t_f}{\frac{\delta_2}{\lambda_2} + \frac{1}{h}}$ and then we can find t_2.

讨论：

(1) 热阻分析时，为什么从界面温度 t_1 开始，而不是从 t_0 开始？这是因为燃料层有内热源，不同截面处的热流量不相等。因此，热阻分析法不适用于有内热源的导热问题。

(2) 能否用热阻分析法先求出界面温度 t_1，再求出 t_2？当然可以。由 $q = \frac{t_1 - t_f}{\frac{\delta_2}{\lambda_2} + \frac{1}{h}}$ 可求出 t_1，然后就可以求出 t_2。

Summary This section illustrates the solution of one-dimensional steady-state heat conduction problem with internal heat source.

小结 本节阐述了有内热源一维稳态导热问题的求解方法。

Question What are the characteristics of temperature distribution in a plane wall with a uniform internal heat source?

思考题 有均匀内热源的平壁内温度分布有什么特点?

Self-tests

自测题

1. When the geometric structure and boundary conditions of the thermal conductor are symmetric, how to select the geometric model?

1. 导热体的几何结构和边界条件都具有对称性时,几何模型如何选取?

2. Why can't thermal resistance analysis be used for plate heat conduction problem with internal heat source?

2. 有内热源平板导热问题为什么不能用热阻分析法?

3. Under what conditions can the third boundary condition of a thermal conductor be transformed into the first boundary condition?

3. 导热物体的第三类边界什么条件下可转化第一类边界条件?

Exercises

习题

2.1 Boil water in a pan, the temperature of the bottom of the pan in contact with the water is 111℃, and the heat flow density is $42400W/m^2$. After a period of use, a layer of scale with an average thickness of 3mm formed on the bottom of the pot. Assuming that the surface temperature and heat flux density of the scale in contact with water at this time are respectively equal to the original values, try to calculate the temperature of the contact surface between the scale and the metal pot bottom. The thermal conductivity of scale is taken as $1W/(m \cdot K)$.

2.1 用平底锅烧开水,与水相接触的锅底温度为111℃,热流密度为$42400W/m^2$。使用一段时间后,锅底结了一层平均厚度为3mm的水垢。假设此时与水相接触水垢的表面温度及热流密度分别等于原来的值,试计算水垢与金属锅底接触面的温度。水垢的导热系数为$1W/(m \cdot K)$。

2.2 The wall of a refrigerating room is composed of three layers of steel slag, slag wool and asbestos board. The thickness of each layer is 0.794mm, 152mm and 9.5mm in turn, and the thermal conductivity is $45W/(m \cdot K)$, $0.07W/(m \cdot K)$ and $0.1W/(m \cdot K)$, respectively. The effective heat exchange area of the refrigerating room is $37.2m^2$, the indoor and outdoor air temperatures are −2℃ and 30℃ respectively, and the surface heat transfer coefficients of the indoor and outdoor walls can be calculated according to $1.5W/(m^2 \cdot K)$ and $2.5W/(m^2 \cdot K)$,

2.2 一冷藏室的墙由钢皮、矿渣棉及石棉板三层叠合构成,各层的厚度依次为0.794mm、152mm及9.5mm,导热系数分别为$45W/(m \cdot K)$、$0.07W/(m \cdot K)$及$0.1W/(m \cdot K)$。冷藏室的有效传热面积为$37.2m^2$,室内外气温分别为−2℃及30℃,室内外壁面的表面传热系数可分别按$1.5W/(m^2 \cdot K)$及$2.5W/(m^2 \cdot K)$计算。为维持冷藏室温度恒定,试

respectively. In order to maintain a constant temperature in the refrigerator, try to determine the amount of heat that the cooling pipes in the refrigerator need to remove per hour.

确定冷藏室内的冷却排管每小时需带走的热量。

2.3 For the one-dimensional heat conduction problem in an infinite flat plate, try to explain which combinations of boundary conditions on both sides can obtain a definite solution to the temperature field in the flat plate among the three types of boundary conditions?

2.3 对于无限大平板内的一维导热问题，试说明在三类边界条件中，两侧边界条件的哪些组合可以使平板中的温度场获得确定的解？

2.4 The principle of an apparatus for measuring thermal conductivity by a comparative method is shown in Fig. 2.33. The standard material with known thermal conductivity and the tested material are made into cylinders with the same diameter, and the two cylinders of the standard material are pressed and placed on both ends of the tested material. Three pairs of thermocouples for measuring the temperature difference between two points at equal intervals are respectively arranged on the three samples. The samples were well insulated around the edges (not shown in the figure). The temperatures at both ends of the sample are known to be:

2.4 一种用比较法测定导热系数的装置原理如图 2.33 所示。将导热系数已知的标准材料与被测材料做成相同直径的圆柱，且标准材料的两段圆柱分别压紧置于被测材料的两端。在三段试样上分别布置三对测定相等间距两点间温差的热电偶。试样的四周绝热良好（图中未示出）。已知试样两端的温度分别为：

$$t_h = 400℃, t_c = 300℃, \Delta t_r = 2.49℃, \Delta t_{t,1} = 3.56℃, \Delta t_{t,2} = 3.60℃$$

Try to determine the thermal conductivity of the tested material and discuss which factors will affect the inequality between $\Delta t_{t,1}$ and $\Delta t_{t,2}$?

试确定被测材料的导热系数，并讨论哪些因素会影响 $\Delta t_{t,1}$、$\Delta t_{t,2}$ 不相等？

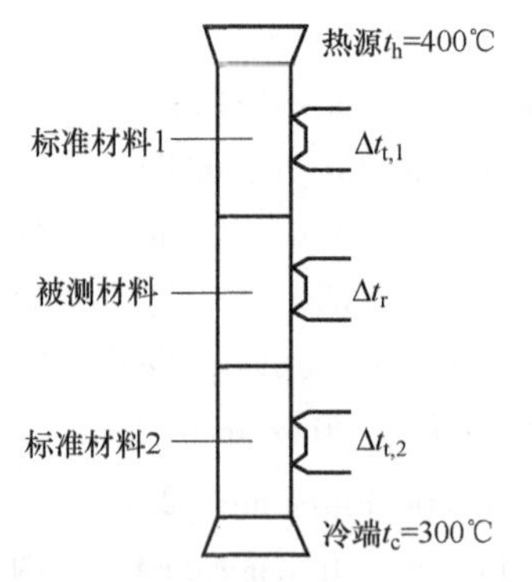

Fig. 2.33 Figure for exercise 2.4

图 2.33 习题 2.4 图

2.5 Raising the gas inlet temperature is an effective way to improve the efficiency of aero-engines. In order to allow the blades of the engine to withstand higher temperatures without damage, the blades are made of high-temperature resistant alloys. At the same time, a method of coating a thin layer of ceramic material on the surface of the blades in contact with high-temperature gas is also proposed, as shown in Fig. 2.34, the internal passages of the blades are cooled by air from the compressor. The thermal conductivity of the ceramic layer is 1.3W/(m · K), the maximum temperature that the high-temperature resistant alloy can withstand is 1250K, and its thermal conductivity is 25W/(m · K). There is a thin layer of bonding material between the high-temperature resistant alloy and the ceramic layer, and the contact thermal resistance caused by it is 10^{-4}m^2 · K/W. If the average temperature of the gas is 1700K, the surface heat transfer coefficient with the ceramic layer is 1000W/(m^2 · K), the average temperature of the cooling air is 400K, and the surface heat transfer coefficient with the inner wall is 500W/(m^2 · K). Try to analyze whether the superalloy can work safely at this time? (The thickness of the ceramic layer is 1.3mm, and the thickness of the alloy layer is 25mm)

2.5 提高燃气进口温度是提高航空发动机效率的有效方法。为了使发动机的叶片能承受更高的温度而不至于损坏，叶片均用耐高温的合金制成，同时还提出了在叶片与高温燃气接触的表面上涂以陶瓷材料薄层的方法，如图 2.34 所示，叶片内部通道由从压气机来的空气予以冷却。陶瓷层的导热系数为 1.3W/(m · K)，耐高温合金能承受的最高温度为 1250K，其导热系数为 25W/(m · K)。在耐高温合金与陶瓷层之间有一薄层黏结材料，其造成的接触热阻为 10^{-4}m^2 · K/W。如果燃气的平均温度为 1700K，与陶瓷层的表面传热系数为 1000W/(m^2 · K)，冷却空气的平均温度为 400K，与内壁间的表面传热系数为 500W/(m^2 · K)。试分析此时耐高温合金是否可以安全地工作（陶瓷层厚度为 1.3mm，合金层厚度为 25mm）?

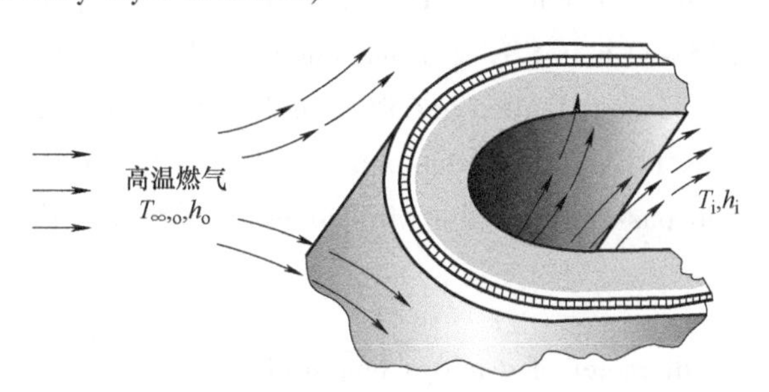

Fig. 2.34 Figure for exercise 2.5

图 2.34 习题 2.5 图

2.6 During the manufacturing process of a certain product, a transparent film with a thickness of 0.2mm is attached to the substrate with a thickness of 1.0mm. There is a cooling air flow on the surface of the film, its temperature is 20℃, and the surface heat transfer coefficient of the convective heat exchange is 40W/(m^2 · K). At the

2.6 在某一产品的制造过程中，厚为 1.0mm 的基板上紧贴了一层透明的薄膜，其厚度为 0.2mm。薄膜表面上有一股冷却气流流过，其温度为 20℃，对流传热的表面传热系数为 40W/(m^2 · K)。同时，

same time, there is a radiation energy projected through the film to the bonding surface of the film and the substrate, as shown in Fig. 2. 35. The other side of the substrate is maintained at temperature t_1 = 30℃. The production process requires the temperature t_0=60℃ of the bonding surface between the film and the substrate. Try to determine what the radiation heat flux q should be? The thermal conductivity of the film λ_f=0. 02W/(m · K), and the thermal conductivity of the substrate λ_s=0. 06W/(m · K). The radiant heat flow projected on the bonding surface is all absorbed by the bonding surface. The film is opaque to thermal radiation at 60℃.

有一股辐射能透过薄膜投射到薄膜与基板的结合面上，如图 2. 35 所示。基板的另一面维持在温度 t_1 = 30℃。生产工艺要求薄膜与基板结合面的温度 t_0 = 60℃。试确定辐射热流密度 q 应为多大？薄膜的导热系数 λ_f = 0. 02W/(m · K)，基板的导热系数 λ_s = 0. 06W/(m · K)。投射到结合面上的辐射热流全部为结合面所吸收。薄膜对 60℃ 的热辐射是不透明的。

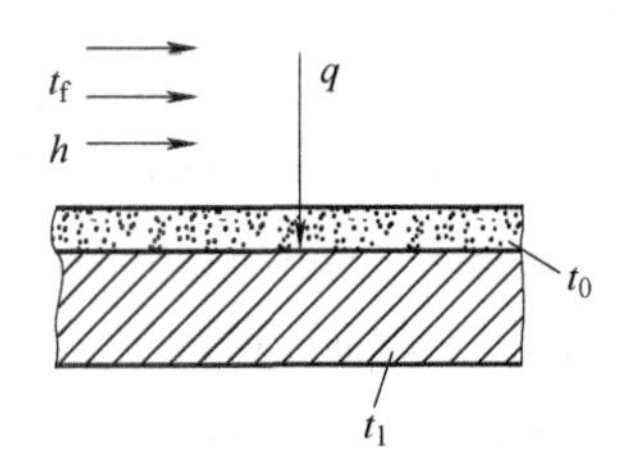

Fig. 2. 35 Figure for exercise 2. 6

图 2. 35 习题 2. 6 图

2. 7 A copper wire with a diameter of 3mm has a resistance per meter of length 2. 22×10^{-3}Ω. The wire is covered with an insulation layer with a thickness of 1mm and a thermal conductivity of 0. 15W/(m · K). The maximum temperature of the insulation layer is limited to 65℃, and the minimum temperature is 0℃. Try to determine the maximum current allowed to pass through the wire under this condition.

2. 7 一根直径为 3mm 的铜导线，每米长的电阻为 2. 22×10^{-3}Ω。导线外包有厚为 1mm、导热系数为 0. 15W/(m · K) 的绝缘层。限定绝缘层的最高温度为 65℃，最低温度为 0℃。试确定在这种条件下导线中允许通过的最大电流。

2. 8 A circular rod with a diameter of d and a length of l, both ends of which are in contact with surfaces at temperatures t_1 and t_2 respectively, and the thermal conductivity λ of the rod is constant. Try to list the differential equations and boundary conditions for the temperature in the rod for the following two situations, and solve them.

2. 8 一直径为 d 长为 l 的圆杆，两端分别与温度为 t_1 及 t_2 的表面接触，杆的导热系数 λ 为常数。试对下列两种情形列出杆中温度的微分方程式及边界条件，并求解。

(1) The side of the rod is insulated.

(1) 杆的侧面是绝热的。

(2) There is stable convection heat transfer between the

(2) 杆的侧面与四周流体间有稳定

side of the rod and the surrounding fluid, the average surface heat transfer coefficient is h, and the fluid temperature t_f is less than t_1 and t_2.

的对流传热，平均表面传热系数为 h，流体温度 t_f 小于 t_1 及 t_2。

2.9 A hollow cylinder, $t = t_1$ at $r = r_1$, $t = t_2$ at $r = r_2$. $\lambda(t) = \lambda_0(1+bt)$, t is the local temperature, try to derive the expression of temperature distribution in the cylinder and the calculation formula of the heat conduction quantity.

2.9 一空心圆柱，在 $r=r_1$ 处 $t=t_1$，$r=r_2$ 处 $t=t_2$。$\lambda(t)=\lambda_0(1+bt)$，$t$ 为局部温度，试导出圆柱中温度分布的表达式及导热量计算式。

2.10 Aluminum ring ribs of equal thickness are installed on the tube wall with an outer diameter of 25mm. The distance between the centerlines of the adjacent ribs is $s = 9.5$mm, and the height of the ring rib is $H = 12.5$mm and the thickness $\delta = 0.8$mm. The temperature of the tube wall $t_w = 200$℃, the temperature of the fluid $t_f = 90$℃, the surface heat transfer coefficient between the tube wall, the ribs and the fluid is 110W/(m^2 · K). Try to determine the heat dissipation per meter of finned tube (including ribs and base tube).

2.10 在外径为 25mm 的管壁上装有铝制的等厚度环肋，相邻肋片中心线之间的距离 $s = 9.5$mm，环肋高 $H = 12.5$mm，厚 $\delta = 0.8$mm。管壁温度 $t_w = 200$℃，流体温度 $t_f = 90$℃，管壁及肋片与流体之间的表面传热系数为 110W/(m^2 · K)。试确定每米长肋片管（包括肋片及基管部分）的散热量。

Chapter 3 Unsteady-state Heat Conduction

第3章 非稳态热传导

3.1 The Characteristics of Unsteady-state Heat Conduction

3.1 非稳态导热的特点

Unsteady-state heat conduction refers to the heat conduction process in which the temperature varies with time in a heat conductor. Temperature t is not just a function of the spatial coordinates x, y and z but also of time τ. In the case of unsteady-state heat conduction, not only the temperature of each point in the object varies with time, but also the heat conduction quantity and internal energy vary with time.

非稳态导热是指在导热体中温度随时间而变化的导热过程。温度 t 不仅是空间坐标 x、y、z 的函数，还是时间 τ 的函数。非稳态导热时，不仅物体中各点的温度随时间变化，其导热量和内能也在随时间而变化。

In nature and engineering, many heat conduction processes are unsteady-state heat conduction. For example, the process of heating or cooling a workpiece in metallurgy, heat treatment and heat processing; or the operation of startup, shutdown or variable operating conditions of the thermal equipment such as boilers, internal combustion engines, heat exchangers and so on; or the heating or shutdown process of the building, in which the temperature of walls and indoor air vary; etc.

在自然界和工程中，许多导热过程都属于非稳态导热。例如，在冶金、热处理和热加工领域中物体被加热或冷却的过程；热力设备如锅炉、内燃机、换热器等在起动、停机或变工况运行时；建筑物的供暖或停暖过程中墙壁及室内空气的温度变化；等等。

In the regenerative heat exchanger, the endothermic and exothermic processes of heat accumulator are also the typical unsteady-state heat conduction processes, as shown in Fig. 3. 1. Its working principle is that the heat accumulator in the heat exchanger cylinder alternately contacts the hot fluid and the cold fluid for heat exchange, so as to indirectly transfer heat from the hot fluid to the cold fluid.

在蓄热式换热器中，蓄热体的吸热和放热过程也是典型的非稳态导热过程，如图3.1所示。其工作原理是：换热器筒体内的蓄热体与热流体及冷流体交替接触换热，从而把热量从热流体间接地传给冷流体。而管壳式换热器在正常运行时，其

When the shell and tube heat exchanger is in normal operation, the heat conduction process of the tube wall is steady-state heat conduction.

管壁的导热过程为稳态导热。

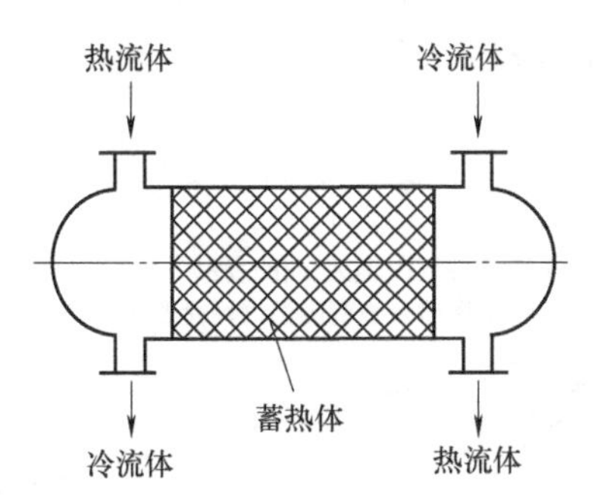

Fig. 3.1 The regenerative heat exchanger
图 3.1 蓄热式换热器

According to the characteristics of the object's temperature changes with time, unsteady-state heat conduction can be divided into two categories: periodic unsteady-state heat conduction and aperiodic unsteady-state heat conduction. In the process of unsteady-state heat conduction, the temperature and heat flux of each point in an object vary periodically with time, which is called periodic unsteady-state heat conduction. For example, the periodic change of solar radiation causes the temperature fields of walls and roofs of houses to vary with time periodically 24 hours per day. The temperature of the earth's surface also varies with the seasons in a one-year cycle.

根据物体温度随时间的变化特点，非稳态导热可分为周期性非稳态导热和非周期性非稳态导热两大类。在非稳态导热过程中，物体中各点温度和热流密度都随着时间做周期性变化，称为周期性非稳态导热。例如太阳辐射的周期性变化，引起房屋墙壁和屋顶的温度场随时间每天 24h 做周期性变化。地球表面层的温度也会随着季节更替，而做以一年为周期的变化。

In the process of heat conduction, if the temperature of an object increases continuously such as heating process or decreases continuously such as cooling process, after a long time, the heat conduction of the object gradually reaches a steady state or thermal equilibrium. This unsteady-state heat conduction process is called aperiodic unsteady-state heat conduction, also known as transient heat conduction.

在导热过程中，如果物体的温度不断升高（如加热过程），或者不断降低（如冷却过程），在经历相当长时间以后，物体导热逐渐达到稳态或热平衡。这种非稳态导热过程称为非周期性非稳态导热，又称为瞬态导热。

For example, put the hot iron block into cold water, the temperature of the iron block gradually decreases, and the temperature of water gradually increases, and they eventually reach thermal equilibrium. Then, the temperature

例如将高温铁块投入凉水中，铁块温度逐渐降低，而水的温度逐渐升高，它们之间最终达到热平衡。那么，水和铁块的温度变化过程都属

change processes of water and the iron block are transient heat conduction. Now let's discuss in detail the characteristics of temperature changes in the objects during unsteady-state heat conduction, as shown in Fig. 3. 2.

于瞬态导热。下面详细讨论一下非稳态导热过程中物体内温度变化的特点，如图 3. 2 所示。

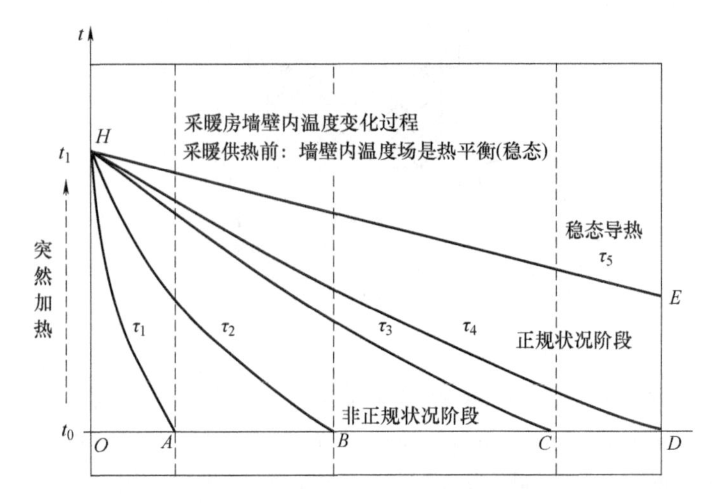

Fig. 3. 2 Temperature changes in a house wall during unsteady-state heat conduction

图 3. 2 非稳态导热过程中墙壁内温度变化

Take a heating house wall, for example. The house wall can be regarded as an infinite plane plate, which means that the length and width are infinite relative to the thickness, its initial temperature is t_0, and its outer wall is always in contact with the air at temperature t_0. After heating starts, the indoor air heats up rapidly and heats the wall surface, the temperature of the inner wall surface suddenly rises from t_0 to t_1 and remains unchanged. Due to the temperature difference inside the wall, heat begins to transfer from the inside to the outside.

以采暖房的墙壁为例。墙壁可看作一块无限大平板，这里是指长度和宽度相对于厚度为无限大，初始温度为 t_0，外壁面始终与温度为 t_0 的空气接触。开始供暖后，室内空气快速升温并加热墙壁面，内壁面温度由 t_0 突然升高到 t_1 并保持不变。由于墙壁内存在温差，热量开始从内侧向外侧传递。

At the moment of τ_1, the heat is transferred to the cross-section corresponding to point A, the temperature distribution is curve HA. At the moment of τ_2, the heat is transferred to the cross-section corresponding to point B, the temperature distribution is curve HB. At the moment of τ_3, the heat is transferred to the cross-section corresponding to point C, the temperature distribution is curve HC. In the process of heat transfer, the cold part of the wall absorbs heat constantly and converts it into internal

在 τ_1 时刻，热量传递到 A 点对应的截面，温度分布为曲线 HA；在 τ_2 时刻，热量传递到 B 点对应的截面，温度分布为曲线 HB；在 τ_3 时刻，热量传递到 C 点对应的截面，温度分布为曲线 HC。在热量的传递过程中，墙壁的低温部分不断吸收热量并转化为内能，各点的温度不断升高。越靠近内壁面的截

energy, and the temperature of each point keeps rising. The section closer to the inner wall surface heats up faster, so the heat transferred to the right gradually decreases. Within a certain heat transfer time, the heat reaching the outer wall surface is 0.

面升温越快，所以，向右侧传递的热量会逐渐减少。在一定的传热时间内，到达外壁面的热量为0。

The heat is transferred to point *D* after the moment of τ_4, that is, the heat reaches the outer wall surface, and the corresponding temperature distribution is curve *HD*. Before the moment of τ_4, heat is not transferred to the outer wall surface, the wall surface still remains the initial temperature t_0. After the moment of τ_4, the outer wall surface starts to absorb heat and its temperature rises gradually. Therefore, from the time point of view, τ_4 is the time demarcation point of whether heat is transmitted to the outer wall surface. According to this time demarcation point, the unsteady-state heat conduction process can be divided into two stages: the non-normal state stage and the normal state stage.

直到 τ_4 时刻之后，热量传递到 *D* 点，也就是到达了外壁面，对应的温度分布为曲线 *HD*。在 τ_4 时刻之前，热量没有传递到外壁面，外壁面仍保持初始温度 t_0。在 τ_4 时刻之后，外壁面开始吸收热量，温度逐渐升高。所以从时间上来看，τ_4 时刻就是热量是否传递到外壁面的时间分界点。根据这个时间分界点，可把非稳态导热过程分为非正规状况阶段和正规状况阶段。

The non-normal state phase refers to the phase in which heat has not been transferred to the outer wall surface, also known as the initial phase. The temperature distribution is mainly controlled by the initial temperature distribution, such as the phase of the curves *HA*, *HB*, *HC* in Fig. 3. 2. The normal state phase refers to the phase after the heat has been transferred to the outer wall surface, also known as the full development phase. The effect of the initial temperature distribution gradually disappears, and the temperature distribution mainly depends on the thermal boundary conditions and the thermal properties of the thermal conductor.

非正规状况阶段是指热量尚未传递到外壁面的阶段，也称为初始阶段；温度分布主要受初始温度分布控制，如图 3. 2 中曲线 *HA*、*HB*、*HC* 所处的阶段。正规状况阶段是指热量已经传递到外壁面之后的阶段，也称为充分发展阶段。初始温度分布的影响逐渐消失，温度分布主要取决于热边界条件和导热体的热物性。

Then, after another period of time, the temperature of each point in the wall will no longer change, and the heat conduction process reaches a steady state, forming a new temperature distribution, such as straight line *HE*. Generally speaking, the whole unsteady-state heat conduction process

然后，再经历一段时间后，墙壁内各点的温度不再变化，导热过程达到稳态，形成新的温度分布，如直线 *HE*。一般来说，物体的整个非稳态导热过程主要处于正规状况阶

of an object is mainly in the normal state phase. Therefore, the temperature variation regularity in the normal state phase of unsteady-state heat conduction is the focus of learning.

段。因此，非稳态导热正规状况阶段的温度变化规律是学习的重点。

Now let's talk about the changes of heat transfer rate on both sides of a plane plate during unsteady-state heat conduction. For an unsteady-state heat conduction process, heat transfer rate varies everywhere in different sections perpendicular to the direction of heat flow. This is also the main characteristic of unsteady-state heat conduction different from steady-state heat conduction, as shown in Fig. 3. 3.

接下来讨论一下非稳态导热过程中平板两侧表面的热流量变化。对于非稳态导热过程，在与热流方向垂直的不同截面上，热流量各处不相等。这也是非稳态导热区别于稳态导热的主要特点，如图 3. 3 所示。

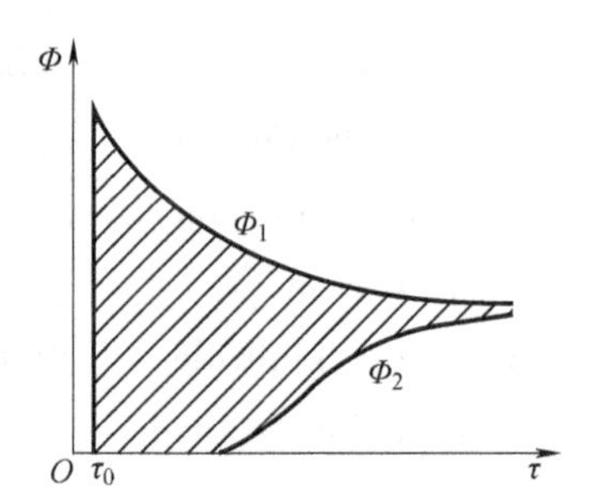

Fig. 3. 3 The main characteristics of unsteady-state heat conduction different from steady-state heat conduction

图 3. 3 非稳态导热区别于稳态导热的主要特点

Where Φ_1 is the heat imported from the left side of the plane plate, while Φ_2 is the heat exported from the right side of the plane plate. Therefore, the shaded area represents the total heat accumulated by the plane plate during heating, which is the increment of internal energy.

图中，Φ_1 是从平板左侧面导入的热量，而 Φ_2 是从右侧面导出的热量。因此，阴影区域代表了平板在升温过程中所积聚的总热量，也就是内能的增量。

It can be seen from Fig. 3. 3 that the imported heat Φ_1 and exported heat Φ_2 are not equal during the whole unsteady-state heat conduction process. As the temperature of each point in the plane plate rises continuously, the temperature difference with the left side wall surface continuously decreases, so the heat Φ_1 imported into the plane plate gradually decreases. Before the heat is transferred to the right side wall surface, the heat Φ_2 exported from the right side wall surface is equal to 0. In other words, the

从图 3. 3 中可以看出，在整个非稳态导热过程中，导入与导出的热量 Φ_1 和 Φ_2 是不相等的。随着平板内各点温度不断升高，与左侧面的温差不断减小，因此导入平板的热量 Φ_1 逐渐减少。在热量传递到右侧面之前，从右侧面导出的热量 Φ_2 等于 0。也就是说，非稳态导热处于非正规状况阶段，即初始阶段。

unsteady-state heat conduction is in the non-normal state phase, that is, the initial phase.

Heat has been transferred to the right side wall surface from the moment of τ_4, heat Φ_2 exported from the right side gradually increases, and the unsteady-state heat conduction enters the normal state phase, that is, the full development phase. With the passage of time, the difference between Φ_1 and Φ_2 gradually decreases until they're equal, which indicates that the plane plate no longer absorbs heat and enters the steady-state heat conduction phase.

从 τ_4 时刻开始，热量已经传递到右侧面，从右侧面导出的热量 Φ_2 逐渐增大，非稳态导热进入正规状况阶段，即充分发展阶段。随着时间的推移，Φ_1 和 Φ_2 的差值逐渐减小，直到两者相等，说明平板不再吸收热量，进入了稳态导热阶段。

It can be seen that, there are also two important characteristics of the unsteady-state heat conduction process: ① The temperature distribution of the unsteady-state heat conduction process is a curve, while the temperature distribution of the steady-state heat conduction process is a straight line. ② The internal energy of the object is also constantly changing during the process of unsteady-state heat conduction.

由此可见，非稳态导热过程还有两个重要特点：① 非稳态导热过程的温度分布是曲线，而稳态导热过程的温度分布是直线；② 非稳态导热过程中物体的内能也在不断变化。

Summary This section illustrates the characteristics and classification of the unsteady-state heat conduction process and focuses on learning the variation characteristics of temperature distribution and heat transfer rate in the non-normal state phase and the normal state phase.

小结 本节阐述了非稳态导热过程的特点及分类，重点学习了非正规状况阶段与正规状况阶段中温度分布和热流量的变化特点。

Question What are the main characteristics of unsteady-state heat conduction different from steady-state heat conduction?

思考题 非稳态导热区别于稳态导热的主要特点有哪些？

Self-tests

自测题

1. Briefly describe the definition and characteristics of unsteady-state heat conduction.

1. 简述非稳态导热的定义及特点。

2. Briefly describe the two stages of transient thermal conduction process and their characteristics of heat transfer rate variation.

2. 简述瞬态导热过程的两个阶段及其热流量变化特点。

3.2 Biot Number and Characteristic Numbers

3.2 毕渥数及特征数

The third type of boundary condition is the focus of unsteady-state heat conduction problems, which is the known convective heat transfer coefficient h and the temperature t_f of the fluid.

非稳态导热问题一般以第三类边界条件为研究重点，也就是已知表面传热系数 h 和流体的温度 t_f。

First, let's analyze the unsteady-state heat conduction problem of an infinite plane plate, as shown in Fig. 3.4. There is an infinite plane plate of thickness 2δ. It is known that the thermal conductivity λ and the thermal diffusivity a are constants, the temperature is t_0 at initial moment, the plane plate is suddenly put into a fluid at temperature t_f, the convective heat transfer coefficient between the wall and the fluid is h. Try to find the temperature distribution inside the plate.

首先，分析一个无限大平板的非稳态导热问题，如图 3.4 所示。厚度为 2δ 的无限大平板，已知导热系数 λ、热扩散率 a 均为常数，初始时刻 $\tau=0$ 时，其温度为 t_0，突然放置于温度为 t_f 的流体中，壁面与流体之间的表面传热系数为 h。求平板内的温度分布。

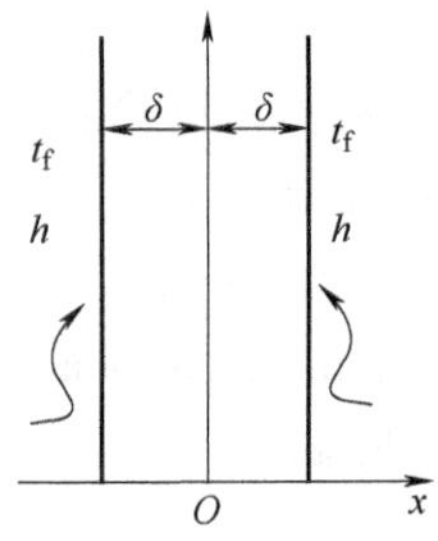

Fig. 3.4 Schematic diagram of unsteady-state heat conduction in an infinite plane plate

图 3.4 无限大平板非稳态导热示意图

Analysis: The plane plate is geometrically symmetric about the center plane, the cooling heat transfer conditions on both sides are the same. The center plane is a plane of symmetry and can also be regarded as an adiabatic plane. It can be inferred that the internal temperature distribution must be symmetrical about the center plane. So, we only need to solve for the temperature field of half a plane plate. Remember the pithy formula: a plane of symmetry, take half of it, which can simplify the problem.

分析：平板关于中心面几何对称，两侧的冷却换热情况相同。中心面就是对称面，也可看作是绝热面。由此可以推测，平板内部的温度分布必然关于中心面对称。因此，只需要求解半个平板的温度场。请记住口诀：对称面，取一半。这样可使问题简化。

Take the center of the plane plate as the origin of the coordinates, the thickness of half a plane plate is δ. The heat transfer process of the plane plate includes two heat transfer links: ① Convection heat transfer between the fluid and the plane plate surface, the thermal resistance is $r_h = 1/h$; ② Heat conduction inside the plane plate, the thermal resistance is $r_\lambda = \delta/\lambda$. Then we can define a new dimensionless number, i. e. Biot number. Biot number represents the ratio of the thermal conduction resistance to convective heat transfer resistance. So, it's a dimensionless number. Its expression is:

取平板中心面处为坐标原点，半个平板的厚度为 δ。该平板的传热过程包括两个传热环节：① 流体与平板表面的对流传热，热阻为 $r_h = 1/h$；② 平板内部的导热，热阻为 $r_\lambda = \delta/\lambda$。于是可定义一个新的无量纲数，即毕渥数。毕渥数表示了导热热阻与对流传热热阻的比值，所以它是一个无量纲数，其表达式为：

$$Bi = \frac{r_\lambda}{r_h} = \frac{\delta/\lambda}{1/h} = \frac{\delta h}{\lambda} \tag{3.1}$$

Now let's discuss about the influence of Biot number on the temperature distribution.

下面讨论一下毕渥数对温度分布的影响。

The first extreme case, when Bi tends to ∞, that is to say, the thermal conduction resistance r_λ of the plane plate is much larger than the convective heat transfer resistance r_h, so the convective heat transfer resistance can be ignored. Which means that the convective heat transfer coefficient of the fluid is infinite, and the wall temperature of the plane plate is equal to the fluid temperature, as shown in Fig. 3. 5a. There is no temperature difference between the plane plate and the fluid near the wall, while the temperature field inside the object is distributed symmetrically about the center plane. Through the above analysis, we can know when Bi tends to ∞, the third type of boundary condition can be transformed into the first type of boundary condition in the unsteady-state heat conduction problem.

第一种极端情况：当 Bi 趋于 ∞ 时，也就是平板的导热热阻 r_λ 远大于对流传热热阻 r_h，因此对流传热热阻可以忽略不计。这意味着流体的表面传热系数为无穷大，平板的壁面温度等于流体温度，如图 3. 5a 所示。在壁面附近流体与平板没有温差，而物体内部的温度场关于中心面对称分布。通过上述分析可知，当 Bi 趋于 ∞ 时，非稳态导热问题的第三类边界条件可转化为第一类边界条件。

The second extreme case, when Bi tends to zero, the thermal conduction resistance r_λ is much smaller than the convective heat transfer resistance r_h, so, the thermal conduction resistance can be ignored, which means that the temperature of each point inside the object is uniform at any time, as shown in Fig. 3. 5b. The temperature distri-

第二种极端情况，当 Bi 趋于 0 时，导热热阻 r_λ 远小于对流传热热阻 r_h，所以导热热阻可以忽略不计，意味着物体内各点的温度在任意时刻都均匀一致，如图 3. 5b 所示。在不同时刻，物体内部的温度分布

bution inside the object is the horizontal lines of different values at different moments, namcly, the temperature field inside the object has nothing to do with the coordinates x, y and z. So, the object can be regarded as a mass point. This is a typical zero-dimensional unsteady-state heat conduction problem.

为不同数值的水平线，也就是物体内部的温度场与坐标 x、y、z 无关。因此，可以把物体看作是一个质点。这就是典型的零维非稳态导热问题。

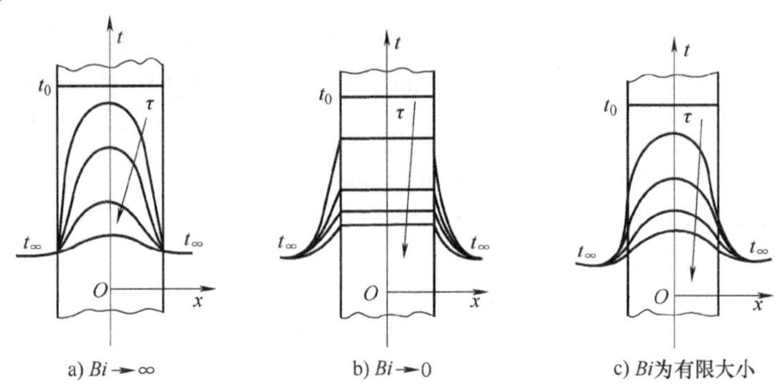

Fig. 3. 5 Influence of Bi number on temperature distribution in plate

图 3. 5 Bi 数对板内温度分布的影响

The general case in engineering is that the Biot number is limited which is between 0 and ∞, neither thermal conduction resistance nor convective heat transfer resistance can be ignored. The temperature distribution of the heat transfer process is shown in Fig. 3. 5c. The temperature distribution is approximately a parabolic curve at any time, and the temperature field is symmetrical about the center plane of the plane plate.

工程中的一般情况是：Bi 为有限大小，也就是介于 0 与 ∞ 之间，导热热阻与对流传热热阻均不能忽略。传热过程的温度分布如图 3. 5c 所示。任意时刻的温度分布近似为抛物线，并且温度场关于平板的中心面对称。

The above analysis methods of the unsteady-state heat conduction problem are summarized as follows: When the research problem is very complex and involves many parameters, in order to reduce the number of parameters in the problem, people combine some parameters organically to characterize the main features of a certain physical phenomenon or process, and there is no dimension in combined parameters. Then, the dimensionless number that characterizes a certain physical phenomenon or process is called the characteristic number, also known as the criterion number such as Reynolds number, Biot number, etc.

下面总结一下非稳态导热问题的上述分析方法：当研究的问题非常复杂且涉及参数较多时，为减少问题所涉及的参数，人们将一些参数有机地组合起来，来表征某一类物理现象或物理过程的主要特征，并且没有量纲。那么，表征某一类物理现象或物理过程特征的无量纲数就称为特征数，又称为准则数，如雷诺数、毕渥数等。

The geometric scale involved in the characteristic number is called the characteristic length, which is generally denoted

特征数涉及的几何尺度称为特征长度，一般用 l 表示，例如以半个平

by l, for example, the thickness of half a plane plate δ is taken as the characteristic length, namely $l=\delta$; for cylinder $l=R$. For each characteristic number, we must master its definition formula and physical meaning, and the meaning of each physical quantity in the definition formula.

板的厚度δ作为特征长度，也就是$l=\delta$；对于圆柱体$l=R$。对每一个特征数，必须掌握其“定义式和物理意义”，以及定义式中各个物理量的含义。

In addition, we have to master the uniqueness theorem of solution: If the independent variables of a function t are x, y, z and τ, and the function satisfies the heat conduction differential equation and certain initial and boundary conditions, then this function is the only solution to the specific heat conduction problem.

另外，要掌握解的唯一性定理：如果某一函数t的自变量是x、y、z和τ，且该函数满足导热微分方程及一定的初始条件与边界条件，那么此函数就是该特定导热问题的唯一解。

Summary This section illustrates the Biot number and characteristic numbers, which are also dimensionless numbers and criterion numbers, and their definitions and physical meanings, and analyzes the influence of Biot number on the temperature distribution inside the plane plate under the third type of boundary condition.

小结 本节阐述了毕渥数及特征数，也是无量纲数、准则数，以及它们的定义与物理意义；分析了第三类边界条件下Bi对平板内温度分布的影响。

Question What is the influence of Biot number on the temperature distribution inside the plane plate under the third type of boundary condition?

思考题 第三类边界条件下，Bi对平板内温度分布有什么影响？

Self-tests

自测题

1. Explain dimensionless number, characteristic number, criterion number and Biot number, master “definition + physical meaning” and parameter composition of the characteristic numbers.

1. 解释无量纲数、特征数、准则数、毕渥数，掌握特征数的“定义式+物理意义”及参数组成。

2. What is the effect of Biot number on unsteady-state thermal conductivity temperature distribution?

2. 毕渥数对非稳态导热温度分布有什么影响？

授课视频

3.3 The Analysis Method of Zero-dimensional Problem

3.3 零维问题的分析法

The analysis method of zero-dimensional problem is the

零维问题的分析法即集中参数法。

lumped parameter method. The lumped parameter method is a simplified analysis method which ignores the thermal conduction resistance inside the object and considers the temperature distribution inside the object to be uniform. The physical meaning is that the thermal conduction resistance of the object is relatively small compared with the convective heat transfer resistance and can be ignored. Bi tends to 0 in this case, and the physical parameters of the object such as m, c and ρ etc. seem to be concentrated at one point. The temperature distribution inside the object is only related to time and has nothing to do with spatial coordinates. So, this kind of problem is called zero-dimensional problem.

集中参数法是忽略物体内部导热热阻，认为物体内温度均匀一致的简化分析方法。其物理意义是：物体的传热热阻相对于对流传热热阻很小，可以忽略，此时 Bi 趋于 0，物体的物性参数（如 m、c、ρ 等）好像都集中于一点。物体内温度分布只与时间有关，而与空间坐标无关。因此这一类问题称为零维问题。

For example, when the thermocouple is used to measure temperature, the thermal conduction process of thermocouple can be regarded as a zero-dimensional problem. The diameter of the temperature measuring ball of the thermocouple is very small and the thermal conductivity of the material is very large, so its thermal conduction resistance can be ignored.

比如用热电偶测温时，热电偶的导热过程即可看作是零维问题。因为热电偶的测温小球直径很小，材料的导热系数却很大，所以其导热热阻可以忽略不计。

Now we solve the zero-dimensional heat conduction problem by the lumped parameter method. As shown in Fig. 3. 6, for an object with arbitrary shape, geometric parameters and physical parameters are known. At initial time $\tau = 0$, the temperature of the object is t_0, suddenly put it into a fluid at temperature t_∞, the convective heat transfer coefficient h between the object and the fluid is a constant. Try to find the relationship between the temperature of the object and time.

下面用集中参数法求解零维导热问题。如图 3. 6 所示，对于任意形状的物体，几何参数和物性参数均为已知。初始时刻 $\tau = 0$ 时，物体的温度为 t_0，突然将其放置于温度为 t_∞ 的流体中，物体与流体的表面传热系数 h 为常数。试求物体温度随时间变化的关系。

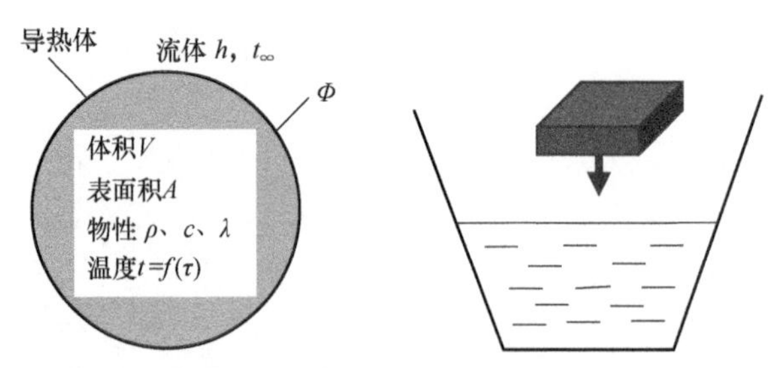

Fig. 3. 6 Schematic diagram of geometric model

图 3. 6 几何模型示意图

The following introduces the solution method of the unsteady-state heat conduction problem. Actually, it is basically the same as the solution method of the steady-state heat conduction problem.

下面介绍一下非稳态导热问题的求解方法。实际上，与稳态导热问题的求解方法基本是一样的。

(1) List the differential equation for the unsteady-state heat conduction problem, which is the basic heat conduction equation derived in Chapter 2.

(1) 列出非稳态导热问题的导热微分方程：也就是第 2 章推导得到的导热基本方程。

$$\rho c \frac{\partial t}{\partial \tau}=\frac{\partial}{\partial x}\left(\lambda \frac{\partial t}{\partial x}\right)+\frac{\partial}{\partial y}\left(\lambda \frac{\partial t}{\partial y}\right)+\frac{\partial}{\partial z}\left(\lambda \frac{\partial t}{\partial z}\right)+\dot{\Phi}$$

(2) Give the definite solution conditions including geometric conditions, physical conditions, boundary conditions and initial condition. Where the initial condition is that at $\tau=0$, temperature t is a function of the coordinates x, y and z, or temperature t is equal to the constant t_0.

(2) 给出定解条件：包括几何条件、物理条件、边界条件+初始条件。这里，初始条件是：$\tau=0$ 时温度 t 对坐标 x、y、z 的函数，或者等于常数 t_0。

$$t(x,y,z,0)=f(x,y,z) \quad \text{或} \quad t(x,y,z,0)=t_0$$

(3) Solve it by mathematical methods, and the concrete solution methods include: ① Analytical methods such as separate variable method, integral transformation etc.; ② Approximate analysis methods such as lumped parameter method, integral method; ③ Numerical methods such as finite difference method, finite element method, etc.

(4) The variation rules of temperature distribution and heat transfer rate with time and space can be obtained.

(3) 采用数学方法等进行求解，具体求解方法包括：① 分析解法：有分离变量法、积分变换等；② 近似分析法：有集中参数法、积分法；③ 数值解法：包括有限差分法、有限元法等。

(4) 得到温度分布和热流量随时间及空间的变化规律。

Next, solve the above zero-dimensional unsteady-state heat conduction by using the lumped parameter method. The thermal conduction resistance of the object can be ignored, then the temperature distribution has nothing to do with the spatial coordinates. According to the definite solution conditions, firstly simplify the differential equation of heat conduction and eliminate the thermal diffusion terms in the three coordinate directions.

下面采用集中参数法求解上述零维的非稳态导热问题。物体的导热热阻可以忽略，那么温度分布与空间坐标无关。根据定解条件，首先简化导热微分方程，消去三个坐标方向的热扩散项。

To obtain the governing equation of the zero-dimensional problem:

可得到零维问题的控制方程：

$$\rho c \frac{\mathrm{d}t}{\mathrm{d}\tau} = \dot{\Phi} \tag{3.2}$$

The following key issue needs to be solved: the boundary where convection heat transfer occurs is not the calculation boundary. Because the zero-dimensional problem has no boundary at all, then how to deal with the heat transfer quantity through the boundary? We can adopt a special treatment method: the heat transfer quantity at the interface is converted into a volumetric heat source of the entire object, which is recorded as a negative value when the object is cooled. Thus, the expression of the heat transfer converted into volumetric heat source is as follows:

下面需要解决一个关键问题：发生对流传热的边界不是计算边界。因为零维问题根本无边界，那么通过边界的传热量该怎么处理呢？可以采用一种特殊的处理方法：界面上的传热量折算成整个物体的体积热源（当物体被冷却时记作负值），从而得到传热量折算成体积热源的表达式：

$$hA(t-t_\infty) = -\dot{\Phi}V$$

By replacing the heat source term in the formula, the governing equation of the zero-dimensional problem can be obtained:

把式中热源项代换掉，即可得到零维问题的控制方程：

$$\rho c V \frac{\mathrm{d}t}{\mathrm{d}\tau} = -hA(t-t_\infty)$$

Obviously, this is a nonhomogeneous differential equation, which needs to be homogenized. Define the excess temperature as $\theta = t - t_\infty$, then the governing equation and the definite solution condition can be written as:

显然，这是一个非齐次微分方程，需要齐次化处理。定义过余温度 $\theta=t-t_\infty$，那么控制方程和定解条件可写为：

$$\rho c V \frac{\mathrm{d}\theta}{\mathrm{d}\tau} = -hA\theta$$

$$\theta(\tau=0) = t_0 - t_\infty = \theta_0$$

Using the separate variable method, the governing equation can be written as:

采用分离变量法，控制方程可写为：

$$\frac{\mathrm{d}\theta}{\theta} = -\frac{hA}{\rho c V}\mathrm{d}\tau$$

When the time is τ, the corresponding excess temperature is θ, the simultaneous integral of both sides of the equation can be written as:

当时间为 τ 时，对应的过余温度为 θ，对方程两边同时积分，可写为：

$$\int_{\theta_0}^{\theta} \frac{\mathrm{d}\theta}{\theta} = -\frac{hA}{\rho c V}\int_0^{\tau} \mathrm{d}\tau$$

The following formula can be obtained by integration:

积分可得:

$$\ln \frac{\theta}{\theta_0}=-\frac{hA}{\rho cV}\tau$$

That is the ratio of excess temperatures:

也就是过余温度的比值:

$$\frac{\theta}{\theta_0}=\frac{t-t_\infty}{t_0-t_\infty}=\exp\left(-\frac{hA}{\rho cV}\tau\right) \tag{3.3}$$

It can be seen that the relationship between the ratio of excess temperature and time τ is exponential. This is the analytical solution to the zero-dimensional problem. In the above analytical solution, V/A has a dimension of length. Therefore in the exponent of Eq. (3.3), the numerator and denominator are multiplied by λ, V and A, and V/A is regarded as the characteristic length. After the mathematical form changes, the exponent can be written as $Bi_V \cdot Fo_V$.

可以看出，过余温度的比值与时间 τ 为指数关系。这就是零维问题的分析解。上述分析解中，V/A 具有长度的量纲。因此，式（3.3）的指数中，分子、分母同乘以 λ、V 和 A，并把 V/A 看作特征长度，经过数学形式变化，该指数即可写为 $Bi_V \cdot Fo_V$。

$$\frac{hA}{\rho cV}\tau=\frac{hV}{\lambda A}\cdot\frac{\lambda A^2}{\rho cV^2}\tau=\frac{h(V/A)}{\lambda}\cdot\frac{a\tau}{(V/A)^2}=Bi_V\cdot Fo_V \tag{3.4}$$

Therefore, the solution of the differential equation can be written as the dimensionless criterion number:

因此，微分方程的解可用无量纲准则数写为:

$$\frac{\theta}{\theta_0}=\exp(-Bi_V\cdot Fo_V) \tag{3.5}$$

It is known from the dimensional analysis of the exponent in the equation that the dimension of the exponent is the same as the dimension of $1/\tau$.

通过对方程中指数的量纲分析可知，指数的量纲与 $1/\tau$ 量纲相同。

$$\frac{hA}{\rho cV}=\frac{\left[\dfrac{\mathrm{W}}{\mathrm{m^2K}}\right]\cdot[\mathrm{m^2}]}{\left[\dfrac{\mathrm{kg}}{\mathrm{m^3}}\right]\cdot\left[\dfrac{\mathrm{J}}{\mathrm{kg\cdot K}}\right]\cdot[\mathrm{m^3}]}=\frac{\mathrm{W}}{\mathrm{J}}=\frac{1}{s} \tag{3.6}$$

Eq. (3.3) shows that when the heat transfer time $\tau=\frac{\rho cV}{hA}$, the excess temperature of the object has dropped to 36.8% of the initial excess temperature (reduced by 63.2%, the excess temperature drops quickly during this time). So, it is called time constant expressed by τ_c. By using the lumped parameter method, the curve of the excess tem-

式（3.3）表明，当传热时间 $\tau=\frac{\rho cV}{hA}$，物体的过余温度已经降到了初始过余温度的 36.8%（降了 63.2%，这段时间内过余温度下降较快）。因此，$\frac{\rho cV}{hA}$ 称为时间常数，

perature of the object with $Bi_V \cdot Fo_V$ can also be obtained, as shown in Fig. 3. 7.

用 τ_c 表示。应用集中参数法，还可以得到物体过余温度随 $Bi_V \cdot Fo_V$ 的变化曲线，如图 3. 7 所示。

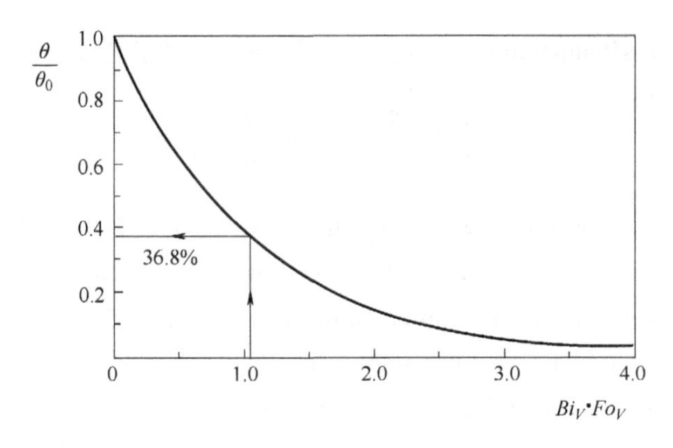

Fig. 3. 7　Diagram of excess temperature variation

图 3. 7　过余温度变化示意图

It can be seen that when $Bi_V \cdot Fo_V = 1$, $\dfrac{\theta}{\theta_0} = e^{-1} = 36.8\%$. The excess temperature of the object has dropped to 36. 8% of the initial excess temperature. When $\tau = 4\dfrac{\rho cV}{hA}$, $\dfrac{\theta}{\theta_0} = 1.83\%$, which is close to 0, which shows that t is approximately equal to t_∞. Therefore, it can be considered in engineering that the heat conductor has reached thermal equilibrium.

可以看出，当 $Bi_V \cdot Fo_V = 1$ 时，$\dfrac{\theta}{\theta_0} = e^{-1} = 36.8\%$。也就是说，物体的过余温度已经降到了初始过余温度的 36. 8%。而当 $\tau = 4\dfrac{\rho cV}{hA}$ 时，$\dfrac{\theta}{\theta_0} = 1.83\%$，已接近于 0，说明 t 约等于 t_∞。因此，工程上认为导热体已达到热平衡状态。

If the heat capacity (ρcV) of the heat conductor is small, heat transfer conditions are good and convective heat transfer coefficient is large, then the time constant ($\rho cV/hA$) is small and the heat transfer quantity per unit time is more, and the temperature of the heat conductor changes quickly, and the time for the object to reach thermal equilibrium is short. For temperature-measured thermocouples, the smaller the time constant, the faster the thermocouple responds to fluid temperature change. This is what the temperature measurement technology requires. Therefore, thermocouples are often made into fine-wire thermocouples and thin film thermocouples.

如果导热体的热容量（ρcV）较小，传热条件较好，表面传热系数较大，时间常数（$\rho cV/hA$）就较小，那么单位时间所传递的热量较多，导热体的温度变化较快，物体达到热平衡的时间就短。对于测量温度用的热电偶，时间常数越小，说明热电偶对流体温度变化的响应越快，这是测温技术所需要的。为此，热电偶常做成微细热电偶、薄膜热电偶。

Now let's discuss about the physical meaning of Bi_V and Fo_V. The Bi_V is the ratio of the thermal conduction resistance inside the object to the convective heat transfer resistance. The smaller the thermal conduction resistance or the larger the convective heat transfer resistance, then the smaller the Bi_V, and the result solved by the lumped parameter method is closer to the real situation.

下面讨论一下 Bi_V 和 Fo_V 的物理意义。Bi_V 是物体内部导热热阻与对流传热热阻的比值。导热热阻越小或对流热阻越大，Bi_V 越小，用集中参数法求解的结果越接近实际情况。

Fo_V is the ratio of the time of the unsteady-state heat conduction process to the time for the boundary thermal disturbance diffusing to a certain depth of the cross-sectional area l^2. Fo_V reflects the dimensionless time for the depth of the unsteady-state process. The larger Fo_V, the deeper thermal disturbance can travel into the object. Thus the temperature of each point inside the object is closer to the temperature of the surrounding medium.

Fo_V 是非稳态导热过程的传热时间与边界热扰动扩散到一定深度的截面面积 l^2 上所需时间的比值。Fo_V 反映了非稳态过程进行深度的无量纲时间。Fo_V 越大，热扰动就能越深入地传播到物体内部。因而，物体各点的温度就越接近周围介质的温度。

Given the temperature distribution, according to the governing equation $\rho cV\dfrac{dt}{d\tau}=-hA(t-t_\infty)$, and the expression of temperature distribution $\dfrac{\theta}{\theta_0}=\exp\left(-\dfrac{hA}{\rho cV}\tau\right)$, the instantaneous heat transfer rate of the heat conductor can be calculated as:

有了温度分布之后，根据控制方程 $\rho cV\dfrac{dt}{d\tau}=-hA(t-t_\infty)$ 和温度分布表达式 $\dfrac{\theta}{\theta_0}=\exp\left(-\dfrac{hA}{\rho cV}\tau\right)$，可求出导热体的瞬时热流量：

$$\Phi(\tau)=hA[t(\tau)-t_\infty]=hA\theta=hA\theta_0\exp\left(-\frac{hA}{\rho cV}\tau\right) \tag{3.7}$$

The total heat transferred by the heat conductor to the fluid in the time period from 0 to τ can be solved by integral.

导热体在 $0\sim\tau$ 时间段内传给流体的总热量，可用积分求解。

$$Q_\tau=\int_0^\tau \Phi(\tau)\,d\tau=\rho cV\theta_0\left[1-\exp\left(-\frac{hA}{\rho cV}\tau\right)\right] \tag{3.8}$$

Summary This section illustrates the analysis method of zero-dimensional problem, namely the lumped parameter method, and analyzes the physical meaning of Biot number and Fourier number.

小结 本节阐述了零维问题的分析法——集中参数法，并分析了毕渥数和傅里叶数的物理意义。

Question When the object is heated ($t<t_\infty$), is the calculation formulas of the temperature distribution inside the object the same as when the object is cooled ($t>t_\infty$)? Why?

思考题 当物体被加热时（$t<t_\infty$），与物体被冷却时（$t>t_\infty$），物体内温度分布计算式相同吗？为什么？

Self-tests

1. Briefly introduce the physical significance and application conditions of the lumped parameter method.

2. How to deal with the convective heat transfer boundary of zero-dimensional problem in calculation?

自测题

1. 简述集中参数法的物理意义及应用条件。

2. 零维问题的对流传热边界在计算时如何处理?

授课视频

3.4 The Application of the Lumped Parameter Method

3.4 集中参数法的应用

According to the previous qualitative analysis, when the Biot number is very small, the lumped parameter method can be used. Then, how small is it suitable for the lumped parameter method?

由前面的定性分析可知，当 Bi 很小时，可以采用集中参数法。那么，要小到什么程度才适合采用集中参数法呢?

When solving the unsteady-state heat conduction problem, the following criterion can be used to determine whether the lumped parameter method can be used. In this way, the difference between the maximum and minimum excess temperature of the object can be controlled within 5%. In engineering calculation, it is accurate enough to consider the whole temperature of the object is uniform. The criterion for application of the lumped parameter method is:

在求解非稳态导热问题时，可根据以下判据确定能否采用集中参数法。这样，可使物体中最大与最小过余温度之差控制在 5% 以内。作为工程计算，认为物体的温度整体上均匀，已足够精确。应用集中参数法的判据:

$$Bi_V = \frac{h(V/A)}{\lambda} < 0.1M$$

Where V/A represents the characteristic length of an object; M is the geometric shape factor of an object. It can be seen that the Biot number is related to the geometric shape factor M. The geometric shape factor M is determined as follows: for an infinite plane plate with a thickness of 2δ, M is 1 and the characteristic length is δ; For an infinite cylinder with radius R, M is 1/2 and the characteristic length is $R/2$; For a sphere with radius R, M is 1/3 and the characteristic length is $R/3$.

式中，V/A 表示物体的特征长度；M 为物体的几何形状因子。可见，Bi 与几何形状因子 M 有关。几何形状因子 M 的确定方法如下: 对厚度为 2δ 的无限大平板，M 取 1，特征长度为 δ；对半径为 R 的无限长圆柱，M 取 1/2，特征长度为 $R/2$；对半径为 R 的球体，M 取 1/3，特征长度为 $R/3$。

Example 3.1 A steel ball with a diameter of 5cm, its initial temperature t_0 is 450℃, suddenly put it into the air at temperature t_∞ of 30℃. Suppose that the convective heat transfer coefficient is 24W/(m^2 · K) between the steel ball surface and the surrounding environment. The specific heat capacity $c = 0.48$kJ/(kg · K), density $\rho = 7753$kg/m^3 and thermal conductivity $\lambda = 33$W/(m · K) of the steel ball are known. Try to calculate the time required for the steel ball to cool to 300℃.

例 3.1 一个直径为 5cm 的钢球，初始温度 t_0 为 450℃，突然放置于温度 t_∞ 为 30℃ 的空气中。设钢球表面与周围环境间的表面传热系数为 24W/(m^2 · K)。已知钢球的比热容 $c = 0.48$kJ/(kg · K)、密度 $\rho = 7753$kg/m^3 和导热系数 $\lambda = 33$W/(m · K)。试计算钢球冷却到 300℃所需的时间。

Assumptions: ① During the cooling process of the steel ball, convection and radiation heat transfer occur with the air and the surrounding cold surfaces, as the surface temperature drops, radiation heat transfer capacity is reduced. The convective heat transfer coefficient takes an average value and is treated as a constant. ② It is considered as constant physical property.

假设：① 钢球冷却过程中，与空气及四周冷表面发生对流与辐射传热，随着表面温度的降低，辐射传热量减少。这里取一个平均值，表面传热系数按常数处理。② 认为是常物性。

Solution: When solving this kind of problem, firstly, it is necessary to check whether the lumped parameter method is applicable to the problem. For the sphere, its shape factor M is equal to 1/3, then first figure out the Biot number.

解：在解决此类问题时，首先必须检验该问题是否可用集中参数法。对于球体，几何形状因子 M 等于 1/3，为此先计算出 Bi 数。

$$Bi_V = \frac{h(V/A)}{\lambda} = \frac{h(R/3)}{\lambda} = \frac{24\text{W}/(\text{m}^2\cdot\text{K}) \times \dfrac{0.025\text{m}}{3}}{33\text{W}/(\text{m}\cdot\text{K})} = 0.00606 < 0.0333$$

Therefore, the lumped parameter method can be used. According to the known conditions, the exponent in the expression of temperature distribution can be calculated:

因此，可用集中参数法。根据已知条件，可算出温度分布表达式中的指数：

$$\frac{hA}{\rho c V} = \frac{24\text{W}/(\text{m}^2\cdot\text{K}) \times 4\pi \times (0.025\text{m})^2}{7753\text{kg}/\text{m}^3 \times 480\text{J}/(\text{kg}\cdot\text{K}) \times \dfrac{\pi}{3} \times (0.025\text{m})^3} = 7.74 \times 10^{-4}\text{s}^{-1}$$

Calculate the excess temperature ratio:

求出过余温度比：

$$\frac{\theta}{\theta_0} = \frac{t - t_\infty}{t_0 - t_\infty} = \frac{300℃ - 30℃}{450℃ - 30℃} = 0.6429$$

According to the expression of excess temperature ratio of zero dimensional problem:

根据零维问题的过余温度比表达式：

$$\frac{\theta}{\theta_0}=\exp\left(-\frac{hA}{\rho cV}\tau\right)=\exp\ (-7.74\times10^{-4}\mathrm{s}^{-1}\tau)$$

Where only τ is unknown, and it can be found that τ is equal to 570s. The convective heat transfer coefficient h is given in this example, its set value has great influence on the calculation results.

式中，只有 τ 是未知量，可求出 τ 等于 570s。本例中给出了表面传热系数 h，其设定值对计算结果影响很大。

Example 3. 2 The mercury bubble of a thermometer is cylindrical, its length l is 20mm, its inner diameter R is 4mm, and its initial temperature is t_0, insert it into the high temperature air tank to measure the gas temperature. Assume the convective heat transfer coefficient $h=11.63\mathrm{W/(m^2 \cdot K)}$ between the mercury bubble and the gas, the effect of thin glass layer outside the mercury bubble is negligible. Try to calculate the time constant of the thermometer under this condition, and determine that percentage of the excess temperature of thermometer reading to the initial excess temperature after 5min. The physical parameters of mercury are known: specific heat capacity $c=0.138\mathrm{kJ/(kg \cdot K)}$, density $\rho=13110\mathrm{kg/m^3}$, and thermal conductivity $\lambda=10.36\mathrm{W/(m \cdot K)}$.

例 3. 2 温度计的水银泡呈圆柱形，长度 l 为 20mm，内径 R 为 4mm，初始温度为 t_0，将其插入温度较高的储气罐中，测量气体温度。设水银泡与气体间的表面传热系数 $h=11.63\mathrm{W/(m^2 \cdot K)}$，水银泡外薄层玻璃的作用可忽略不计。试计算此条件下温度计的时间常数，并确定 5min 后，温度计读数的过余温度为初始过余温度的百分比。已知水银的物性参数：比热容 $c=0.138\mathrm{kJ/(kg \cdot K)}$，密度 $\rho=13110\mathrm{kg/m^3}$，导热系数 $\lambda=10.36\mathrm{W/(m \cdot K)}$。

Solution: First check whether the lumped parameter method can be used. Considering that the upper end face of the mercury bubble cylinder is not directly heated, therefore when calculating the characteristic length, the surface area of the cylinder is the sum of the side area of the cylinder and the area of one end face.

解： 首先检验是否可用集中参数法。考虑到水银泡柱体的上端面不直接受热，因此计算特征长度时，柱体的表面积取圆柱侧面积与一个端面面积之和。

$$\frac{V}{A}=\frac{\pi R^2 l}{2\pi Rl+\pi R^2}=\frac{Rl}{2(l+0.5R)}=\frac{0.002\mathrm{m}\times0.02\mathrm{m}}{2\times(0.020\mathrm{m}+0.001\mathrm{m})}=0.953\times10^{-3}\mathrm{m}$$

$$Bi=\frac{h(V/A)}{\lambda}=\frac{11.63\mathrm{W/(m^2 \cdot K)}\times0.953\times10^{-3}\mathrm{m}}{10.36\mathrm{W/(m \cdot K)}}=1.07\times10^{-3}<0.05$$

Therefore, the lumped parameter method can be used. According to known conditions, the time constant can be obtained:

因此，可以采用集中参数法。根据已知条件，可求出时间常数：

$$\tau_c=\frac{\rho cV}{hA}=\frac{13110\mathrm{kg/m^3}\times138\mathrm{J/(kg \cdot K)}\times0.953\times10^{-3}\mathrm{m}}{11.63\mathrm{W/(m^2 \cdot K)}}=148\mathrm{s}$$

When the time τ is equal to 5min, the Fourier number can be obtained.

时间 $\tau=5\text{min}$ 时，可求出 Fo。

$$Fo=\frac{a\tau}{\left(\frac{V}{A}\right)^2}=\frac{\lambda}{\rho c}\frac{\tau}{(V/A)^2}=\frac{10.36\text{W}/(\text{m}\cdot\text{K})}{13110\text{kg/m}^3\times138\text{J}/(\text{kg}\cdot\text{K})}\times\frac{5\times60\text{s}}{(0.953\times10^{-3}\text{m})^2}=1.89\times10^3$$

Then, the excess temperature ratio can be obtained.

然后，可求出过余温度比。

$$\frac{\theta}{\theta_0}=\exp(-Bi\cdot Fo)=\exp(-1.07\times10^{-3}\times1.89\times10^3)=\exp(-2.02)=0.1333$$

After five minutes, the excess temperature of the thermometer reading is 13.3% of the initial excess temperature. During this time, the thermometer reading increased 86.7% of the temperature jump (i. e. from t_0 to t_∞) in this measurement.

过5min以后，温度计读数的过余温度是初始过余温度的13.3%。在这段时间内，温度计读数上升了此次测定中温度跃升（从 t_0 到 t_∞）的86.7%。

Discussion: It can be seen from the example that, when measuring fluid temperature with mercury thermometer, the thermometer must be placed in the fluid for a long enough time to reach heat balance between the thermometer and the fluid. This measurement is fine for the steady-state process, but for the measurement of the unsteady-state fluid temperature, the thermal capacity of the mercury thermometer is so large that its measurement temperature change can not keep up with the fluid temperature change, so its response characteristics are very poor. In this case, it is necessary to use a temperature sensor with small time constant, such as the thermocouple with small diameter, and its diameter is generally only 0.05mm, which can accurately measure the dynamic change of fluid temperature.

讨论： 通过此例可知，用水银温度计测量流体温度时，必须在流体中放置足够长的时间，以使温度计与流体之间达到热平衡。对于稳态过程这样测量没问题，但是对于非稳态流体温度的测量，由于水银温度计的热容量过大，其测量温度变化将无法跟上流体温度的变化，其响应特性很差。此时需要采用时间常数很小的感温元件，如直径很小的热电偶，一般直径只有0.05mm，才能准确测量流体温度的动态变化。

Example 3.3 There is a steel cylinder with a diameter of 5cm and a length of 30cm. The initial temperature t_0 is 30℃. It is put into the heating furnace at temperature t_∞ of 1200℃ and can be taken out after heating to 800℃. Assuming that the combined convective heat transfer coefficient between steel cylinder and flue gas is 140W/(m^2·K), the physical parameters of steel are the same as that of example 3.1. Try to calculate how long it will take to reach the temperature requirement.

例 3.3 有一直径为5cm、长为30cm的钢制圆柱体，初始温度 t_0 为30℃，将其放入炉温 t_∞ 为1200℃的加热炉中，加热升温到800℃方可取出。设钢制圆柱体与烟气间的复合传热的表面传热系数为140W/(m^2·K)，钢的物性参数同例题3.1，试计算需要多长时间才能达到温度要求。

Assumptions: The combined convective heat transfer coefficient is constant and physical properties are constant.

假设：复合表面传热系数为常数并且常物性。

Solution: First of all, it is necessary to check whether the lumped parameter method can be used. First the Biot number should be calculated:

解：首先还是要检验是否可用集中参数法。先计算出 Bi：

$$Bi=\frac{h(V/A)}{\lambda}=\frac{h[(\pi d^2l/4)/(\pi dl+2\pi d^2/4)]}{\lambda}=\frac{h}{\lambda}\frac{dl/4}{l+d/2}=\frac{140\text{W}/(\text{m}^2\cdot\text{K})}{33\text{W}/(\text{m}\cdot\text{K})}\times\frac{0.05\text{m}\times0.3\text{m}/4}{0.3\text{m}+0.025\text{m}}$$
$$=0.049<0.05$$

Therefore, the lumped parameter method can be used reluctantly.

因此，勉强可以采用集中参数法。

The exponent in the expression of temperature distribution can be obtained from the known parameters:

由已知参数可求出温度分布表达式中的指数：

$$\frac{hA}{\rho cV}=\frac{h}{\rho c}\left(\frac{V}{A}\right)^{-1}=\frac{h}{\rho c}\frac{4(l+d/2)}{dl}=\frac{140\text{W}/(\text{m}^2\cdot\text{K})\times4\times0.325\text{m}}{7753\text{kg/m}^3\times480\text{J}/(\text{kg}\cdot\text{K})\times0.05\text{m}\times0.3\text{m}}=0.326\times10^{-2}\text{s}^{-1}$$

Then, the excess temperature ratio can be obtained:

然后可求出过余温度比：

$$\frac{\theta}{\theta_0}=\frac{t-t_\infty}{t_0-t_\infty}=\frac{800℃-1200℃}{30℃-1200℃}=0.342$$

According to the expression of excess temperature ratio, substitute the known parameters into the expression to find that the time $\tau=3.29\text{s}$.

根据过余温度比的表达式，代入已知参数，可求出时间 $\tau=3.29\text{s}$。

Discussion: In this example V/A is used as the characteristic length, the Biot number has reached 0.049, which is close to the criterion of 0.05, so the central temperature has a certain calculation error. Estimate it based on the excess temperature ratio of 95%, the center temperature t_m is about 779℃. In order to accurately predict the time required for the center temperature to reach 800℃, the analytical solution of one-dimensional problem is needed to be used.

讨论：本例中以 V/A 作为特征长度，Bi 已经达到了 0.049，该 Bi 数值已接近判据 0.05，因此，中心温度具有一定的计算误差。根据过余温度比 95%加以估算，中心温度 t_m 约为 779℃。要准确预测中心温度达到 800℃所需时间，需采用一维问题的分析解。

Summary This section illustrates the application of the lumped parameter method through three examples.

小结 本节通过三个例题学习了集中参数法的应用方法。

Question Please start from the temperature distribution expression of zero-dimensional problem, and analyze the

思考题 请从零维问题的温度分布表达式出发，分析小直径的热电偶

reason why the thermocouple with small diameter can reduce the time constant.

能减小时间常数的原因。

Self-tests

自测题

1. The physical parameters ρ, λ, c and geometric parameters are known, try to design a method by measuring the temperature variation of a metal ball in the unsteady-state heat conduction process to obtain the convective heat transfer coefficient h between the metal ball and the cooling liquid.

1. 已知物性参数ρ、λ、c和几何参数，试设计一种方法，通过测定金属球非稳态导热过程中的温度变化，从而获得金属球与冷却液体间的表面传热系数h。

2. Illustrate definition and physical significance of *Bi* criterion number and *Fo* criterion number.

2. 说明*Bi*准则数、*Fo*准则数的定义及物理意义。

授课视频

3.5 The Analytical Solution of One-dimensional Unsteady-state Heat Conduction of an Infinite Plane Plate

3.5 无限大平板一维非稳态导热分析解

The research objects of typical one-dimensional unsteady-state heat conduction include plane plate, cylinder and sphere. Where one-dimension means that the temperature varies only along the thickness direction for plane plates, the temperature varies only along the radius direction for cylinders and spheres. Now we only study the plane plate, as shown in Fig. 3.8.

典型的一维非稳态导热研究对象有平板、圆柱和球。这里一维是指：对于平板，温度仅沿厚度方向变化；对于圆柱和球，温度仅沿半径方向变化。这里只研究平板，如图3.8所示。

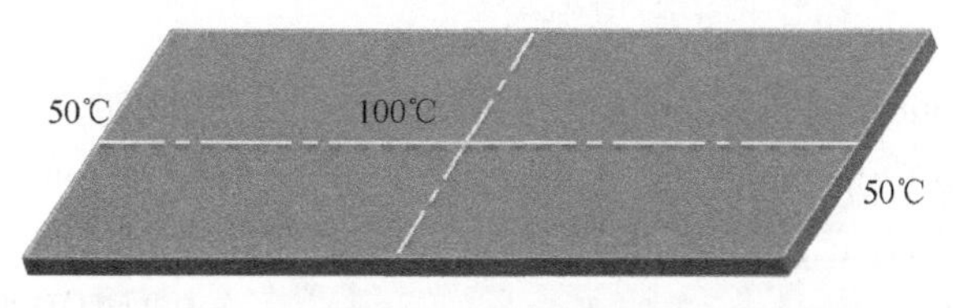

Fig. 3.8 Diagram of plane plate

图3.8 平板示意图

For the one-dimensional unsteady-state heat conduction

对于平板一维非稳态导热问题，几

problem of a plane plate, its geometric model can be simplified to an infinite plane plate, that is, the length and width of the plane plate are much larger than its thickness, so only one-dimensional heat conduction in the thickness direction is considered. The physical model is that the heat dissipation from the edge of the length and the width of the plane plate has little effect on the temperature distribution inside the plane plate. Therefore, the temperature distribution inside the plane plate can be regarded as a function of thickness and time. Namely, $t=f(x, \tau)$. Boundary conditions: Upper side boundary is 100℃, the other boundaries are 50℃, as shown in Fig. 3. 8.

何模型可简化为“无限大”平板，即平板的长度和宽度远大于其厚度，因此只考虑厚度方向的一维导热。物理模型是，平板长度和宽度的边缘向四周的散热，对平板内的温度分布影响很小。因此，可把平板内温度分布看作是厚度和时间的函数，即 $t=f(x,\tau)$。边界条件：上侧边界为 100℃，其他边界为 50℃，如图 3. 8 所示。

This is the simulation result of temperature distribution of an infinite plane plate under steady state. Fig. 3. 9a shows two mutually perpendicular cross-sections through the center of the plane plate, and Fig. 3. 9b shows the local magnification of the end of the cross-section. It can be seen that the temperature distribution is the same in the length and width cross-sections, and the temperature influence area at the end is very small, so the above simplified model is reasonable.

这是“无限大”平板在稳态状况下温度分布模拟结果。图 3. 9a 是通过板中心的两个相互垂直的截面，图 3. 9b 是截面端部的局部放大。可以看出，在长度和宽度方向的截面上温度分布相同，端部的温度影响区域很小，因此上述简化模型是合理的。

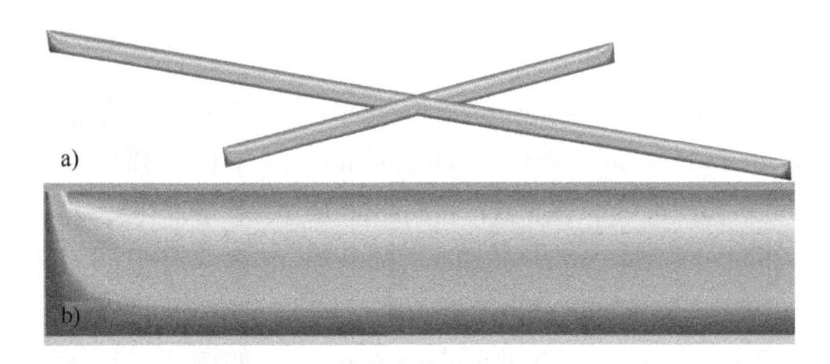

Fig. 3. 9 Simulation diagram of temperature distribution of plane plate

图 3. 9 平板截面温度分布模拟图

Next, we use the theoretical analysis method to derive the analytical solution of the temperature distribution of an infinite plane plate. There is an infinite plane plate with a thickness of 2δ, λ and a are constants, the temperature is t_0 at $\tau=0$, the plane plate is suddenly put into a fluid at temperature t_∞, the convective heat transfer coefficient between the wall and the fluid is h.

下面用理论分析法推导无限大平板温度分布的分析解。已知厚度为 2δ 的无限大平板，λ、a 均为常数，$\tau=0$ 时温度为 t_0，突然放置于温度为 t_∞ 的流体中，壁面与流体间的表面传热系数为 h。

Through simple analysis, we know that the heat transfer conditions on both sides of the plane plate are the same, then the internal temperature distribution must be symmetrical, so just study half a plane plate. The mathematical description for heat conduction of half a plane plate is the general form of the heat conduction differential equation. According to the definite solution conditions, a simplified heat conduction differential equation can be obtained:

通过简单分析可知，平板两侧传热状况相同，则内部温度分布必然对称，所以研究半块平板即可。半块平板导热的数学描写就是导热微分方程的一般形式。根据定解条件，可得到简化的导热微分方程：

$$\frac{\partial t}{\partial \tau}=a\frac{\partial^2 t}{\partial x^2}\quad(0<x<\delta,\tau>0)\tag{3.9}$$

Initial condition:

初始条件：

$$\tau=0,t(x,0)=t_0\quad(0\leqslant x\leqslant\delta)\tag{3.10}$$

Boundary conditions: When $x=0$, the boundary surface is a symmetric surface, the symmetric surface is also an adiabatic surface, $\frac{\partial t}{\partial x}=0$.

边界条件：$x=0$，边界面为对称面，对称面也就是绝热面，$\frac{\partial t}{\partial x}=0$。

When $x=\delta$, it is the third type of boundary condition.

$x=\delta$，为第三类边界条件。

$$-\lambda\frac{\partial t}{\partial x}=h(t-t_\infty)\tag{3.11}$$

A new variable, i. e. excess temperature, is introduced as follows.

下面引入新的变量过余温度。

$$\theta(x,\tau)=t(x,\tau)-t_\infty\tag{3.12}$$

Where x and τ are the independent variables of functions θ and t.

式中，θ 和 t 的自变量都是 x 和 τ。

So, the governing equation and the definite solution conditions become:

于是，控制方程和定解条件变为：

$$\frac{\partial\theta}{\partial\tau}=a\frac{\partial^2\theta}{\partial x^2},\quad 0<x<\delta,\tau>0$$

$$\theta=\theta_0,\quad\tau=0$$

$$\frac{\partial\theta}{\partial x}=0,\quad x=0$$

$$-\lambda\frac{\partial\theta}{\partial x}=h\theta,\quad x=\delta$$

Solve with the variable separation method, the analytical solution of an infinite plane plate can be obtained, which is an infinite series:

用分离变量法求解，可以得到无限大平板的分析解，是一个无穷级数：

$$\frac{\theta(\eta,\tau)}{\theta_0}=\sum_{n=1}^{\infty}C_n\exp(-\mu_n^2 Fo)\cos(\mu_n\eta) \tag{3.13}$$

Where μ_n is a discrete value, also known as a characteristic value, $n=1$, 2, 3 and so on. μ_n is the root of the transcendental equation: $\cot\mu_n=\mu_n/Bi$. So, the characteristic value μ_n is related to Biot number.

式中，μ_n 是离散值，也称为特征值，$n=1$，2，3，…。μ_n 为超越方程的根：$\cot\mu_n=\mu_n/Bi$。可见，特征值 μ_n 与 Bi 有关。

The unsteady-state heat conduction analytical solution is simplified for the normal state phase. For the unsteady-state heat conduction of an infinite plane plate:

下面对非稳态导热分析解针对正规状况阶段进行简化。对于无限大平板非稳态导热时：

$$Fo=a\tau/\delta^2 \tag{3.14}$$

When Fourier number>0.2, take the first term of the infinite series, that is, $n=1$, the temperature error is less than 1% at the center of the plane plate, i.e. $x=0$, it can be seen that this calculation method is accurate enough. So, the equation of dimensionless excess temperature can be simplified as:

当 $Fo>0.2$ 时，取无穷级数的首项，也就是 $n=1$，平板中心（即 $x=0$）处温度误差小于 1%，可见，这样计算已足够精确。因此，无量纲过余温度的表达式可简化为：

$$\frac{\theta(\eta,\tau)}{\theta_0}=\frac{2\sin\mu_1}{\mu_1+\sin\mu_1\cos\mu_1}\exp(-\mu_1 Fo)\cos(\mu_1\eta) \tag{3.15}$$

At the center of the plane plate, i.e. $x=0$ or $\eta=0$, the equation of dimensionless excess temperature is:

平板中心处（即 $x=0$ 或 $\eta=0$）无量纲过余温度的表达式：

$$\frac{\theta_m(\tau)}{\theta_0}=\frac{2\sin\mu_1}{\mu_1+\sin\mu_1\cos\mu_1}\exp(-\mu_1^2 Fo) \tag{3.16}$$

By dividing the above two equations and removing the same term, the excess temperature ratio of any point to the center of the plate can be obtained when the $Fo>0.2$.

上述两式相除，约去相同的项，可以得到 $Fo>0.2$ 以后，任意点与平板中心处的过余温度比。

$$\frac{\theta(x,\tau)}{\theta_m(\tau)}=\cos(\mu_1\eta) \tag{3.17}$$

The excess temperature ratio is independent of time, but related to the boundary conditions. Then, when $Fo>0.2$, the dimensionless excess temperature θ/θ_0 at any point can be calculated by Eq. (3.18):

该过余温度比与时间无关，而与边界条件有关。那么 $Fo>0.2$ 时，任意点的无量纲过余温度 θ/θ_0 可用式（3.18）来计算：

$$\frac{\theta}{\theta_0}=\frac{\theta}{\theta_m}\cdot\frac{\theta_m}{\theta_0} \tag{3.18}$$

The equation can be regarded as a two-stage calculation of dimensionless excess temperature to simplify the problem. When $Fo>0.2$, the excess temperature everywhere is related to time τ, but their ratio is independent of the time τ, only depends on the geometric position (η) and the boundary conditions (Bi). So, an important conclusion can be drawn that no matter what the initial temperature t_0 distribution is, as long as $Fo>0.2$, the excess temperature ratio is independent of the initial temperature t_0. It also indicates that the unsteady-state heat conduction enters the normal state phase when $Fo>0.2$.

该计算式可看作通过将无量纲过余温度分成两段计算，使问题得到简化。当 $Fo>0.2$ 时，各处的过余温度均与时间 τ 有关，但其比值与时间 τ 无关，仅取决于几何位置（η）及边界条件（Bi）。由此可得出一个重要结论，无论初始温度 t_0 分布怎样，只要 $Fo>0.2$，过余温度比值就与初始温度 t_0 无关。也说明 $Fo>0.2$ 时，标志着非稳态导热进入了正规状况阶段。

Next, learn about the heat transfer quantity in the process of unsteady-state heat conduction. The maximum heat transfer quantity from the initial moment until the plate reaches heat balance with the surrounding medium is defined as:

下面考察非稳态导热过程中的传热量，从初始时刻到平板与周围介质达到热平衡传递的最大热量为：

$$Q_0=\rho cV(t_0-t_\infty) \tag{3.19}$$

Suppose Q is the heat transfer quantity in any period of $[0,\tau]$.

令 Q 为 $[0,\tau]$ 任意时间段内所传递热量。

$$\frac{Q}{Q_0}=\frac{\rho c\int_V(t_0-t)\,\mathrm{d}V}{\rho cV(t_0-t_\infty)}=1-\frac{\bar{\theta}}{\theta_0} \tag{3.20}$$

Where θ denotes the average excess temperature at time τ, which is equal to the volume integral of θ.

式中，θ 是 τ 时刻的平均过余温度，等于 θ 对体积积分。

$$\bar{\theta}=\frac{1}{V}\int_V\theta\mathrm{d}v=\theta_0\frac{2\sin\mu_1}{\mu_1+\sin\mu_1\cos\mu_1}\mathrm{e}^{-(\mu_1^2Fo)}\frac{\sin\mu_1}{\mu_1} \tag{3.21}$$

By substituting the average excess temperature into the above heat calculation equation, the ratio of heat Q and Q_0 transferred from the initial time τ_0 to a certain time τ can be obtained. The expression is as follows:

把平均过余温度代入上述热量计算式，可得到从初始时刻 τ_0 到某一时刻 τ 所传递的热量 Q 与 Q_0 之比，表达式为：

$$\frac{Q}{Q_0}=1-\frac{\sin\mu_1}{\mu_1}\frac{2\sin\mu_1}{\mu_1+\sin\mu_1\cos\mu_1}\exp(-\mu_1^2Fo) \tag{3.22}$$

Summary　This section illustrates the analytical solution method of one-dimensional unsteady-state heat conduction of an infinite plane plate.

小结　本节阐述了无限大平板一维非稳态导热的分析解法。

Question　What is the judgment condition for the unsteady-state heat conduction process to reach the normal state phase?

思考题　非稳态导热达到正规状况阶段的判断条件是什么？

Self-tests

自测题

1. What are the characteristics of the normal state phase of unsteady-state heat conduction?

1. 非稳态导热的正规状况阶段有什么特点？

2. Under what conditions is the analytical solution of an infinite plane plate an infinite series? Under what conditions take the first term of series?

2. 无限大平板的分析解在什么条件下为无穷级数？在什么条件下取级数首项？

授课视频

3.6　The Analytical Solution of Unsteady-state Heat Conduction of Semi-infinite Objects

3.6　半无限大物体非稳态导热分析解

For the plane plate with finite thickness, the initial temperature is uniform, and one side of the plane plate is suddenly subjected to thermal disturbance, the effect of the disturbance is limited to the near area of the surface, and is not transmitted to another side of the plane plate, namely, the unsteady-state heat conduction is in the initial phase. At this time the plane plate can be regarded as a semi-infinite object.

对于有限厚度的平板，初始时刻温度均匀，其一侧表面突然受到热扰动，扰动的影响局限在表面附近，而未传递到平板的另一个侧面，也就是非稳态导热处于初始阶段。此时该平板可看作半无限大物体。

Its geometric characteristics are as follows: From the interface of $x=0$, the plane plate can extend infinitely to the positive x-axis and the y, z-axis directions. Find the analytical solution of the temperature field of semi-infinite objects under the first type of boundary condition. Firstly, the governing equation and the definite solution conditions of the unsteady-state heat conduction problem are given.

其几何特点：从 $x=0$ 的界面开始，平板可以向 x 轴正向及 y、z 轴方向无限延伸。求半无限大物体在第一类边界条件下温度场的分析解。首先给出非稳态导热问题的控制方程和定解条件。

$$\frac{\partial t}{\partial \tau}=a\frac{\partial^2 t}{\partial x^2}$$

$$\tau=0,\quad t=t_0$$

$$x=0,\quad t=t_w$$

$$x\to\infty,\quad t=t_0$$

Next, introduce the excess temperature $\theta=t-t_w$, then to find out the analytical solutions of the unsteady-state heat conduction of semi-infinite objects.

而后引入过余温度 $\theta=t-t_w$，可求出半无限大物体非稳态导热的分析解。

$$\frac{\theta}{\theta_0}=\text{erf}\left(\frac{x}{\sqrt{4a\tau}}\right)=\text{erf}(\eta) \tag{3.23}$$

Where erf is the error function, which can be found in the relevant table; $\frac{x}{\sqrt{4a\tau}}$ is a dimensionless variable, denoted as η.

式中，erf 为误差函数，可查相关用表；$\frac{x}{\sqrt{4a\tau}}$为无量纲变量，记作 η。

$$\eta=\frac{x}{\sqrt{4a\tau}}=2,\quad \text{erf}(2)=0.9953$$

$$\frac{\theta}{\theta_0}=99.53\%$$

At this time the temperature of the point x can be still considered as t_0. So, the condition for regarding an object as semi-infinite is:

此时，可认为 x 处的温度仍为 t_0。因此，把物体看作半无限大的条件是：

$$\eta=\frac{x}{\sqrt{4a\tau}}\geqslant 2 \tag{3.24}$$

Two important parameters can be inferred from this condition: ① Geometric position. If $\eta\geqslant 2$, then we can get $x\geqslant 4\sqrt{a\tau}$. For a plane plate of 2δ thickness, if $\delta\geqslant 4\sqrt{a\tau}$, then it can be treated as a semi-infinite object. ② Inertia time. If $\eta\geqslant 2$, then we can also get $\tau\leqslant\frac{x^2}{16a}$.

由此条件可推出两个重要参数：① 几何位置。如果 $\eta\geqslant 2$，可推出 $x\geqslant 4\sqrt{a\tau}$。对于一厚度为 2δ 的平板，若 $\delta\geqslant 4\sqrt{a\tau}$，那么该平板就可看作半无限大物体来处理。② 惰性时间。如果 $\eta\geqslant 2$，还可推出 $\tau\leqslant\frac{x^2}{16a}$。

For real objects of finite size, the concept of semi-infinite object is only applicable to the initial phase of the unsteady-state heat conduction, that is within the $\tau=\frac{x^2}{16a}$

对于有限大的实际物体，半无限大物体的概念只适用于物体非稳态导热的初始阶段，即在惰性时间 $\tau=$

inertia time, the temperature of the point x is still t_0. According to the above temperature distribution equation:

$\frac{x^2}{16a}$以内，x 处的温度仍为 t_0。根据上述温度分布表达式：

$$\frac{\theta}{\theta_0}=\mathrm{erf}\left(\frac{x}{\sqrt{4a\tau}}\right) \tag{3.25}$$

Using Fourier's law, the heat flux through any section x can be obtained:

应用傅里叶定律，可求出通过任一截面 x 处的热流密度：

$$q_x=-\lambda\frac{\partial\theta}{\partial x}=-\lambda\theta_0\frac{1}{\sqrt{\pi a\tau}}\exp\left(\frac{-x^2}{4a\tau}\right) \tag{3.26}$$

Suppose $x=0$ to get the instantaneous heat flux on the boundary surface:

令 $x=0$，即可得到边界面上的瞬时热流密度：

$$q_{\mathrm{w}}=-\lambda\frac{\theta_0}{\sqrt{\pi a\tau}} \tag{3.27}$$

Within the time period $[0,\tau]$, we can find the total heat transfer quantity Q through area A:

在 $[0,\ \tau]$ 时间段内，可求出流过面积 A 的总热量 Q：

$$Q=A\int_0^{\tau}q_{\mathrm{w}}\mathrm{d}\tau=-2A\sqrt{\frac{\tau}{\pi}}\sqrt{\rho c\lambda}\,\theta_0 \tag{3.28}$$

Where $\sqrt{\rho c\lambda}$ is called the Endothermic coefficient, which represents the ability of an object to absorb heat from a high-temperature object in contact with it.

式中，$\sqrt{\rho c\lambda}$ 称为吸热系数，代表了物体从与其接触的高温物体吸热的能力。

Example 3.4 It is known that pouring a plate type casting by moulding sand with the thermal diffusivity of $a=0.89\times10^{-6}\ \mathrm{m^2/s}$, the temperature of the moulding sand is 20℃ before pouring, and pouring time is short, the surface temperature of the casting maintained at solidification temperature of 1450℃ after pouring. Calculate the temperature of the moulding sand at a distance of 80mm from the bottom of the casting after 2h of pouring.

例 3.4 浇注一平板型铸件，型砂热扩散率为 $a=0.89\times10^{-6}\mathrm{m^2/s}$，浇注前型砂温度为 20℃，浇注时间很短，浇注后，铸件表面温度维持在凝固温度 1450℃。计算浇注 2h 后，离铸件底面 80mm 处型砂的温度。

The unsteady-state heat conduction process of the moulding sand below the bottom surface of the casting can be solved according to the heat conduction of a semi-infinite object. First figure out η based on known parameters.

求解铸件底面以下型砂的非稳态导热过程，可以按照半无限大物体导热来处理。根据已知参数先求出 η。

$$\eta = \frac{x}{\sqrt{4a\tau}} = 0.5$$

Refer to the error function table, the error function is 0.5205, then according to the temperature distribution equation, we can find that $t = 705.7℃$.

查误差函数表得误差函数为0.5205，再由温度分布表达式，可求出 $t = 705.7℃$。

$$\frac{\theta}{\theta_0} = \frac{t - t_w}{t_0 - t_w} = \text{erf}(\eta)$$

For the one-dimensional unsteady-state heat conduction calculation of a plane plate in summary, there are four cases:

对于平板的一维非稳态导热计算，概括起来有以下四种情况：

(1) When $Bi_V \to 0$, that is, $Bi_V = \frac{h(V/A)}{\lambda} < 0.1$, as a zero-dimensional problems to calculate.

(1) $Bi_V \to 0$ 即 $Bi_V = \frac{h(V/A)}{\lambda} < 0.1$，按零维问题处理。

(2) When $Bi_V > 0.1$ and $Fo_V < 0.06$ or $\eta = \frac{x}{\sqrt{4a\tau}} \geqslant 2$, the unsteady-state heat conduction is in the initial phase, as a semi-infinite object to calculate.

(2) $Bi_V > 0.1$ 且 $Fo_V < 0.06$ 或 $\eta = \frac{x}{\sqrt{4a\tau}} \geqslant 2$ 时，非稳态导热处于初始阶段，按半无限大物体计算。

(3) When $Bi_V > 0.1$ and $0.06 < Fo_V < 0.2$, as the complete series solution of a one-dimensional plane plate to calculate.

(3) $Bi_V > 0.1$ 且 $0.06 < Fo_V < 0.2$ 时，按一维平板的完整级数解计算。

(4) When $Bi_V > 0.1$ and $Fo_V > 0.2$, the unsteady-state heat conduction is in the normal state phase, as the first term of series solution of a one-dimensional plane plate to calculate.

(4) $Bi_V > 0.1$ 且 $Fo_V > 0.2$ 时，非稳态导热处于正规状况阶段，按一维平板的级数首项解计算。

Summary This section illustrates the analytical solution of the unsteady-state heat conduction of semi-infinite objects.

小结 本节阐述了半无限大物体非稳态导热的分析解法。

Question What is the difference between the geometric and physical models of one-dimensional plane plate and semi-infinite plane plate?

思考题 一维平板和半无限大平板的几何模型及物理模型有何区别？

Self-tests

自测题

1. What is the physical significance of semi-infinite objects?

1. “半无限大”物体的物理意义是什么？

2. Try to illustrate the simplified calculation method for unsteady-state thermal conduction of infinite rectangular cylinder.

2. 试述无限长方柱体非稳态导热的简化计算方法。

Exercises

习题

3.1 Assume that a bar with a length of l have a uniform initial temperature t_0, after that, the two ends are kept constant at $t_1(x=0)$ and $t_2(x=l)$ respectively, and $t_2>t_1>t_0$, and keep insulation around the bar. Try to draw the schematic curve of the temperature distribution in the bar with time and the final temperature distribution curve.

3.1 假设一根长为 l 的棒有均匀初温度 t_0，此后使其两端各维持在恒定的 $t_1(x=0)$ 及 $t_2(x=l)$，并且 $t_2>t_1>t_0$。棒的四周保持绝热。试画出棒中温度分布随时间变化的示意性曲线及最终的温度分布曲线。

3.2 The principle of modern microwave oven heating an object is to use high-frequency electromagnetic waves to polarize molecules in the object, resulting in oscillation. The result is equivalent to an internal heat source which is nearly evenly distributed in the object, while the typical oven heats the surface of the object with near a constant heat flow. Taking a piece of beef as an infinite plate with a thickness of 2δ, try to qualitatively draw the temperature distribution curve of beef during heating (from room temperature to the lowest temperature of 85℃) by microwave oven and oven (before heating, at a certain time during heating and at the end of heating).

3.2 现代微波炉加热物体的原理是利用高频电磁波使物体中的分子极化，从而产生振荡，其结果相当于物体中产生了一个接近于均匀分布的内热源，而一般的烘箱是从物体的表面上进行接近恒热流的加热。设把一块牛肉当作厚为 2δ 的无限大平板，试定性地画出采用微波炉及烘箱对牛肉加热（从室温到最低温度为85℃）过程中牛肉的温度分布曲线（加热开始前、加热过程中某一时刻及加热终了三个时刻）。

3.3 A solid with an initial temperature of t_0 is placed in a room with a room temperature of t_∞. The emissivity of the object surface is ε, and the surface heat transfer coefficient between the surface and the air is h. The volume of the object is V, the area involved in heat transfer is A, and the specific heat capacity and density are c and ρ respectively. The internal thermal resistance of the object can be neglected. Try to list the differential equation of the temperature of the object changing with time. Reminder: The radiation heat transfer capacity per unit area of an object is $\varepsilon\sigma(T^4-T_\infty^4)$.

3.3 一初始温度为 t_0 的固体，被置于室温为 t_∞ 的房间中。物体表面的发射率为 ε，表面与空气间的表面传热系数为 h。物体的体积为 V，参与传热的面积为 A，比热容和密度分别为 c 及 ρ。物体的内热阻可忽略不计，试列出物体温度随时间变化的微分方程式。提示：物体单位面积上的辐射传热量为 $\varepsilon\sigma(T^4-T_\infty^4)$。

3.4 The value of a thermocouple $\rho cV/A$ is 2.094kJ/(m^2 · K), the initial temperature is 20℃, and then it is

3.4 一热电偶的 $\rho cV/A$ 值为2.094kJ/(m^2 · K)，初始温度为20℃，

placed in the airflow of 320℃. Try to calculate the time constant of the thermocouple in two cases where the surface heat transfer coefficient between air flow and thermocouple is 58W/(m^2 · K) and 116W/(m^2 · K), and draw the curve of the excess temperature of thermocouple reading with time in two cases.

后将其置于320℃的气流中。试计算在气流与热电偶之间的表面传热系数为58W/(m^2 · K) 及116W/(m^2 · K)两种情况下热电偶的时间常数，并画出两种情况下热电偶读数的过余温度随时间变化的曲线。

3.5 A flat plate with a unilateral surface area of A and an initial temperature of t_0, one side surface is suddenly heated by a constant heat flux q_0, the other side surface is cooled by an airflow with an initial temperature of t_∞, and the surface heat transfer coefficient is h. List the differential equation of the temperature change of the object over time and solve it. The internal resistance can be neglected, and other geometric and physical parameters are known.

3.5 一块单侧表面积为A、初温为t_0的平板，一侧表面突然受到恒定热流密度q_0的加热，另一侧表面受到初温为t_∞的气流冷却，表面传热系数为h。试列出物体温度随时间变化的微分方程式并求解。设内阻可以不计，其他的几何、物性参数均已知。

3.6 In the heat treatment process, the cooling capacity of quenching medium under different conditions is measured by silver ball sample. Now there are two silver balls with a diameter of 20mm, which are heated to 600℃ and then placed in a large container containing static water at 20℃ and in circulating water at 20℃ respectively. Measured by thermocouple, when the center temperature of the ball changes from 650℃ to 450℃, its cooling rate is 180℃/s and 360℃/s respectively. Try to determine the surface heat transfer coefficient between the silver ball surface and water under two conditions. It is known that the physical parameters of silver in the above temperature range are $c = 2.62 \times 10^2$J/(kg · K)、ρ = 10500kg/m^3、λ = 360W/(m · K).

3.6 在热处理工艺中，用银球试样来测定淬火介质在不同条件下的冷却能力。现有两个直径为20mm的银球，加热到600℃后被分别置于20℃的盛有静止水的大容器及20℃的循环水中。用热电偶测得，当银球中心温度从650℃变化到450℃时，其降温速率分别为180℃/s及360℃/s。试确定两种情况下银球表面与水之间的表面传热系数。已知在上述温度范围内银的物性参数为$c = 2.62 \times 10^2$J/(kg · K)、ρ = 10500kg/m^3、λ = 360W/(m · K)。

Chapter 4 Numerical Solution of a Heat Conduction Problem

第 4 章 热传导问题的数值解法

4.1 The Numerical Solution Steps of a Heat Conduction Problem

4.1 导热问题数值求解步骤

There are usually four methods to solve the heat conduction problems:

求解导热问题通常有以下四种方法：

(1) The theoretical analysis method: On the basis of simplification and hypothesis, establish the differential equation and give the definite conditions, then separate the variables and integrate to obtain the analytical or theoretical solution.

(1) 理论分析法：该方法在简化和假设的基础上，建立微分方程式，并给出定解条件，然后分离变量、积分求解，进而得到分析解或理论解。

The characteristics of this method are as follows: ① The accurate solutions to the research problems can be obtained to provide a basis for comparison between experimental results or numerical calculation. ② The mathematical analysis is more rigorous in the solving process, and the result is a function expression, which can clearly show the temperature distribution law and its influencing factors. ③ It has great limitations and is unable to solve complex problems.

理论分析法的特点是：① 能获得所研究问题的精确解，可为实验结果或数值计算提供对比依据；② 求解过程中数学分析较严谨，结果是函数表达式，能清晰显示出温度分布规律及其影响因素；③ 局限性较大，对复杂问题无法求解。

(2) The numerical calculation method, also called semi-theoretical method: This method firstly simplifies and assumes the geometric and physical models reasonably, and then establishes a mathematical description of the physical problem being studied, that is, a mathematical model. Because geometric models and physical models are complicated, the mathematical models are also complicated, and mathematicians are powerless. In this case,

(2) 数值计算法，也称为半理论解法：该方法首先对几何、物理模型进行合理的简化与假设，然后建立所研究物理问题的数学描写，即数学模型。因为几何模型、物理模型比较复杂，导致数学模型也很复杂，数学家无能为力。在这种情况下，只好借助计算机来解决，从而

we have to use computers to solve the problem, thus forming the numerical calculation method.

形成了数值计算法。

The basic idea of the numerical calculation method is that the fields of continuous physical quantities in time and space, such as the temperature fields of heat conductor, are replaced with a set of values at a finite number of discrete points. The algebraic equation group about these points is established by a certain method, the values of the physical quantities to be solved on the discrete points are obtained by computer solution, which are called numerical solution or approximate solution.

数值计算法的基本思想是，把时间和空间上连续物理量的场，如导热物体的温度场，用有限个离散点上的值的集合来代替。按一定方法建立起关于这些点的代数方程组，通过计算机求解获得离散点上被求物理量的值，称为数值解或近似解。

The characteristics of the numerical calculation method are as follows: ① It makes up for the shortcomings of the analysis method and has strong adaptability and obvious superiority especially for complex problems. ② Compared with the experimental method, it is more convenient, faster and lower cost. ③ It is an approximate method with a certain accuracy. The common numerical methods include finite difference method, finite element method and boundary element method, and so on.

数值计算法的特点：① 弥补了分析法的缺点，适应性强，特别对于复杂问题更显其优越性；② 与实验法相比，方便快捷，成本低；③ 是具有一定精度的近似方法。常见的数值解法有有限差分法、有限元法和边界元法等。

(3) The experimental method: Under the guidance of the basic theory of Heat Transfer, the method of testing the heat transfer process using the similarity principle is called the experimental method. It is the most basic and reliable method to study Heat Transfer. Its characteristics are as follows: ① The adaptability to the research object is not good, and it is limited by the experiment and other conditions. ② The process is complicated, time-consuming and expensive.

(3) 实验法：在传热学基本理论指导下，采用相似原理对传热过程进行测试的方法就叫作实验法。实验法是研究传热学的最基本、最可靠的方法。其特点是：① 对研究对象的适应性不好，受实验条件等限制；② 过程繁琐、时间长、费用高。

(4) The analogy method. For example, the heat conduction process is likened to the electricity conduction process. We all know that the relationship between current, voltage and resistance, which is similar to the relationship between heat flux, temperature difference and thermal resistance of heat conduction, and the mechanism of heat conduction and electrical conduction is the same, which can be studied and analyzed by comparison.

(4) 比拟法。比如，把导热过程与导电过程比拟。我们都知道电流、电压、电阻的关系，与导热的热流、温差、热阻的关系是类似的，且导热与导电的机理相同，可以对比着研究分析。

$$I=\frac{U}{R} \qquad q=\frac{\Delta t}{r_\lambda}$$

The specific steps of numerical solution of heat conduction problem are shown in Fig. 4. 1.

导热问题数值求解的具体步骤如图 4. 1 所示。

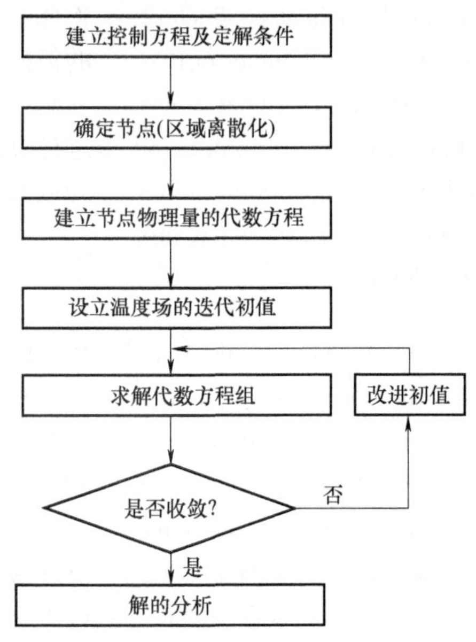

Fig. 4. 1 The specific steps of numerical solution of heat conduction problem

图 4. 1 导热问题数值求解的具体步骤

(1) Establish the control equations and definite solution conditions.

(1) 建立控制方程及定解条件。

(2) Determine nodes, that is, region discretization.

(2) 确定节点，即区域离散化。

(3) Establish the algebraic equations of node physical quantities.

(3) 建立节点物理量的代数方程。

(4) Set up the initial value of the temperature field iteration.

(4) 设立温度场的迭代初值。

(5) Solve algebraic equation group and judge whether the result converges, that is, whether the relative error between the calculated value and the initial value meets the accuracy requirements. If accuracy requirements are met, proceed to step 6.

(5) 求解代数方程组，判断是否收敛，即计算值与初值之间的相对误差是否达到了精度要求。如果达到了精度要求，则进行第 6 步。

(6) Analysis of the solutions. If the accuracy requirements are not met, then improve the initial value and return to step 5 and iterate again.

(6) 解的分析。如果没达到精度要求，那么改进初值，返回第 5 步重新迭代计算。

Next, we will introduce the steps of numerical solution in detail with a specific heat conduction problem.

下面，结合一个具体的导热问题来详细介绍数值求解的步骤。

Example 4.1 Taking a steady-state heat conduction problem with no internal heat source and constant physical properties in a two-dimensional rectangular domain as an example. The boundary conditions are shown in Fig. 4.2.

例 4.1 以一个二维矩形域内稳态无内热源、常物性的导热问题为例，边界条件如图 4.2 所示。

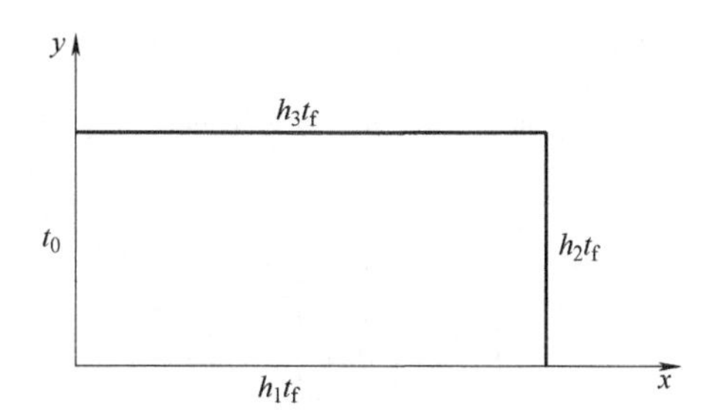

Fig. 4.2 A two-dimensional rectangular region and its boundary conditions

图 4.2 二维矩形域及其边界条件

Numerical calculation step 1: Establish the mathematical model. Analyze the heat conduction problem, and establish control equation and definite conditions according to geometric and physical models. The equation group is shown as follows:

数值计算第 1 步：建立数学模型。 分析导热问题，根据几何与物理模型，建立控制方程及定解条件。方程组如下：

$$\frac{\partial^2 t}{\partial x^2}+\frac{\partial^2 t}{\partial y^2}=0 \tag{4.1a}$$

$$t(0,y)=t_0 \tag{4.1b}$$

$$-\lambda(\partial t/\partial n)_w=h(t_w-t_f) \tag{4.1c}$$

Numerical calculation step 2: Region discretization. The temperature field of heat conduction problem is a continuous function of space. In the numerical solution, firstly the studied region is spatially divided into a finite number of small regions. The original continuous temperature field is discretized into a temperature distribution of step change. This is the basic idea of numerical calculation. For example, the total temperature difference between left and right boundaries is 10℃, and is divided into 10, 100 and 1000 grids. The more grids, the smaller the temperature difference between adjacent nodes.

数值计算第 2 步：区域离散化。 导热问题的温度场是空间的连续函数。进行数值求解时，首先对所研究的区域在空间上分割成有限个小区域。原来连续的温度场被离散成阶跃变化的温度分布。这就是数值计算的基本思想。比如，左右边界总温差 10℃，分成 10 格、100 格、1000 格。格越多，相邻节点之间温差越小。

Note: Several very important basic concepts are defined here. As shown in Fig. 4. 3, the interface line on the interface, the grid line that dividing the grid, the inner nodes where the grid lines intersect, the distance between two adjacent nodes in x or y directions is called the step length, every studied micro-element also known as the control volume or element, and the boundary nodes on the boundary. These are the important basic concepts.

注意：这里规定了几个非常重要的基本概念。如图 4. 3 所示，在界面上的界面线，划分网格的网格线，网格线相交的内节点，x 或 y 方向相邻两个节点的距离称为步长，研究的每个微元也叫作控制容积或元体，还有边界上的边界节点。这些都是重要的基本概念。

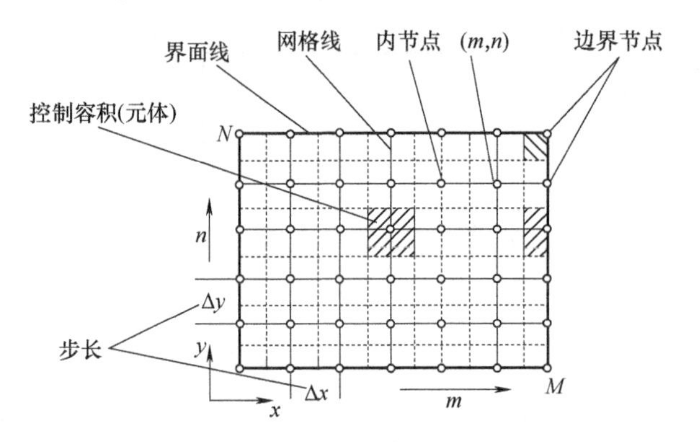

Fig. 4. 3 Example of numerical solution of heat conduction problem

图 4. 3 导热问题数值求解示例

Numerical calculation step 3: Establish the algebraic equations of node physical quantities, namely, the discrete equations. There are four common methods to establish the discrete equations: ① Taylor series expansion method, that is, discretization of the differential equation. ② Heat balance method, also called the control volume method. ③ Polynomial fitting method. ④ Control volume integral method. In this chapter, we mainly use Taylor series expansion method and heat balance method.

数值计算第 3 步：建立节点物理量的代数方程，即离散方程。建立离散方程的常用方法有四种：① 泰勒级数展开法，即微分方程离散化；② 热平衡法，即控制容积法；③ 多项式拟合法；④ 控制容积积分法。本章主要采用泰勒级数展开法和热平衡法。

Numerical calculation step 4: Solve algebraic equation group. In addition to the boundary nodes of known temperature, each node corresponds to an algebraic equation, that is, $(m-1)\times n$ nodes correspond to $(m-1)\times n$ algebraic equations. These $(m-1)\times n$ algebraic equations are solved by iterative method. In some cases, these $(m-1)\times n$ algebraic equations can also be solved directly, such as matrix inversion and Gauss elimination method.

数值计算第 4 步：求解代数方程组。除了已知温度的边界节点外，每个节点对应一个代数方程，即 $(m-1)\times n$ 个节点对应有 $(m-1)\times n$ 个代数方程。对这 $(m-1)\times n$ 个代数方程进行迭代法求解。有些情况下，也可以对这 $(m-1)\times n$ 个代数方程进行直接求解，如矩阵求逆、高斯消元法。

Summary This section illustrates four basic methods to solve heat conduction problems and basic steps for numerical solution of heat conduction problems.

小结 本节阐述了求解导热问题的四种基本方法，以及导热问题数值求解的基本步骤。

Question What is the basic idea of numerical calculation method?

思考题 数值计算法的基本思想是什么？

Self-tests

自测题

1. What are the basic steps of numerical solution of a heat conduction problem?

2. What is convergence? Discrete?

1. 导热问题数值求解的基本步骤是什么？

2. 什么是收敛？什么是离散？

授课视频

4.2 Establishment of the Discrete Equations of Internal Nodes—Taylor Series Method

4.2 内节点离散方程的建立——泰勒级数法

The methods of establishing internal node discrete equation include Taylor series expansion method and heat balance method. In this section, we will study Taylor series expansion method.

建立内节点离散方程的方法有泰勒级数展开法及热平衡法。本节学习泰勒级数展开法。

Taylor series expansion method is also called discretization method of differential equation. The temperature of node (m,n) is denoted by $t_{m,n}$, and the temperature of node $(m+1,n)$ is denoted by $t_{m+1,n}$. According to Taylor series expansion, the value of node $t_{m+1,n}$ can be denoted by the value of node $t_{m,n}$, and the expansions are as follows:

泰勒级数展开法也称为微分方程离散化法。节点（m,n）的温度用$t_{m,n}$来表示，节点（$m+1,n$）的温度用$t_{m+1,n}$来表示。根据泰勒级数展开式，可以用节点$t_{m,n}$的值来表示节点$t_{m+1,n}$的值，展开式如下：

$$t_{m+1,n}=t_{m,n}+\left.\frac{\partial t}{\partial x}\right|_{m,n}\Delta x+\left.\frac{\partial^2 t}{\partial x^2}\right|_{m,n}\frac{\Delta x^2}{2!}+\left.\frac{\partial^3 t}{\partial x^3}\right|_{m,n}\frac{\Delta x^3}{3!}+\cdots \tag{4.2a}$$

Similarly, the temperature $t_{m,n}$ of node (m,n) is used to denote the temperature $t_{m-1,n}$ of node $(m-1,n)$ and the expression is as follows:

同样，用节点（m,n）的温度$t_{m,n}$来表示节点（$m-1,n$）的温度$t_{m-1,n}$，表达式如下：

$$t_{m-1,n}=t_{m,n}-\left.\frac{\partial t}{\partial x}\right|_{m,n}\Delta x+\left.\frac{\partial^2 t}{\partial x^2}\right|_{m,n}\frac{\Delta x^2}{2!}-\left.\frac{\partial^3 t}{\partial x^3}\right|_{m,n}\frac{\Delta x^3}{3!}+\cdots \tag{4.2b}$$

Add the left and right sides of the equal sign of the two equations respectively, remove the same item and move the items and organize to obtain the central difference expression of second derivative along x-axis direction:

把两式等号左边与右边分别相加，约去相同的项，移项整理可得 x 轴方向二阶导数的中心差分表达式：

$$\left.\frac{\partial^2 t}{\partial x^2}\right|_{m,n}=\frac{t_{m+1,n}-2t_{m,n}+t_{m-1,n}}{\Delta x^2}+O(\Delta x^2) \tag{4.2c}$$

Where $O(\Delta x^2)$ is called truncation error, which represents the unspecified high-order remainder, and the lowest order of Δx is 2 in the remainder. Similarly, the central difference expression of second derivative in the y-axis direction can be obtained:

式中，$O(\Delta x^2)$ 称为截断误差，表示未明确写出的高阶余项，余项中 Δx 的最低阶数为 2。同样，可得 y 轴方向二阶导数的中心差分表达式为：

$$\left.\frac{\partial^2 t}{\partial y^2}\right|_{m,n}=\frac{t_{m,n+1}-2t_{m,n}+t_{m,n-1}}{\Delta y^2}+O(\Delta y^2) \tag{4.2d}$$

Ignoring the truncation error, that is, the high-order remainder, the central difference approximation of second derivative can be obtained as follows:

忽略截断误差，即高阶余项，得到二阶导数的中心差分近似表达式为：

$$\left.\frac{\partial^2 t}{\partial x^2}\right|_{m,n}=\frac{t_{m+1,n}-2t_{m,n}+t_{m-1,n}}{\Delta x^2} \tag{4.3a}$$

$$\left.\frac{\partial^2 t}{\partial y^2}\right|_{m,n}=\frac{t_{m,n+1}-2t_{m,n}+t_{m,n-1}}{\Delta y^2} \tag{4.3b}$$

For the two-dimensional steady-state heat conduction problem with no internal heat source, the differential equation of heat conduction in rectangular coordinates is as follows:

对于二维稳态、无内热源的导热问题，在直角坐标中其导热微分方程为：

$$\frac{\partial^2 t}{\partial x^2}+\frac{\partial^2 t}{\partial y^2}=0 \tag{4.4}$$

Substituting the central difference approximate expression of the above second-order derivative into the above equation, the discrete equation of the node (m,n) can be obtained as follows:

将上述二阶导数的中心差分近似表达式代入，可以得到节点 (m,n) 的离散方程为：

$$\frac{t_{m+1,n}-2t_{m,n}+t_{m-1,n}}{\Delta x^2}+\frac{t_{m,n+1}-2t_{m,n}+t_{m,n-1}}{\Delta y^2}=0 \tag{4.5a}$$

If $\Delta x=\Delta y$, and merge similar items, then Eq. (4.5a) becomes:

如果 $\Delta x=\Delta y$，并合并同类项，式（4.5a）变为：

$$t_{m,n}=\frac{1}{4}(t_{m-1,n}+t_{m+1,n}+t_{m,n+1}+t_{m,n-1}) \tag{4.5b}$$

In the whole computational region, there are $(m-1)\times n$ nodes, and $(m-1)\times n$ algebraic equations of the above form can be established, thus the discrete equation group of the inner nodes can be obtained. In the governing equations of heat transfer problems, the first and second derivatives are mainly encountered. In the mean grids, the common differential expressions of the first and second derivatives are listed in Table 4.1.

在整个计算区域内，有 $(m-1)\times n$ 个节点，就可以相应地建立 $(m-1)\times n$ 个上述形式的代数方程，从而得到了内节点的离散方程组。在传热学问题的控制方程中，主要遇到的是一阶和二阶导数。在均分网格中，一阶、二阶导数常用的差分表示式列于表 4.1 中。

Table 4.1 The common differential expressions of the first and second derivatives

表 4.1 一阶、二阶导数常用的差分表示式

导数	差分表示式	截断误差	备注
$\left(\frac{\partial t}{\partial x}\right)_i$	$\frac{t_{i+1}-t_i}{\Delta x}$	$O(\Delta x)$	称为 i 点的向前差分（forward difference）
	$\frac{t_i-t_{i-1}}{\Delta x}$	$O(\Delta x)$	称为 i 点的向后差分（backward difference）
	$\frac{t_{i+1}-t_{i-1}}{2\Delta x}$	$O(\Delta x^2)$	称为 i 点的中心差分
$\left(\frac{\partial^2 t}{\partial x^2}\right)_i$	$\frac{t_{i+1}-2t_i+t_{i-1}}{\Delta x^2}$	$O(\Delta x^2)$	称为 i 点的中心差分

Summary This section illustrates Taylor series method to establish the discrete equations of internal nodes.

小结 本节阐述了建立内节点离散方程的泰勒级数法。

Question What is the basic idea of establishing the discrete equations by Taylor series method?

思考题 用泰勒级数法建立离散方程的基本思路是什么？

Self-tests

自测题

1. Write the Taylor series expansion of temperature for two adjacent nodes.

1. 写出相邻两节点的温度泰勒级数展开式。

2. Write the central difference expression of second derivative.

2. 写出二阶导数的中心差分表达式。

4.3 Establishment of the Discrete Equations of Internal Nodes—The Heat Balance Method

4.3 内节点离散方程的建立——热平衡法

The heat balance method is also called the control volume method. Its basic idea is to apply Fourier's law of heat conduction to the control volume represented by each node, and directly write the energy conservation expression, so as to obtain the algebraic equation group of temperature field. $m\times n$ nodes correspond to write $m\times n$ algebraic equations. This method is based on the basic physical phenomenon and does not need to establish the governing equations in advance.

热平衡法也叫作控制容积法。其基本思想是对每个节点代表的控制容积应用傅里叶导热定律，直接写出能量守恒表达式，从而得到温度场的代数方程组。$m\times n$ 个节点就对应写出 $m\times n$ 个代数方程。这种方法从基本物理现象出发，不必事先建立控制方程。

According to the law of energy conservation, for any element, the total heat flux imported and exported from all directions of the control body, and plus the heat generated by the internal heat source in the control body, is equal to the internal energy increment of the control body. The equation can be written as:

根据能量守恒定律可知，对于任一个元体，从所有方向导入导出控制体的总热流，加上控制体内热源的发热，等于控制体内能的增量。方程式可写为：

$$\Phi_i+(-\Phi_o)+\Phi_v=\Phi_\tau \tag{4.6a}$$

Note: Eq. (4.6a) is applicable to both internal nodes and boundary nodes. For two-dimensional steady-state heat conduction with an internal heat source, the internal energy increment of the control body is 0. The equation can be written as:

注意：式（4.6a）对内部节点和边界节点均适用。对二维稳态导热有内热源时，控制体内能增量为 0。方程式可写为：

$$\Phi_i+(-\Phi_o)+\Phi_v=0 \tag{4.6b}$$

According to the above energy conservation equation, for the internal node (m,n), the equation can be written as following:

依据上述能量守恒方程，对于内部节点（m,n），方程可写成以下形式：

$$\Phi_{左}+\Phi_{右}+\Phi_{上}+\Phi_{下}+\Phi_v=0 \tag{4.6c}$$

Suppose that each heat transfer rate is positive in the direction of the input element (m,n). Assume that the temperature distribution in the element is piecewise linear, then,

设各项热流量均以导入元体 (m,n) 的方向为正。假定元体内温度呈分段线性分布，那么，

$$\Phi_{左}=\lambda\Delta y\frac{t_{m-1,n}-t_{m,n}}{\Delta x},\quad \Phi_{右}=\lambda\Delta y\frac{t_{m+1,n}-t_{m,n}}{\Delta x}$$

$$\Phi_{上}=\lambda\Delta x\frac{t_{m,n+1}-t_{m,n}}{\Delta y},\quad \Phi_{下}=\lambda\Delta x\frac{t_{m,n-1}-t_{m,n}}{\Delta y}$$

$$\Phi_{\mathrm{v}}=\dot{\Phi}V=\dot{\Phi}\Delta x\Delta y$$

Substitute them into the energy conservation equation, it can be obtained:

代入能量守恒式中，可得：

$$\lambda\Delta y\frac{t_{m-1,n}-t_{m,n}}{\Delta x}+\lambda\Delta y\frac{t_{m+1,n}-t_{m,n}}{\Delta x}+\lambda\Delta x\frac{t_{m,n+1}-t_{m,n}}{\Delta y}+\lambda\Delta x\frac{t_{m,n-1}-t_{m,n}}{\Delta y}+\dot{\Phi}\Delta x\Delta y=0 \quad (4.7\mathrm{a})$$

For uniform grid, that is, $\Delta x=\Delta y$, it can be obtained:

对于均匀网格，即 $\Delta x=\Delta y$ 时，可得：

$$t_{m-1,n}+t_{m+1,n}+t_{m,n+1}+t_{m,n-1}-4t_{m,n}+\frac{\Delta x^2}{\lambda}\dot{\Phi}=0$$

$$t_{m,n}=\frac{1}{4}\left(t_{m-1,n}+t_{m+1,n}+t_{m,n+1}+t_{m,n-1}+\frac{\Delta x^2}{\lambda}\dot{\Phi}\right) \quad (4.7\mathrm{b})$$

This is an explicit difference equation. When there is no internal heat source, the internal heat source term is 0, it can be obtained:

这是一个显式差分方程式。当无内热源时，内热源项为 0，可得：

$$t_{m,n}=\frac{1}{4}(t_{m-1,n}+t_{m+1,n}+t_{m,n+1}+t_{m,n-1}) \quad (4.7\mathrm{c})$$

Summary This section illustrates the method for establishing the discrete equations of internal nodes, that is the heat balance method.

小结 本节阐述了内节点离散方程的建立方法——热平衡法。

Question What is the basic idea of the heat balance method for establishing the discrete equations of internal nodes?

思考题 热平衡法建立内节点离散方程的基本思想是什么？

Self-test What is the heat balance method (the control volume method)?

自测题 什么是热平衡法（控制容积法）？

4.4 Establishment of the Discrete Equations of Boundary Nodes

4.4 边界节点离散方程的建立

For the heat conduction of the first type of boundary condition, it is easy to deal with boundary nodes. Given the temperature t_w of the boundary, it can be added to the discrete equations of the inner nodes in numerical form to form a closed algebraic equation group as follows:

对于第一类边界条件的导热问题，边界节点处理较简单。已知边界的温度为 t_w，可将其以数值形式加入内节点的离散方程中，从而组成封闭的代数方程组如下：

$$t_{m,n}=\frac{1}{4}(t_{m+1,n}+t_{m,n+1}+t_{m,n-1}+t_w) \tag{4.8}$$

For the heat conduction of the second or third type of boundary conditions, the equation group composed of the discrete equations of the inner nodes is not closed because the temperature of the boundary nodes is unknown. So, the discrete equations of boundary nodes must be established by the heat balance method, then the closed algebraic equation group jointly composed by the discrete equations of inner nodes and boundary nodes can be solved. For the convenience of solution, the second and third types of boundary conditions are considered together, and can be written as:

对于第二类或第三类边界条件的热传导问题，因为边界节点温度未知，由内节点的离散方程组成的方程组不封闭。所以，必须用热平衡法建立边界节点的离散方程，由内节点与边界节点的离散方程共同组成封闭的代数方程组才能求解。为求解方便，将第二类、第三类边界条件合并起来考虑，并写成：

$$q_w=h(t_f-t_{m,n}) \tag{4.9}$$

Where q_w is the heat flux on the boundary, as shown in Fig. 4.4. $\dot{\Phi}$ is the intensity of the internal heat source.

式中，q_w 表示边界上的热流密度，如图4.4所示。用 $\dot{\Phi}$ 表示内热源强度。

From figure it can be seen the difference between the boundary nodes and the inner nodes, and the elements represented by the boundary nodes are incomplete. According to different types of boundary nodes, the corresponding discrete equations can be established.

从图中可以看出边界节点与内节点的区别，边界节点代表的元体都不完整。根据不同类型的边界节点，可建立相应的离散方程。

(1) Nodes on a straight boundary, as shown in Fig. 4.5.

(1) 平直边界上的节点，如图4.5所示。

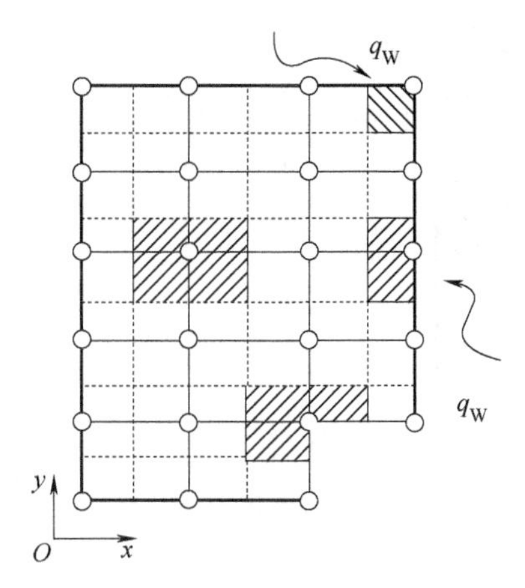

Fig. 4.4 The boundary nodes and the inner nodes
图 4.4 边界节点和内节点

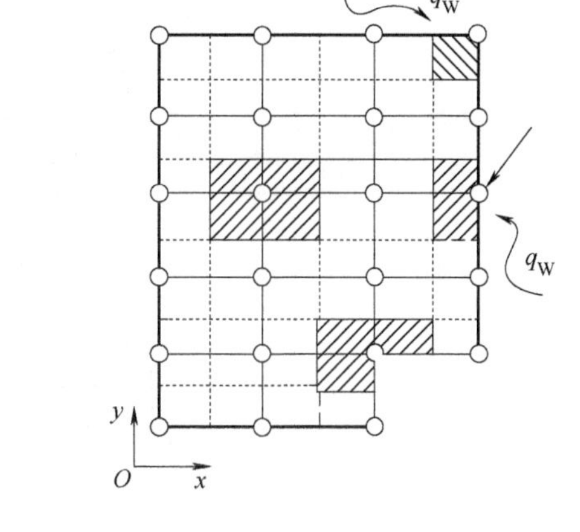

Fig. 4.5 Nodes on a straight boundary
图 4.5 平直边界上的节点

According to the same idea as the internal nodes, the discrete equation of boundary node (m,n) is established by the heat balance method as follows:

仍然按照与内节点一样的思路，对边界节点 (m,n) 用热平衡法建立离散方程如下：

$$\lambda\Delta y\frac{t_{m-1,n}-t_{m,n}}{\Delta x}+\Delta yq_{w}+\lambda\frac{\Delta x}{2}\frac{t_{m,n+1}-t_{m,n}}{\Delta y}+\lambda\frac{\Delta x}{2}\frac{t_{m,n-1}-t_{m,n}}{\Delta y}+\dot{\Phi}_{m,n}\frac{\Delta x}{2}\Delta y=0 \quad (4.10a)$$

When $\Delta x=\Delta y$, namely the step size is equal, then:

当 $\Delta x=\Delta y$ 时，即步长相等，则有：

$$4t_{m,n}=2t_{m-1,n}+\frac{2\Delta x}{\lambda}q_{w}+t_{m,n+1}+t_{m,n-1}+\dot{\Phi}_{m,n}\frac{\Delta x^{2}}{\lambda} \quad (4.10b)$$

(2) **External corner node**, as shown in Fig. 4.6.

(2) **外部角点**，如图 4.6 所示。

Establish the discrete equation by the heat balance method as follows:

用热平衡法建立离散方程如下：

$$\lambda\frac{\Delta y}{2}\frac{t_{m-1,n}-t_{m,n}}{\Delta x}+\frac{\Delta y}{2}q_{w}+\frac{\Delta x}{2}q_{w}+\lambda\frac{\Delta x}{2}\frac{t_{m,n-1}-t_{m,n}}{\Delta y}+\dot{\Phi}_{m,n}\frac{\Delta x}{2}\frac{\Delta y}{2}=0 \quad (4.11a)$$

When $\Delta x=\Delta y$, then:

当 $\Delta x=\Delta y$ 时，有：

$$2t_{m,n}=t_{m-1,n}+t_{m,n-1}+\frac{2\Delta x}{\lambda}q_{w}+\dot{\Phi}_{m,n}\frac{\Delta x^{2}}{2\lambda} \quad (4.11b)$$

(3) **Internal corner node**, as shown in Fig. 4.6.

(3) **内部角点**，如图 4.6 所示。

In the same way, the algebraic equation of the internal corner can be written as follows:

同样的方法可写出内部角点的代数方程如下：

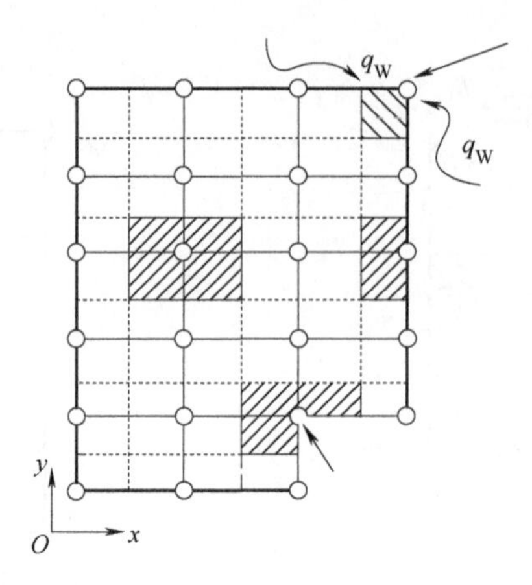

Fig. 4.6 External corner node and internal corner node

图 4.6 外部角点和内部角点

$$\lambda \Delta y \frac{t_{m-1,n}-t_{m,n}}{\Delta x}+\left(\lambda \frac{\Delta y}{2}\frac{t_{m+1,n}-t_{m,n}}{\Delta x}+\frac{\Delta y}{2}q_{w}\right)+\lambda \Delta x \frac{t_{m,n+1}-t_{m,n}}{\Delta y}+$$
$$\left(\lambda \frac{\Delta x}{2}\frac{t_{m,n-1}-t_{m,n}}{\Delta y}+\frac{\Delta x}{2}q_{w}\right)+\dot{\Phi}_{m,n}\frac{3\Delta x\Delta y}{4}=0 \quad (4.12a)$$

When $\Delta x=\Delta y$, then:

当 $\Delta x=\Delta y$ 时，有：

$$t_{m,n}=\frac{1}{6}\left(2t_{m-1,n}+t_{m+1,n}+2t_{m,n+1}+t_{m,n-1}+\frac{3\Delta x^{2}}{2\lambda}\dot{\Phi}+\frac{2\Delta x^{2}}{\lambda}q_{w}\right) \quad (4.12b)$$

If there are curved or inclined boundaries in the calculation region, as shown in Fig. 4.7. The stepped broken line is often used to simulate the real boundary, then the discrete equations of boundary nodes are established by the above method.

如果计算区域出现曲线边界或倾斜边界，如图 4.7 所示。常用阶梯形的折线来模拟真实边界，然后再用上述方法建立边界节点的离散方程。

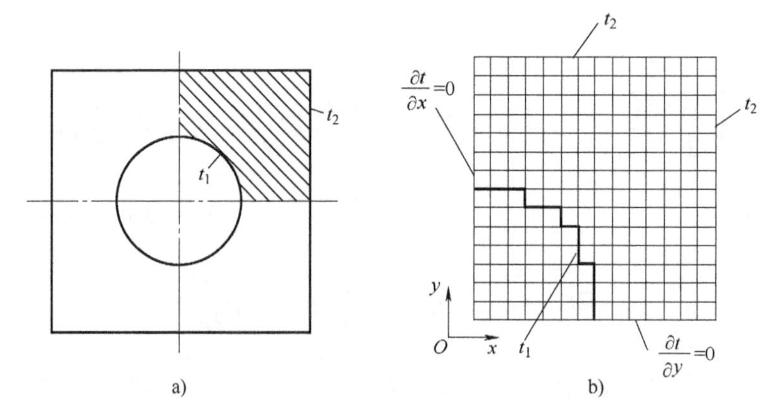

Fig. 4.7 Curved or inclined boundaries

图 4.7 曲线边界或倾斜边界

Summary This section illustrates the method for establishing the discrete equations of boundary nodes.

小结 本节阐述了边界节点离散方程建立的方法。

Question What are the different characteristics of the nodes on the straight boundary, the external corner nodes and the internal corner nodes compared with the internal nodes?

思考题 平直边界上的节点、外部角点、内部角点与内节点相比有什么不同特点？

Self-tests

自测题

1. What is the method of establishing the discrete equations of boundary nodes?
2. What is the difference between the boundary nodes and the center nodes?

1. 边界节点离散方程的建立方法是什么？
2. 边界节点与中心节点有什么区别？

授课视频

4.5 The Iteration Solution Method of Algebraic Equation Group

4.5 求解代数方程组的迭代法

Firstly, according to the methods learned in the previous sections, write out the difference equations of temperature of all internal nodes and boundary nodes in the calculation region to form an algebraic equation group. There must be n algebraic equations for n unknown node temperatures. As shown in the following equation group.

首先，根据前面几节所学的方法，写出计算区域内所有内节点和边界节点的温度差分方程，从而构成代数方程组。n 个未知节点温度，一定对应有 n 个代数方程式。如下面方程组所示。

$$\begin{cases} a_{11}t_1+a_{12}t_2+\cdots+a_{1n}t_n+b_1=0 \\ a_{21}t_1+a_{22}t_2+\cdots+a_{2n}t_n+b_2=0 \\ \qquad\qquad\vdots \\ a_{n1}t_1+a_{n2}t_2+\cdots+a_{nn}t_n+b_n=0 \end{cases}$$

Where a_{mn} and b_n are known coefficients or constants and are not all zeros. The solution methods for algebraic equation group include direct solution method and iterative solution method.

式中，a_{mn}、b_n 为已知系数或常数，且不全为0。代数方程组的求解方法有直接解法和迭代解法两种。

(1) Direct solution method. The accurate solution of algebraic equation group can be obtained by finite calculation. The calculation methods such as matrix inversion and

(1) 直接解法。通过有限次运算获得代数方程组的精确解。运算方法如矩阵求逆、高斯消元法等。

Gaussian elimination method etc.

The disadvantages of the direct solution method include large memory required, inconvenience to use when there are too many equations, and no suitability for nonlinear problems. If the physical properties are the function of temperature, then the coefficients in the difference equation of temperature are also the function of temperature, and these coefficients are no longer constants but constantly change with temperature. This type of problem is called a nonlinear problem.

直接解法的缺点是所需内存较大，方程数目太多时使用不便，不适用于非线性问题。若物性为温度的函数，温度差分方程中的系数也是温度的函数，这些系数不再是常数，而是随温度变化而不断更新。这类问题称为非线性问题。

(2) **Iterative solution method.** Firstly, set the initial field or initial value for the calculated field, and continuously improve it in the iterative calculation process, until the difference between the two adjacent calculation results is less than the allowable value, which is called iterative calculation convergence, otherwise known as divergence.

(2) 迭代解法。先对计算的场设定初场或者叫作初值，在迭代计算过程中不断改进，直到相邻两次计算结果相差小于允许值，称为迭代计算收敛，否则称为发散。

The common iterative solution methods include Gauss-Seidel iteration and simple iteration (Jacobian iteration) and so on. The characteristics of Gauss-Seidel iteration method: The latest value of the node temperature is always used in each iteration, thus it converges faster. For example, the following equation group (a) has n equations, rewrite all algebraic equations into explicit form (b).

常见的迭代解法有高斯-赛德尔迭代和简单迭代（雅可比迭代）等。高斯-赛德尔迭代法的特点：每次迭代时总是使用节点温度的最新值，这样收敛更快。例如，下面这个方程组（a）有 n 个这样的方程式，将 n 个代数方程式均改写成显式形式（b）。

$$\begin{cases} a_{11}t_1+a_{12}t_2+\cdots+a_{1n}t_n+b_1=0 \\ a_{21}t_1+a_{22}t_2+\cdots+a_{2n}t_n+b_2=0 \\ \vdots \\ a_{n1}t_1+a_{n2}t_2+\cdots+a_{nn}t_n+b_n=0 \end{cases} \tag{a}$$

$$\begin{cases} t_1=-\dfrac{1}{a_{11}}(a_{12}t_2+\cdots+a_{1n}t_n+b_1) \\ t_2=-\dfrac{1}{a_{22}}(a_{21}t_1+\cdots+a_{2n}t_n+b_2) \\ \vdots \\ t_n=-\dfrac{1}{a_{nn}}(a_{n1}t_1+\cdots+a_{nn-1}t_{n-1}+b_n) \end{cases} \tag{b}$$

For the above equations, first assume the initial value of iteration:

对于上面这组方程，首先假设迭代的初值：

$$t_1^{(0)}, t_2^{(0)}, \cdots, t_n^{(0)}$$

Where the superscript with parentheses of temperature t denotes the number of iterations, the subscript denotes the variable sequence number. Substitute into equation (b) to get the first iteration value:

式中，温度 t 带括号的上标表示迭代轮次数，下标表示变量序号。代入式（b）得到第一轮迭代数值：

$$t_1^{(1)}, t_2^{(1)}, \cdots, t_n^{(1)}$$

In the following equation, the superscript (k) denotes the value of the previous iteration, ($k+1$) denotes the value of the next iteration.

下式中，上标（k）表示前一轮数值，（$k+1$）表示后一轮迭代数值。

$$t_1^{(k+1)} = a_{12}t_2^{(k)} + \cdots + a_{1n}t_n^{(k)} + b_1^{(k)}$$

And then we can use the value of (k) iteration to figure out the value of ($k+1$) iteration. So when calculating the temperature t of the following nodes, the latest value $t^{(k+1)}$ is used to substitute into the equation, then to get the following equation.

这样就可以用（k）轮迭代的值算出（$k+1$）轮迭代的值。在计算后续节点的温度 t 时，都采用最新值 $t^{(k+1)}$ 代入计算式，可得到以下公式。

$$\begin{aligned}
t_2^{(k+1)} &= a_{21}t_1^{(k+1)} + a_{23}t_3^{(k)} + \cdots + a_{2n}t_n^{(k)} + b_2^{(k)} \\
t_3^{(k+1)} &= a_{31}t_1^{(k+1)} + a_{32}t_2^{(k+1)} + \cdots + a_{3n}t_n^{(k)} + b_3^{(k)} \\
&\vdots \\
t_n^{(k+1)} &= a_{n1}t_1^{(k+1)} + a_{n2}t_2^{(k+1)} + \cdots + a_{nn-1}t_{n-1}^{(k+1)} + b_n^{(k)}
\end{aligned}$$

Where the new value t_1 of ($k+1$) iteration is used when calculating t_2. Similarly, the criterion for judging whether the iteration converges in above equations is that: the temperature deviation of each node in the two adjacent rounds of iteration is less than the specified allowable deviation ε. The temperature deviation includes absolute deviation and relative deviation. The maximum value of absolute deviation between ($k+1$) and (k) iterations should be less than or equal to the allowable deviation ε.

这里，计算 t_2 时用了（$k+1$）轮的 t_1 新值。同样地，在上述方程中，判断迭代是否收敛的准则是：所有节点相邻两轮的温度偏差小于指定允许的偏差 ε。这里的偏差有绝对偏差和相对偏差两种。（$k+1$）轮和（k）轮的绝对偏差的最大值要小于或等于允许偏差 ε。

$$\max \left| t_i^{(k+1)} - t_i^{(k)} \right| \leqslant \varepsilon \tag{4.13a}$$

There are two forms of relative deviation:

相对偏差有两种形式：

$$\max\left|\frac{t_i^{(k+1)}-t_i^{(k)}}{t_i^{(k)}}\right|\leqslant\varepsilon_i \tag{4.13b}$$

$$\max\left|\frac{t_i^{(k+1)}-t_i^{(k)}}{t_{\max}^{(k)}}\right|\leqslant\varepsilon_i' \tag{4.13c}$$

The allowable deviation ε in absolute and relative deviations takes generally $\varepsilon=10^{-3}\sim10^{-6}$. (k) and $(k+1)$ in the equation denote two adjacent iterations, where $t_{\max}^{(k)}$ is the maximum value obtained in iteration k. When there is a node with temperature close to zero, the third criterion is more reliable.

绝对偏差和相对偏差中的允许偏差 ε，一般取 $10^{-3}\sim10^{-6}$。式中，(k) 及 $(k+1)$ 表示的是相邻的两次迭代。其中，$t_{\max}^{(k)}$ 表示第 k 次迭代得到的最大值。当有温度接近于零的节点时，采用第三个准则更可靠。

Example 4.2 The Gauss-Seidel iteration method is used to solve the following equation group (c).

例 4.2 用高斯-赛德尔迭代法求解下列方程组（c）。

$$\begin{aligned}8t_1+2t_2+t_3&=29\\ t_1+5t_2+2t_3&=32\\ 2t_1+t_2+4t_3&=28\end{aligned} \tag{c}$$

Analysis: First rewrite the above equations into the following iteration form, which is the explicit form.

分析：先将式（c）改写成以下迭代形式，即显示形式。

$$\begin{aligned}t_1&=\frac{1}{8}(29-2t_2-t_3)\\ t_2&=\frac{1}{5}(32-t_1-2t_3)\\ t_3&=\frac{1}{4}(28-2t_1-t_2)\end{aligned} \tag{d}$$

Note: The above rewritten equation group meets the conditions of iteration convergence. Assume a set of initial values, such as taking:

注意：对上述改写后的方程组，迭代收敛的条件是满足的。假设一组初值，例如取：

$$t_1^{(0)}=t_2^{(0)}=t_3^{(0)}=0$$

The results of the first iteration can be obtained by using the above iteration method and substituting the initial values into the equations:

利用上述迭代方法，把初值代到方程中，可得出第一次迭代的结果：

$$t_1^{(1)},t_2^{(1)},t_3^{(1)}$$

After several iterations, the required solution can be obtained. After 7 iterations, the results are consistent with

经过数次迭代后，可获得所需的解。经过了 7 次迭代后，在 4 位有效数

the exact solution in 4 significant digits. The intermediate values of the iterative process are listed in Table 4. 2.

字内得到了与精确解一致的结果。迭代过程的中间值列在表4. 2中。

Table 4. 2 The intermediate values of the iterative process

表4. 2 迭代过程的中间值

迭代次数	t_1	t_2	t_3
0	0	0	0
1	3. 625	5. 675	3. 769
2	1. 735	4. 545	4. 996
3	1. 864	4. 038	5. 058
4	1. 983	3. 980	5. 013
5	2. 003	3. 994	5. 000
6	2. 0001	4. 000	5. 000
7	2. 000	4. 000	5. 000

Discussion: If we establish the iterative equation of equation group (c) in the following way, and write as the form of equation group (e).

讨论：如果按下列方式来构造方程组（c）的迭代方程，并写成方程组（e）的形式。

$$\begin{aligned} t_1 &= 32-5t_2-2t_3 \\ t_2 &= 28-2t_1-4t_3 \\ t_3 &= 29-8t_1-2t_2 \end{aligned} \tag{e}$$

Then we iteratively solve again, the results of iteration are shown in Table 4. 3.

然后再迭代求解，迭代结果见表4. 3。

Table 4. 3 The results of iteration

表4. 3 迭代结果

迭代次数	0	1	2	3	4
t_1	0	32	522	8722	143522
t_2	0	-36	-396	-3996	-39996
t_3	0	-155	-3355	-61755	-1068075

Obviously, the iteration method of equation (e) cannot obtain a convergent solution, which is called the iterative process divergence. This example shows that for the same algebraic equation group, if the iteration method is not appropriate, it may cause iteration divergence.

显然，按式（e）的迭代方式得不到收敛解，这称为迭代过程发散。这一例子说明同一个代数方程组，如果选用的迭代方式不合适，则可能导致迭代发散。

Conclusion: The differential equation system for heat conduction problems with constant physical properties (i. e. linear problem), the selection of the iteration formula should make the coefficient of each iterative variable always be greater than or equal to the algebraic sum of the absolute values of the coefficients of the other variables in the equation. In this case, the algebraic equation group solved by the iteration method must converge. This condition is mathematically called the dominance of the main diagonal, and written as the equation as follows:

结论：对于常物性导热问题（线性问题）的差分方程组，迭代公式的选择，应使每一个迭代变量的系数总是要大于或等于该式其他变量系数绝对值的代数和。此时，迭代法求解的代数方程组一定收敛。这一条件在数学上称为主对角线占优，写成公式如下：

$$\frac{|a_{12}|+|a_{13}|}{a_{11}}\leqslant 1,\frac{|a_{21}|+|a_{23}|}{a_{22}}\leqslant 1,\frac{|a_{31}|+|a_{32}|}{a_{33}}\leqslant 1$$

Example 4.3 A square body with an isotropic material, its thermal conductivity is constant. The temperatures of each boundary are known as shown in Fig. 4.8. Try to use Gauss-Seidel iteration method to find the temperature of the internal grid nodes 1, 2, 3 and 4.

例 4.3 有一各向同性材料的方形物体，其导热系数为常量。已知各边界的温度如图 4.8 所示，试用高斯-赛德尔迭代法求其内部网格节点 1、2、3 和 4 的温度。

Solution: This is a two-dimensional steady-state heat conduction problem. For the temperatures of each grid node inside the object, the relation of Eq. (4.5b) is applicable. At the beginning suppose:

解：这是一个二维稳态导热问题。对于物体内部每个网格节点的温度，式（4.5b）的关系均适用。开始计算时假设：

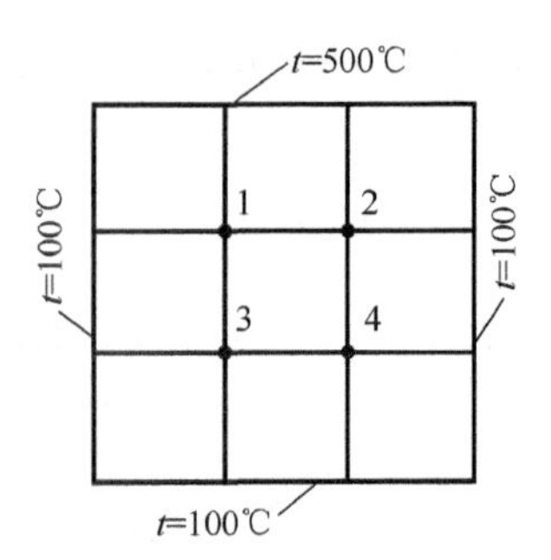

Fig. 4.8 Schematic diagram of a grid of the square object

图 4.8 方形物体的网格示意图

$$t_1^{(0)}=t_2^{(0)}=300℃,t_3^{(0)}=t_4^{(0)}=200℃$$

According to the Gauss-Seidel iterative method, the initial values are substituted into the iterative equations for calculation:

按照高斯-赛德尔迭代法，把初值代入迭代公式中进行计算：

$$t_1^{(1)} = \frac{1}{4} \times (500℃ + 100℃ + t_2^{(0)} + t_3^{(0)})$$

$$= \frac{1}{4} \times (500+100+300+200)℃ = 275℃$$

$$t_2^{(1)} = \frac{1}{4} \times (500℃ + 100℃ + t_1^{(1)} + t_4^{(0)})$$

$$= \frac{1}{4} \times (500+100+275+200)℃ = 268.75℃$$

$$t_3^{(1)} = \frac{1}{4} \times (100℃ + 100℃ + t_1^{(1)} + t_4^{(0)})$$

$$= \frac{1}{4} \times (100+100+275+200)℃ = 168.75℃$$

$$t_4^{(1)} = \frac{1}{4} \times (100℃ + 100℃ + t_2^{(1)} + t_3^{(1)})$$

$$= \frac{1}{4} \times (100+100+268.75+168.75)℃$$

$$= 159.38℃$$

And so on to get other values for each iteration, the first to fifth iteration values are summarized in Table 4.4.

以此类推，可得其他各次迭代值，第1~第5次迭代值汇总于表4.4中。

Table 4.4 The first to fifth iteration values

表 4.4 第 1~第 5 次迭代值

迭代次数	t_1/℃	t_2/℃	t_3/℃	t_4/℃
0	300	300	200	200
1	275	268.75	168.75	159.38
2	259.38	254.69	154.69	152.35
3	252.35	251.18	151.18	150.59
4	250.59	250.30	150.30	150.15
5	250.15	250.07	150.07	150.04
6	250.04	250.02	150.02	150.01

Substitute the results of the fifth and sixth iterations into the equation of relative deviation (4.13c), it is found that the relative deviation is already less than 2×10^{-4}, then the iteration is over.

把第 5 与第 6 次迭代的结果代入式（4.13c）相对偏差的计算式，发现相对偏差已小于 2×10^{-4}，迭代终止。

Only 4 nodes are taken for the convenience of manual cal-

例题 4.3 为了手工计算方便，仅取

culation in Example 4. 3. We could take a large number of nodes in numerical calculation in engineering. The principle of taking the number of nodes is that: further increasing the number of nodes, the effect on the main results of numerical calculation is within the allowable range, and the calculation results are basically independent of the grid at the time, which is called grid independence.

4 个节点。工程数值计算时，节点可能取很多。取节点数的原则：再进一步增加节点数，对数值计算主要结果的影响已在允许范围内，此时计算结果基本与网格无关，称为网格独立。

Summary This section illustrates the iteration solution method for algebraic equation group and the convergence criteria of iterative calculations.

小结 本节阐述了求解代数方程组的迭代解法，以及迭代计算收敛的准则。

Question When solving algebraic equations by Gauss-Seidel iteration method, can it be sure to get a convergent solution? If a convergent solution cannot be got, try to analyze the reasons.

思考题 用高斯-赛德尔迭代法求解代数方程时，是否一定可以得到收敛的解？如不能得到收敛解，试分析其原因。

Self-tests

自测题

1. What is a linear problem? A non-linear problem?

1. 什么是线性问题？什么是非线性问题？

2. What is the convergence criterion? Grid independence?

2. 什么是收敛判据？什么是网格独立性？

Exercises

习题

4. 1 Try to calculate the values of t_1, t_2, t_3 and t_4 using the Gauss-Seidel iteration method for the two-dimensional steady-state heat conduction problem with constant physical properties and no internal heat source as shown in Fig. 4. 9.

4. 1 试对图 4. 9 所示的常物性、无内热源的二维稳态导热问题，用高斯-赛德尔迭代法计算 t_1、t_2、t_3、t_4 值。

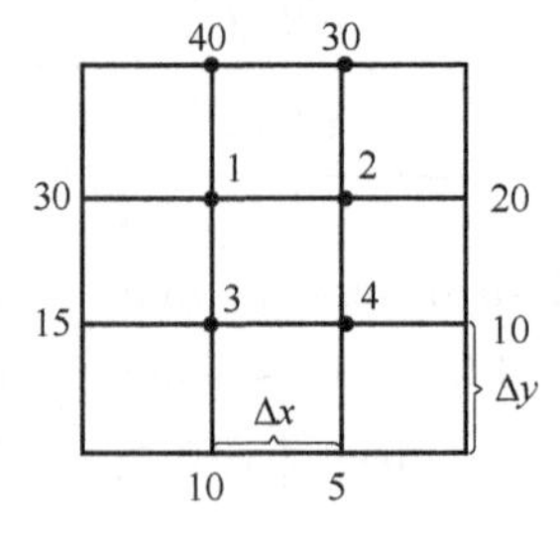

Fig. 4. 9 Attached figure of exercise 4. 1 ($\Delta x=\Delta y$)

图 4. 9 习题 4. 1 附图（$\Delta x=\Delta y$）

4.2 In the two-dimensional heat conduction region with an internal heat source as shown in Fig. 4.10, one interface is adiabatic, the other interface is constant temperature of t_0 (including node 4), and the other two interfaces have convective heat transfer with a fluid at a temperature of t_f. λ and h is uniform, and the intensity of internal heat source is $\dot{\Phi}$. List the discrete equations of nodes 1, 2, 5, 6, 9 and 10.

4.2 在图 4.10 所示的有内热源的二维导热区域中，一个界面绝热，另一个界面恒温温度为 t_0（包括节点 4），其余两个界面与温度为 t_f 的流体对流传热，λ、h 均匀，内热源强度为 $\dot{\Phi}$。试列出节点1、2、5、6、9、10 的离散方程式。

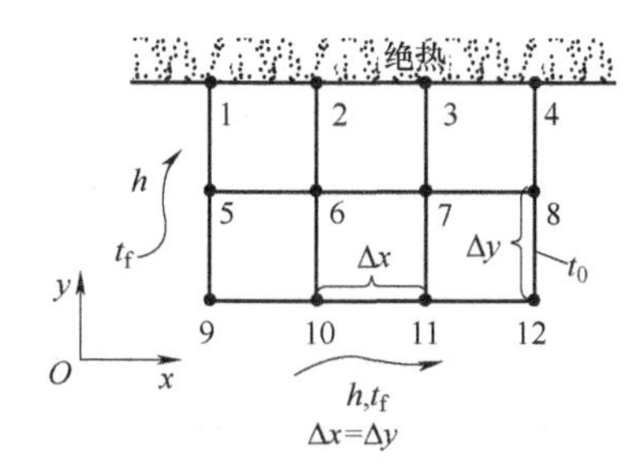

Fig. 4.10 Attached figure of exercise 4.2

图 4.10 习题 4.2 附图

4.3 A straight rib with equal cross-section with height H and thickness δ, the rib root temperature is t_0, the side fluid temperature is t_f, the surface heat transfer coefficient is h, and the thermal conductivity of the rib is λ. Divide it into four nodes (Fig. 4.11), and for two cases where the rib end is adiabatic and the convective boundary condition (h is on the same side), list the discrete equations for nodes 2, 3, and 4. Assuming $H = 45\text{mm}$, $\delta = 10\text{mm}$, $t_0 = 100℃$, $t_f = 20℃$, $h = 50\text{W}/(\text{m}^2 \cdot \text{K})$, $\lambda = 50\text{W}/(\text{m} \cdot \text{K})$, calculate the temperature of nodes 2, 3, and 4 (for the two boundary conditions at the rib end).

4.3 一等截面直肋，高为 H，厚为 δ，肋根温度为 t_0，侧面流体温度为 t_f，表面传热系数为 h，肋片导热系数为 λ。将它均分成 4 个节点（图 4.11），并对肋端为绝热及为对流边界条件（h 同侧面）的两种情况列出节点 2、3、4 的离散方程式。假设 $H = 45\text{mm}$，$\delta = 10\text{mm}$，$t_0 = 100℃$，$t_f = 20℃$，$h = 50\text{W}/(\text{m}^2 \cdot \text{K})$，$\lambda = 50\text{W}/(\text{m} \cdot \text{K})$，计算节点2、3、4 的温度（对于肋端的两种边界条件）。

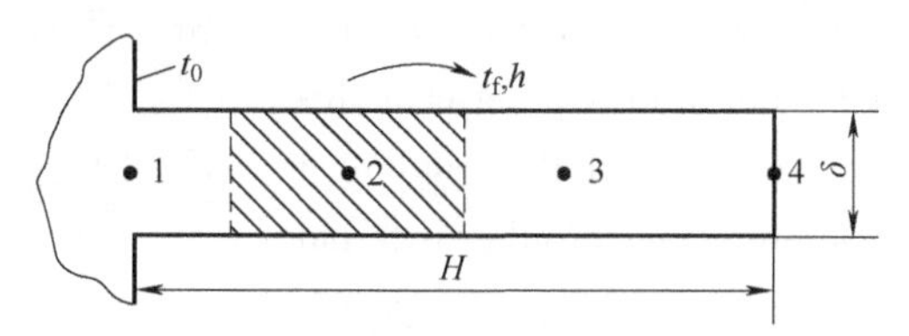

Fig. 4.11 Attached figure of exercise 4.3

图 4.11 习题 4.3 附图

Chapter 5 Theoretical Basis of Convection Heat Transfer

第5章 对流传热的理论基础

授课视频

5.1 The Influencing Factors of Convection Heat Transfer

5.1 对流传热的影响因素

The influencing factors of convection heat transfer include two factors that affect fluid flow and heat transfer, and specifically include five aspects: ① The causes of fluid flow; ② The states of fluid flow; ③ Whether the phase transition of fluid or not; ④ The thermophysical properties of fluid; ⑤ The geometric factors of heat transfer surface etc. It can be seen that convection heat transfer is related to both fluid and heat transfer wall.

对流传热的影响因素包括影响流动和影响传热两方面的因素，具体包括五个方面：① 流体流动的起因；② 流体流动的状态；③ 流体有无相变；④ 流体的热物理性质；⑤ 传热表面的几何因素等。可见，对流传热既和流体有关，又和传热壁面有关。

According to the influencing factors of convection heat transfer, the convective heat transfer phenomenon is classified as follows:

根据对流传热的影响因素，对流传热现象的分类如下：

(1) According to the causes of fluid flow, convection heat transfer can be divided into natural convection heat transfer and forced convection heat transfer. Natural convection heat transfer: Fluid flow and heat transfer are caused by the density difference which is caused by the temperature difference between the various parts of a fluid. Forced convection heat transfer: Fluid flow and heat transfer are caused by water pumps, fans or other external power sources.

(1) 按照流体流动的起因，对流传热可分为自然对流传热和强制对流传热。自然对流传热：流体各部分之间温度不同而引起密度差，由密度差所引起的流动和传热。强制对流传热：由水泵、风机或其他外部动力源所引起的流动和传热。

The causes of fluid flow determine the velocity field of fluid, and the velocity field determines the surface heat transfer effect. Therefore, the surface heat transfer coefficient of forced convection is much larger than that of natural convection.

流体流动的起因决定了流体的速度场，速度场决定了传热效果。因此，强制对流传热的表面传热系数远大于自然对流传热的表面传热系数。

(2) According to the states of fluid flow, convection heat transfer can be divided into laminar convection heat transfer and turbulent convection heat transfer. Laminar convection heat transfer: Fluid microcluster flows in regular stratification along the main flow direction, and the streamlines are a cluster of parallel lines. Turbulent convection heat transfer: Fluid particles move in complex and irregular ways, and fluid is more fully mixed with each other. Therefore, the surface heat transfer coefficient of turbulent convection is much larger than that of laminar convection.

(2) 按照流体流动的状态，对流传热可分为层流对流传热和湍流对流传热。层流对流传热：流体微团沿主流方向做规则的分层流动，流线为一簇平行线。湍流对流传热：流体质点做复杂无规则的运动，流体相互掺混更加充分。因此，湍流对流传热的表面传热系数远大于层流对流传热。

(3) Depending on whether the fluid has a phase transition, convection heat transfer can be divided into single-phase convection heat transfer and phase transition convection heat transfer. Single-phase convective heat transfer relies on fluid sensible heat change to transfer heat. Phase transition convection heat transfer is mainly by release or absorption of phase transition heat (i. e. latent heat) of fluid, such as condensation and boiling.

(3) 按照流体有无相变，对流传热可分为单相对流传热和相变对流传热。单相对流传热靠流体显热变化传递热量。相变对流传热以流体相变热（即潜热）释放或吸收为主，如冷凝和沸腾。

The vaporization of liquid, i. e. phase transition, needs to absorb a large amount of latent heat of vaporization. Similarly, steam condensation also needs to release a large amount of latent heat of vaporization. For example, 1kg of water at 100℃ needs to absorb 2256kJ of heat capacity to turn into steam at 100℃, while the temperature of 1kg of water rises by 1℃, the absorbed heat capacity is 4.186kJ. As we can see that latent heat of phase transition is 540 times of sensible heat. Therefore, the convective heat transfer coefficient of phase transition is much larger than that of single phase.

液体汽化（相变）需吸收大量的汽化潜热。同样，蒸汽冷凝需要放出大量的汽化潜热。比如1kg温度为100℃的水变成100℃的蒸汽需要吸收2256kJ的热量，而1kg的水温度升高1℃吸收的热量是4.186kJ。可见，相变潜热是显热的540倍。因此，相变对流传热的表面传热系数远大于单相对流传热。

Fig. 5.1 is a schematic diagram of four heat transfer processes: forced convection, natural convection, boiling and condensation. In Fig. 5.1a and b, the squares represent the hot components or printed circuit boards. In Fig. 5.1a, the hot components are placed horizontally. Under the action of fan, forced convection heat transfer occurs between air and hot components. In Fig. 5.1b, the hot

图5.1所示为强制对流、自然对流、沸腾、凝结四种传热过程的示意图。图5.1a和b中，方块表示热元件或印制电路板。图5.1a中热元件水平放置。在风扇的作用下，空气与热元件进行强制对流传热。图5.1b中热元件竖直放置。附

components are placed vertically. After the nearby air is heated, its volumc cxpands and the density decreases, then the air naturally rises. Natural convection heat transfer occurs between air and hot components. In Fig. 5. 1c, a heating plate is used to heat water in the pot. Water bubbles constantly generate when water boils, which belongs to convection heat transfer of boiling phase transition. In Fig. 5. 1d, the hot humid air meets the tube wall at a lower temperature. Water vapor in the air condenses into droplets on the surface of the tube wall, which belongs to the convection heat transfer of condensed phase transition.

近空气受热后，体积膨胀而密度减小，空气自然上升。空气与热元件之间进行自然对流传热。图 5. 1c 中用加热板加热锅里的水。水沸腾后不断产生气泡，属于沸腾相变对流传热。图 5. 1d 中湿热空气遇到温度较低的管壁。空气中的水蒸气在管壁表面冷凝为液滴，属于冷凝相变对流传热。

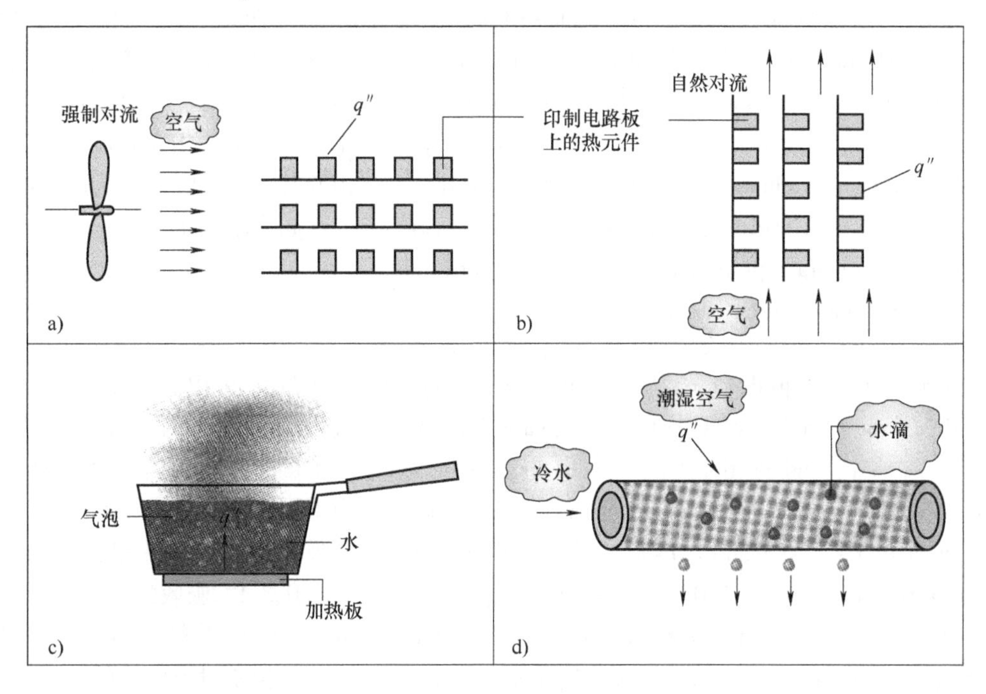

Fig. 5. 1 A schematic diagram of four heat transfer processes

图 5. 1 四种传热过程的示意图

(4) Thermophysical properties of the fluid: Including density, viscosity, thermal conductivity, specific heat capacity. Viscosity is divided into dynamic viscosity and kinematic viscosity. With the increase of thermal conductivity, convective heat transfer coefficient also increases. The convective heat transfer resistance inside the fluid and the heat conduction resistance between the fluid and the wall decrease. As density and specific heat capacity increase, convective heat transfer coefficient also increases.

(4) 流体的热物理性质：包括密度、黏度、导热系数、比热容。黏度分为动力黏度、运动黏度。随着导热系数增大，表面传热系数也增大，流体内部对流传热热阻及流体与壁面间导热热阻都变小。随着密度和比热容增大，表面传热系数也增大。单位体积流体能携带更多热量。随着黏度增大，表面传热系数

Fluid can carry more heat per unit volume. As viscosity increases, convective heat transfer coefficient decreases. Because the high viscosity hinders fluid flow, and it is not conducive to convection heat transfer.

减小。因为黏度大有碍于流体流动，不利于对流传热。

(5) Geometric factors on heat transfer surfaces refer to the shape, size and state (such as smooth or rough) of the heat transfer surface, the relative positions between the heat transfer surface and the fluid flow, etc. According to the relative positions, fluid flow is divided into internal flow and external flow, as shown in Fig. 5. 2.

(5) 传热表面的几何因素：是指传热表面的形状、大小、状态（如光滑或粗糙），传热表面与流体运动的相对位置等。按照相对位置，流体流动分为内部流动和外部流动，如图 5. 2 所示。

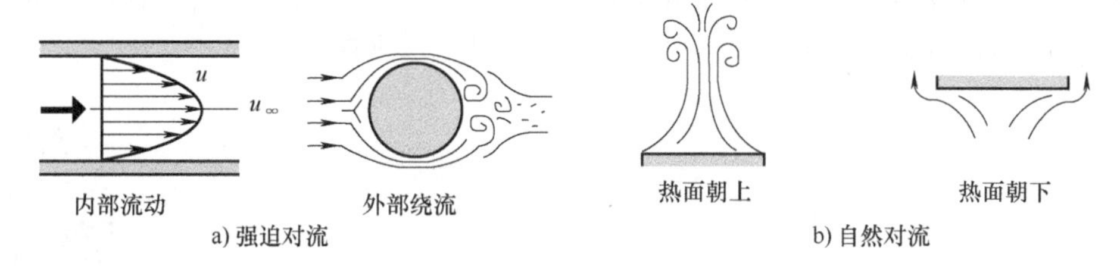

Fig. 5. 2 Classification of convection heat transfer and influence of geometric factors
图 5. 2 对流传热分类和几何因素的影响

Internal flow: When the boundary layer develops fully, its outer boundary converges in the center of the flow channel, such as fluid flow in a tube or a groove. **External flow**: The boundary layer develops infinitely, such as fluid externally flowing through a plane plate or circular tube. In addition, the geometric factors of heat transfer surface also include the heat transfer surface state, such as the extended surface and rough surface.

内部流动：边界层充分发展后，其外边界汇聚于流道中心，如管内、槽内流体流动。**外部流动**：边界层无限发展，如流体外掠平板/圆管。另外，传热表面的几何因素还包括传热表面状态，如扩展表面、粗糙表面。

To sum up, there are six main factors affecting the convective heat transfer coefficient. Therefore, the convective heat transfer coefficient can be expressed as:

综上所述，表面传热系数的主要影响因素有六个。因此，表面传热系数可表示为：

$$h=f(\boldsymbol{u},l,\lambda,c_p,\rho,\eta)$$

The core problem of convective heat transfer research is to determine the convective heat transfer coefficient. Usually, the velocity field (i. e. flow field) is solved first, then the temperature field can be obtained from the velocity field, and the convective heat transfer coefficient can be obtained from the temperature field.

对流传热研究的核心问题就是确定表面传热系数。通常，先求出速度场（流场），由速度场可求出温度场，由温度场可求出表面传热系数。

Summary This section illustrates the influencing factors of convective heat transfer coefficient.

小结 本节阐述了表面传热系数的影响因素。

Question Why is the heat transfer coefficient of turbulent convection much larger than that of laminar convection?

思考题 为什么湍流表面传热系数远远大于层流表面传热系数?

Self-tests

1. What are the influencing factors of convection heat transfer? What are the types of convective heat transfer phenomena?

2. What is the influence of geometric factors of heat transfer surface on convection heat transfer?

自测题

1. 对流传热的影响因素有哪些?对流传热现象有哪些类型?

2. 传热表面的几何因素对对流传热有什么影响?

授课视频

5.2 The Differential Equation of Convective Heat Transfer Process

5.2 对流传热过程微分方程式

When the viscous fluid flows on the wall, there is a velocity boundary layer due to the viscosity of fluid. Namely, with the decrease of the distance between the fluid and the wall, the fluid velocity gradually decreases near the wall; The fluid close to the wall, i. e. $y=0$, is stagnant and in a non-slip state, the velocity along the x-direction $u=0$. The velocity boundary layer is shown in Fig. 5. 3.

黏性流体在壁面上流动时，由于黏性的作用，存在速度边界层。即在壁面附近，随着流体与壁面距离的减小，流体的流速逐渐降低；在紧贴壁面处（$y=0$）流体被滞止，处于无滑移状态，沿着 x 方向的速度 $u=0$。速度边界层示意图如图 5.3 所示。

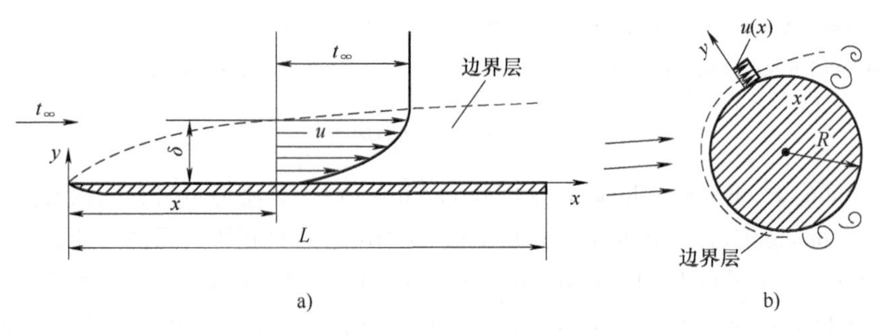

Fig. 5. 3 Schematic diagram of velocity boundary layer

图 5. 3 速度边界层示意图

In the extremely thin layer of fluid that sticks to the wall, the fluid adheres to the wall surface and the heat can be transferred only by heat conduction. Therefore, according to Fourier's law of heat conduction, the heat conduction quantity of the thin layer fluid can be calculated with the following formula:

在这个极薄的贴壁流体层中，流体黏附在壁面上，热量只能以导热方式传递。因此，根据傅里叶导热定律，此薄层流体的导热量可以用式（5.1）进行计算：

$$q_{w,x} = -\lambda\left(\frac{\partial t}{\partial y}\right)_{w,x} \tag{5.1}$$

Where λ is the thermal conductivity of the fluid, $\left(\frac{\partial t}{\partial y}\right)_{w,x}$ is the temperature gradient of the fluid at the coordinate $(x,0)$. The convective heat transfer quantity of the fluid outside the thin layer is calculated according to Newton cooling formula:

式中，λ 是流体的导热系数；$\left(\frac{\partial t}{\partial y}\right)_{w,x}$ 是指坐标（$x,0$）点处流体的温度梯度。薄层以外的流体对流传热量按照牛顿冷却公式计算：

$$q_{w,x} = h_x(t_w - t_\infty) \tag{5.2}$$

Where h_x is the localized convective heat transfer coefficient at the x site of the wall, also known as the local convective heat transfer coefficient. In a steady state, the convective heat transfer quantity is equal to the heat conduction quantity of the fluid layer that sticks to the wall. That is to say, the heat transfer quantity of the above two formulas is equal. The differential equation of convective heat transfer process can be obtained:

式中，h_x 是壁面 x 处局部表面传热系数，也称为当地表面传热系数。稳态情况下，对流传热量等于贴壁流体层的导热量，即上述两式的传热量相等。由此可得对流传热过程微分方程式为：

$$h_x = -\frac{\lambda}{t_w - t_\infty}\left(\frac{\partial t}{\partial y}\right)_{w,x} \tag{5.3}$$

Eq. (5.3) reveals the relationship between the local convective heat transfer coefficient and the local temperature alteration ratio. Therefore, we find out the temperature field of fluid, then the convective heat transfer coefficient can be obtained from this equation. Mathematically speaking, the equation is the same as the third type of boundary condition of the heat conduction problem; But their physical meanings are completely different. Where h_x in this formula is the unknown quantity and λ is the thermal conductivity of the fluid.

式（5.3）揭示了局部表面传热系数 h_x 与当地的温度变化率 $\frac{\partial t}{\partial y}$ 之间的关系。因此，只要求出流体的温度场，通过该方程式即可求出表面传热系数。从数学形式上，该方程式与导热问题的第三类边界条件相同，但是其物理意义完全不同。式中，h_x 为待求量，λ 为流体的导热系数。

From the differential equation of the convective heat transfer process, we can find that the local convective heat transfer coefficient h_x depends on the thermal conductivity of the fluid, the temperature difference between the fluid and the wall, and the temperature gradient of the fluid that sticks to the wall. Where the temperature gradient or temperature field depends on the thermophysical properties of the fluid, the flow state, i. e. laminar flow or turbulent flow, the velocity magnitude and its distribution, and surface roughness, etc. In conclusion, the temperature field depends on the velocity field.

从对流传热过程微分方程式可以发现，局部表面传热系数 h_x 取决于流体的导热系数、流体与壁面的温差及贴壁流体的温度梯度。其中，温度梯度（或温度场）取决于流体热物性、流动状况（层流或湍流）、流速大小及其分布，以及表面粗糙度等因素。总之，温度场取决于速度场。

Summary This section illustrates the derivation process of the differential equation of the convective heat transfer process and analyzes its physical meaning.

小结 本节阐述了对流传热过程微分方程式的推导过程，并分析了其物理意义。

Question What is the key problem of convection heat transfer research?

思考题 对流传热研究的核心问题是什么？

Self-tests

自测题

1. What is the difference between the differential equation for convective heat transfer process and the third type of boundary condition for heat conduction problem?

1. 对流传热过程微分方程式与导热问题的第三类边界条件有什么不同？

2. To solve the local surface heat transfer coefficient through the differential equation of convective heat transfer process, what characteristics of the fluid must be determined?

2. 通过对流传热过程微分方程式求解当地表面传热系数，必须要确定流体的什么特性？

授课视频

5.3 The Differential Equation Group of Convection Heat Transfer

5.3 对流传热微分方程组

To solve the convective heat transfer coefficient, we should obtain the velocity field and temperature field first. While the velocity field and the temperature field

要求解表面传热系数，首先需要求出速度场和温度场。而速度场和温度场取决于对流传热微分方程组。

depend on the differential equation group for convection heat transfer. The equation group includes three types of equations: mass conservation equation, momentum conservation equation, and energy conservation equation. Now we study the derivation process of the differential equation group of convection heat transfer.

该方程组包括三类方程：质量守恒方程、动量守恒方程和能量守恒方程。下面来学习对流传热微分方程组的推导过程。

First, make simplifications and assumptions: ① The fluid is a continuous medium and two-dimensional flow; ② The fluid is incompressible Newtonian fluid. The so-called Newtonian fluid refers to a fluid subject to Newton's law of viscosity; ③ All physical parameters are constant and there is no internal heat source; ④ The heat dissipation generated by viscous dissipation is neglected.

首先做出简化和假设：① 流体为连续性介质，并且为二维流动；② 流体为不可压缩的牛顿流体。所谓牛顿流体是指服从牛顿黏性定律的流体；③ 所有物性参数均为常数，并且无内热源；④ 黏性耗散所产生的耗散热忽略不计。

For two-dimensional convection heat transfer, there are four unknown quantities: velocity u and v; temperature t; pressure p. So, four equations are required to solve, including one continuity equation, two momentum conservation equations, and one energy conservation equation. We first derive the continuity equation, i. e. the mass conservation equation. The continuous flow of fluid complies with the law of mass conservation, as shown in Fig. 5. 4.

对于二维对流传热问题，有 4 个未知量：速度 u、v，温度 t，压力 p。因此，需要 4 个方程才能求解：也就是 1 个连续性方程、2 个动量守恒方程和 1 个能量守恒方程。首先推导出连续性方程（即质量守恒方程）。流体的连续流动遵循质量守恒规律，如图 5. 4 所示。

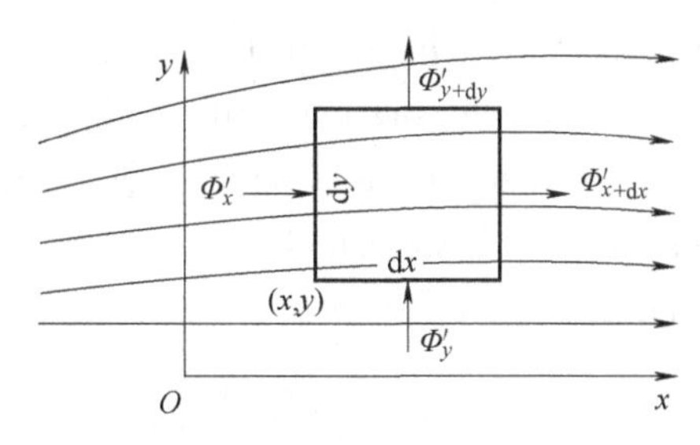

Fig. 5. 4 The micro-element body in the derivation of energy differential equations

图 5. 4 能量微分方程推导中的微元体

For any point (x, y) in the flow field, take out the micro-element body with edge lengths of dx and dy, and M is the mass flow rate. The mass flowing into the micro-element body through the x-surface along the x-axis direction per unit time:

从流场中任一点 (x, y) 处，取出边长为 dx、dy 的微元体，M 为质量流率。单位时间内，沿 x 轴方向，经 x 表面流入微元体的质量为：

$$M_x = \rho u dy \cdot 1 \tag{5.4}$$

The thickness of the micro-element body is 1. The mass flowing out of the micro-element body through the $x+\mathrm{d}x$ surface along the x-axis direction per unit time:

1是指微元体的厚度。单位时间内，沿x轴方向，经$x+\mathrm{d}x$表面流出微元体的质量为：

$$M_{x+\mathrm{d}x}=M_x+\frac{\partial M_x}{\partial x}\mathrm{d}x \tag{5.5}$$

Then, the net mass flowing into and out of the micro-element body along the x-axis direction per unit time is equal to subtracting Eq. (5.4) and Eq. (5.5):

那么，单位时间内，沿x轴方向，流入、流出微元体的净质量可用式（5.4）和式（5.5）相减：

$$M_x-M_{x+\mathrm{d}x}=-\frac{\partial M_x}{\partial x}\mathrm{d}x=-\frac{\partial(\rho u)}{\partial x}\mathrm{d}x\mathrm{d}y \tag{5.6}$$

The same method can be used to obtain the net mass flowing into and out of the micro-element body along the y-axis direction per unit time:

同样方法可求出，单位时间内，沿y轴方向，流入、流出微元体的净质量为：

$$M_y-M_{y+\mathrm{d}y}=-\frac{\partial M_y}{\partial y}\mathrm{d}y=-\frac{\partial(\rho v)}{\partial y}\mathrm{d}x\mathrm{d}y \tag{5.7}$$

The fluid mass change rate in the micro-element body per unit time can be calculated as follows:

单位时间内微元体内流体质量的变化率可计算如下：

$$\frac{\partial(\rho\mathrm{d}x\mathrm{d}y)}{\partial\tau}=\frac{\partial\rho}{\partial\tau}\mathrm{d}x\mathrm{d}y \tag{5.8}$$

According to the mass conservation of fluid in the micro-element body, we can get the net mass flowing into and out of the micro-element body is equal to the fluid mass change rate in the micro-element body. So, we can obtain the two-dimensional continuity equation:

根据微元体内流体质量守恒关系可得：流入流出微元体的净质量=微元体内流体质量的变化率。由此可得到二维连续性方程为：

$$\frac{\partial\rho}{\partial\tau}+\frac{\partial(\rho u)}{\partial x}+\frac{\partial(\rho v)}{\partial y}=0 \tag{5.9}$$

For the two-dimensional steady-state flow, and the constant density, i. e. incompressible fluid:

对于二维稳态流动，密度为常数，也就是不可压缩流体：

$$\frac{\partial\rho}{\partial\tau}=0$$

The two-dimensional continuity equation is:

二维连续性方程为：

$$\frac{\partial u}{\partial x}+\frac{\partial v}{\partial y}=0 \tag{5.10}$$

The momentum conservation equation of the fluid micro-element body describes the velocity field of fluid, also known as equations of motion. According to Newton's second law, the sum of all the external forces acting on the micro-element body is equal to the fluid momentum change rate of the micro-element body. In hydrodynamics, we have obtained the momentum conservation equations, i. e. Navier-Stokes equations. Here is no longer detailed.

流体微元体的动量守恒方程描述了流体的速度场，也称为运动方程。根据牛顿第二定律，作用在微元体上所有外力的总和等于微元体中流体动量的变化率。在流体力学中已经得到了动量守恒方程，即 Navier-Stokes 方程。这里不再详述。

Next, we look at the energy conservation equation. The energy conservation equation of the fluid micro-element body describes the temperature field of the fluid. According to the law of energy conservation: the net heat imported to and exported from the micro-element body, plus the net heat transferred by heat convection, plus the heat generated by the internal heat source, are equal to the increment of internal energy plus the expansion work done externally.

接下来是能量守恒方程。流体微元体的能量守恒方程描述了流体的温度场。根据能量守恒定律：[导入与导出微元体的净热量]+[热对流传递的净热量]+[内热源发热量]=[内能的增量]+[对外做的膨胀功]。

Generally, the heat generated by the internal heat source and the work done externally is neglected. So, the expression is as follows: heat conduction quantity plus heat convection quantity is equal to the increment of thermodynamic internal energy. Along the direction of the x-axis and y-axis, the net heat imported to and exported from the micro-element body is derived in Chapter 2:

一般情况下，不考虑内热源发热量，以及对外做功。所以，有表达式：导热量+热对流传热量=热力学内能增量。沿 x 轴、y 轴方向，导入导出微元体的净热量在第 2 章已推导过：

$$Q_{导热}=\lambda\frac{\partial^2 t}{\partial x^2}\mathrm{d}x\mathrm{d}y\mathrm{d}\tau+\lambda\frac{\partial^2 t}{\partial y^2}\mathrm{d}x\mathrm{d}y\mathrm{d}\tau=\lambda\left(\frac{\partial^2 t}{\partial x^2}+\frac{\partial^2 t}{\partial y^2}\right)\mathrm{d}x\mathrm{d}y\mathrm{d}\tau \tag{5.11}$$

Within time $\mathrm{d}\tau$, the heat imported to the micro-element body by heat convection along the x-axis direction is:

在 $\mathrm{d}\tau$ 时间内，沿 x 方向热对流传入微元体的热量为：

$$Q_x=\rho c_p ut\mathrm{d}y\mathrm{d}\tau$$

The heat exported from the micro-element body is:

传出微元体的热量为：

$$Q_{x+\mathrm{d}x}=Q_x+\frac{\partial Q_x}{\partial x}\mathrm{d}x$$

The net heat transferred into the micro-element body by heat convection is equal to subtracting the above two formulas:

热对流传递到微元体的净热量等于上面两式相减：

$$Q_x - Q_{x+dx} = -\frac{\partial Q_x}{\partial x}dx = -\rho c_p \frac{\partial (ut)}{\partial x}dxdyd\tau$$

Similarly, within time dt, the net heat transferred into the micro-element body by heat convection along the y-axis direction is:

同样，dτ 时间内，沿 y 方向热对流传递到微元体的净热量为：

$$Q_y - Q_{y+dy} = -\rho c_p \frac{\partial (vt)}{\partial y}dydxd\tau$$

Therefore, within time dτ, the total heat transferred by heat convection is equal to the sum of the above two formulas. And merge the same items to get:

因此，在 dτ 时间内，热对流传递的总热量等于上面两式相加。将相同的项合并可以得到：

$$Q_{对流} = -\rho c_p \left[u\frac{\partial t}{\partial x} + v\frac{\partial t}{\partial y} + t\frac{\partial u}{\partial x} + t\frac{\partial v}{\partial y}\right]dxdyd\tau$$

According to the continuity equation, in the equation:

根据连续性方程，式中：

$$t\frac{\partial u}{\partial x} + t\frac{\partial v}{\partial y} = 0$$

Then, the total heat transferred by heat convection is:

那么，热对流传递的总热量为：

$$Q_{对流} = -\rho c_p \left[u\frac{\partial t}{\partial x} + v\frac{\partial t}{\partial y}\right]dxdyd\tau \tag{5.12}$$

The net heat conduction quantity and total heat convection quantity are obtained above. And considering the increment of internal energy $\Delta U = \rho c_p dxdy\frac{\partial t}{\partial \tau}d\tau$, substitute them into the law of energy conservation to obtain the energy conservation equation:

以上求出了净导热量、热对流传递的总热量，再考虑到内能增量 $\Delta U = \rho c_p dxdy\frac{\partial t}{\partial \tau}d\tau$，代入能量守恒定律，可得能量守恒方程为：

$$\rho c_p\left(\frac{\partial t}{\partial \tau} + u\frac{\partial t}{\partial x} + v\frac{\partial t}{\partial y}\right) = \lambda\left(\frac{\partial^2 t}{\partial x^2} + \frac{\partial^2 t}{\partial y^2}\right) \tag{5.13}$$

This is the energy conservation equation. Where the first term on the left side of the equal sign is unsteady-state term, the second and third terms are convection terms, and the two items on the right side of the equal sign are thermal diffusion terms. When the velocity of the fluid is 0, Eq. (5.13) can be transformed into a differential equation of heat conduction.

这就是能量守恒方程。式中，等号左侧第一项为非稳态项，第二、三项为对流项，等号右侧两项为热扩散项。当流体的速度为 0 时，式（5.13）可转化为导热微分方程式。

To sum up, for constant physical properties, no internal heat source, two-dimensional, incompressible Newtonian

综上可得，对于常物性、无内热源、二维、不可压缩牛顿流体，对

fluid, the differential equation group of convection heat transfer includes one continuity equation, two momentum equations, and one energy equation.

流传热微分方程组包括一个连续性方程、两个动量方程和一个能量方程。

$$\frac{\partial u}{\partial x}+\frac{\partial v}{\partial y}=0 \tag{5.14}$$

$$\rho\left(\frac{\partial u}{\partial \tau}+u\frac{\partial u}{\partial x}+v\frac{\partial u}{\partial y}\right)=F_x-\frac{\partial p}{\partial x}+\eta\left(\frac{\partial^2 u}{\partial x^2}+\frac{\partial^2 u}{\partial y^2}\right) \tag{5.15}$$

$$\rho\left(\frac{\partial v}{\partial \tau}+u\frac{\partial v}{\partial x}+v\frac{\partial v}{\partial y}\right)=F_y-\frac{\partial p}{\partial y}+\eta\left(\frac{\partial^2 v}{\partial x^2}+\frac{\partial^2 v}{\partial y^2}\right) \tag{5.16}$$

$$\rho c_p\left(\frac{\partial t}{\partial \tau}+u\frac{\partial t}{\partial x}+v\frac{\partial t}{\partial y}\right)=\lambda\left(\frac{\partial^2 t}{\partial x^2}+\frac{\partial^2 t}{\partial y^2}\right) \tag{5.17}$$

The equation group applies to both laminar and turbulent flows of the incompressible viscous fluids. As we know, the complete mathematical formulation of the heat transfer problem includes the differential equation group and the definite conditions. There are also four types of definite conditions: geometric conditions, physical conditions, time condition, and boundary conditions.

该方程组适用于不可压缩黏性流体的层流及湍流流动。我们知道传热问题的完整数学描述包括微分方程组和定解条件。定解条件同样包括四类：几何条件、物理条件、时间条件和边界条件。

Now we focus on the boundary conditions of convection heat transfer, namely, the conditions related to velocity, pressure and temperature, etc. at the boundary. The first type of boundary condition: It is given the wall temperature, to calculate the temperature change rate in the normal direction of the wall. The second type of boundary condition: It is given the heat flux on the wall, namely, the temperature change rate, to calculate the wall temperature. There is no third type of boundary condition for the convective heat transfer problem because the convective heat transfer coefficient h is an unknown quantity.

下面重点讨论一下对流传热的边界条件，即边界上与速度、压力和温度等相关的条件。第一类边界条件：已知壁温，求壁面法向的温度变化率；第二类边界条件：已知壁面上热流密度，也就是温度变化率，求壁温。对流换热问题没有第三类边界条件，因为表面传热系数 h 是一个待求量。

There are four equations in the equation group, corresponding to four unknown quantities, the velocity field (u,v), temperature field (t), and pressure field (p) can be obtained theoretically. However, for practical convective heat transfer problems, it is still difficult to solve the above equation group mathematically. After the above four

方程组有 4 个方程，对应有 4 个未知量，从理论上可求得速度场 (u,v) 和温度场 (t) 以及压力场 (p)。但是，针对实际对流传热问题，数学上求解上述方程组仍然很困难。利用上述 4 个方程及定解条件求出

equations and the definite conditions are used to obtain the temperature field, the local convective heat transfer coefficient can be calculated by using the differential equation of the convective heat transfer process.

温度场之后，用对流传热过程微分方程式就可计算出当地的表面传热系数。

Summary This section illustrates the derivation of the differential equation group of convection heat transfer, and gives the definite conditions for the convective heat transfer process.

小结 本节阐述了对流传热微分方程组的推导过程，并给出了对流传热过程的定解条件。

Question What is the difference between the boundary conditions of convection heat transfer and heat conduction?

思考题 对流传热问题与导热问题的边界条件有什么区别？

Self-tests

自测题

1. Which conservation equations and variables does the differential equation group of convection heat transfer include?

1. 对流传热微分方程组包括哪些守恒方程式？哪几个变量？

2. Under what conditions can the energy conservation equation of convective heat transfer be transformed into the differential equation of heat conduction?

2. 对流传热能量守恒方程在什么条件下可转化为导热微分方程式？

授课视频

5.4 The Boundary Layer

5.4 边界层

In 1904, Prandtl discovered that when the viscous fluid flows over the surface of an object, a velocity boundary layer with a large velocity gradient will be formed. In 1921, Polhausen discovered that when there is a temperature difference between the fluid and the wall, a temperature boundary layer with a large temperature gradient will also be produced. Two boundary layers are defined as follows.

1904 年，普朗特发现，当黏性流体流过物体表面时，会形成速度梯度很大的速度边界层。1921 年，波尔豪森发现，当流体与壁面之间有温差时，也会产生温度梯度很大的温度边界层。下面给出两个边界层的定义。

A thin layer where the fluid velocity changes drastically near the solid surface is called the velocity boundary layer,

在固体表面附近流体速度发生剧烈变化的薄层，称为速度边界层，或

or called the flow boundary layer. Its physical meaning is that the viscous effect of the fluid is just restricted to the thin layer of fluid near the wall.

称为流动边界层。其物理意义是，流体的黏滞作用仅局限于靠近壁面的薄层流体内。

The characteristics of fluid velocity distribution near the wall: At $y=0$, that is, on the wall, the fluid velocity $u=0$; As the y value increases, the velocity u increases rapidly; At $y=\delta$, the velocity u is close to the main flow velocity. Then, the thin fluid layer of $y=\delta$ is called the velocity boundary layer. The distance from site of $u/u_\infty=99\%$ to the wall is defined as the boundary layer thickness δ.

流体在壁面附近速度分布的特点是：$y=0$ 时，即壁面上，流体速度 $u=0$；随着 y 值增大，速度 u 迅速增大；在 $y=\delta$ 处，速度 u 接近主流速度。那么，$y=\delta$ 的流体薄层，称为速度边界层。$u/u_\infty=99\%$处离壁面的距离定义为边界层厚度 δ。

In the boundary layer, the average velocity gradient is large, and the velocity gradient is maximum at $y=0$. As shown in Fig. 5. 5, the boundary layer thickness δ is small compared to the plate length L. It can be seen from the figure when the air externally flows through a plane plate at a velocity of 10m/s, the velocity boundary layer thickness at different locations is different. For example, $\delta_{x=100\text{mm}}=1.8\text{mm}$; $\delta_{x=200\text{mm}}=2.5\text{mm}$.

在边界层内，平均速度梯度很大，$y=0$ 处速度梯度最大。如图 5. 5 所示，边界层厚度 δ 与板长 L 相比很小。从图中可以看出，当空气以速度 10m/s 外掠平板时，不同位置的速度边界层厚度不同。例如，$\delta_{x=100\text{mm}}=1.8\text{mm}$、$\delta_{x=200\text{mm}}=2.5\text{mm}$。

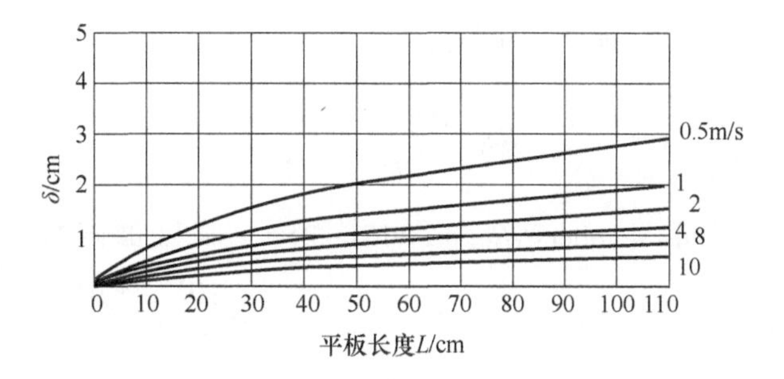

Fig. 5. 5 Boundary layer thickness versus plate length as air flows along a plate

图 5. 5 空气沿平板流动时边界层厚度与板长的关系

It can be seen that the velocity boundary layer thickness gradually increases with the extension of fluid flow. But as the fluid velocity increases, the boundary layer thickness gradually decreases. The variation of fluid flow state and convective heat transfer coefficient in the boundary layer is shown in Fig. 5. 6.

可见，随着流体流程的延伸，速度边界层厚度逐渐增大。但是随着流体的速度增大，边界层厚度逐渐减小。边界层内部流动状态及表面传热系数的变化情况如图 5. 6 所示。

When the fluid externally flows through a plane plate, with the development of the velocity boundary layer, the

当流体外掠平板时，随着速度边界层的发展，边界层内会由层流逐渐

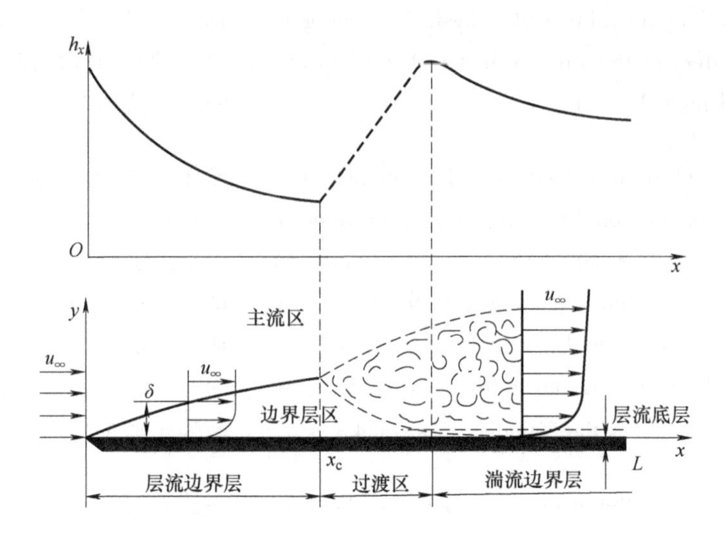

Fig. 5.6 The variation of fluid flow state and convective heat transfer coefficient in the boundary layer

图 5.6 边界层内部流动状态及表面传热系数的变化情况

laminar flow gradually becomes turbulent flow in the boundary layer. The distance from the point of transition from laminar flow to turbulent flow to the starting point is called the critical distance, written as x_c, and usually denoted by the corresponding critical Reynolds number:

变为湍流。由层流开始向湍流过渡的点到起始点的距离称为临界距离，记作 x_c，通常用对应的临界雷诺数表示：

$$Re_c = \frac{\rho u_\infty x_c}{\eta} = \frac{u_\infty x_c}{\nu} \tag{5.18}$$

The range of critical Reynolds number: $Re_c = 2\times10^5 \sim 3\times10^6$, take $Re_c = 5\times10^5$.

临界雷诺数的范围是：$Re_c = 2\times10^5 \sim 3\times10^6$，取 $Re_c = 5\times10^5$。

In the turbulent boundary layer there exists a laminar bottom layer, also called a viscous bottom layer. The viscous force is still dominant near the wall, so that the very thin layer of fluid adjacent to the wall maintains laminar flow characteristics and has the maximum velocity gradient. Inside the boundary layer, the fluid satisfies Newton's law of viscosity:

在湍流边界层内存在层流底层，也称为黏性底层。在紧靠壁面处，黏滞力仍占绝对优势，使紧邻壁面的极薄层流体保持层流特征，并具有最大的速度梯度。边界层内，流体满足牛顿黏性定律：

$$\tau = \eta \frac{\partial u}{\partial y} \tag{5.19}$$

Outside the boundary layer, the fluid velocity is basically constant in the y-direction, and its velocity gradient is 0,

边界层外，流体速度在 y 方向基本不变，速度梯度为 0，因此黏滞力

so, the viscous force is zero, it is called the main flow region. In other words, the boundary layer divides the flow field, i. e. velocity field, into two regions, boundary layer region and the main flow region.

为零，称为主流区。也就是说，边界层把流场（速度场）分成边界层区和主流区两个区。

In the boundary layer region, the viscosity of the fluid is dominant, and fluid flow can be described by the differential equation of viscous fluid motion, i. e. N-S equation. In the main flow region, the velocity gradient is 0, and the viscous shear stress $\tau=0$, and the fluid can be regarded as an ideal fluid without viscosity, described with the Euler equation.

在边界层区，流体的黏性起主导作用，流体流动可用黏性流体运动微分方程，即 N-S 方程描述。在主流区，速度梯度为 0，黏性切应力 $\tau=0$，流体可视为无黏性的理想流体，用欧拉方程描述。

The thin layer near the solid surface where the fluid temperature changes drastically is called the temperature boundary layer, or thermal boundary layer.

固体表面附近流体温度发生剧烈变化的薄层称为温度边界层，或热边界层。

The development trend of the temperature boundary layer, i. e. thermal boundary layer, is similar to the velocity boundary layer basically. The thermal boundary layer can be divided into laminar thermal boundary layer and turbulent thermal boundary layer, as shown in Fig. 5. 7.

温度边界层（热边界层）的发展趋势基本与速度边界层相似。热边界层可分为层流热边界层、湍流热边界层，如图 5. 7 所示。

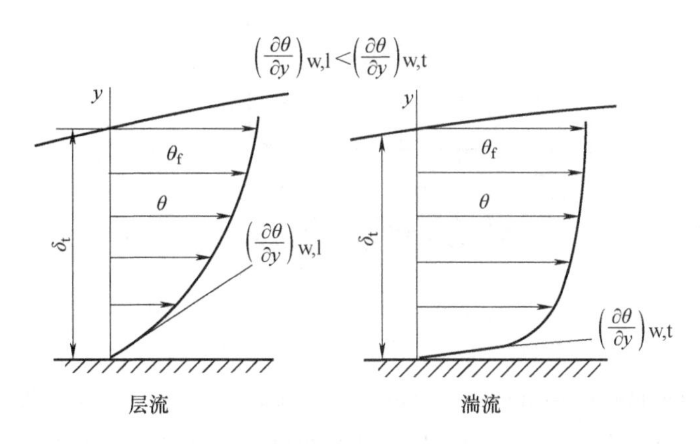

Fig. 5. 7 Laminar thermal boundary layer and turbulent thermal boundary layer

图 5. 7 层流热边界层和湍流热边界层

$$y=0,\quad \theta_w=T-T_w=0$$

$$y=\delta_t,\quad \theta=T-T_w=0.99\theta_\infty$$

$$\theta_\infty=T_\infty-T_w$$

The temperature distribution of the laminar thermal boundary layer is parabolic, and the temperature distribution of the turbulent thermal boundary layer is a power function. The temperature gradient of the laminar boundary layer adjacent to the wall is significantly smaller than that of the turbulent boundary layer. So, turbulent heat transfer is stronger than laminar heat transfer.

层流热边界层温度分布呈抛物线形式，湍流热边界层温度分布呈幂函数形式。层流边界层贴壁处的温度梯度明显小于湍流边界层。所以，湍流传热比层流传热的强度更大。

According to the above analysis, we can obtain the following conclusion: The conditions of the flow boundary layer and thermal boundary layer determine the temperature distribution and heat transfer in the boundary layer. The relationship between δ and δ_t is $\frac{\delta_t}{\delta} \approx Pr^{1/3}$, as shown in Fig. 5. 8. Laminar flow: $0.36 \leqslant Pr \leqslant 50$. Prandtl number is a characteristic number, $Pr = \frac{v}{a} = \frac{c_p \eta}{\lambda}$, which characterizes the relative thickness of the flow boundary layer and the thermal boundary layer, and also reflects the ability of momentum diffusion and heat diffusion in the fluid.

通过以上分析，可得出结论：流动边界层与热边界层的状况决定了边界层内的温度分布和热量传递。δ 与 δ_t 的关系：$\frac{\delta_t}{\delta} \approx Pr^{\frac{1}{3}}$，如图 5. 8 所示。层流时：$0.36 \leqslant Pr \leqslant 50$。普朗特数是一个特征数，$Pr = \frac{v}{a} = \frac{c_p \eta}{\lambda}$。它表征了流动边界层与热边界层的相对厚度，也反映了流体中动量扩散与热量扩散的能力。

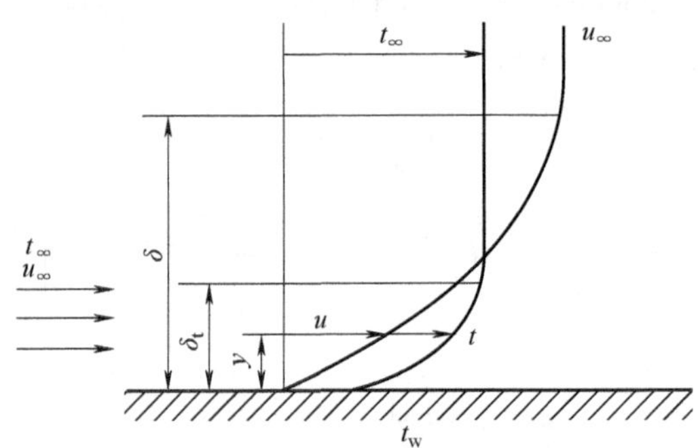

Fig. 5. 8 Velocity boundary layer and temperature boundary layer

图 5. 8 速度边界层和温度边界层

Summary This section illustrates the definition and characteristics of the boundary layer, and the relationship between the flow boundary layer and the thermal boundary layer, and defines a new dimensionless number, i. e. Prandtl number.

小结 本节阐述了边界层的定义和特性，以及流动边界层与热边界层的关系，定义了一个新的无量纲数，即普朗特数。

Question What are the differences between the development states of the fluid velocity boundary layer outside and inside the tube?

思考题 管外与管内流体速度边界层发展状态有何区别？

Self-tests

1. What are the velocity boundary layer and thermal boundary layer? What is the relationship between them?

2. What are the characteristics of the flow boundary layer? What is its function?

自测题

1. 什么是速度边界层？什么是热边界层？两者之间是什么关系？

2. 流动边界层有什么特性？其作用是什么？

授课视频

5.5 The Differential Equation Group of Heat Transfer in Laminar Boundary Layer

5.5 层流边界层传热微分方程组

The incompressible, constant physical property, two-dimensional differential equation group of steady-state convection heat transfer is difficult to solve, so the equation group needs to be simplified. Here, we will adopt the order of magnitude analysis method to simplify.

不可压缩、常物性、二维稳态对流传热微分方程组难于求解，还需要再简化。下面采用数量级分析法进行简化。

The order of magnitude analysis method refers to comparing the relative magnitude of each item in the equation, retaining the terms of large magnitude, and eliminating the terms of small magnitude, so that the equation can be reasonably simplified. Here, $O(1)$ means the order of magnitude is 1, and $O(\delta)$ means the order of magnitude is δ. Note that δ in brackets is a mathematical symbol that represents a number much less than 1.

数量级分析法：通过比较方程式中各项的数量级相对大小，保留数量级大的项，舍去数量级小的项，从而使方程式得到合理简化。这里，用 $O(1)$ 表示数量级为 1，用 $O(\delta)$ 表示数量级为 δ。注意，括号中的 δ 是一个数学符号，代表一个远小于 1 的数。

There are five basic quantities related to the boundary layer, and their magnitudes are: main flow direction coordinates $x \sim O(1)$, coordinates $y \sim O(\delta)$, main flow velocity $u \sim O(1)$, coordinates y direction velocity $v \sim O(\delta)$, and fluid temperature $t \sim O(1)$. The magnitude of the thickness of the velocity/temperature boundary layer is $O(\delta)$.

边界层相关的基本量有 5 个，它们的数量级分别是：主流方向坐标 $x \sim O(1)$、坐标 $y \sim O(\delta)$，主流速度 $u \sim O(1)$，坐标 y 方向速度 $v \sim O(\delta)$，流体温度 $t \sim O(1)$。速度/温度边界层厚度的数量级均为 $O(\delta)$。

The order of magnitude of the momentum equation is analyzed as follows. According to the above rules, list the order of magnitude of each physical quantity and keep the

下面进行动量方程的数量级分析。按照上面的规则，列出每个物理量的数量级，并注意等式两边的数量

same order of magnitude on both sides of the equation. First look at the momentum equation in the x direction. Both the order of magnitude of u and x in the first term on the left of the equation is 1, then the order of magnitude of this term is 1.

级一致。先看 x 方向的动量方程。等式左侧第一项中 u、x 的数量级均为 1，则这一项的数量级为 1。

The order of magnitude of v and y in the second term is δ, δ in the numerator and the denominator cancels out, then the order of magnitude of this term is also 1. Thus the order of magnitude of the density ρ is also 1. Therefore, the order of magnitude of the left term of the equation is 1.

第二项中，v、y 数量级为 δ，分子与分母的 δ 相约，则这一项的数量级也是 1，因而，密度 ρ 的数量级也是 1，所以等式左侧的数量级为 1。

$$\rho\left(u\frac{\partial u}{\partial x}+v\frac{\partial u}{\partial y}\right)=-\frac{\partial p}{\partial x}+\eta\left(\frac{\partial^2 u}{\partial x^2}+\frac{\partial^2 u}{\partial y^2}\right) \tag{5.20}$$

On the right of the equation, in the first term in the brackets, the order of magnitude of u and x is 1, then the order of magnitude of the first term is 1 temporarily, the order of magnitude of the second term y squared is δ squared and the order of magnitude of the second term is one over δ squared. To keep the same order of magnitude on both sides of the equation, the order of magnitude of the viscosity must be δ squared. Thus, the order of magnitude of the first term in brackets becomes δ squared, so this term can be neglected. According to the rule that it's the same order of magnitude on both sides of the equation, the order of magnitude of $\frac{\partial p}{\partial x}$ should be 1.

等式右侧，括号中第一项，u、x 的数量级为 1，则第一项的数量级暂时为 1，第二项 y^2 的数量级为 δ^2，第二项的数量级是 $1/\delta^2$。要使等式两边数量级一致，黏度的数量级必须为 δ^2。因而，括号中第一项的数量级就变成了 δ^2，所以这一项可忽略。按照“等式两边数量级一致”来判断，$\frac{\partial p}{\partial x}$的数量级应该为 1。

By the same method, analyze the momentum equation in the y-direction. For the first term on the left of the equation, the order of magnitude of u and x is 1 and the order of magnitude of v is δ, so the order of magnitude of the first term is δ. The order of magnitude of v and y in the second term is δ, then the order of magnitude of this term is also δ. Therefore, the order of magnitude of the left term of the equation is δ.

同样方法，分析 y 方向的动量方程：等式左侧第一项中，u、x 的数量级均为 1，v 的数量级为 δ，则第一项的数量级为 δ。第二项中，v、y 的数量级均为 δ，则这一项的数量级也是 δ。因此，等式左侧的数量级为 δ。

$$\rho\left(u\frac{\partial v}{\partial x}+v\frac{\partial v}{\partial y}\right)=-\frac{\partial p}{\partial y}+\eta\left(\frac{\partial^2 v}{\partial x^2}+\frac{\partial^2 v}{\partial y^2}\right) \tag{5.21}$$

On the right of the equation, the order of magnitude of the viscosity is δ squared, times the two items in brackets, so the magnitude is δ. According to the rule that it's the same order of magnitude on both sides of the equation, the order of magnitude of $\frac{\partial p}{\partial y}$ should also be δ. The order of magnitude of the left of the equation is δ, and the order of magnitude of the right is also δ. Therefore, the entire equation can be neglected.

等式右侧，黏度的数量级为 δ^2，乘上括号中两项，则数量级均为 δ。按照“等式两边数量级一致”来判断，$\frac{\partial p}{\partial y}$的数量级应该也为 δ；等式左侧的数量级为 δ，右侧的数量级也为 δ。因此整个式子可忽略。

From the above magnitude analysis, we can know $\left(\frac{\partial p}{\partial x}\sim O(1), \frac{\partial p}{\partial y}\sim O(\delta)\right)$, these two formulas show that the pressure gradient changes only along the x direction in the boundary layer, and the normal pressure gradient is extremely small in the boundary layer. So, the velocity u in the x direction is much greater than the velocity v in the y direction. The pressure at any section in the boundary layer is independent of the y coordinate and equal to the main flow pressure. The momentum equation of the boundary layer is simplified as follows:

通过上述数量级分析可知，$\left[\frac{\partial p}{\partial x}\sim O(1), \frac{\partial p}{\partial y}\sim O(\delta)\right]$这两个式子表明，边界层内的压力梯度仅沿 x 方向变化，边界层内法向的压力梯度极小。因此，x 方向的速度 u 远大于 y 方向的速度 v。边界层内任一截面上的压力与 y 坐标无关，而等于主流压力。边界层动量方程简化为：

$$\rho\left(u\frac{\partial u}{\partial x}+v\frac{\partial u}{\partial y}\right)=-\frac{\mathrm{d}p}{\mathrm{d}x}+\eta\frac{\partial^2 u}{\partial y^2} \tag{5.22}$$

By the same method, analyze the order of magnitude of the energy equation. Listing the order of magnitude of each physical quantity, it can be seen that the order of magnitude of the first item on the right of the equation is δ squared, which can be neglected. The order of magnitude of the other items is 1. The boundary layer energy equation can be simplified as:

同样方法，分析能量方程的数量级。列出各个物理量的数量级以后，可见，等式右侧第一项的数量级为 δ^2，可忽略。其他项的数量级均为 1。边界层能量方程可简化为：

$$\rho c_p\left(u\frac{\partial t}{\partial x}+v\frac{\partial t}{\partial y}\right)=\lambda\frac{\partial^2 t}{\partial y^2} \tag{5.23}$$

The differential equation group of convection heat transfer in the laminar boundary layer ends up with three simplified equations as follows:

层流边界层对流传热微分方程组最终剩下三个简化的方程：

Continuity equation:

连续方程：

$$\frac{\partial u}{\partial x}+\frac{\partial v}{\partial y}=0$$

Momentum equation:

动量方程:

$$u\frac{\partial u}{\partial x}+v\frac{\partial u}{\partial y}=-\frac{1}{\rho}\frac{\mathrm{d}p}{\mathrm{d}x}+\nu\frac{\partial^2 u}{\partial y^2}$$

Energy equation:

能量方程:

$$u\frac{\partial t}{\partial x}+v\frac{\partial t}{\partial y}=a\frac{\partial^2 t}{\partial y^2}$$

According to Bernoulli equation of the ideal fluid outside the boundary layer we can obtain:

由边界层外理想流体的伯努利方程可求出:

$$-\frac{\mathrm{d}p}{\mathrm{d}x}=\rho u_\infty\frac{\mathrm{d}u}{\mathrm{d}x} \tag{5.24}$$

3 equations, and 3 unknown quantities u, v and t, the equation group is closed, and the equation group can be solved by adding the definite conditions.

3个方程，3个未知量u、v、t，方程组封闭，加上定解条件即可求解。

Summary This section illustrates the simplified method for the order of magnitude analysis for the differential equation group of convection heat transfer, and obtains the differential equation group of heat transfer in laminar boundary layer.

小结 本节阐述了对流传热微分方程组进行数量级分析的简化方法，得到了层流边界层传热微分方程组。

Question What are the fundamental characteristics of the boundary layer type flow and heat transfer?

思考题 边界层型流动与传热的根本特点是什么?

Warm prompt: Which coordinate direction is the viscous force and heat conduction neglected, namely, the second derivative of this direction is neglected.

温馨提示：忽略了哪个坐标方向的黏性力与导热，即略去了该方向的二阶导数。

Self-tests

自测题

1. How to simplify the differential equation group for convection heat transfer using boundary layer theory?

1. 如何使用边界层理论简化对流传热微分方程组?

2. Briefly describe the basic idea of the order of magnitude analysis method.

2. 简述数量级分析方法的基本思想。

5.6 Local Convective Heat Transfer Coefficient in Laminar Boundary Layer

5.6 层流边界层局部表面传热系数

Take the laminar heat transfer of fluid externally flowing through the isothermal plane plate as an example. For the main flow field with constant velocity and uniform temperature, in the case of constant wall temperature, the heat transfer of the fluid longitudinally flowing through the plane plate, the boundary conditions of the boundary layer are as follows:

以流体外掠等温平板层流传热为例。对于主流场匀速、均温，在恒定壁温情况下流体纵掠平板传热，边界层的边界条件：

$$y=0\text{时}, u=0, v=0, t=t_w$$

$$y=\delta\text{时}, u=u_\infty, t=t_\infty$$

Solve the differential equation group of convection heat transfer in laminar boundary layer to obtain the dimensionless temperature distribution in the temperature boundary layer:

求解层流边界层对流传热微分方程组，可得温度边界层内无量纲温度分布：

$$\frac{\theta}{\theta_0}=\frac{t-t_w}{t_\infty-t_w}=\frac{3}{2}\frac{y}{\delta_t}-\frac{1}{2}\left(\frac{y}{\delta_t}\right)^3 \tag{5.25}$$

The solution process is rather complicated, no more details here. The expression of local convective heat transfer coefficient in laminar boundary layer can also be obtained:

求解过程比较繁琐，这里不再详述。还可得到层流边界层局部表面传热系数的表达式为：

$$\frac{h_x x}{\lambda}=0.332\left(\frac{u_\infty x}{\nu}\right)^{\frac{1}{2}}\left(\frac{\nu}{a}\right)^{\frac{1}{3}} \tag{5.26}$$

Eq. (5.26) can be denoted by characteristic numbers:

式（5.26）可用特征数表示：

$$Nu_x=0.332Re_x^{1/2}Pr^{1/3} \tag{5.27}$$

The equation is called the characteristic number equation or criterion equation. It is necessary to pay attention to the applicable conditions of the criterion equation: Fluid externally flowing through an isothermal plane plate without an internal heat source, steady state, and laminar flow. In this formula, Nusselt number:

该方程称为特征数方程或准则方程。需注意准则方程的适用条件：流体外掠等温平板、无内热源、稳态、层流。式中努塞尔数：

$$Nu_x = \frac{h_x x}{\lambda}$$

Reynolds number:

雷诺数:

$$Re_x = \frac{u_\infty x}{\nu}$$

Prandtl number:

普朗特数:

$$Pr = \frac{\nu}{a}$$

Where the subscript x of the characteristic number denotes that the characteristic number is a local value, or the characteristic length is the local coordinate.

式中,特征数的下标 x 表示该特征数为局部值,或特征长度为当地坐标。

According to the formula of local convective heat transfer coefficient, integrate x to obtain the average of the convective heat transfer coefficient on the entire plane plate with length of L:

由局部表面传热系数的计算式对 x 积分可得整个平板长度 L 的表面传热系数平均值:

$$\overline{Nu} = \frac{\bar{h}L}{\lambda} = 0.664 Re^{\frac{1}{2}} Pr^{\frac{1}{3}} \tag{5.28}$$

The thickness of velocity boundary layer at x away from the front edge:

离平板前缘 x 处的速度边界层厚度为:

$$\frac{\delta}{x} = \frac{5.0}{\sqrt{Re_x}} \tag{5.29}$$

The ratio of the thickness of flow boundary layer to thermal boundary layer:

流动边界层与热边界层厚度之比:

$$\frac{\delta}{\delta_t} \approx Pr^{\frac{1}{3}} \tag{5.30}$$

The velocity distribution in the boundary layer:

边界层内的速度分布:

$$\frac{u}{u_\infty} = \frac{3}{2}\frac{y}{\delta} - \frac{1}{2}\left(\frac{y}{\delta}\right)^3 \tag{5.31}$$

The local shear stress at the wall surface x is:

壁面 x 处局部切应力为:

$$\tau_w = \eta \left.\frac{\mathrm{d}u}{\mathrm{d}y}\right|_{y=0} = \frac{0.332\rho u_\infty^2}{\sqrt{Re_x}} \tag{5.32}$$

In engineering, the ratio of local shear stress to fluid dynamic head is often used, called Fanning friction coefficient, which is abbreviated as friction coefficient, and it is a dimensionless parameter:

在工程中,常使用局部切应力与流体动压头之比,称为范宁摩擦系数,简称摩擦系数,是一个无量纲量:

$$c_f = \frac{\tau_w}{\frac{1}{2}\rho u_\infty^2} = \frac{0.664}{\sqrt{Re_x}} \tag{5.33}$$

When applying the above formula, we should pay attention to the following five points: ① Prandtl number is greater than or equal to 1; ② Note the difference between two pairs of variables (i. e. local, average); ③ Don't get confused when picking x and L; ④ The above formula is only applicable to the laminar boundary layer, namely, the critical Reynolds number $Re_c \leqslant 5\times10^5$; ⑤ Reference temperature of the fluid in the boundary layer $t=(t_\infty+t_w)/2$.

应用以上公式计算时，要注意以下五点：① $Pr\geqslant1$；② 注意两对变量的差别（局部、平均）；③ x 与 L 的选取不要混淆；④ 上述公式仅适用于层流边界层，即临界雷诺数 $Re_c \leqslant 5\times10^5$；⑤ 边界层中流体的定性温度 $t=(t_\infty+t_w)/2$。

The reference temperature is used to determine the fluid physical parameters in the characteristic numbers. The experimental results of laminar convection heat transfer of air longitudinally flowing through a plane plate are shown in Fig. 5. 9. It can be seen from the figure that the boundary layer is in a laminar flow state when $Re\leqslant2\times10^5$, the criterion correlation of Nusselt number is in good agreement with the experimental data. When the Reynolds number exceeds the critical value, the error of the calculation result is large.

定性温度：用以确定特征数中流体物性参数的温度。空气纵掠平板层流对流传热的实验结果如图 5. 9 所示。由图可见，当 $Re\leqslant2\times10^5$ 时，边界层处于层流状态，Nu 准则数关联式与实验数据吻合较好。超过该 Re 时，计算结果误差较大。

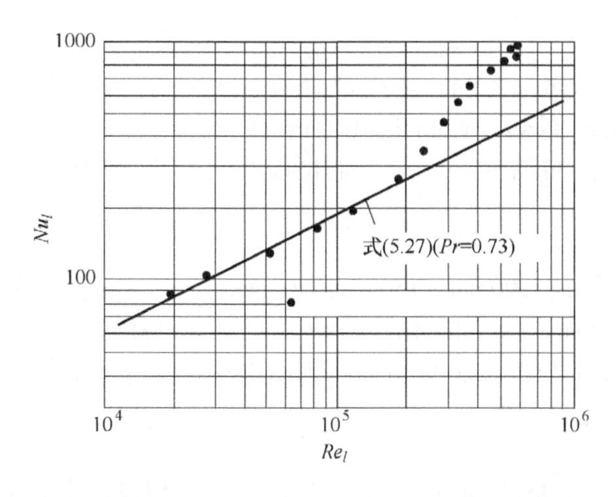

Fig. 5. 9 Experimental results of laminar convection heat transfer of air longitudinally flowing through a plane plate

图 5. 9 空气纵掠平板层流对流传热的实验结果

Summary　This section solves the differential equation group of convection heat transfer in the laminar boundary layer to obtain the temperature and velocity distributions in the boundary layer, and the calculation formula and criterion equation of local convective heat transfer coefficient in the laminar boundary layer.

小结　本节通过求解层流边界层对流传热微分方程组，得到了边界层内的温度分布和速度分布，以及层流边界层局部表面传热系数的计算公式和准则方程式。

Question　Why does the convective heat transfer coefficient of the laminar boundary layer depend on the thickness of the velocity boundary layer?

思考题　为什么层流边界层表面传热系数取决于速度边界层厚度？

Tips: The thickness of the velocity boundary layer depends on the velocity and physical parameters of the fluid.

提示：速度边界层厚度取决于流体的速度和物性参数。

Self-test　What are the definition and physical meaning of Nusselt number?

自测题　*Nu* 的定义和物理意义是什么？

授课视频

5.7　The Calculation Examples of Heat Transfer in Laminar Boundary Layer

5.7　层流边界层传热计算实例

Example 5.1　The air with a temperature of 20℃ and one atmospheric pressure flows longitudinally through a 320mm long plane plate with a temperature of 40℃, its velocity is 10m/s.

例 5.1　压力为大气压的 20℃ 空气，纵向掠过一块长 320mm、温度为 40℃ 的平板，流速为 10m/s。

At 7 points that are 50mm, 100mm, 150mm, 200mm, 250mm, 300mm and 320mm away from the front edge of the plane plate, find the thickness of the velocity boundary layer and the thermal boundary layer.

求离平板前缘 50mm、100mm、150mm、200mm、250mm、300mm、320mm 七个点处，流动边界层和热边界层的厚度。

Suppose that the flow is in steady state.

假设：流动为稳态。

Calculation is as follows: The physical parameters of air are calculated according to the average of the surface tem-

下面进行计算：空气的物性参数按照板的表面温度和空气来流温度的

perature of the plate and the temperature of the air incoming flow, namely 30℃. Kinematic viscosity and Pr number of air at 30℃ are $\nu=16\times10^{-6}\text{m}^2/\text{s}$, $Pr=0.701$.

平均值计算，即 30℃。30℃ 时空气的运动黏度 $\nu=16\times10^{-6}\text{m}^2/\text{s}$ 和 $Pr=0.701$。

For a 320mm long plane plate,

对长 320mm 的平板，

$$Re=\frac{uL}{\nu}=\frac{10\text{m/s}\times0.32\text{m}}{16\times10^{-6}\text{m}^2/\text{s}}=2\times10^5<5\times10^5$$

Therefore, it belongs to the laminar boundary layer, and the analytical solution of the laminar boundary layer can be used for calculation. The thickness of the velocity boundary layer is calculated according to the following formula:

因此，属于层流边界层，可用层流边界层的分析解进行计算。其流动边界层厚度按下式计算：

$$\frac{\delta}{x}=\frac{5.0}{\sqrt{Re_x}}$$

As a result:

由此可得：

$$\delta=\frac{5.0x}{\sqrt{\frac{u_\infty x}{\nu}}}=5.0\sqrt{\frac{\nu x}{u_\infty}}=5.0\times\sqrt{\frac{16\times10^{-6}\text{m}^2/\text{s}}{10\text{m/s}}}\sqrt{x}=6.36\times10^{-3}\sqrt{x}\,\text{m}^{\frac{1}{2}}$$

Note that the units of x and δ are m.

注意：x 和 δ 的单位均为 m。

The thickness of the thermal boundary layer can be calculated according to the following formula:

热边界层厚度可按下式计算：

$$\frac{\delta}{\delta_t}\approx Pr^{\frac{1}{3}},\quad \delta_t=\frac{\delta}{\sqrt[3]{Pr}}=\frac{\delta}{\sqrt[3]{0.701}}=1.13\delta$$

The calculation results of δ and δ_t are shown in Fig. 5.10.

δ 及 δ_t 的计算结果如图 5.10 所示。

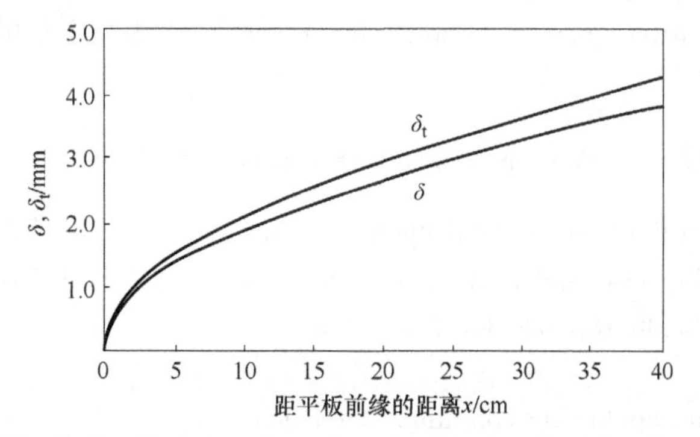

Fig. 5.10 Variation of δ and δ_t along the length of the plate

图 5.10 δ 和 δ_t 沿平板长度的变化

It can be seen from Fig. 5.10 that the thermal boundary layer of air on the plane plate is slightly thicker than the velocity boundary layer.

由图 5.10 可见，平板上空气的热边界层略厚于流动边界层。

Example 5.2 In the above example, the other conditions remain unchanged, if the width of the plane plate is 1m, calculate the heat transfer quantity between the plane plate and the air.

例 5.2 上例中，其他条件不变，如果平板的宽度为 1m，求平板与空气的传热量。

Analysis: Convection heat transfer must occur between the plane plate and the air, so the heat transfer should be calculated by Newton cooling formula.

分析一下：平板与空气之间必然是对流传热，需用牛顿冷却公式计算。

$$\Phi = hA(t_w - t_f)$$

Suppose that it is a steady-state heat transfer process and the radiation heat dissipation of the plane plate is ignored.

假设：为稳态传热过程，并且不计平板的辐射散热。

According to the analytical solution of laminar heat transfer of fluid externally flowing through the isothermal plane plate:

由流体外掠等温平板层流传热的分析解：

$$Nu = 0.664Re^{1/2}Pr^{1/3} = 0.664 \times (2.0 \times 10^5)^{1/2} \times 0.701^{1/3} = 263.7$$

Then, the average convective heat transfer coefficient of the plane plate can be obtained:

由此，可求出平板的平均表面传热系数为：

$$\bar{h} = \frac{\lambda}{L}Nu = \frac{2.67 \times 10^{-2}}{0.32} \times 263.7\text{W/(m}^2 \cdot \text{K)} = 22.0\text{W/(m}^2 \cdot \text{K)}$$

Where λ is the thermal conductivity of air at 30℃.

式中，λ 是 30℃时空气的导热系数。

The convective heat transfer quantity between the plane plate and the air is:

平板与空气的对流传热量为：

$$\Phi = \bar{h}A\Delta t = 22.0\text{W/(m}^2 \cdot \text{K)} \times 1\text{m} \times 0.32\text{m} \times (40℃ - 20℃) = 140.8\text{W}$$

Summary This section illustrates the calculation method of the boundary layer thickness and heat transfer quantity of fluid flowing longitudinally through the plane plate.

小结 本节阐述了流体纵掠平板的边界层厚度和传热量的计算方法。

Question What are the applicable conditions for calculating the local convective heat transfer coefficient of the laminar boundary layer?

思考题 层流边界层局部表面传热系数计算式的适用条件是什么？

Exercises

习题

5.1 For oil, air, and liquid metals, $Pr \gg 1$, $Pr \approx 1$ and $Pr \ll 1$. Regarding the laminar flow of fluid sweeping over an isothermal plate, try to draw rough images of the velocity and temperature distributions in the boundary layer of the three fluids (To be able to show the relative magnitude of δ and δ_t).

5.1 对于油、空气及液态金属，分别有 $Pr \gg 1$、$Pr \approx 1$、$Pr \ll 1$。试就流体外掠等温平板的层流流动，画出三种流体边界层中速度分布和温度分布的大致图像（要能显示出 δ 与 δ_t 的相对大小）。

5.2 The fluid carries out fully developed laminar convective heat transfer between two parallel plates (Fig. 5.11). Try to draw the temperature distribution curves of the fluid on the cross section of the fully developed area under the following three conditions: (1) $q_{w1}=q_{w2}$; (2) $q_{w1}=2q_{w2}$; (3) $q_{w1}=0$。

5.2 流体在两平行平板间进行充分发展的层流对流传热（图 5.11）。试画出下列三种情形下充分发展区域截面上的流体温度分布曲线：(1) $q_{w1}=q_{w2}$；(2) $q_{w1}=2q_{w2}$；(3) $q_{w1}=0$。

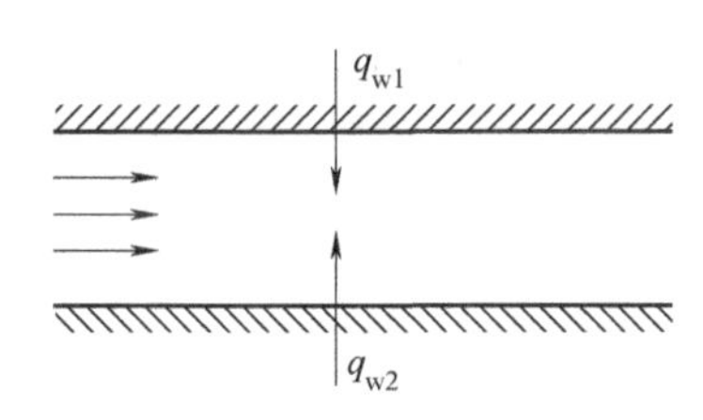

Fig. 5.11 Figures for exercise 5.2

图 5.11 习题 5.2 图

5.3 A plate with a temperature of 80℃ was placed in the air flow with an inflow temperature of 20℃. Assuming that the temperature gradient at a point on the surface of the plate in the direction perpendicular to the wall is 40℃/mm, try to calculate the heat flux at that point.

5.3 温度为 80℃ 的平板置于来流温度为 20℃ 的气流中。假设平板表面某点在垂直于壁面方向的温度梯度为 40℃/mm，试确定该处的热流密度。

5.4 The water at 20℃ flows through a flat plate in parallel with a velocity of 2m/s, and the velocity in the boundary layer is a cubic polynomial distribution. Try to calculate the thickness of the flow boundary layer at 10cm and 20cm away from the front edge of the plate and the mass flow rate of the fluid in the boundary layer on the two sections (Based on the unit width perpendicular to the flow direction).

5.4 20℃ 的水以 2m/s 的流速平行地流过一块平板，边界层内的流速为三次多项式分布。试计算离开平板前缘 10cm 及 20cm 处的流动边界层厚度及两截面上边界层内流体的质量流量（以垂直于流动方向的单位宽度计）。

5.5 The air of 1.013×10^5Pa and 100℃ flows through a plate at the speed of 100m/s, and the plate temperature

5.5 1.013×10^5Pa、100℃ 的空气以 100m/s 的速度流过一块平板，平

is 30℃. Try to calculate the normal velocity, thicknesses of flow boundary layer and thermal boundary layer, local shear stress and local surface heat transfer coefficient, average drag coefficient, and average surface heat transfer coefficient on the outer boundary of the boundary layer at 3cm and 6cm away from the front edge of the plate.

板温度为 30℃。试计算离开平板前缘 3cm 及 6cm 处边界层外边界上的法向速度、流动边界层及热边界层厚度、局部切应力和局部表面传热系数、平均阻力系数和平均表面传热系数。

5.6 An aircraft flies at an altitude of 10000m at a speed of 600km/h. The temperature is -40℃. Think of the wing as a flat plate, try to determine how far away from the leading edge of the wing is the flow of air fully developed turbulence? The air is treated as dry air.

5.6 一架飞机在 10000m 高空飞行，时速为 600km/h，该处温度为-40℃。把机翼当成一块平板，试确定离开机翼前沿点多远的位置上，空气的流动为充分发展的湍流？空气当作干空气处理。

Chapter 6　The Experimental Correlations of Convection Heat Transfer

第 6 章　对流传热的实验关联式

6.1　The Similarity Principle

6.1　相似原理

When studying convective heat transfer problems, experiment is a common and reliable method. However, there are many factors that affect convective heat transfer coefficient, such as flow velocity, characteristic length, density, viscosity, thermal conductivity, specific heat capacity and so on, so there are generally three key problems encountered in experimental research.

研究对流传热问题时，实验是一种常用的、可靠的研究方法。然而，表面传热系数的影响因素很多，如流体的流速、特征长度、密度、黏度、导热系数、比热容等，所以在实验研究时一般会遇到三个关键问题。

① Before the experiment, there are too many variables, which physical quantities need to be measured? ② After the experiment, how to organize the date, which kind of functional relationship should it be organized into? ③ The physical equipment is bulky, and the experiment is difficult and expensive, what to do?

① 实验前，变量太多，哪些物理量需要测量？② 实验后，数据如何整理？应该整理成怎样的函数关系？③ 实物设备体积庞大，实验困难且昂贵，怎么办？

Fortunately, the similarity principle can solve the above three problems well. The similarity principle mainly studies the relationship between similar physical phenomena. First of all, two concepts should be clarified: same kind of physical phenomena and similar physical phenomena.

幸好相似原理能够很好地解决上述三个问题。相似原理主要研究相似物理现象之间的关系。首先要明确同类物理现象和相似物理现象两个概念。

Same kind of physical phenomena is a phenomenon described by differential equations of the same form and content. Take two counter examples: ① Heat conduction and electrical conduction. Although the form of the differential equations is the same, the content is different. So the two can only be analogized or compared, but they are not the

同类物理现象是用相同形式和内容的微分方程式所描述的现象。举两个反例：① 导热和导电，虽然微分方程的形式相同，但内容不同，所以两者之间只能“类比”或“比拟”，而不是同类物理现象。

same kind of physical phenomenon. ② The velocity field and the temperature field. They have different content, so they are not the same kind of physical phenomenon.

② 速度场和温度场。它们的内容不同，所以也不是同类物理现象。

Similar physical phenomena are two physical phenomena of the same kind, and the physical quantities related to the phenomenon are proportional to each other at the corresponding time and the corresponding place. For example, two steady-state convective heat transfer phenomena, if they are similar to each other, then the geometry of heat transfer surface, velocity field, temperature field, thermal property field and so on must be similar. For another example, for two unsteady-state heat conduction problems, the spatial distribution of each physical quantity at the corresponding moment should be similar.

相似物理现象是两个同类物理现象，并且在相应的时刻与相应的地点上与现象有关的物理量一一对应成比例。比如，两个稳态的对流传热现象，如果彼此相似，那么必有传热面几何形状相似、速度场相似、温度场相似、热物性场相似等。又比如，对于两个非稳态导热问题，在相应时刻各物理量的空间分布要相似。

Similar physical phenomena have the following two characteristics: ① The homonymic characteristic numbers describing the phenomenon are equal correspondingly (or physical quantities are proportional correspondingly); ② There is a functional relationship among the characteristic numbers. For example, the characteristic number equation of convection heat transfer of a constant physical fluid externally flowing through a plane plate:

相似的物理现象具有以下两个特性：① 描述该现象的同名特征数对应相等（或者物理量对应成比例）；② 特征数之间存在着函数关系。比如，常物性流体外掠平板对流传热的特征数方程：

$$Nu_x = 0.332 Re_x^{1/2} Pr^{1/3} \tag{6.1}$$

The characteristic number equation expresses the functional relationship among the characteristic numbers.

特征数方程表示了特征数之间的函数关系。

Judgment conditions for similar physical phenomena are as follows: ① The homonymic determined characteristic numbers are equal. For example, *Re* and *Pr* are determined characteristic numbers in the convective heat transfer phenomenon, while *Nu* is an unknown quantity, so it is an undetermined characteristic number; ② The conditions for unique solution are similar, that is, the geometric condition, the physical condition, the time condition and the boundary condition should be similar.

物理现象相似的判定条件如下：① 同名的已定特征数相等，如对流传热现象中 *Re*、*Pr* 为已定特征数，而 *Nu* 是未知量，所以是待定特征数；② 单值性条件相似，即几何条件、物理条件、时间条件及边界条件都要相似。

The following is a summary of the basic contents of the

归纳一下相似原理的基本内容：

similarity principle: ① The differential equations of two phenomena have the same form and the same content (i. e. same kind); ② The homonymic characteristic numbers describing two phenomena are equal correspondingly (That is, the relevant physical quantities should be proportional); ③ There is a functional relationship among the characteristic numbers; ④ The conditions for unique solution are similar.

① 两个现象的微分方程形式相同，并具有相同内容（即同类）；② 描写两个现象的同名特征数对应相等（即有关物理量要对应成比例）；③ 各特征数间存在着函数关系；④ 单值性条件相似。

The similarity principle solves the following problems in experiments: ① The physical quantities contained in the characteristic numbers should be only measured according to the characteristic numbers in the experiment. —The problem of which physical quantities to measure is solved, and measurement blindness is avoided. ② According to the functional relationship among characteristic numbers, the experimental data are organized to get experimental correlation formula. —The problem of organizing experimental data is solved, which is beneficial to reduce the number of experiments and enhance the universality of experimental results. ③ Modeling experiments can be carried out under the guidance of the similarity principle. —The problem of bulky physical equipment, difficult experiment and high-cost is solved.

相似原理解决了实验中存在的以下问题：① 实验中只需根据特征数测量特征数所包含的物理量——解决了测量哪些物理量的问题，避免了测量盲目性；② 按特征数之间的函数关系整理实验数据，得到实验关联式——解决了实验数据整理的问题，有利于减少实验次数，增强实验结果的通用性；③ 可以在相似原理的指导下，采用模化实验——解决了实物设备体积庞大、实验难度大、成本高的问题。

After the three key problems of experiment are solved, the next step is to determine which characteristic numbers are involved in a physical phenomenon, such as Nu, Pr, Re, Gr, Bi, Fo..., what is the functional relationship among them?

实验的三个关键问题解决之后，下一步需要确定一个物理现象中涉及哪些特征数，比如 Nu、Pr、Re、Gr、Bi、Fo……它们之间的函数关系是怎样的？

Summary This section illustrates the content of the similarity principle, and characteristics and judgment conditions of similar physical phenomena.

小结 本节阐述了相似原理的内容，以及物理现象相似的特性和判定条件。

Question Are the velocity field and temperature field similar physical phenomena? Why?

思考题 速度场和温度场是相似物理现象吗？为什么？

Self-tests

自测题

1. What are the judging conditions for two physical phenomena to be similar?

1. 两个物理现象相似的判定条件是什么？

2. What are the basic contents of the similarity principle?

2. 相似原理的基本内容包括哪些?

3. What problems does the similarity principle solve in the experiment?

3. 相似原理解决了实验中存在的哪些问题?

6.2 The Similarity Analysis Method

6.2 相似分析法

On the basis of the similarity principle, we organize the experimental data according to the functional relationship among the characteristic numbers in the experiment. Therefore, the first step is to determine that which characteristic numbers are involved in a physical phenomenon. There are two methods to obtain the characteristic numbers: ① similarity analysis method; ② dimensional analysis method.

根据相似原理，在实验中按照特征数之间的函数关系来整理实验数据。因此，首先需要确定某一物理现象涉及了哪些特征数。获得特征数有两种方法：① 相似分析法；② 量纲分析法。

The similarity analysis method (i. e. the equation analysis method): Based on known mathematical descriptions of physical phenomena, a series of scale coefficients or similarity multiples are established between two phenomena, and the relationship between these similarity multiples is derived to obtain the characteristic numbers.

相似分析法（即方程分析法）：在已知物理现象的数学描述基础上，建立两个现象之间的一系列比例系数（或相似倍数），并导出这些相似倍数之间的关系，从而获得特征数。

The two convective heat transfer phenomena are shown in Fig. 6.1, their mathematical descriptions are:

图 6.1 所示的两个对流传热现象，其数学描述为：

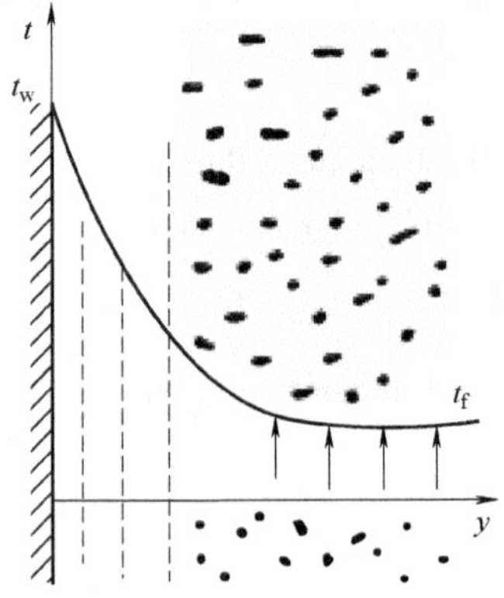

Fig. 6.1 Fluid temperature distribution near the wall

图 6.1 壁面附近的流体温度分布

Phenomenon 1:

现象 1：

$$h' = -\frac{\lambda'}{\Delta t'}\frac{\partial t'}{\partial y'}\bigg|_{y'=0} \tag{6.2a}$$

Phenomenon 2:

现象 2：

$$h'' = -\frac{\lambda''}{\Delta t''}\frac{\partial t''}{\partial y''}\bigg|_{y''=0} \tag{6.2b}$$

It can be seen that the two convective heat transfer phenomena have the identical expressions, so they are the same kind of physical phenomenon. Next, create similarity multiples:

可见，两个对流传热现象的表达式完全相同，所以属于同类物理现象。接着建立相似倍数：

$$\frac{h'}{h''} = C_h, \frac{\lambda'}{\lambda''} = C_\lambda, \frac{t'}{t''} = C_t, \frac{y'}{y''} = C_y \tag{6.2c}$$

Then, substitute them into the differential equation of convection heat transfer for phenomenon 1, the relationship between similar multiples can be obtained:

然后，代入现象 1 的对流传热微分方程式，得到相似倍数之间的关系：

$$\frac{C_h C_y}{C_\lambda}h'' = -\frac{\lambda''}{\Delta t''}\frac{\partial t''}{\partial y''}\bigg|_{y''=0} \tag{6.2d}$$

Compare this equation with the differential equation of convection heat transfer for phenomenon 2 to get:

将该式与现象 2 的对流传热微分方程式比较，可得：

$$\frac{C_h C_y}{C_\lambda} = 1 \tag{6.2e}$$

This equation reflects the constraint relationship between the similarity multiples when the two heat transfer phenomena are similar. Substitute the expression of similarity multiples into the constraint relationship, we can get the dimensionless quantities and their relationship, that is, Nusselt numbers of the two phenomena are equal.

该式反映了两个传热现象相似时相似倍数间的制约关系。把相似倍数表达式代入该制约关系式，得到无量纲量及其关系，即两个现象的努塞尔数相等。

$$\frac{h'y'}{\lambda'} = \frac{h''y''}{\lambda''} \Rightarrow Nu_1 = Nu_2 \tag{6.3}$$

Eq. (6.3) also proves that "corresponding equality of characteristic numbers with the same name" is a property of similar physical phenomena. Similarly, the analysis of momentum differential equation also shows that Reynolds numbers of the two phenomena are equal.

式（6.3）也证明了"同名特征数对应相等"是相似物理现象的特性。类似地，通过动量微分方程分析，也可以得到两个现象的雷诺数相等。

$$\frac{u'l'}{\nu'}=\frac{u''l''}{\nu''}\Rightarrow Re_1=Re_2 \tag{6.4}$$

The analysis of energy differential equation also shows that Peclet numbers of the two phenomena are equal.

通过能量微分方程分析，也可以得到两个现象的贝克莱数相等。

$$\frac{u'l'}{a'}=\frac{u''l''}{a''}\Rightarrow Pe_1=Pe_2 \tag{6.5}$$

Peclet number characterizes the relative magnitude of convection and diffusion in heat transfer process. As long as two heat transfer phenomena are similar, their Peclet numbers must be equal. If $Pe=Pr \cdot Re$, it can be also deduced that Prandtl numbers of the two phenomena are equal, namely, $Pr_1=Pr_2$.

贝克莱数表征了传热过程中对流作用与扩散作用的相对大小。只要两个热量传递现象相似，其贝克莱数一定相等。若 $Pe=Pr \cdot Re$，则可以推出两个现象的普朗特数相等，即 $Pr_1=Pr_2$。

For the momentum equation of the forced convective boundary layer:

根据强制对流边界层动量方程：

$$\rho\left(u\frac{\partial u}{\partial x}+v\frac{\partial u}{\partial y}\right)=-\frac{\mathrm{d}p}{\mathrm{d}x}+\eta\frac{\partial^2 u}{\partial y^2} \tag{6.6}$$

The momentum equation of the natural convective boundary layer is obtained by adding the volumetric force term to the right side of the equation:

在等式右侧增加体积力项，可以得到自然对流边界层动量方程：

$$u\frac{\partial u}{\partial x}+v\frac{\partial u}{\partial y}=g\alpha_V\theta+\nu\frac{\partial^2 u}{\partial y^2} \tag{6.7}$$

A similarity analysis of the momentum equation of the natural convective boundary layer can get Graschof number, which characterizes the ratio of buoyancy force to viscous force of a fluid.

对自然对流边界层动量方程进行相似分析，可以得到**格拉晓夫数**，它表征流体浮升力与黏性力的比值。

$$Gr=\frac{g\alpha_V\Delta t l^3}{\nu^2} \tag{6.8}$$

Summary This section illustrates the principles and steps of the similarity analysis method.

小结 本节阐述了相似分析法的原理和步骤。

Question Why organize experimental data by the characteristic number equation?

思考题 为什么按特征数方程整理实验数据？

Self-tests

自测题

1. What is the similarity analysis method (i. e. the equation analysis method)?

1. 什么是相似分析法（即方程分析法）？

2. What are the prerequisites to obtain characteristic numbers by the similarity analysis method?

2. 使用相似分析法获得特征数的前提条件是什么？

授课视频

6.3 The Dimensional Analysis Method

6.3 量纲分析法

The dimensional analysis method is the second method to obtain characteristic number. The characteristic numbers are obtained according to π theorem and the dimensional harmony principle on the premise that the relevant physical quantities are known.

量纲分析法是获得特征数的第二种方法。在已知相关物理量的前提下，按 π 定理及量纲和谐原理而获得特征数。

π theorem: An equation containing n physical quantities (the dimensions are the same on both sides of the equation) can definitely be converted to a relationship consisting of $(n-r)$ mutually independent dimensionless quantities. Where r is the number of basic dimensions contained in n physical quantities.

π 定理：一个包含 n 个物理量的方程式（等式两侧的量纲一致），一定可以转换为（$n-r$）个相互独立的无量纲量所组成的关系式。其中，r 是 n 个物理量中包含的基本量纲数。

The advantages of the dimensional analysis method are as follows: ① Simple method; ② Dimensionless quantities can still be obtained without knowing the differential equation. The following takes single-phase fluid forced convection heat transfer in a circular tube as an example to conduct dimensional analysis.

量纲分析法的优点是：① 方法简单；② 在不知道微分方程的情况下，仍可以获得无量纲量。下面以圆管内单相流体强制对流传热为例，进行量纲分析。

Step 1, determine the relevant physical quantities. According to the previous analysis of convection heat transfer, there are 7 physical quantities related to convection heat transfer, that is, $n=7$, including h, u, d, λ, η, ρ, c_p.

第 1 步，确定相关的物理量。由前面的对流传热分析可知：对流传热的相关物理量有 7 个，即 $n=7$，包括 h、u、d、λ、η、ρ、c_p。

Step 2, determine the number of basic dimensions r. There are 7 basic dimensions in International System of Units: length [L], quality [M], time [T], temperature [θ], current [I], amount of substance [N], luminous intensity [J]. Take convective heat transfer coefficient as an example

第 2 步，确定基本量纲数 r。国际单位制中有 7 个基本量纲：长度 [L]、质量 [M]、时间 [T]、温度 [θ]、电流 [I]、物质的量 [N]、发光强度 [J]。以表面传热系数为例进行量

to conduct dimensional analysis, the units of each physical quantity are expressed in the units corresponding to the basic dimensions:

纲分析，把每个物理量的单位都用“基本量纲”对应的单位表示：

$$h:\frac{\mathrm{W}}{\mathrm{m}^2\cdot\mathrm{K}}=\frac{\mathrm{J/s}}{\mathrm{m}^2\cdot\mathrm{K}}=\frac{\mathrm{N}\cdot\mathrm{m}}{\mathrm{s}\cdot\mathrm{m}^2\cdot\mathrm{K}}=\frac{\mathrm{kg}\cdot\mathrm{m/s}^2}{\mathrm{s}\cdot\mathrm{m}\cdot\mathrm{K}}=\frac{\mathrm{kg}}{\mathrm{s}^3\cdot\mathrm{K}}\Rightarrow\frac{\mathrm{M}}{\mathrm{T}^3\Theta}$$

$$u:\frac{\mathrm{m}}{\mathrm{s}},\quad d:\mathrm{m},\quad \lambda:\frac{\mathrm{W}}{\mathrm{m}\cdot\mathrm{K}}=\frac{\mathrm{kg}\cdot\mathrm{m}}{\mathrm{s}^3\cdot\mathrm{K}}$$

$$\eta:\mathrm{Pa}\cdot\mathrm{s}=\frac{\mathrm{kg}}{\mathrm{m}\cdot\mathrm{s}},\quad \rho:\frac{\mathrm{kg}}{\mathrm{m}^3},\quad c_p:\frac{\mathrm{J}}{\mathrm{kg}\cdot\mathrm{K}}=\frac{\mathrm{m}^2}{\mathrm{s}^2\cdot\mathrm{K}}$$

It can be seen that the seven related physical quantities involve four basic dimensions: length [L], mass [M], time [T], temperature [θ], that is, $r=4$. Therefore, $n-r=3$, that is, 7 physical quantities can form 3 dimensionless quantities.

由此可见，相关的 7 个物理量共涉及 4 个基本量纲：长度［L］、质量［M］、时间［T］、温度［θ］，即 $r=4$。所以，$n-r=3$，即 7 个物理量可组建 3 个无量纲量。

The specific method of forming: Select 4 basic physical quantities from 7 physical quantities, such as u, d, λ and η. The basic requirement for selection is that the above four basic dimensions must be included in the four dimensions of basic physical quantities, then three dimensionless quantities are formed with other physical quantities respectively.

组建的具体方法：从 7 个物理量中选定 4 个“基本物理量”，如选 u、d、λ 和 η。选择的基本要求是：4 个“基本物理量”的量纲中必须包含上述 4 个“基本量纲”，然后分别与其他物理量组成 3 个无量纲量。

Step 3, three dimensionless quantities are represented by π_1, π_2 and π_3, and add exponents a, b, c and d to the 4 selected basic physical quantities u, d, λ and η, respectively, then multiply the four basic physical quantities to each other, and then multiply by the other 3 physical quantities respectively, we can get the following expressions:

第 3 步，用 π_1、π_2、π_3 代表三个无量纲量，对所选取的 4 个“基本物理量”u、d、λ 和 η 分别添加指数 a、b、c、d，然后四个“基本物理量”相乘，再分别乘以其他 3 个物理量，就可以得到以下的表达式：

$$\pi_1=hu^{a_1}d^{b_1}\lambda^{c_1}\eta^{d_1}$$

$$\pi_2=\rho u^{a_2}d^{b_2}\lambda^{c_2}\eta^{d_2}$$

$$\pi_3=c_p u^{a_3}d^{b_3}\lambda^{c_3}\eta^{d_3}$$

Where the exponents a, b, c and d are all undetermined constants.

式中，指数 a、b、c、d 均为待定常数。

Step 4, solve for undetermined constants, take π_1 as an example, substitute the dimension of each physical quantity into π_1 and merge similar items to get:

第 4 步，求解待定常数，以 π_1 为例，把每个物理量的量纲代入 π_1，合并同类项可得到：

$$\begin{aligned}\pi_1 &= hu^{a_1}d^{b_1}\lambda^{c_1}\eta^{d_1}\\ &= \mathrm{M}^{1+c_1+d_1}\mathrm{T}^{-3-a_1-3c_1-d_1}\Theta^{-1-c_1}\mathrm{L}^{a_1+b_1+c_1-d_1}\end{aligned}$$

According to the principle of dimensional harmony, the left side of the equal sign is a dimensionless quantity, then the exponent of every dimension on the right side of the equal sign must be zero. Therefore, we can get a quaternary linear equation group, then solve the equation group to get four undetermined constants.

根据量纲和谐原理，等号左边为无量纲量，等号右边各量纲的指数必然为零。于是，可得到一个四元一次方程组，解方程组即可求出 4 个待定常数。

$$\begin{cases}1+c_1+d_1=0\\ -3-a_1-3c_1-d_1=0\\ -1-c_1=0\\ a_1+b_1+c_1-d_1=0\end{cases} \Rightarrow \begin{cases}a_1=0\\ b_1=1\\ c_1=-1\\ d_1=0\end{cases}$$

Substitute four undetermined constants into the expression of π_1, sort it out to obtain:

把 4 个待定常数代入 π_1 的表达式，整理后就可以得到：

$$\pi_1=\frac{hd}{\lambda}=Nu$$

In the same way, we can also get:

同样方法也可以得到：

$$\pi_2=\frac{\rho ud}{\eta}=Re,\quad \pi_3=\frac{\eta c_p}{\lambda}=Pr$$

Therefore, the dimensionless correlation formula of forced convection heat transfer for a single-phase fluid in a circular tube can be expressed as:

于是，圆管内单相流体强制对流传热的无量纲关联式可表示为：

$$Nu=f(Re,Pr) \tag{6.9}$$

Summary This section illustrates the method to obtain characteristic numbers by dimensional analysis.

小结 本节阐述了通过量纲分析而获得特征数的方法。

Question What is the basic method of deriving the correlation formula of characteristic numbers by using π theorem?

思考题 使用 π 定理推导特征数关联式的基本方法是什么？

Self-tests

自测题

1. What is the dimensional analysis method? What is the principle of dimensional harmony?

1. 什么是量纲分析法？什么是量纲和谐原理？

2. For the forced convection heat transfer of single-phase fluid in a circular tube, which characteristic numbers can be obtained by dimensional analysis?

2. 对于圆管内单相流体强制对流传热，通过量纲分析可得到哪几个特征数？

授课视频

6.4 The Application of the Similarity Principle

6.4 相似原理的应用

The application of the similarity principle mainly has three aspects: ① Guide the arrangement of experiment; ② Organization of experimental data; ③ Guide the modeling experiments.

相似原理的应用主要有三个方面：① 指导实验安排；② 实验数据整理；③ 指导模化实验。

The so-called modeling experiment refers to adopting the models that are different from physical geometric scale (most of which are miniaturized models) to study the physical process in the actual devices. In the modeling experiment, the experiment should be arranged according to similar characteristic number. Its purpose is that the number of experiments is greatly reduced on the one hand, and the obtained experimental correlation formula is universal on the other hand.

所谓模化实验，是指用不同于实物几何尺度的模型（多数是缩小模型）来研究实际装置中所进行物理过程的实验。在模化实验时，要按照相似特征数安排实验。其目的一方面大幅度减少实验次数，另一方面得到的实验关联式具有一定的通用性。

To conduct modeling experiment, the first thing to know is the two principles that modeling experiments should follow: ① The convective heat transfer process in both the model and the prototype must be similar; ② When changing the experimental conditions, i. e. the relevant parameters, during the experiment, all the physical quantities contained in similar characteristic numbers should be measured.

开展模化实验首先要知道模化实验应遵循的两个原则：① 模型与原型中的对流传热过程必须相似；② 实验时，改变实验条件即相关参数，需测量相似特征数中所包含的全部物理量。

To calculate the characteristic numbers, the first thing is to identify reference temperature, characteristic length and characteristic velocity in the characteristic numbers.

要计算特征数，首先要确定特征数中定性温度、特征长度和特征速度。

The reference temperature refers to the temperature used

定性温度：计算特征数中物性参数

to calculate the physical parameters in the characteristic numbers. The physical parameters contained in the similar characteristic numbers, such as λ, ν, Pr, etc., all depend on the reference temperature.

所用的温度。相似特征数中所包含的物性参数（比如 λ、ν、Pr 等）都取决于定性温度。

The reference temperature of fluid: When the fluid externally flowing through a plane plate to exchange heat, the reference temperature is the incoming temperature of the fluid; When the fluid flows in a tube for heat transfer, the reference temperature is the average of the inlet and outlet temperatures of the fluid; For the thermal boundary layer, the reference temperature is the average of wall temperature and fluid temperature.

流体的定性温度：当流体沿平板流动传热时，流体的定性温度为来流温度；当流体在管内流动传热时，定性温度为流体进出口温度的平均值；对于热边界层，定性温度为壁面温度与流体温度的平均值。

In the characteristic number correlations of convection heat transfer, the subscript of characteristic number denotes the value at a certain reference temperature. For example, Nu_f, Re_f, Pr_f, the subscript f denotes the value at the reference temperature of the fluid, Nu_m, Re_m, Pr_m, the subscript m denotes the value at the reference temperature of the boundary layer. Pay particular attention to the fact that: When the characteristic number correlation is used, it must correspond to the reference temperature.

在对流传热特征数关联式中，用特征数的下标表示某种定性温度下的值。比如，Nu_f，Re_f，Pr_f，下标 f 表示流体定性温度下的值，Nu_m，Re_m，Pr_m 下标 m 表示边界层定性温度下的值。特别注意：使用特征数关联式时，必须与定性温度相对应。

The characteristic length is the geometric scale which is contained in the characteristic numbers and has a great effect on fluid flow and heat transfer. For example, when the fluid flows in a tube for heat transfer, the characteristic length is taken as the inner diameter of the tube; When the fluid longitudinally flows through a plane plate, the characteristic length is taken as the length of the plane plate; When the fluid flows in a channel with irregular cross-section, the characteristic length is taken as the equivalent diameter. The equivalent diameter is denoted by d_e, its calculation formula is as follows:

特征长度：是包含在特征数中，对流动和传热影响较大的几何尺度。比如管内流动传热时，特征长度取管内径；流体纵掠平板时，特征长度取板长；流体在流通截面形状不规则的槽道内流动时，特征长度取当量直径。当量直径用 d_e 表示，计算公式为：

$$d_e = \frac{4A_c}{P} \tag{6.10}$$

That is 4 times the area of the flow cross-section divided by the wetted perimeter, where A_c is the area of the flow

即 4 倍的流通截面面积除以润湿周长，式中，A_c 表示流通截面面积；

cross-section, P is the wetted perimeter.

P 为润湿周长。

The characteristic velocity refers to the fluid velocity in Reynolds number. When fluid externally flows through a plane plate or around a cylinder, it is taken as the velocity of the incoming fluid u_∞; When fluid flows in a tube, it is taken as the average velocity u_m of the fluid on the cross-section of a tube; When the fluid externally flows around the tube bundle, it is taken as the maximum velocity u_{max} at the minimum flow cross-section.

特征速度：是指 Re 中的流体速度。当流体外掠平板或绕流圆柱时，取来流速度 u_∞；当流体在管内流动时，取管截面上的平均速度 u_m；当流体绕流管束时，取最小流通截面处的最大速度 u_{max}。

The determination of the reference temperature, characteristic length and characteristic velocity lays a foundation for the calculation of characteristic numbers. We must expertly master the expressions and the physical meanings of common characteristic numbers. In particular, it is important to master Reynolds number, Prandtl number and Nusselt number.

定性温度、特征长度和特征速度的确定，为计算特征数奠定了基础。必须熟练掌握常用特征数的表达式及其物理意义，尤其是要重点掌握雷诺数、普朗特数和努塞尔数。

Reynolds number: The physical meaning is the ratio of inertia force to viscous force.

雷诺数：物理意义是惯性力和黏性力之比。

$$Re=\frac{ul}{\nu} \tag{6.11}$$

Prandtl number: The physical meaning is the ratio of the momentum diffusion capacity to the heat diffusion capacity.

普朗特数：物理意义是动量扩散能力和热量扩散能力之比。

$$Pr=\frac{\eta c_p}{\lambda}=\frac{\nu}{a} \tag{6.12}$$

Nusselt number: The physical meaning is the dimensionless temperature gradient of the fluid on the wall, which also indicates the intensity of convection heat transfer, and λ is the thermal conductivity of the fluid in the formula.

努塞尔数：物理意义是壁面上流体的无量纲温度梯度，也反映了对流传热的强度，式中，λ 为液体的导热系数。

$$Nu=\frac{hl}{\lambda} \tag{6.13}$$

Biot number: The physical meaning is the ratio of the thermal conduction resistance of the solid to the convective heat transfer resistance of the fluid on the surface, and λ is the thermal conductivity of the solid in the formula.

毕渥数：物理意义是固体的导热热阻和面上流体的对流传热热阻之比，式中，λ 为固体的导热系数。

$$Bi = \frac{hl}{\lambda} \tag{6.14}$$

Fourier number: The physical meaning is the dimensionless time of unsteady-state heat transfer process, which characterizes the depth of the process.

傅里叶数：物理意义是非稳态传热过程的无量纲时间，表征了过程进行的深度。

$$Fo = \frac{a\tau}{l^2} \tag{6.15}$$

Grashof number: The physical meaning is the ratio of buoyancy force to viscous force.

格拉晓夫数：物理意义是浮升力与黏性力之比。

$$Gr = \frac{gl^3 \alpha_V \Delta t}{\nu^2} \tag{6.16}$$

After obtaining the characteristic numbers, we know how to organize the experimental data, or what kind of functional relationship to organize it. The purpose of organizing experimental data is to express the regularity of the experimental data so as to facilitate application.

有了特征数之后，就知道了实验数据如何整理，或者整理成什么样的函数关系。实验数据整理的目的是：为了表达实验数据的规律性，以便于应用。

We need to know that it is empirical to determine the specific function form of characteristic number correlation, as well as the reference temperature, characteristic length and characteristic velocity, etc. The characteristic number correlation is usually organized into the power function of known characteristic number. For example, $Nu = cRe^n Pr^m$, where the undetermined constants c, m and n need to be determined by experimental data, and the graphical method and the least square method are usually used.

要知道，确定特征数关联式的具体函数形式，以及定性温度、特征长度、特征速度等，都具有一定的经验性。特征数关联式通常整理成已定特征数的幂函数形式。比如 $Nu = cRe^n Pr^m$，式中，待定常数 c、m、n 都需通过实验数据确定，通常用图解法和最小二乘法。

The graphical method: According to the characteristic that the power function is a straight line in logarithmic coordinate system, the graphical method is as follows: For power functions $Nu = cRe^n$, take the logarithm of both sides to get $\lg Nu = \lg c + n\lg Re$, this expression is equivalent to a linear function $y = a + bx$, as shown in Fig. 6.2. Where $\lg c$ is the intercept of the ordinate in the figure, n is the slope of the straight line, the deviation between the characteristic number correlation and experimental data is generally less than 25%.

图解法： 根据“幂函数在对数坐标图中是直线”的特点，图解法的处理方法如下：对于幂函数 $Nu = cRe^n$，两边取对数就可以得到 $\lg Nu = \lg c + n\lg Re$，这个表达式相当于线性函数 $y = a + bx$，如图 6.2 所示。式中，$\lg c$ 是图中纵坐标的截距，n 就是直线的斜率，特征数关联式与实验数据的偏差一般要小于 25%。

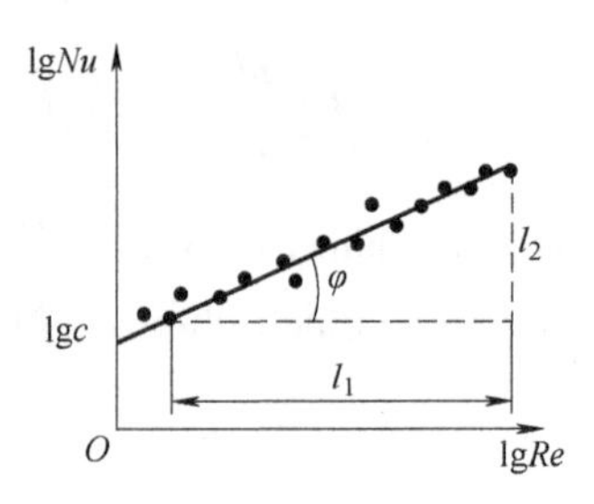

Fig. 6. 2 $Nu = cRe^n$ double logarithmic coordinate diagram

图 6. 2 $Nu = cRe^n$ 双对数坐标图示

Summary This section illustrates the application of similarity principle in guiding experimental arrangement, organizing data and modeling experiment, etc.

小结 本节阐述了相似原理在指导实验安排、数据整理和模化实验等方面的应用。

Question What is the purpose of arranging experiments according to the similar characteristic number?

思考题 按照相似特征数安排实验的目的是什么？

Self-tests

自测题

1. What principle should be followed in the modeling test?

2. What are the expressions and physical meanings of Nu, Pr, Re, Bi, Fo?

1. 模化实验应遵循的原则是什么？

2. Nu、Pr、Re、Bi、Fo 的表达式和物理意义是什么？

授课视频

6. 5 The Application Examples of the Similarity Principle

6. 5 相似原理应用实例

Example 6. 1 The working conditions of a heat exchange equipment are as follows: The wall temperature $t_w = 120℃$, the temperature of the heated air $t_f = 80℃$, the air velocity $u = 0.5\text{m/s}$. Use a model which is fully reduced to 1/5 of the original equipment to study its heat transfer condition. The air is also heated in the model, air temperature $t_f' = 10℃$, wall temperature $t_w' = 30℃$. What should the fluid velocity u' be in the model to ensure that its heat transfer phenomenon is similar to that in the original equipment

例 6. 1 一台换热设备的工作条件是：壁温 $t_w = 120℃$，加热 $t_f = 80℃$ 的空气，空气流速 $u = 0.5\text{m/s}$。采用一个全面缩小为原设备 1/5 的模型来研究它的换热状况。在模型中也是对空气加热，空气温度 $t_f' = 10℃$，壁面温度 $t_w' = 30℃$。试问模型中流速 u'需要多大才能保证与原设备中的换热现象相似（模型中的

(All parameters in the model are indicated by a superscript′).

各参数用上标“′”标明)。

Before solving the problem, we first make simplifications and hypotheses. It is considered as a steady-state heat transfer process. The same kind of phenomena is studied in the model as in the original equipment, both of which is convection heat transfer, and the conditions for unique solution are also similar. Therefore, as long as the determined characteristic numbers, such as *Re* and *Pr*, in the prototype and model correspond equally, their similarity can be realized.

在求解之前，先做出简化和假设：认为是稳态传热过程，模型与原设备中研究的是同类现象（对流传热），单值性条件也相似。所以，只要原型和模型中已定的特征数（如 *Re*、*Pr*）对应相等，即可实现相似。

Solution: Because *Pr* number of air changes very little with temperature, *Pr* number of the fluid in the model can be considered unchanged from that in the original equipment. As long as *Re* number is equal, the similarity can be realized. In the calculation, according to the definition of Reynolds number and its similarity relation:

解：因为空气的 *Pr* 随温度变化很小，所以可以认为模型与原设备中流体 *Pr* 不变。只要保证 *Re* 相等，即可实现相似。在计算时根据雷诺数的定义式及其相似关系：

$$\frac{u'l'}{\nu'}=\frac{ul}{\nu}\Rightarrow u'=u\frac{\nu'}{\nu}\frac{l}{l'}$$

The reference temperature is the average value of fluid temperature and wall temperature, the kinematic viscosity of air at two reference temperatures can be found in the appendix.

取定性温度为流体温度与壁面温度的平均值，从附录中可查出两个定性温度下空气的运动黏度。

It is known that $l/l'=5$, the fluid velocity required $u'=1.63\text{m/s}$ in the model can be obtained by substituting the known parameters.

已知 $l/l'=5$，代入已知参数，可求出模型中要求的流体流速 $u'=1.63\text{m/s}$。

Scaling the model down to 1/5 of the size of the real object, it would be wrong to think that the flow velocity in the model should be 1/5 of that in the real object without considering the similarity principle. In fact, the opposite is true, the velocity in the model should be about 5 times that in the real object. As for not strictly 5 times, that is because the physical properties of the fluid vary with temperature.

模型的尺度缩小到实物的 1/5，如果不考虑相似原理，以为模型中的流速应该是实物中的 1/5，那就错了。实际上恰恰相反，模型中的流速应该是实物中的 5 倍左右。至于不是严格的 5 倍，那是因为流体物性随温度变化而引起的。

The test is carried out according to the parameters calculated in this example, the processes in the model and in the real object belong to the same group of similarity, then the test results of the model can represent the whole group of similarity.

按照本例计算所得的参数进行实验，模型与实物中的过程就属于同一个相似组，那么模型的实验结果就可以代表整个相似组。

Example 6.2 The air with an average temperature of 50℃ is used to simulate the convective heat transfer of flue gas with an average temperature of 400℃ across tube bundles, the velocity range of flue gas is 10 ~ 15 m/s in the real object. The model adopts the same tube diameter as the real object. What is the velocity range of the air in the model?

例 6.2 用平均温度为 50℃ 的空气来模拟平均温度为 400℃ 的烟气横掠管束的对流传热，实物中烟气流速的变化范围是 10 ~ 15m/s。模型采用与实物一样的管径，问模型中空气的流速应在多大范围内变化?

Simplification and assumptions first: ① It is considered as a steady-state heat transfer process; ② The physical properties of simulated gas (i. e. air) are calculated at 50℃, and the physical properties of the actual working gas (i. e. flue gas) are calculated at 400℃.

先做出简化和假设：① 认为是稳态传热过程；② 以 50℃ 计算模拟气体（即空气）的物性，以 400℃ 计算实际工作气体（即烟气）的物性。

In order to calculate the velocity range of air in the model, according to the principle of similarity, *Re* number in the model and in the real object should be equal, so there is a similarity relation:

要计算模型中空气流速的变化范围，根据相似原理，模型与实物中 *Re* 应该相等，于是有相似关系式：

$$u' = u\frac{\nu'}{\nu}\frac{l}{l'}$$

Where l and l' are the characteristic lengths of the real object and the model respectively, and they are the same tube diameter, so its ratio is 1. According to the appendix: The kinematic viscosity of flue gas at 400℃, and the kinematic viscosity of air at 50℃.

式中，l 和 l' 分别为实物与模型的特征长度，而它们为相同的管径，因此比值是 1。由附录可以查出：400℃ 烟气的运动黏度和 50℃ 空气的运动黏度。

According to the similarity relation, when arranging the test, the air velocity in the model should be $u' = 2.94 \sim 4.46$m/s.

根据相似关系式，在安排实验时，模型中空气流速应该为 $u' = 2.94 \sim 4.46$m/s。

Discussion: $Pr = 0.64$ for flue gas at 400℃ and $Pr = 0.698$ for air at 50℃. It can be seen that *Pr* numbers of the two are not equal. But considering that *Pr* number is not the main affecting factor of heat transfer and they' re

讨论：400℃ 烟气的 $Pr = 0.64$，50℃ 空气的 $Pr = 0.698$，可见两者的 *Pr* 并不相等。但是考虑到 *Pr* 不是影响传热的主要因素，而且两者

not much different. Therefore, the results of the modeling tests are still of practical engineering value.

数值相差不大，因而模化实验的结果仍有工程实用价值。

Summary This section illustrates the application method of similarity principle through two examples.

小结 本节通过两个例子阐述了相似原理的应用方法。

Question What is the modeling test?

思考题 什么是模化实验?

授课视频

6.6 The Characteristics of Forced Convection Heat Transfer in Tubes

6.6 管内强制对流传热的特征

According to the relative relationship between fluid flow and channel wall, there are two states of internal flow and external flow. Their main differences are as follows: The relative relationship between the flow boundary layer and the channel wall is different in different flow states.

按照流体流动时与流道壁面之间的相对关系，有内部流动与外部流动两种状态。它们的主要区别是：在不同的流动状态下，流动边界层与流道壁面之间的相对关系不同。

The development of the fluid boundary layer on the heat transfer wall is restricted by the wall in the internal flow, such as fluid flow in a tube, the outer boundary of the boundary layer will converge to the center line of the flow channel. The fluid boundary layer on the heat transfer wall can develop freely and is not restricted by the wall in the external flow, such as the fluid externally flowing across a plane plate and a circular tube, they belong to external flow.

内部流动时，传热壁面上流体边界层的发展受到壁面的限制，如管内流动，边界层的外边界将汇聚到流道的中心线上。外部流动时，传热壁面上流体边界层可自由发展，没有壁面的限制，如流体外掠平板、外掠圆管时，都属于外部流动。

The forced convection heat transfer in tubes has five main characteristics:

管内强制对流传热有以下 5 个主要特征：

(1) The flow can be divided into laminar flow and turbulent flow. From the knowledge of fluid mechanics, according to Reynolds number of fluid flow, the flow states of a fluid can be divided into laminar flow and turbulent flow and transition flow between them. It is generally believed that it is laminar flow when Reynolds number $Re<2300$,

(1) 流动有层流和湍流之分。由流体力学知识，按照流体流动的雷诺数不同，流体的流动状态可分为层流和湍流，以及介于两者之间的过渡流。一般认为，当 $Re<2300$ 时为层流，当 $2300<Re<10000$ 时为过渡

it is transition flow, when $2300<Re<10000$, and it is strong turbulent flow (fully developed turbulent flow) when $Re>10000$. The flow state of a fluid directly affects the convective heat transfer coefficient, the convective heat transfer coefficient in turbulent flow is much larger than that in laminar flow for the same fluid

流，当 $Re>10000$ 时为旺盛湍流（充分湍流）。流体的流动状态直接影响表面传热系数的大小，同种流体湍流时的表面传热系数会远大于层流。

(2) Entrance effect. According to the change of the fluid flow boundary layer in the tube in Fig. 6. 3, it can be found that the thermal boundary layer at the entrance of the tube is thin, so the convective heat transfer coefficient is large. The length of the entrance section can be estimated by $l/d\approx 0.05Re\ Pr$ in laminar flow; $l/d\approx 60$ in turbulent flow; The entrance effect is negligible when the length-diameter ratio is greater than 60.

(2) 入口效应。根据图 6.3 中管内流体流动的边界层变化，可以发现管子的入口段热边界层较薄，所以表面传热系数较大。层流时，入口段长度可用 $l/d\approx 0.05Re\ Pr$ 估算；湍流时，$l/d\approx 60$；l/d 大于 60 时，入口效应可忽略不计。

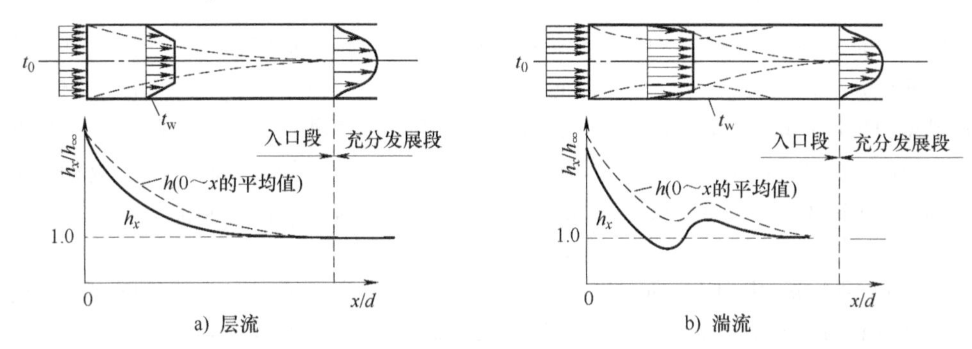

Fig. 6. 3 The change of the fluid flow boundary layer and convective heat transfer coefficient in the tube

图 6.3 管内流体流动的边界层及表面传热系数的变化

Fig. 6. 4 shows the numerical simulation results of laminar flow and turbulent flow in a tube. The lines along the axis in the figure are called isokinetic lines. As can be seen from the figure, the fluid velocity changes mainly near the wall, the velocity distribution on the cross-section of a tube is parabolic in laminar flow, the velocity distribution is approximately trapezoidal in turbulent flow.

图 6.4 所示为管内流体层流和湍流流动时的数值模拟结果，图中沿轴向的线为等速线。从图中可以看出，流体的速度变化主要在壁面附近，层流时管内横截面上速度分布呈抛物线形，湍流时速度分布近似为梯形。

(3) Thermal boundary conditions of a wall. There are two kinds of uniform wall temperature and uniform heat flux, which are also called constant wall temperature and constant heat flux. For example, the temperature of fluid is constant during vapor condensation or liquid boiling, the wall temperature can also be considered to be uniform.

(3) 壁面的热边界条件。有均匀壁温和均匀热流两种，也称为恒壁温和恒热流。如蒸气冷凝或液体沸腾时，流体温度不变，也可认为壁温均匀。又如采用均匀缠绕的电热丝加热壁面时，可认为均匀热流。

When the wall is heated by evenly wound electric wire, it can be considered as uniform heat flux.

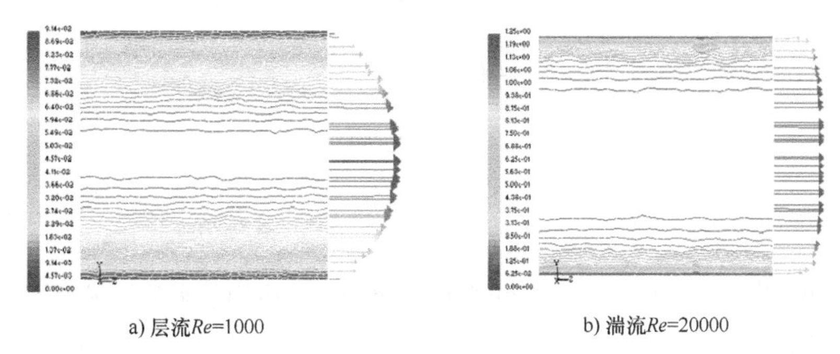

a) 层流Re=1000　　b) 湍流Re=20000

Fig. 6.4　The numerical simulation results of laminar flow and turbulent flow states in a tube

图 6.4　管内流体层流和湍流流动时的数值模拟结果

Except for liquid metal, the influence of two thermal boundary conditions on convective heat transfer coefficient h can be negligible in turbulent flow. The influence of two kinds of thermal boundary conditions on the average temperature difference is very obvious in laminar flow.

湍流时，除液态金属外，两种热边界条件对表面传热系数 h 的影响可忽略不计。层流时，两种热边界条件对平均温差的影响非常明显。

(4) Determination of reference temperature and characteristic velocity. The reference temperature (the bulk temperature) takes the average temperature of the fluid on the cross-section, or the average temperature of the fluid at the inlet and outlet sections. The average temperature of the fluid can be measured accurately after full mixing by the mixer. The characteristic velocity generally takes the average velocity of the fluid on cross-section of the channel.

(4) 定性温度及特征速度的确定。定性温度（整体温度）取截面上流体的平均温度，或者进出口截面的平均温度。通过混合器充分混合后可以准确测出流体的平均温度。特征速度一般取流道截面的平均流速。

(5) Calculation of average temperature difference in Newton cooling formula. For the condition of constant heat flux, the temperature difference between the wall and the fluid in the fully developed section, i. e. $t_w - t_f$ can be taken as the average temperature difference. For the condition of constant wall temperature, the local temperature difference on the section is constantly changing, the following heat balance formula is required for calculation:

(5) 牛顿冷却公式中平均温差的计算。对于恒热流条件，取充分发展段的壁面温度与流体温度之差，即 $t_w - t_f$ 作为平均温差。对于恒壁温条件，截面上的局部温差不断变化，需用以下热平衡式计算：

$$h_m A \Delta t_m = q_m c_p (t_f'' - t_f') \tag{6.17}$$

Where q_m is the mass flow rate, t_f'', t_f' are the average temperature of fluid on the outlet and inlet sections respec-

式中，q_m 是质量流量；t_f''、t_f' 分别是出口、进口截面上的平均温度；

tively, Δt_m is calculated according to the logarithmic mean temperature difference:

Δt_m 按对数平均温差计算：

$$\Delta t_m=\frac{t_f''-t_f'}{\ln\dfrac{t_w-t_f'}{t_w-t_f''}} \tag{6.18}$$

When the ratio of temperature differences $\frac{t_w-t_f'}{t_w-t_f''}$ is between 0.5-2, the arithmetic mean temperature difference can be used to calculate.

当温差之比$\frac{t_w-t_f'}{t_w-t_f''}$在 0.5～2 之间时，可用算术平均温差来计算。

Summary This section illustrates the difference between internal and external flows, and the characteristics of forced convection heat transfer in tubes.

小结 本节阐述了内部流动和外部流动的区别，以及管内强制对流换热的特征。

Question Why is the convective heat transfer coefficient of turbulent flow far greater than that of laminar flow?

思考题 为什么湍流表面传热系数远大于层流表面传热系数？

Self-tests

自测题

1. What is the difference between internal flow and external flow?

1. 内部流动与外部流动的区别是什么？

2. What are the velocity distribution characteristics of fluid laminar flow and turbulent flow in the cross section of the tube?

2. 管内横截面上流体层流与湍流流动的速度分布特征是什么？

3. What are the two thermal boundary conditions for fluid convection heat transfer? What effect do they have on heat transfer?

3. 流体对流传热的热边界条件有哪两种？对传热有什么影响？

授课视频

6.7 The Experimental Correlations of Turbulent Heat Transfer in Tubes

6.7 管内湍流传热实验关联式

For the calculation of turbulent heat transfer in tubes, Dittus-Boelter formula is the most widely used in engineering:

对于管内流体湍流传热计算，工程上使用最广的是迪图斯-贝尔特公式：

$$Nu_f = 0.023Re_f^{0.8}Pr_f^n \quad (6.19)$$

n takes 0.4 when the fluid is heated, n takes 0.3 when the fluid is cooled. Where the reference temperature adopts the average of fluid temperature t_f, and the characteristic length is the inner diameter of the tube. The scope of experimental verification is also the applicable scope of the formula. So are the following formulas.

流体被加热时，n 取 0.4，流体被冷却时，n 取 0.3。式中，定性温度采用流体平均温度，特征长度为管内径。实验验证范围也是该公式的适用范围，下面的公式也一样。

$$Re_f = 10^4 \sim 1.2\times10^5, Pr_f = 0.7 \sim 120, l/d \geqslant 60$$

This formula is applicable to the occasions where the temperature difference between fluid and wall surface is below medium, which is no more than 50℃ for gases, and no more than 30℃ for water.

该式适用于流体与壁面具有中等以下温差的场合，对于气体不超过50℃，对于水不超过30℃。

The velocity distribution of the fluid in a tube will be distorted with the change of temperature. The dotted line 1 in Fig. 6.5 is the velocity distribution of isothermal flow. In fact, the fluid temperature on the cross-section is not uniform, which affects the change of fluid thermophysical properties, leading to the distortion of velocity distribution, as shown in the figure of solid lines 2 and 3.

管内流体的速度分布随着温度的变化会发生畸变。图6.5中虚线1为等温流动时速度分布。实际上，横截面上流体温度并不均匀，并影响流体热物性的变化，导致速度分布发生畸变，如图中的实线2和实线3。

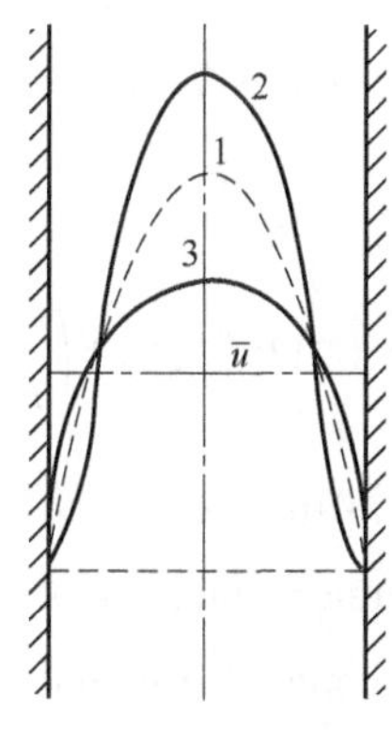

Fig. 6.5 Distortion of the velocity distribution of the fluid in a tube with changes in temperature

图6.5 管内流体速度分布随温度变化的畸变

1—等温流动 2—液体冷却或气体加热 3—液体加热或气体冷却

The correction coefficient should be generally introduced in the correlation formula, i.e. $(\eta_f/\eta_w)^n$ or $(Pr_f/Pr_w)^n$, to modify the effect of non-uniform physical field on heat transfer. Where the subscript f denotes the value at the fluid temperature, the subscript w denotes the value at the wall temperature.

一般在关联式中需引进修正系数，即 $(\eta_f/\eta_w)^n$ 或 $(Pr_f/Pr_w)^n$ 来修正不均匀物性场对传热的影响。式中，下标 f 表示流体温度下的值，下标 w 表示壁面温度下的值。

For the case of large temperature difference, the following correction formulas can be used for calculation.

对于大温差情形，可采用下列修正式计算。

(1) Dittus-Boelter correction formula

(1) 迪图斯-贝尔特修正式

$$Nu_f = 0.023 Re_f^{0.8} Pr_f^n c_t \tag{6.20}$$

When the gas is heated, $c_t = \left(\frac{T_f}{T_w}\right)^{0.5}$; when the gas is cooled, $c_t = 1$; for liquids, $c_t = \left(\frac{\eta_f}{\eta_w}\right)^m$, m takes 0.11 when the fluid is heated, and m takes 0.25 when the fluid is cooled.

对于气体被加热时，$c_t = \left(\frac{T_f}{T_w}\right)^{0.5}$；对于气体被冷却时，$c_t = 1$；对于液体，$c_t = \left(\frac{\eta_f}{\eta_w}\right)^m$，流体受热时，$m = 0.11$，流体被冷却时，$m = 0.25$。

(2) Sider-Tate formula

(2) 齐德-泰特公式

$$Nu_f = 0.027 Re_f^{0.8} Pr_f^{1/3} \left(\frac{\eta_f}{\eta_w}\right)^{0.14} \tag{6.21}$$

The reference temperature is the average of fluid temperature t_f, η_w is determined by the wall temperature t_w, and the characteristic length is the inner diameter of the tube. The experimental verification scope of Re and Pr is

定性温度为流体平均温度 t_f，η_w 按壁温 t_w 确定，特征长度为管内径。Re 和 Pr 的实验验证范围为：

$$Re_f \geqslant 10^4, Pr_f = 0.7 \sim 16700, l/d \geqslant 60$$

(3) Mihaiev formula

(3) 米海耶夫公式

$$Nu_f = 0.021 Re_f^{0.8} Pr_f^{0.43} \left(\frac{Pr_f}{Pr_w}\right)^{0.25} \tag{6.22}$$

The reference temperature is the average of fluid temperature t_f, the characteristic length is the inner diameter of the tube. The experimental verification scope of Re and Pr is

定性温度为流体平均温度 t_f，特征长度为管内径。Re 和 Pr 的实验验证范围为：

$$10^4 \leqslant Re_f \leqslant 1.75 \times 10^6, Pr_f = 0.6 \sim 700, l/d \geqslant 50$$

(4) Gnielinski formula

(4) 格尼林斯基公式

$$Nu_f = \frac{(f/8)(Re - 1000) Pr_f}{1 + 12.7\sqrt{f/8}(Pr_f^{2/3} - 1)} \left[1 + \left(\frac{d}{l}\right)^{2/3}\right] c_t \tag{6.23}$$

At present, it is generally believed that the formula has the highest accuracy, but the formula form is relatively

目前大家普遍认为，该式的计算准确度最高，但是公式形式比较复杂。

complex. Different modification coefficients c_t are given respectively for liquids and gases.

对液体和气体分别给出了不同的修改系数 c_t。

For liquids,

对于液体，

$$c_t=\left(\frac{Pr_f}{Pr_w}\right)^{0.01},\frac{Pr_f}{Pr_w}=0.05\sim20$$

For gases,

对于气体，

$$c_t=\left(\frac{T_f}{T_w}\right)^{0.45},\frac{T_f}{T_w}=0.5\sim1.5$$

The Experimental verification scope of Re and Pr is

Re 和 Pr 的实验验证范围为：

$$Re_f=2300\sim10^6,Pr_f=0.6\sim10^5$$

Where f is the resistance coefficient of turbulent flow in a tube:

式中，f 为管内湍流流动的阻力系数：

$$f=(1.82\lg Re-1.62)^{-2} \tag{6.24}$$

The above criterion equations are obtained for circular section flow channels, their application scope can be further expanded. Some corrections need to be made. For non-circular section flow channels, the equivalent diameter can be used as the characteristic length and applied to the above criterion equations. Note: for the flow region with sharp angles on the cross-section, the calculation with equivalent diameter will lead to a large error.

上述准则方程都是针对圆形截面流道得到的，其应用范围可以进一步扩大，当然也需要做一些修正。对于非圆形截面流道，可用当量直径作为特征长度，应用到上述准则方程中。注意：对截面上有尖角的流动区域，用当量直径计算会导致较大的误差。

For entrance region, the thermal boundary layer in the entrance region is thin, so it has a high convective heat transfer coefficient. The following entrance effect correction coefficient is available:

对于入口段，由于入口段热边界层较薄，因而具有较高的表面传热系数。可用以下入口效应修正系数：

$$c_l=1+(d/l)^{0.7} \tag{6.25a}$$

For spiral tube, the fluid in the tube may cause secondary circulation in the cross-section and enhance heat transfer, as shown in Fig. 6.6, so, the spiral tube correction coefficients can be used:

对于螺旋管，管内流体在横截面上会引起二次环流而产生强化传热作用，如图 6.6 所示，因此可以采用螺旋管修正系数：

For gases:

对于气体：

$$c_r=1+1.77\frac{d}{R} \tag{6.25b}$$

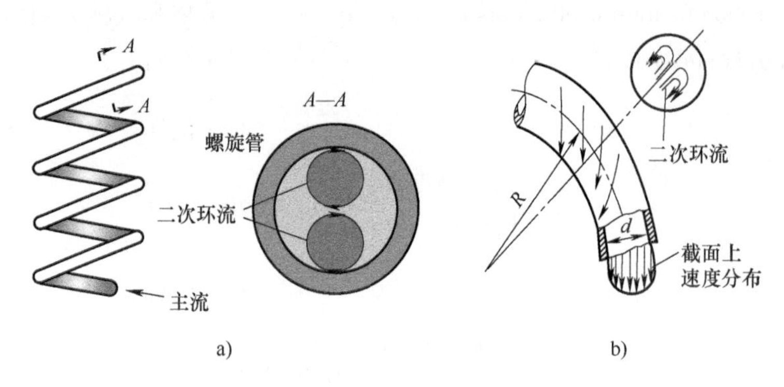

Fig. 6. 6 Flow state of fluid in a spiral tube

图 6. 6 螺旋管内流体的流动状态

For liquids:

对于液体:

$$c_r = 1 + 10.3\left(\frac{d}{R}\right)^3 \tag{6.25c}$$

Where d is the inner diameter of the tube, and R is the spiral radius. All the above equations are applicable only to gases or liquids with Pr number greater than 0. 6. For liquid metals with very small Pr number, the law of heat transfer is completely different. In the smooth circular tube, the criterion equations of the fully developed turbulent heat transfer for liquid metals are as follows:

式中,d 是管内径;R 是螺旋半径。以上所有方程仅适用于 Pr 大于 0. 6 的气体或液体。对 Pr 很小的液态金属,其传热规律完全不同。在光滑圆管内,液态金属充分发展湍流传热的准则式为:

For the uniform heat flux boundary:

对于均匀热流边界:

$$Nu_f = 4.82 + 0.0185 Pe_f^{0.827} \tag{6.26}$$

The scope of experimental verification is

实验验证范围为

$$Re_f = 3.6 \times 10^3 \sim 9.05 \times 10^5, \quad Pe_f = 10^2 \sim 10^4$$

For the uniform wall temperature boundary: The experimental verification range is $Pe_f > 100$. Where the characteristic length is the inner diameter of the tube, the reference temperature is the average of fluid temperature.

对于均匀壁温边界:实验验证范围是 $Pe_f > 100$。这里特征长度为管内径,定性温度为流体平均温度。

Summary This section illustrates the experimental correlations of turbulent heat transfer in tubes in different working conditions, and the correction methods of the experimental correlation with the change of the working conditions.

小结 本节阐述了不同工况下管内湍流传热计算的实验关联式,以及随着工况变化实验关联式的修正方法。

Question What are the variation characteristics of fluid velocity distribution in tubes under the condition of non-uniform fluid temperature on the cross-section?

思考题 在横截面上流体温度不均匀的条件下，管内流体速度分布变化的特点是什么？

Self-tests

1. What is the most widely used formula for turbulent heat transfer coefficient in tubes?
2. How to calculate the characteristic scale of non-circular section channel?
3. Why can the spiral tube enhance heat transfer?

自测题

1. 使用最广的管内湍流传热的表面传热系数公式是什么？
2. 非圆形截面槽道的特征尺度怎么计算？
3. 螺旋管为什么能强化传热？

授课视频

6.8 Nusselt Number of Laminar Heat Transfer in Tubes

6.8 管内层流传热的 *Nu*

For the fully developed laminar convection heat transfer in tubes, its theoretical research is sufficient, and there are many research results available. As can be seen from Table 6.1: ① *Nu* numbers of fully developed laminar heat transfer in the tubes with different cross-section shapes are independent of *Re* numbers; ② Among them, *Nu* number of the circular section channel is the maximum, and *Nu* number of the triangular section channel is the minimum; ③ For channels with the same cross-section shape, *Nu* number of uniform heat flux is always greater than that of uniform wall temperature.

管内层流充分发展对流传热的理论研究比较充分，有许多研究结果可选用。从表6.1中可以看出：① 不同截面形状的管内，层流充分发展传热的 *Nu* 均与 *Re* 无关；② 其中，圆形截面流道的 *Nu* 最大，三角形截面流道的 *Nu* 最小；③ 对于同一截面形状的流道，均匀热流的 *Nu* 总是大于均匀壁温的 *Nu*。

Table 6.1 *Nu* numbers of fully developed laminar heat transfer in tubes with different cross-section shapes

表6.1 不同截面形状的管内层流充分发展传热的 *Nu*

截面形状	$Nu=hd_e/\lambda$		$fRe\left(Re=\dfrac{ud_e}{\nu}\right)$
	均匀热流	均匀壁温	
正三角形 △	3.11	2.47	53

（续）

截面形状	$Nu=hd_e/\lambda$		$fRe\left(Re=\frac{ud_e}{\nu}\right)$
	均匀热流	均匀壁温	
正方形	3.61	2.98	57
正六边形	4.00	3.34	60
圆	4.36	3.66	64
长方形 $\left(\frac{b}{a}=2\right)$	4.12	3.39	62

As can be seen from Table 6.2: For the rectangular channels, as the aspect ratio of the rectangular section increases, Nu number of fully developed laminar heat transfer also increases. Similarly Nu numbers are independent of Re numbers, Nu number of uniform heat flow is always greater than that of uniform wall temperature.

从表 6.2 中可以看出：对于长方形流道，随着长方形截面的长宽比增大，层流充分发展传热的 Nu 也随之增大；同样，Nu 与 Re 无关，均匀热流的 Nu 总是大于均匀壁温的 Nu。

Table 6.2 Nu numbers of fully developed laminar heat transfer in the rectangular channels

表 6.2 长方形流道内层流充分发展传热的 Nu

截面形状	$Nu=hd_e/\lambda$		$fRe\left(Re=\frac{ud_e}{\nu}\right)$
	均匀热流	均匀壁温	
$\frac{b}{a}=3$	4.79	3.96	69
$\frac{b}{a}=4$	5.33	4.44	73
$\frac{b}{a}=8$	6.49	5.60	82
$\frac{b}{a}=\infty$	8.23	7.54	96

As can be seen from Table 6.3: For fully developed laminar heat transfer in the annular space, when one side of the wall is adiabatic and the other is isothermal, Nu number of the inner wall decreases and Nu number of the outer

从表 6.3 中可以看出：对于环形空间内层流充分发展传热，当一侧壁面绝热，另一侧壁面等温时，随着内外径之比的增大，内壁面的 Nu

wall increases as the ratio of inner and outer diameters increases, and Nu number of inner wall changes faster.

随之减小，外壁面的 Nu 随之增大，且内壁面的 Nu 变化更快。

Table 6.3 Nu numbers of fully developed laminar heat transfer in the annular space

表 6.3 环形空间内层流充分发展传热的 Nu

内、外径之比 d_i/d_o	内壁 Nu_i（外壁绝热）	外壁 Nu_o（内壁绝热）
0	—	3.66
0.05	17.46	4.06
0.10	11.56	4.11
0.25	7.37	4.23
0.50	5.74	4.43
1.00	4.86	4.86

Table 6.4 also shows that for the fully developed laminar heat transfer in the annular space, when the inner and outer walls maintain uniform heat flux, Nu number of the inner wall decreases and Nu number of the outer wall increases as the ratio of inner and outer diameters increases, and Nu number of the inner wall changes faster.

表 6.4 也表明，对于环形空间内层流充分发展传热，当内外侧壁面均维持均匀热流时，随着内外径之比的增大，同样是内壁面的 Nu 随之减小，外壁面的 Nu 随之增大，且内壁面的 Nu 变化更快。

Table 6.4 Nu numbers of fully developed laminar heat transfer in the annular space

表 6.4 环形空间内层流充分发展传热的 Nu

内外径之比 d_i/d_o	内壁 Nu_i	外壁 Nu_o
0	—	4.364
0.05	17.81	4.792
0.10	11.91	4.834
0.20	8.499	4.833
0.40	6.583	4.979
0.60	5.912	5.099
0.80	5.580	5.240
1.00	5.385	5.385

The above research results show that the fully developed laminar convection heat transfer in tubes has the following common characteristics:

上述研究结果表明，管内层流充分发展对流传热具有以下共同特点：

(1) For channels of the same cross-section shape, Nu number of uniform heat flux is always greater than that of

(1) 对于同一截面形状的流道，均匀热流的 Nu 总是大于均匀壁温

uniform wall temperature, and more than 19 percent for the circular tubes.

的 Nu，对于圆管超出 19%。

(2) For the straight channels with constant cross-section, no matter how the shape of the section changes, Nu number of the fully developed laminar flow is independent of Re number.

(2) 对于等截面直流道，不论截面形状如何，层流充分发展时，Nu 均与 Re 无关。

(3) Even if the equivalent diameter is taken as the characteristic length and the equivalent diameters are equal, Nu numbers of the fully developed laminar heat transfer in the different section channels are not equal, that is, Nu number is related to the shape of the channel cross-section.

(3) 即使把当量直径作为特征长度，且当量直径相等，不同截面流道层流充分发展传热的 Nu 也不相等，即 Nu 与流道截面形状有关。

Summary This section illustrates the characteristics of Nu number of fully developed laminar convection heat transfer in tubes with the shape of the channel cross-section.

小结 本节阐述了管内层流充分发展对流传热的 Nu 随流道截面形状变化的特点。

Question Why do all the heat transfer tubes used in engineering choose the circular section?

思考题 为什么工程中使用的传热管都选择圆形截面？

Self-tests

自测题

1. Why is Nusselt number independent of Reynolds number when laminar flow is fully developed?

1. 为什么层流充分发展时努塞尔数与雷诺数无关？

2. The laminar flow in tubes with different section shapes fully develops and transfers heat, and the equivalent diameters are equal. Why is the Nu number of circular section channels the largest?

2. 不同截面形状的管内层流充分发展传热，且当量直径相等，为什么圆形截面流道的 Nu 最大？

授课视频

6.9 An Application Example of the Experimental Correlations in Tubes

6.9 管内实验关联式的应用实例

Example 6.3 When water flows through a straight tube with a length l of 5m at a uniform wall temperature, it is heated from 25.3℃ to 34.6℃, the inner diameter of the

例 6.3 水流过长 $l=5\text{m}$ 壁温均匀的直管时，从 25.3℃ 被加热到 34.6℃，管子的内径 $d=20\text{mm}$，水

tube d is 20mm, the flow velocity of water in the tube is 2m/s. Find the convective heat transfer coefficient.

在管内的流速为 2m/s，求表面传热系数。

Solution: Once the convective heat transfer coefficient is calculated, it may first be thought of using Newton cooling formula. However, can Newton cooling formula be directly used for calculation?

解： 一旦计算表面传热系数，可能首先会想到用牛顿冷却公式。但是，能否直接用牛顿冷却公式计算呢?

According to the known conditions, we can get the heat transfer quantity of water by the formula $\Phi=\rho u\frac{\pi d^2}{4}c_p(t_f''-t_f')$. But the temperature of the tube wall is unknown, so Newton cooling formula can't be directly used for calculation. Therefore, it needs to be calculated using Dittus-Boelter formula.

根据已知条件，由公式 $\Phi=\rho u\frac{\pi d^2}{4}c_p(t_f''-t_f')$，可求出水的传热量。但是管壁温度未知，所以无法直接采用牛顿冷却公式来计算。因此，需要采用迪图斯-贝尔特公式计算。

Assume that: ① The ratio of length to diameter is greater or equal to 60; ② Heat transfer is in the range of small temperature difference. After the convective heat transfer coefficient is calculated, the average wall temperature can be calculated, and the assumptions will be verified. If the assumptions are not valid, the next calculation will be performed on the basis of the first calculation result.

假定：① 长径比大于或等于 60；② 传热处于小温差的范围。在下面计算出表面传热系数以后，再推算平均壁温，并校核假定条件是否成立。如果不成立，则在第一次计算结果的基础上再进行计算。

First find the average temperature of the water, and take it as a reference temperature. It can be found in the appendix that the thermal conductivity, kinematic viscosity and Prandtl number of water at the reference temperature. So we can obtain:

下面先求出水的平均温度，以此作为定性温度。从附录中可查出该定性温度下水的导热系数、运动黏度和普朗特数。由此可以求得：

$$Re_f=\frac{ud}{\nu_f}=\frac{2\text{m/s}\times0.02\text{m}}{0.805\times10^{-6}\text{m}^2/\text{s}}=4.97\times10^4>10^4$$

So, the fluid flow is in fully turbulent region and can be calculated using the Dittus-Boelter formula:

因此，流动处于旺盛湍流区，可采用迪图斯-贝尔特公式计算：

$$Nu_f=0.023Re_f^{0.8}Pr_f^{0.4}=0.023\times(4.97\times10^4)^{0.8}\times5.42^{0.4}=258.5$$

Thus, the convective heat transfer coefficient can be calculated:

由此可以计算出表面传热系数：

$$h_m=\frac{\lambda_f}{d}Nu_f=\frac{0.618\text{W}/(\text{m}\cdot\text{K})}{0.02\text{m}}\times258.5=7988\text{W}/(\text{m}^2\cdot\text{K})$$

Now let's check it: density ρ and specific heat capacity c_p of water at 30℃ can be found in the appendix. Then, the heat absorption per second of the heated water can be calculated as:

下面进行校核：从附录中可查出，30℃时水的密度 ρ、比热容 c_p，则被加热水每秒钟的吸热量为：

$$\Phi = \rho u \frac{\pi d^2}{4} c_p (t_f'' - t_f') = 2.43 \times 10^4 \mathrm{W}$$

The wall temperature can be calculated by Newton cooling formula:

用牛顿冷却公式可计算出壁温：

$$t_w = t_f + \frac{\Phi}{hA} = 39.7℃$$

Then the temperature difference can be calculated $t_w - t_f$ = 9.7℃, it is much less than 20℃. Therefore, The working condition of this problem is within the applicable scope of Dittus-Boelter formula, and the convective heat transfer coefficient is the answer to this question.

于是可计算出温差 $t_w - t_f$ = 9.7℃，远小于20℃。因此，本题工况在迪图斯-贝尔特公式的适用范围内，所求的表面传热系数即为本题的答案。

Let's make a comparison and discussion. Calculate again according to Gnielinski formula, and approximate t_w = 40℃. From the appendix we can get Prandtl number and viscosity at the wall temperature and viscosity at the fluid temperature, and calculate the resistance coefficient of turbulent flow in the tube.

下面做一个对比和讨论。再按格尼林斯基公式计算，并近似取 t_w = 40℃。由附录可查得壁温下普朗特数、黏度及流体温度下黏度，并计算出管内湍流流动的阻力系数。

$$f = (1.82 \times \lg Re - 1.62)^{-2} = 0.02096$$

Substitute these data into Gnielinski formula, we can get Nu_f = 308.8.

把这些数据代入格尼林斯基公式，就可以得到 Nu_f = 308.8。

It can be seen by comparison that the calculation results of the two correlations differ by about 19.5%. According to the views of most literature, the calculation result of Gnielinski formula is more accurate, so, it can be considered that the calculation result of Dittus-Boelter formula is about 19.5% lower.

通过对比可知，两个关联式的计算结果相差约19.5%。根据大多数文献的观点，格尼林斯基公式计算结果的准确性更高一些。因此，可认为本题按迪图斯-贝尔特公式计算结果大约偏低19.5%。

The calculation difference of the convective heat transfer coefficient is evaluated as follows: Firstly, the uncertainty of the experimental correlation of convection heat transfer is generally about 25%, so, the above differences are all within uncertainty; Secondly, 5% to 10% deviation is

下面对表面传热系数计算差别进行评价：首先，对流传热实验关联式的不确定度一般在25%左右，因此上述差别都在不确定度之内；其次，对于一般工程计算，5%~10%的偏

acceptable for general engineering calculations.

差是可以接受的。

Summary This section illustrates the application method of experimental correlations in tubes by a practical example.

小结 本节通过实例阐述了管内实验关联式的应用方法。

Question What are the methods for calculating turbulent heat transfer coefficient in tubes in engineering?

思考题 工程上计算管内湍流的表面传热系数有哪些方法?

授课视频

6.10 The Experimental Correlation Equations for Heat Transfer of Fluid Flowing Across a Single Tube

6.10 流体横掠单管传热实验关联式

The situation of fluid flowing across a single tube is an external flow. The characteristics of external flow: The flow boundary layer and thermal boundary layer can fully develop freely on the heat transfer wall. Because there is only restriction of the inner wall surface and no outer wall surface in the fluid flow process.

流体横掠单管属于外部流动。外部流动的特征：传热壁面上，流动边界层与热边界层能够充分地自由发展。因为流体在流动过程中，只有内壁面，没有外壁面的限制。

Flowing across a single tube refers to the fluid flowing across the surface of the tube along the direction perpendicular to the tube axis. In addition to the characteristics of the boundary layer, fluid flow can also produce detour flow and body-shedding and resulting in backflow, vortex and vortex beam. Fig. 6.7 shows the separation phenomenon of the boundary layer when the fluid crosses a single tube.

横掠单管：指流体沿着垂直于管轴线方向流过管子表面。流体流动除了具有边界层特征以外，还会发生绕流脱体而产生回流、旋涡和涡束。图 6.7 所示为流体横掠单管时边界层的分离现象。

The phenomenon of detour flow and body-shedding of fluid flowing across a single tube: The starting point of detour flow and body-shedding is called the separation point, and starting from the separation point, the separation of the inner edge of the boundary layer from the wall is called body-shedding. The starting point of fluid body-shedding depends on *Re* number. If *Re* number is less than 10, no body-shedding occurs, because the viscous force is dominant at the time.

流体横掠单管的绕流脱体现象：绕流脱体的起点称为分离点，从分离点开始，边界层内缘脱离壁面称为脱体。流体脱体的起点取决于 *Re* 的大小。如果 *Re* 小于 10，不会出现脱体，因为此时黏性力占主导。

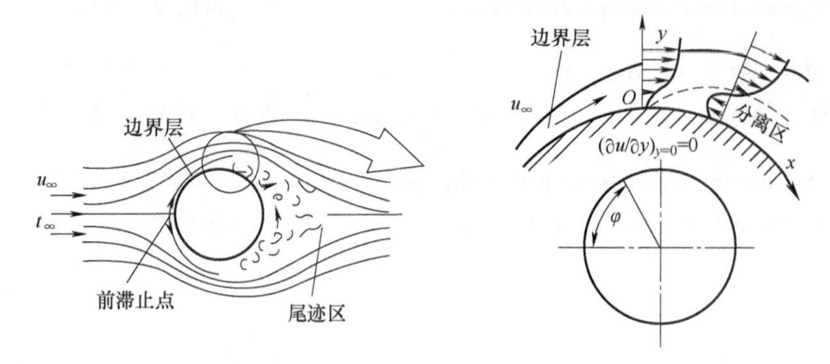

Fig. 6. 7 The separation phenomenon of the boundary layer when the fluid crosses a single tube

图 6. 7 流体横掠单管时边界层的分离现象

If *Re* number is greater than 10 but less than 1.5×10^5, then the boundary layer is in laminar flow, body-shedding occurs at the position of $\varphi=80°\sim85°$, and the inertial force is dominant at this time. If *Re* number is greater than 1.5×10^5, the boundary layer has turned into turbulent flow before body-shedding, body-shedding is pushed back to the position of $\varphi=140°$. Fig. 6. 8 is the variation of the local convective heat transfer coefficient on the circular tube surface under the condition of constant heat flux on the wall.

如果 *Re* 大于 10 而小于 1.5×10^5，边界层为层流，脱体发生在 φ 等于 $80°\sim85°$的位置，此时惯性力占主导。如果 *Re* 大于 1.5×10^5，边界层在脱体前已经转变为湍流，脱体会推后到 $\varphi=140°$的位置。图 6. 8 所示为在壁面恒热流的条件下，圆管表面的局部表面传热系数变化。

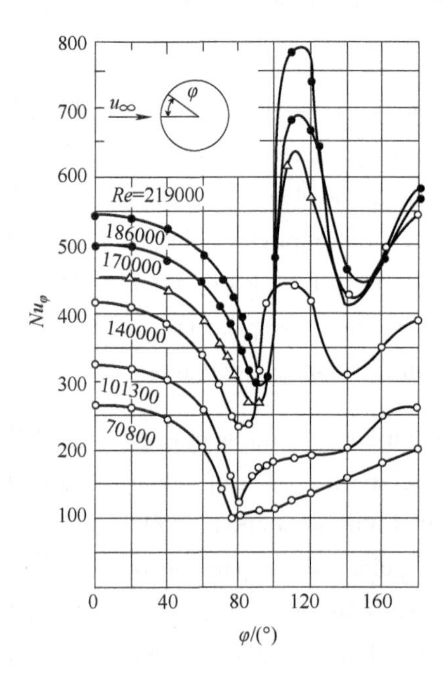

Fig. 6. 8 The variation of the local convective heat transfer coefficient on the circular tube

图 6. 8 圆管表面的局部表面传热系数变化

We can see 4 main characteristics when the fluid flowing across the circular tube for heat transfer: ① When φ is between 0° and 80°, as the laminar boundary layer thickens continuously, then the Nu number curve is going down; ② When Re number is small (i. e. hollow point in Fig. 6. 8), the curve rises at $\varphi = 80°$, the rebound point reflects the starting point of detour flow and body-shedding, which shows that body-shedding enhances surface heat transfer; ③ When the Re number is big (i. e. solid dots in the figure), the first point that the curve goes up, is the transition point from laminar flow to turbulent flow, and the second point that the curve goes up, at the position of about $\varphi = 140°$, is a new starting point of detour flow and body-shedding; ④ As the Re number of fluid increases, Nu number also increases.

可以看到，流体横掠圆管传热时，有4个主要特征：① φ 位于0°~80°之间时，随着层流边界层不断增厚，Nu 曲线逐渐下降；② 低 Re 时（如图6.8中空心点），在 φ 为80°的位置曲线回升，回升点反映了绕流脱体的起点，说明脱体强化了表面传热；③ 高 Re 时（如图中实心点），第一次曲线回升点，是层流向湍流的转变点，第二次曲线回升点，约在 φ 为140°的位置，是绕流脱体的新起点；④ 随着流体 Re 增大，Nu 也随之增大。

It can be seen that the development and body-shedding of the boundary layer determines the convective heat transfer of the fluid across the circular tube. Although the change of local convective heat transfer coefficient is more complicated, the regularity of average convective heat transfer coefficient gradient is very obvious.

由此可见，边界层的发展和脱体决定了流体横掠圆管的对流传热状况。虽然局部表面传热系数变化比较复杂，但是平均表面传热系数渐变的规律性非常明显。

A power correlation can be used to calculate the fluid flowing across a tube:

流体横掠单管可采用幂次关联式计算：

$$Nu = CRe^{n}Pr^{1/3} \tag{6.27}$$

Please look at Table 6. 5 for the values of C and n in the formula. The reference temperature is the average of wall temperature and fluid temperature. The characteristic length is the outer diameter of the tube. The characteristic velocity is the incoming fluid velocity. The scope of experimental verification is:

式中，C 和 n 的值请看表6.5。定性温度为壁面温度与流体温度的平均值，特征长度为管外径，特征速度为来流速度。实验验证范围为：

$$t_{\infty} = 15.5 \sim 980℃, t_{w} = 21 \sim 1046℃$$

The convection heat transfer of gas across non-circular tubes can also be calculated by using the above correlation formula. The values of C and n in the equation are shown in Table 6. 6. The geometric dimension l in the table is the characteristic length.

气体横掠非圆形截面管道的对流传热也可采用上述关联式计算。式中，C 及 n 值见表6.6。表中的几何尺寸 l 为特征长度。

Table 6.5 Values of C and n in Eq. (6.27) (fluid)

表 6.5 式 (6.27) 中 C 和 n 的值 (流体)

Re	C	n
0.4~4	0.989	0.330
4~40	0.911	0.385
40~4000	0.683	0.466
4000~40000	0.193	0.618
40000~400000	0.0266	0.805

Table 6.6 Values of C and n in Eq. (6.27) (gas)

表 6.6 式 (6.27) 中 C 和 n 的值 (气体)

	Re	C	n
正方形 (l)	5×10^3 ~ 10^5	0.246	0.588
(l)	5×10^3 ~ 10^5	0.102	0.675
正六边形 (l)	5×10^3 ~ 1.95×10^4 1.95×10^4 ~ 10^5	0.160 0.0385	0.638 0.782
(l)	5×10^3 ~ 10^5	0.153	0.638
竖直平板 (l)	4×10^3 ~ 1.5×10^4	0.228	0.731

In addition, Churchill and Bonsteadon proposed a criterion correlation formula for fluid flowing across a single tube that was applicable throughout the test range:

另外，邱吉尔与朋斯登对流体横掠单管提出了在整个实验范围内都适用的准则关系式：

$$Nu=0.3+\frac{0.62Re^{1/2}Pr^{1/3}}{[1+(0.4/Pr)^{2/3}]^{1/4}}\left[1+\left(\frac{Re}{282000}\right)^{5/8}\right]^{4/5} \tag{6.28}$$

Where the reference temperature is the average of the wall temperature and the fluid temperature. It is applicable when $Re \cdot Pr$ is greater than 0. 2.

式中，定性温度为壁面温度与流体温度的平均值，适用于 $Re \cdot Pr>0.2$ 的情形。

Example 6. 4 In a low-speed wind tunnel, the convection heat transfer test of air flowing across a horizontal circular tube is carried out by electric heating method. The test tube is placed in the two side walls of the wind tunnel, the part exposed to the air is 100mm long, the outer diameter is 12mm. The incoming fluid temperature is measured experimentally $t_\infty = 15℃$. The average temperature of heat transfer surface $t_w = 125℃$. Electric power is $P = 40.5W$. Due to the radiation on the surface of the heat transfer tube and the heat conduction between both ends of the heat transfer tube and the side walls of the wind tunnel, it is estimated that about 15% of the power is lost. Try to calculate the convective heat transfer coefficient at this time.

例 6. 4 在低速风洞中，用电加热方法进行空气横掠水平圆管的对流传热实验。实验管置于风洞的两个侧壁上，暴露在空气中的部分长 100mm，外径为 12mm。实验测得来流温度 $t_\infty = 15℃$，传热表面平均温度 $t_w = 125℃$，电功率 $P = 40.5W$。由于换热管表面的辐射及换热管两端通过风洞侧壁的导热，估计约有 15%的功率损失。试计算此时对流传热的表面传热系数。

Solution: This is an example of experimentally measuring the convective heat transfer coefficient of fluid flowing across a single tube. It can be calculated with Newton cooling formula. The average surface heat transfer coefficient of the entire heat transfer tube is:

解：这是用实验方法测定横掠单管表面传热系数的例子。可用牛顿冷却公式计算，整个换热管的平均表面传热系数为：

$$h = \frac{\Phi}{A(t_w - t_\infty)}$$

From the known conditions we can get total heat transfer quantity:

由已知条件可得总传热量为：

$$\Phi = 0.85P = 34.43W$$

And the surface area of a single tube is known:

又有单管的表面面积：

$$A = \pi dl = 3.768 \times 10^{-3} m^2$$

Substitute them into the above formula to get:

代入上式就可以得出：

$$h = 83.1W/(m^2 \cdot K)$$

In order to improve the measurement accuracy of the convective heat transfer coefficient, it is of great significance to try to reduce the radiation and heat conduction loss at

为了提高表面传热系数测定的精确度，在本实验中，尽量降低换热管的辐射及两端导热损失具有重要意

both ends of the heat transfer tube in this experiment. To reduce radiation heat transfer, a layer of chromium can be plated on the surface of the heat transfer tube, the surface emissivity can be reduced to 0. 1 ~ 0. 05. In order to reduce the heat conduction loss at both ends, insulation material is used to separate the heat transfer tube from the wall of the wind tunnel.

义。为了减少辐射传热，可在换热管表面镀一层铬，使表面发射率下降到0.1 ~ 0.05。为了减少两端导热损失，在换热管穿过风洞壁面处用绝热材料隔开。

In the natural convection experiment, reducing radiation and heat conduction losses at the ends is more important to improve the accuracy of test results.

在进行自然对流实验时，减少辐射及端部导热损失，对提高测试结果的准确度具有更重要意义。

Summary This section illustrates the variation law of local convective heat transfer coefficient and the experimental correlations for the heat transfer of the fluid flowing across a single tube.

小结 本节阐述了流体横掠圆管时局部表面传热系数的变化规律，以及流体横掠单管传热的实验关联式。

Question Can the example in this section adopt the power correlation equation of a fluid flowing across a single tube to calculate the average convective heat transfer coefficient of heat transfer tubes?

思考题 本节的例题能否采用流体横掠单管幂次关联式计算换热管的平均表面传热系数？

Self-tests

自测题

1. What is the effect of the fluid flow state on the phenomenon of detour flow and body-shedding?

1. 流体的流动状态对绕流脱体现象有何影响？

2. Please use the boundary layer theory to explain the characteristics of the heat transfer of the fluid flowing across a circular tube.

2. 用边界层理论解释流体横掠圆管传热的特征。

授课视频

6.11 The Experimental Correlation Equations for Heat Transfer of Fluid Flowing Across Tube Bundle

6.11 流体横掠管束传热实验关联式

In engineering, in order to increase the heat transfer area, the tube bundle consisting of a large number of heat transfer

在工程中，为了增大传热面积，通常采用大量换热管所组成的管束，

tubes is commonly used, and thus forming a heat transfer component. The phenomenon of fluid externally flowing across tube bundle is most common in shell and tube heat exchangers and boilers, as shown in Fig. 6. 9.

从而构成传热组件。流体外掠管束现象在管壳式换热器和锅炉中最为常见，如图 6. 9 所示。

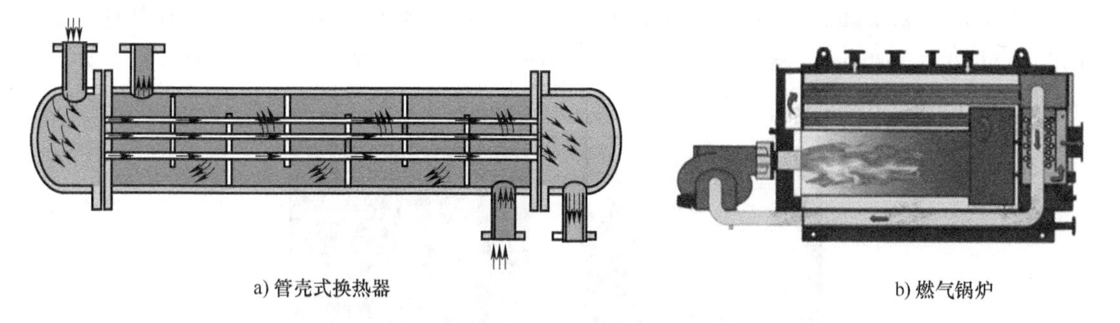

Fig. 6. 9 Equipment for fluid externally flowing across tube bundle

图 6. 9 流体外掠管束的设备

In shell and tube heat exchangers and gas boilers, the tubes are usually arranged in two ways: cross array and in-line array. As shown in Fig. 6. 10, cross array means that the central line of the adjacent tubes has a certain angle with the flow direction of the fluid; In-line array means that the central line of the adjacent tubes is parallel to the flow direction of the fluid.

在管壳式换热器和燃气锅炉中，通常管子的排列方式有叉排和顺排两种。如图 6. 10 所示，叉排是指相邻管子的中心连线与流体流向成一定夹角，顺排是指相邻管子的中心连线与流体流向平行。

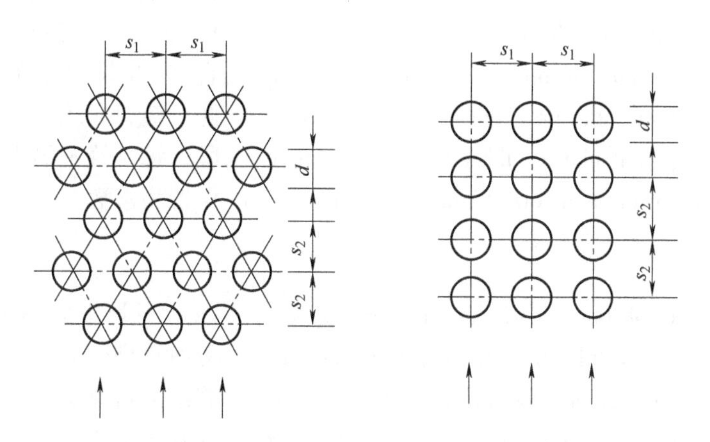

Fig. 6. 10 Cross and in-line tube bundle

图 6. 10 叉排和顺排管束

The different arrangement has great influence on fluid flow and heat transfer outside the tube. Therefore, in addition to *Re* and *Pr*, the factors affecting the heat transfer of tube bundle also include tubes arrangement, transverse and longitudinal tube spacing, tube row number and other factors.

不同的排列方式对管外流体流动和传热的影响很大，因此，影响管束传热的因素，除 *Re*、*Pr* 以外，还有管子排列方式、横向及纵向管间距，以及管束排数等因素。

The flow disturbance effect of the tube bundle on the fluid is very strong in the cross array, so, the heat transfer intensity is also large, but the resistance loss is large and it's difficult to clean. The effect of in-line array on fluid flow disturbance is small, so the heat transfer intensity is small, and the resistance loss is small and it's easy to clean, as shown in Fig. 6. 11.

在叉排时，管束对流体的扰流作用非常大，因而传热强度也较大，但阻力损失较大，并且难以清洗。在顺排时，对流体扰流的作用小，因此传热强度也较小，阻力损失较小，并且清洗方便，如图 6. 11 所示。

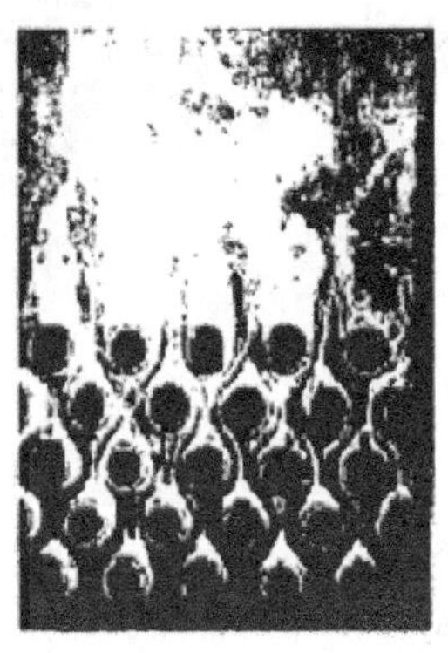

a) 叉排（正三角形布管）

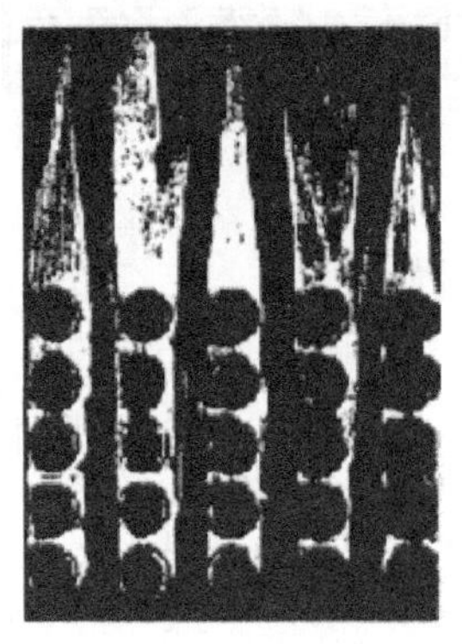

b) 顺排（正方形布管）

Fig. 6. 11 Flow scene when fluid laterally scours tube bundle

图 6. 11 流体横向冲刷管束时的流动景象

Experiments show that:

(1) The rear row tube is disturbed by the wake of the front row tube, the effect on the average convective heat transfer coefficient will not disappear until more than 16 rows of tubes. Since then, the flow and heat transfer have entered a full development phase, the average convective heat transfer coefficient of each row of tubes remains constant;

实验表明：

(1) 后排管受前排管尾流的扰动作用，对平均表面传热系数的影响直到 16 排以上的管子才会消失。此后流动和传热进入充分发展阶段，每排管子的平均表面传热系数保持为常数。

(2) When the number of tube rows is less than 16, ignore the effect of the number of tube rows firstly, then the calculation results of the correlation formula are multiplied by the tube row correction coefficient. The Soviet heat transfer scientist Zhukauskas summarized a set of experimental correlations for heat transfer of fluid flowing across tube bundle, as shown in Table 6. 7 and Table 6. 8. These correlations apply to a wide range of Pr number.

(2) 当管排数小于 16 排时，先不考虑管排数的影响，对关联式的计算结果再乘以管排修正系数即可。苏联传热学家茹卡乌斯卡斯总结出一套流体外掠管束传热实验关联式，见表 6. 7 和表 6. 8。这些关联式适用的 Pr 范围很广。

Where the reference temperature is the average of the inlet and outlet temperature of the fluid, Pr number is deter-

式中，定性温度为流体进出口平均温度；Pr 按管束的平均壁温确定；

mined by the average wall temperature of the tube bundle. Flow velocity in the Re number takes the average flow velocity in the minimum section of the tube bundle. The characteristic length is the outer diameter of the tube. The experimental verification range of Pr number is between 0.6~500.

Re 中的流速取管束中最小截面的平均流速；特征长度为管子外径；实验验证范围是 Pr 在 0.6~500 之间。

Table 6.7 Experimental correlations of the average convective heat transfer coefficient of fluid flowing across the in-line tube bundle (≥16 rows)

表 6.7 流经顺排管束（≥16 排）的流体平均表面传热系数实验关联式

关联式	适用 Re 范围
$Nu_f=0.9Re_f^{0.4}Pr_f^{0.36}(Pr_f/Pr_w)^{0.25}$	$1\sim10^2$
$Nu_f=0.52Re_f^{0.5}Pr_f^{0.36}(Pr_f/Pr_w)^{0.25}$	$10^2\sim10^3$
$Nu_f=0.27Re_f^{0.63}Pr_f^{0.36}(Pr_f/Pr_w)^{0.25}$	$10^3\sim2\times10^5$
$Nu_f=0.033Re_f^{0.8}Pr_f^{0.36}(Pr_f/Pr_w)^{0.25}$	$2\times10^5\sim2\times10^6$

Table 6.8 Experimental correlations of the average convective heat transfer coefficient of fluid flowing across the cross tube bundle(≥16 rows)

表 6.8 流经叉排管束(≥16 排)的流体平均表面传热系数计算公式

关联式	适用 Re 范围
$Nu_f=1.04Re_f^{0.4}Pr_f^{0.36}(Pr_f/Pr_w)^{0.25}$	$1\sim5\times10^2$
$Nu_f=0.71Re_f^{0.5}Pr_f^{0.36}(Pr_f/Pr_w)^{0.25}$	$5\times10^2\sim10^3$
$Nu_f=0.35\left(\frac{s_1}{s_2}\right)^{0.2}Re_f^{0.6}Pr_f^{0.36}(Pr_f/Pr_w)^{0.25},\ \frac{s_1}{s_2}\leqslant2$	$10^3\sim2\times10^5$
$Nu_f=0.40Re_f^{0.6}Pr_f^{0.36}(Pr_f/Pr_w)^{0.25},\ \frac{s_1}{s_2}>2$	$10^3\sim2\times10^5$
$Nu_f=0.031\left(\frac{s_1}{s_2}\right)^{0.2}Re_f^{0.8}Pr_f^{0.36}(Pr_f/Pr_w)^{0.25}$	$2\times10^5\sim2\times10^6$

For tube rows less than 16, on the basis of the above formulas the average convective heat transfer coefficient is multiplied by a tube row correction factor of less than 1, as shown in Table 6.9.

管排数小于 16 时，平均表面传热系数在上述公式基础上乘以一个小于 1 的管排修正系数，见表 6.9。

Table 6.9 Tube row correction factor of Zhukauskas Formulas

表 6.9 茹卡乌斯卡斯公式的管排修正系数

总排数	1	2	3	4	5	6	7	8	9	10	11	12	13	14	15
顺排 $Re>10^3$	0.700	0.800	0.865	0.910	0.928	0.942	0.954	0.965	0.972	0.978	0.983	0.987	0.990	0.992	0.994

(续)

总排数	1	2	3	4	5	6	7	8	9	10	11	12	13	14	15
叉排 $10^2<Re<10^3$	0.832	0.874	0.914	0.939	0.955	0.963	0.970	0.976	0.980	0.984	0.987	0.990	0.993	0.996	0.999
$Re>10^3$	0.619	0.758	0.840	0.897	0.923	0.942	0.954	0.965	0.971	0.977	0.982	0.986	0.990	0.994	0.997

Example 6.5 In a boiler, flue gas flows across a in-line tube bundle consisting of 4 rows of tubes. We know the tube diameter $d = 60\text{mm}$, $s_1/d = 2$, $s_2/d = 2$, and the average flue gas temperature $t_f = 600℃$, the wall temperature $t_w = 120℃$, the average velocity at the narrowest part of flue gas passage $u = 8\text{m/s}$. Try to find the average convective heat transfer coefficient of tube bundle.

例 6.5 一台锅炉中，烟气横掠 4 排管组成的顺排管束。已知管外径 $d=60\text{mm}$，$s_1/d=2$，$s_2/d=2$，烟气平均温度 $t_f=600℃$，壁面温度 $t_w=120℃$，烟气通道最窄处平均流速 $u=8\text{m/s}$。试求管束平均表面传热系数。

Solution: This question directly gives all the parameters required for the Zhukauskas formulas, the physical parameters of the average flue gas composition in the attached table of the textbook can be used for calculation. From the attached table we can obtain, Pr number, kinematic viscosity, thermal conductivity at the average temperature of flue gas, and Pr number at wall temperature.

解： 本题直接给出了茹卡乌斯卡斯公式所需的所有参数，可采用教材附表中平均烟气成分的物性参数进行计算。由附表可查出，烟气平均温度下的普朗特数、运动黏度、导热系数，以及壁面温度下的普朗特数。

According to Re number, We know from Table 6.7 that the following correlation formula should be used.

根据 Re 查表 6.7 可知，应按照下面的关联式计算。

$$Nu_f = 0.27 Re_f^{0.63} Pr_f^{0.36} (Pr_f/Pr_w)^{0.25}$$

From this formula, the convective surface heat transfer coefficient can be obtained:

由此式可以求出表面传热系数：

$$h = Nu \frac{\lambda}{d} = 59.6\text{W}/(\text{m}^2 \cdot \text{K})$$

According to Table 6.9, it can be found that the tube row correction factor is 0.91, so the average convective heat transfer coefficient is $h = 54.2\text{W}/(\text{m}^2 \cdot \text{K})$.

按表 6.9 可以查出管排修正系数为 0.91，因此，平均表面传热系数为 $h=54.2\text{W}/(\text{m}^2 \cdot \text{K})$。

Discussions: ① Similar to the case that there are multiple correlations for convective heat transfer in tubes, there are also different correlations for fluid flowing across tube

讨论： ① 与管内对流传热存在多个关联式的情形相似，流体外掠管束也有不同的关联式，对同一个问

bundle. The calculation results of the same problems are also different from each other; ② As an example, all the conditions required for adoption of the correlation are given directly. But in practical engineering, it is difficult to measure the average surface temperature of the heat transfer tube.

题的计算结果，相互间也有一定的差异；② 作为例题，直接给出了采用关联式所需的所有条件。但在工程实际中，测定换热管表面的平均温度是很困难的。

Summary This section illustrates experimental correlations for heat transfer of fluid flowing across tube bundle, the application method of experimental correlations is introduced by an example.

小结 本节阐述了流体横掠管束传热实验关联式，并通过实例介绍了实验关联式的应用方法。

Question The calculation model for heat transfer of the fluid flowing across tube bundle, which is close to practical application conditions, is that the average temperature of the fluid at the inlet and outlet of the tube bundle, and quantity of flow, are measured, and the geometric conditions of the tube row are given. Try to analyze in this case, how to apply the results of tables 6.7 to 6.9 to determine the average convective heat transfer coefficient of tube bundle?

思考题 接近实际应用条件的流体横掠管束传热计算模型是：测定了流体进出管排处的平均温度、流体的流量，给出管排的几何条件。试分析在这种情形下，该如何应用表 6.7～表 6.9 的结果，来确定管束的平均表面传热系数？

Self-tests

自测题

1. What is the effect of the two arrangements of cross array and in-line array on fluid heat transfer?

1. 叉排和顺排两种排列方式对流体传热有何影响？

2. Why use the correction factor for the row number of tube bundles?

2. 为什么要采用管束排数修正系数？

3. What are the factors affecting heat transfer of fluid flowing across the tube bundle?

3. 流体横掠管束传热时影响因素有哪些？

Exercises

习题

6.1 In a model reduced to 1/8 of the physical object, air at 20℃ is used to simulate the heating process of air with an average temperature of 200℃ in the object. The average velocity of air in the physical object is 6.03m/s. What should be the velocity in the model? If the average surface

6.1 在一缩小为实物 1/8 的模型中，用 20℃ 的空气来模拟实物中平均温度为 200℃ 空气的加热过程。实物中空气的平均流速为 6.03m/s，问模型中的流速应为多少？

heat transfer coefficient in the model is 195W/($m^2 \cdot K$), find the value in the corresponding physical object. In this test, the Pr numbers of the fluid in the model and the object are not strictly equal. Do you think such a modeling test is of practical value?

若模型中的平均表面传热系数为195W/($m^2 \cdot K$)，求相应实物中的值。在这一实验中，模型与实物中流体的 Pr 并不严格相等，你认为这样的模化实验有无实用价值?

6.2 For the convective heat transfer of constant physical fluid flowing across the tube bundle, when the number of tube rows in the flow direction is greater than 10, the test found that the average surface heat transfer coefficient h of the tube bundle depends on the following factors: fluid velocity u, fluid physical properties ρ, η, λ, c_p and geometric parameters d, s_1, s_2. It is proved by the dimensional analysis method that the convective heat transfer relationship at this time can be sorted into $Nu=f(Re, Pr, s_1/d, s_2/d)$ (where s_1 and s_2 are the tube center distances in different directions).

6.2 对于常物性流体横向掠过管束时的对流传热，当流动方向上的管排数大于10时，实验发现，管束的平均表面传热系数 h 取决于下列因素：流体速度 u，流体物性 ρ、η、λ、c_p，几何参数 d、s_1、s_2。试用量纲分析法证明此时的对流传热关系式可整理为 $Nu=f(Re, Pr, s_1/d, s_2/d)$（其中，$s_1$、$s_2$ 为不同方向的管心距）。

6.3 Try to calculate the equivalent diameter in the following four cases: (1) A rectangular channel with side lengths a and b; (2) Same as (1), but $b \ll a$; (3) An annular channel, the outer diameter of the inner tube is d, and the inner diameter of the outer tube is D; (4) n circular tubes with an outer diameter of d are arranged in a circular cylinder with an inner diameter of D, and the fluid flows longitudinally outside the circular tube.

6.3 试计算下列四种情形下的当量直径：(1) 边长为 a 及 b 的矩形通道；(2) 同 (1)，但 $b \ll a$；(3) 环形通道，内管外径为 d，外管内径为 D；(4) 在一个内径为 D 的圆形筒体内布置了 n 根外径为 d 的圆管，流体在圆管外纵向流动。

6.4 The cooling efficiency of the generator can be improved after the cooling medium changes from air to hydrogen. Try to compare the cooling effects of hydrogen and air. The conditions for comparison are as those described below: turbulent convection heat transfer in the pipeline, the channel geometry sizes and flow velocity are same, the reference temperature is 50℃, and the gas under normal pressure, regardless of the temperature difference correction. The physical properties of hydrogen at 50℃ are as follows:

6.4 发电机的冷却介质从空气改为氢气后可以提高冷却效率，试比较氢气与空气的冷却效果。比较的条件是：管道内湍流对流传热，通道几何尺寸、流速均相同，定性温度为50℃，气体均处于常压下，不考虑温差修正。50℃氢气的物性数据如下：

$$\rho=0.0755\text{kg/m}^3, \lambda=19.42\times10^{-2}\text{W/(m}\cdot\text{K)}, \eta=9.41\times10^{-6}\text{Pa}\cdot\text{s}, c_p=14.36\text{kJ/(kg}\cdot\text{K)}$$

6.5 The air at 1.013×10^5Pa flows in a straight pipe with an inner diameter of 76mm, the inlet temperature is 65℃, the inlet volume flow rate is $0.022\text{m}^3/\text{s}$, and the average temperature of the pipe wall is 180℃. How long is the pipe required to heat the air up to 115℃?

6.5 1.013×10^5Pa 下的空气在内径为 76mm 的直管内流动，入口温度为 65℃，入口体积流量为 $0.022\text{m}^3/\text{s}$，管壁的平均温度为 180℃。问管子要多长才能使空气加热到 115℃？

6.6 Water flows through a long straight tube with an inner diameter of 20mm at an average velocity of 1.2m/s. (1) The tube wall temperature is 75℃, and the water is heated from 20℃ to 70℃; (2) The tube wall temperature is 15℃, and the water is cooled from 70℃ to 20℃. Try to calculate the surface heat transfer coefficient for both cases and discuss the reasons for the difference.

6.6 水以 1.2m/s 的平均速度流过内径为 20mm 的长直管。(1) 管子壁温为 75℃，水从 20℃ 加热到 70℃；(2) 管子壁温为 15℃，水从 70℃ 冷却到 20℃。试计算两种情形下的表面传热系数，并讨论造成差别的原因。

6.7 A flat plate is 400mm long and the average wall temperature is 40℃. Air at 20℃ under normal pressure flows longitudinally across the surface of the plate at a speed of 10m/s. Try to calculate the thermal boundary layer thickness, local surface heat transfer coefficient and average surface heat transfer coefficient at 50mm, 100mm, 200mm, 300mm, and 400mm from the front edge of the plate.

6.7 一平板长 400mm，平均壁温为 40℃，常压下 20℃ 的空气以 10m/s 的速度纵向流过板表面，试计算离平板前缘 50mm、100mm、200mm、300mm、400mm 处的热边界层厚度、局部表面传热系数及平均表面传热系数。

6.8 There is a heat sink with a height of 2cm and a length of 12cm on the shell of a motorcycle engine (the length direction is parallel to the body). The heat sink surface temperature $t_w = 150℃$, if the motorcycle travels against the wind in the environment of $t_\infty = 20℃$ at a speed of 30km/h, and the wind speed is 2m/s, the wind speed is parallel to the vehicle speed. Try to calculate the heat dissipation of the heat sink at this time.

6.8 在一摩托车发动机的壳体上有一条高 2cm、长 12cm 的散热片（长度方向与车身平行）。散热片表面温度为 $t_w = 150℃$，如果车子在 $t_\infty = 20℃$ 的环境中逆风前行，车速为 30km/h，而风速为 2m/s，风速与车速平行。试计算此时散热片的散热量。

6.9 A human body can be regarded as a cylinder with a height of 1.75m and a diameter of 0.35m, with a surface temperature of 31℃. A marathon runner runs the entire course (42195m) in 2.5 hours. The air is stationary at 15℃. Excluding the heat dissipation at the two ends of the column and the portion of sweat loss. Try to estimate the heat dissipation of the runner after finishing the race.

6.9 人体可看成高 1.75m、直径为 0.35m 的圆柱体，表面温度为 31℃。一个马拉松运动员在 2.5h 内跑完全程（42195m），空气是静止的，温度为 15℃。不计柱体两端面的散热，不计出汗散失的部分，试估算此运动员跑完全程后的散热量。

6. 10　In the air preheater of a boiler, the air flows transversely across a group of cross tube bundle, s_1 = 80mm, s_2 = 50mm, the outer diameter of the tubes is d_o = 40mm, the air velocity at the minimum section is 6m/s, fluid temperature t_f = 133℃. The number of tube rows in the flow direction is greater than 10, and the average temperature of the tube wall is 165℃. Try to determine the average surface heat transfer coefficient between the air and the tube bundle.

6. 10　在锅炉的空气预热器中，空气横向掠过一组叉排管束，s_1 = 80mm，s_2 = 50mm，管子外径 d_o = 40mm，空气在最小截面处的流速为 6m/s，流体温度 t_f = 133℃。在流动方向上管排数大于 10，管壁平均温度为 165℃。试确定空气与管束间的平均表面传热系数。

Chapter 7 Phase Change Convection Heat Transfer

第 7 章 相变对流传热

7.1 The Modes of Condensation Heat Transfer

Condensation heat transfer is a phase change heat transfer mode in which vapor condenses into liquid. There are many examples of condensation heat transfer in daily life and industry. For example, we can see droplets and ice flowers on the indoor windows in cold winter. This is the phenomenon that the high-temperature water vapor in the indoor air condenses when it meets the low-temperature glass wall surface.

In industrial production, condensers in power station system and condensers in refrigeration system, etc., will have condensation heat transfer during operation. The condensation process takes place when saturated vapor contacts the wall surface below the saturation temperature. There are two modes of condensation heat transfer:

The first mode is film-wise condensation. Taking the vertical wall as an example, if the condensate liquid can wet the wall surface well, a thin film will be formed along the entire wall surface, and the condensate liquid film will flow downwards under the action of gravity, as shown in Fig. 7.1. The latent heat of vaporization released by condensation must be transferred to the wall through the condensate film. So, the thickness of the condensate film directly affects heat transfer.

The second mode is drop-wise condensation as shown in Fig. 7.2. When the condensate liquid cannot wet the wall

7.1 凝结传热的模式

凝结传热是一种由蒸气凝结成液体的相变传热方式。在日常生活和工业生产中，有许多凝结传热的例子。比如在寒冷的冬天，我们看到室内窗户上会形成露珠和冰花等。这是室内空气中温度较高的水蒸气遇到了低温的玻璃壁面发生凝结的现象。

在工业生产中，电站系统中的凝汽器及制冷系统中的冷凝器等运行时都会发生凝结传热。凝结过程发生的条件是饱和蒸气与低于饱和温度的壁面接触。凝结传热有两种模式：

第一种模式是膜状凝结。以竖壁为例，如果凝结液体能很好地浸润壁面，那么沿整个壁面会形成一层薄膜，并在重力的作用下凝结液膜向下流动，如图 7.1 所示。凝结放出的汽化潜热必须通过凝结液膜传给壁面，因此，凝结液膜的厚度直接影响了热量传递。

第二种模式是珠状凝结，如图 7.2 所示。当凝结液体不能很好地浸润

surface well, then many small droplets will be formed on the wall surface. At this time, part of the wall surfacc is in direct contact with the vapor, thus, the heat transfer rate of drop-wise condensation is much higher than that of film-wise condensation, which may be several times or even an order of magnitude larger.

壁面时，则在壁面上会形成许多小液珠。这时壁面的部分表面与蒸气直接接触，因此珠状凝结的传热速率远大于膜状凝结，可能大几倍甚至是大一个数量级。

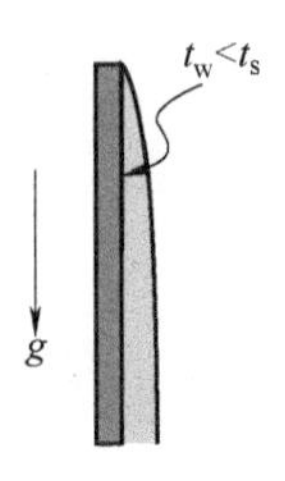

Fig. 7. 1 Film-wise condensation
图 7. 1 膜状凝结

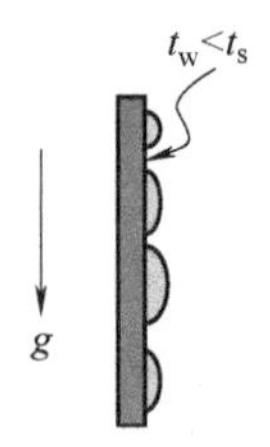

Fig. 7. 2 Drop-wise condensation
图 7. 2 珠状凝结

Both of the above condensation modes mentioned a phenomenon, that is, the condensate liquid wets the wall surface. The wettability can be measured by the contact angle between the condensate and the wall surface. The contact angle θ is formed between the gas-liquid interface and the wall surface, as shown in Fig. 7. 3.

上面两种凝结模式都提到了一个现象，即凝结液体浸润壁面。浸润性的好坏，可以用凝结液与壁面的接触角来衡量。接触角是气液分界面与壁面之间形成的夹角 θ，如图 7. 3 所示。

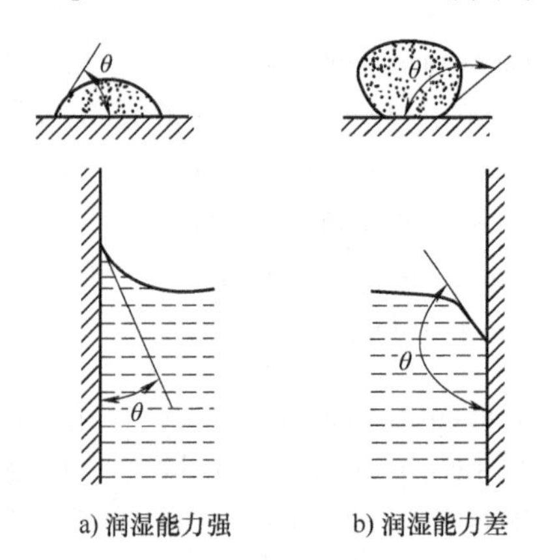

Fig. 7. 3 Contact angles of liquid film on wall surface under different wetting conditions
图 7. 3 不同润湿条件下表面的液膜接触角 θ

The smaller the contact angle θ is, the stronger the wetting ability of the liquid to the wall surface is, and the liquid can spread out on the wall surface and form film-wise condensation. When the contact angle θ is relatively large, a large number of droplets with small diameter will be formed on the wall surface, and grow up gradually.

接触角 θ 越小，则液体润湿壁面能力越强，液体会铺展在壁面上形成膜状凝结。当接触角 θ 较大时，就会在壁面形成大量直径很小的液珠，并逐渐增大。

On the non-horizontal wall surface, the droplets roll down under the action of gravity and combine with other droplets to form larger droplets, so that the wall surface is exposed again for continuous condensation. Whether it is film-wise condensation or drop-wise condensation, condensed liquid is the main thermal resistance carrier for heat transfer between the vapor and the wall surface.

在非水平壁面上，受重力作用液珠向下滚落，与其他液珠结合形成更大的液珠，从而使壁面重新裸露出来，以便继续冷凝。不论是膜状凝结还是珠状凝结，凝结液体都是蒸气与壁面之间传热的主要热阻载体。

Experiments have proved that almost all common vapors, including water vapor, condense in film-wise on the clean surface of commonly used engineering materials under pure conditions. This situation is consistent with our daily experience of cleaning laboratory utensils. Generally, the criterion to judge whether the utensil is clean is whether a uniform film is formed on its surface.

实验证明，几乎所有的常用蒸气（包括水蒸气）在纯净的条件下，在常用工程材料的洁净表面上都发生膜状凝结。这种情况与我们日常清洗实验器皿的经验相符合。通常判断器皿是否洗净的标准是，在其表面上是否形成均匀薄膜。

Although the heat transfer rate of drop-wise condensation is much higher than that of film-wise condensation, most of the condensation encountered in engineering belongs to film-wise condensation. Therefore, from the point of view of engineering design, in order to ensure the condensation effect, only the film-wise condensation can be used as the calculation basis.

尽管珠状凝结传热速率远大于膜状凝结，然而在工程中遇到的大多数凝结都属于膜状凝结。因此，从工程设计的观点出发，为保证凝结效果，只能以膜状凝结为计算依据。

For condensation heat transfer, the following important parameters need to be noticed: ① Condensation driving force: That is the difference between the saturation temperature of vapor and the wall temperature; ② The latent heat of vaporization; ③ Characteristic length; ④ Other normal thermophysical parameters, such as dynamic viscosity, thermal conductivity, specific heat capacity, etc.

对于凝结传热，以下重要参数需要注意：① 凝结推动力，即蒸气的饱和温度与壁面温度之差（t_s-t_w）；② 汽化潜热 r；③ 特征长度；④ 其他标准的热物理性质参数，如动力黏度、导热系数、比热容等。

Summary This section illustrates two modes of condensation heat transfer, that is, film-wise condensation and drop-wise condensation, and their definitions and differences.

小结 本节阐述了凝结传热的两种模式，即膜状凝结和珠状凝结，以及它们的定义和区别。

Question Why are the condensation phenomena in engineering all film-wise condensation?

思考题 为什么工程中凝结现象都是膜状凝结？

Self-tests

自测题

1. What is the difference between film-wise condensation and drop-wise condensation?

1. 膜状凝结和珠状凝结有什么区别?

2. What is the main resistance of the heat transfer process during film-wise condensation?

2. 膜状凝结时热量传递过程的主要阻力是什么?

3. Why is the convective heat transfer coefficient of drop-wise condensation larger than that of film-wise condensation?

3. 为什么珠状凝结比膜状凝结的表面传热系数大?

授课视频

7.2 The Analytical Solution of Laminar Film-wise Condensation

7.2 层流膜状凝结的分析解

In 1916, Nusselt first acquired the analytical solution of the laminar film-wise condensation heat transfer of pure vapor. He grasped the key point of this problem, that is, the thermal conduction resistance of the liquid film is the main thermal resistance in the condensation process, ignoring other secondary factors, and theoretically revealed the influence of relevant physical quantities on condensation heat transfer, which laid a foundation for the modern film-wise condensation theory and heat transfer analysis.

1916年，努塞尔首先得到了纯净蒸气层流膜状凝结传热的分析解。他抓住了这一问题的关键，即液膜的导热热阻是凝结过程的主要热阻，忽略了其他的次要因素，从理论上揭示了有关物理量对凝结传热的影响，为近代膜状凝结理论和传热分析奠定了基础。

Since then, various modifications or developments have been carried out in response to the restrictive assumptions of Nusselt's analysis, thus forming a variety of practical calculation methods. This section focuses on Nusselt's analysis method. Nusselt's analysis was based on the laminar film-wise condensation of pure saturated vapor on a vertical surface with uniform wall temperature.

此后，各种修正或发展都是针对努塞尔分析的限制性假设来进行的，从而形成了各种实用的计算方法。本节重点学习努塞尔的分析方法。努塞尔的分析是对纯净的饱和蒸气在均匀壁温的竖直表面上的层流膜状凝结做出的。

According to the characteristics of the actual condensation process, the following eight assumptions are made to facilitate the mathematical solution:

根据实际凝结过程的特点，为便于数学求解做出以下8个假设:

① The physical properties keep constant; ② The vapor is stationary and the viscous force of the vapor-liquid interface on the liquid film is 0; ③ The inertial force of the liquid film is ignored; ④ There is no temperature difference on the vapor-liquid interface, namely the temperature of the liquid film is equal to the saturation temperature on the interface, written as $t_\delta = t_s$; ⑤ The temperature in the liquid film is linearly distributed, which means that heat transfer is only heat conduction; ⑥ The subcooling degree of liquid film can be ignored, that is, the sensible heat can be ignored; ⑦ The density of vapor can be ignored, because the density of vapor is much smaller than that of liquid; ⑧ The surface of the liquid film is smooth without fluctuation.

① 常物性；② 蒸气静止且气液界面对液膜的黏滞力为 0；③ 忽略液膜的惯性力；④ 气液界面无温差，即界面上液膜温度等于饱和温度，写成 $t_\delta = t_s$；⑤ 液膜内温度线性分布，意味着热量转移只有导热；⑥ 液膜的过冷度可忽略，即显热可忽略；⑦ 忽略蒸气密度，因为蒸气的密度远远小于液体的密度；⑧ 液膜表面平整无波动。

Under the above assumptions, the flow and heat transfer of the condensation film conform to the thin-layer properties of the boundary layer. Taking the laminar film-wise condensation on the vertical wall as an example and the coordinate x as the direction of gravity, as shown in Fig. 7.4.

在以上假设条件下，凝结液膜的流动和传热符合边界层的薄层性质。以竖壁的层流膜状凝结为例，把坐标 x 取为重力方向，如图 7.4 所示。

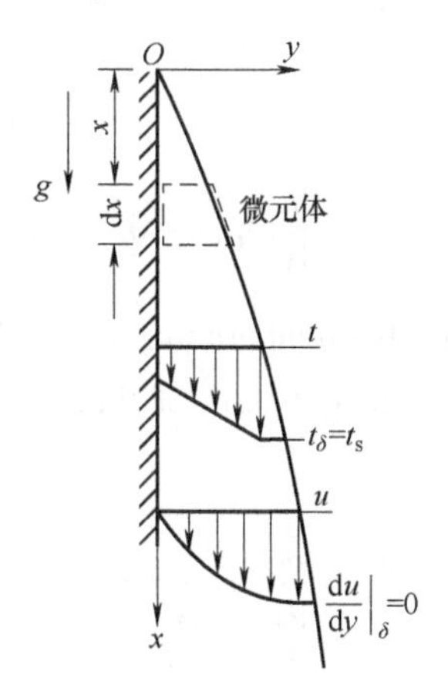

Fig. 7.4 Laminar film-wise condensation on the vertical wall

图 7.4 竖壁的层流膜状凝结

Under steady-state condition, the governing equations of the boundary layer in Chapter 5 are applicable to the flow and heat transfer process of the condensate film. The governing equations include continuity equation, momentum equation and energy equation. According to the above eight assumptions, simplify the equation group:

在稳态条件下，第 5 章的边界层控制方程适用于凝结液膜的流动和传热过程。控制方程包括连续性方程、动量方程和能量方程。根据上面的 8 条假设，对该方程组进行简化：

$$
\begin{cases}
\dfrac{\partial u}{\partial x}+\dfrac{\partial v}{\partial y}=0 \\
\rho_l\left(u\,\dfrac{\partial u}{\partial x}+v\,\dfrac{\partial u}{\partial y}\right)=-\dfrac{\mathrm{d}p}{\mathrm{d}x}+\rho_l g+\eta_l\,\dfrac{\partial^2 u}{\partial y^2} \\
u\,\dfrac{\partial y}{\partial x}+v\,\dfrac{\partial t}{\partial y}=a_l\,\dfrac{\partial^2 t}{\partial y^2}
\end{cases}
\tag{7.1}
$$

Considering the third assumption that the inertial force of the liquid film is ignored, then the left term of the momentum equation in Eq. (7.1) equals 0:

考虑假设 3 忽略液膜的惯性力，式（7.1）中动量方程左边项为 0：

$$\rho_l\left(u\,\frac{\partial u}{\partial x}+v\,\frac{\partial u}{\partial y}\right)=0$$

Considering the second assumption that the vapor is stationary, the first term on the right side of the momentum equation in Eq. (7.1) is:

考虑假设 2 蒸气是静止的，式（7.1）中动量方程右边第一项：

$$\frac{\mathrm{d}p}{\mathrm{d}x}=\rho_v g$$

Considering the seventh assumption that the vapor density is ignored, then

考虑假设 7 忽略蒸气密度，则

$$\rho_v g=0$$

So

因此

$$\frac{\mathrm{d}p}{\mathrm{d}x}=\rho_v g=0$$

Considering the fifth assumption that the temperature in the liquid film is linearly distributed, the left term of the energy equation in Eq. (7.1) is 0:

考虑假设 5 液膜内温度线性分布，式（7.1）中能量方程左边项为 0：

$$u\,\frac{\partial t}{\partial x}+v\,\frac{\partial t}{\partial y}=0$$

At this time, only u and t in Eq. (7.1) are unknown, and they can be solved by using the momentum equation and energy equation. Therefore, the continuity equation can be discarded, and the equation group can be simplified as:

此时，式（7.1）中只有 u 和 t 两个未知量，用动量方程和能量方程即可求解，所以可舍去连续性方程，方程组可简化为：

$$
\begin{cases}
\rho_l g+\eta_l\,\dfrac{\partial^2 u}{\partial y^2}=0 \\
a_l\,\dfrac{\partial^2 t}{\partial y^2}=0
\end{cases}
$$

And its boundary conditions are:

其边界条件为:

$$y=0\text{ 时},\quad u=0,\quad t=t_w$$

$$y=\delta\text{ 时},\quad \left.\frac{du}{dy}\right|_{\delta}=0,\quad t=t_s$$

Solve the equations and directly integrate the above two equations twice to obtain:

求解方程组，对上述两个方程直接做两次积分可得:

$$u=\frac{\rho_l g}{\eta_l}\left(\delta y-\frac{1}{2}y^2\right)\qquad t=t_w+(t_s-t_w)\frac{y}{\delta}\tag{7.2}$$

Note that the variation of liquid film thickness δ with x is unknown in the formula, which means that δ is still unknown.

注意式中液膜厚度 δ 随 x 的变化规律未知，也就是说 δ 还是未知数。

Next, solve δ by applying the sixth assumption, that is, the sensible heat released by liquid film subcooling is not considered. At this time, the heat conduction quantity through the liquid film with thickness δ should be equal to the latent heat released by vapor condensation. Then the following relationship can be obtained:

接下来求解 δ，这里应用假设 6，即不考虑液膜过冷释放显热。此时，通过厚度 δ 的液膜的导热量应等于蒸气凝结释放出来的潜热量。于是可得到以下关系式:

$$\lambda_l\frac{t_s-t_w}{\delta}dx=r\,dq_m=r\frac{g\rho_l^2\delta^2 d\delta}{\eta_l}$$

The expression of liquid film thickness δ can be obtained by integrating the above formula:

对上式进行积分，即得到液膜厚度 δ 的表达式为:

$$\delta=\left[\frac{4\eta_l\lambda_l(t_s-t_w)x}{g\rho_l^2 r}\right]^{1/4}\tag{7.3}$$

Where the physical parameters are determined by the following reference temperature:

式中，物性参数按以下定性温度取值:

$$t_m=\frac{t_s+t_w}{2}$$

Note that the latent heat of vaporization r is determined by t_s.

注意，汽化潜热 r 是由 t_s 确定的。

Next, find the convective heat transfer coefficient h between the vapor and the wall. According to the relationship that the convective heat transfer quantity between the vapor and the wall is equal to the heat conduction quantity of the liquid film, the following formula can be obtained:

接下来求解蒸气与壁面之间的表面传热系数 h。根据蒸气和壁面之间的对流传热量与液膜导热量相等的关系，可得到下式:

$$h_x A(t_s - t_w) = \lambda A \frac{t_s - t_w}{\delta}$$

Thus, we get:

从而得到：

$$h_x = \frac{\lambda}{\delta}$$

Substitute the expression of δ into the above expression, the local convective heat transfer coefficient h_x can be obtained as follows:

把 δ 的表达式代入上式，可得到局部表面传热系数 h_x 的表达式为：

$$h_x = \left[\frac{g r \rho_l^2 \lambda_l^3}{4\eta_l (t_s - t_w) x}\right]^{1/4} \tag{7.4}$$

Eq. (7.4) is the local convective heat transfer coefficient at the position x on the vertical wall.

式（7.4）为竖壁 x 位置处的局部表面传热系数。

If we want to calculate the average convective heat transfer coefficient on the entire vertical wall, the above local convective heat transfer coefficient h_x is integrated on the entire wall to obtain the expression of average convective heat transfer coefficient h_V:

如果要计算整个竖壁的平均表面传热系数，对上面的局部表面传热系数 h_x 在整个壁面上积分，可得到平均表面传热系数 h_V 的表达式为：

$$h_V = \frac{1}{l}\int_0^l h_x \mathrm{d}x = 0.943\left[\frac{g r \rho_l^2 \lambda_l^3}{\eta_l l (t_s - t_w)}\right]^{1/4} \tag{7.5}$$

Eq. (7.5) is Nusselt's theoretical solution of laminar liquid film condensation on a vertical wall. Where the subscript V represents the vertical wall. Experiments show that the condensation heat transfer is enhanced due to the fluctuation of the liquid film surface. So, the experimental value is about 20% higher than the above theoretical value, as shown in Fig. 7.5.

式（7.5）就是竖壁上液膜层流膜状凝结的努塞尔理论解。其中，下标 V 代表竖壁。实验表明，由于液膜表面波动，凝结传热得到强化。因此，实验值比上述理论值高 20%左右，如图 7.5 所示。

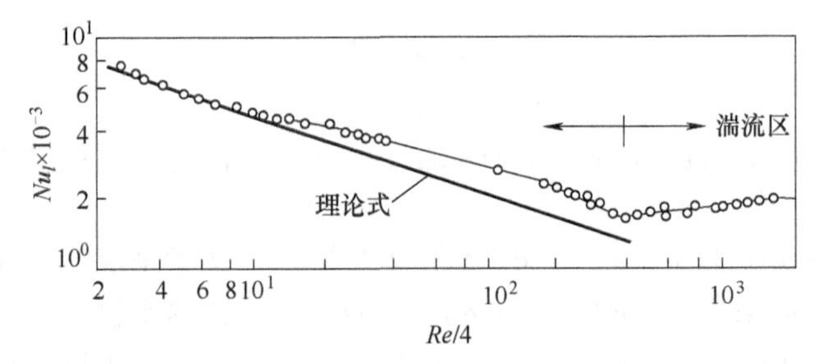

Fig. 7.5 Comparison of theoretical value and experimental value

图 7.5 理论值与实验值的比较

Therefore, the coefficient of the theoretical calculation formula is increased by 20% in engineering application. The revised calculation formula is as follows:

因此，工程应用时，将理论计算公式系数增加 20%。修正后的计算公式为：

$$h_V = 1.13\left[\frac{gr\rho_l^2\lambda_l^3}{\eta_l l(t_s - t_w)}\right]^{1/4} \tag{7.6}$$

If the wall is inclined, $g\sin\phi$ is used instead of g in the above formulas.

如果是倾斜壁，则用 $g\sin\phi$ 代替以上各式中的 g 即可。

Nusselt's theoretical analysis can be extended to laminar film-wise condensation on the surface of a horizontal circular tube and a sphere. The calculation formula for a horizontal circular tube is as follows:

努塞尔的理论分析可推广到水平圆管及球的表面上层流膜状凝结。对于水平圆管，计算公式为：

$$h_H = 0.729\left[\frac{gr\rho_l^2\lambda_l^3}{\eta_l d(t_s - t_w)}\right]^{1/4} \tag{7.7}$$

The calculation formula for a spherical surface is as follows:

对于球表面，计算公式为：

$$h_S = 0.826\left[\frac{gr\rho_l^2\lambda_l^3}{\eta_l d(t_s - t_w)}\right]^{1/4} \tag{7.8}$$

Where the subscript H represents horizontal tube, S represents sphere, d is the diameter of horizontal tube or sphere. The reference temperature is the same as that of the previous formula. Since the horizontal tube and the vertical wall have different characteristic lengths, the coefficients in the formula are also different. And the ratio of the convective heat transfer coefficients is:

式中，下标 H 表示水平管；下标 S 表示球；d 为水平管或球的直径。定性温度与前面的公式相同。水平管与竖壁的特征长度不同，公式中的系数也就不同。其表面传热系数比为：

$$\frac{h_H}{h_V} = 0.77\left(\frac{l}{d}\right)^{1/4} \tag{7.9}$$

When $l/d = 50$, the average convective heat transfer coefficient of the horizontal tube is twice that of the vertical tube. Therefore, the condensers usually adopt a horizontal tube arrangement.

当 $l/d = 50$ 时，水平管的平均表面传热系数是竖管的 2 倍，所以冷凝器通常采用水平管的布置方案。

Summary This section illustrates the solution idea and derivation process of Nusselt's analytical solution of laminar film-wise condensation, and the formulas of theoretical calculation are obtained.

小结 本节阐述了努塞尔层流膜状凝结分析解的求解思路和推导过程，并得到了理论计算式。

Question What is the basic idea of solving the analytical solution of laminar film-wise condensation?

思考题 求解层流膜状凝结分析解的基本思路是什么？

Self-tests

自测题

1. What are the basic assumptions in the derivation of the analytical solution for laminar film-wise condensation heat transfer of pure saturated vapor?

1. 纯净饱和蒸气层流膜状凝结传热分析解的推导中，最基本的假设是什么？

2. Under the same conditions, why is the condensation heat transfer outside the horizontal tube stronger than that outside the vertical tube?

2. 在相同的条件下，为什么水平管外的凝结传热比竖直管强烈？

授课视频

7.3 The Correlation Formula of Film-wise Condensation Calculation

7.3 膜状凝结计算关联式

For film-wise condensation, the flow states of the condensed liquid on the surface of an object can also be classified as laminar flow and turbulent flow. The judgment is based on the Reynolds number of the film layer, which expression is as follows.

对于膜状凝结，物体表面的凝结液体流动状态也有层流和湍流之别，判断依据是膜层雷诺数，其表达式为：

$$Re=\frac{d_e\rho u_l}{\eta}$$

Where u_l is the average flow velocity of the liquid film layer at $x=l$, and d_e is the equivalent diameter of the liquid film layer at this section, as shown in Fig. 7.6.

式中，u_l 是 $x=l$ 处的液膜层的平均流速；d_e 是该截面处液膜层的当量直径，如图 7.6 所示。

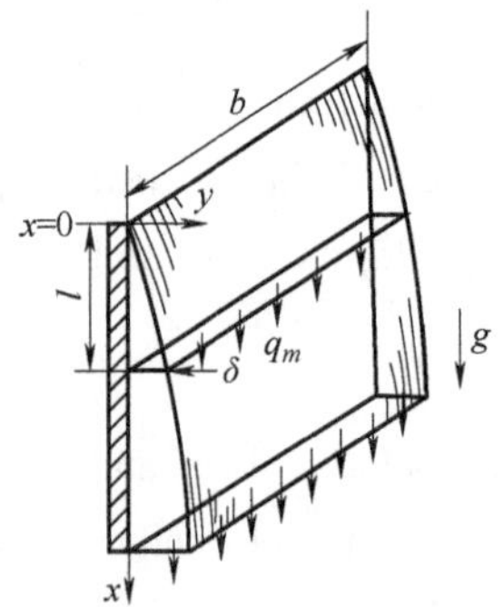

Fig. 7.6 Flow states inside the liquid film layer
图 7.6 液膜层内的流态

According to the definition of equivalent diameter.

根据当量直径的定义

$$d_e = 4A_c/P = 4b\delta/b = 4\delta$$

Then, the Reynolds number of the film layer can be written as:

此时，膜层雷诺数可写为：

$$Re = \frac{4\delta\rho u_l}{\eta} = \frac{4q_{ml}}{\eta} \quad (7.10)$$

Where q_{ml} is the mass flow rate of the condensate on the interface with a width of 1m at $x=l$.

式中，q_{ml}是 $x=l$ 处宽为 1m 的界面上的凝结液的质量流量。

According to the heat balance equation

由热平衡方程式

$$h(t_s - t_w)l = rq_{ml}$$

Substitute q_{ml} in this formula into the expression of Reynolds number to get

将该式中的 q_{ml} 代入雷诺数的表达式，得到

$$Re = \frac{4hl(t_s - t_w)}{\eta r} \quad (7.11)$$

For a horizontal tube, as long as l in Eq. (7.11) is replaced by πr, it is the Reynolds number of the film layer. Research shows that the critical Reynolds number can be set as 1600 for the liquid film to change from laminar flow to turbulent flow. Due to the small diameter of the horizontal tube, Reynolds number is in the laminar flow range in engineering practice.

对于水平管而言，只要式（7.11）中的 l 用 πr 代替，即是其膜层雷诺数。研究表明，液膜从层流转变为湍流的临界雷诺数可定为 1600。水平管因直径较小，在工程实践中 Re 均在层流范围。

For the turbulent liquid film, except that the laminar sub-layer close to the wall still relies on heat conduction to transfer heat, the turbulent transfer is dominant outside the laminar sub-layer, and the heat transfer is greatly enhanced. For the turbulent condensation heat transfer on the vertical wall, the calculation formula of the average convective heat transfer coefficient along the entire wall is as follows:

对于湍流液膜，除了靠近壁面的层流底层仍依靠导热来传递热量外，层流底层之外以湍流传递为主，传热大为增强。对竖壁的湍流凝结传热，沿整个壁面的平均表面传热系数计算公式为：

$$h = h_l \frac{x_c}{l} + h_t\left(1 - \frac{x_c}{l}\right) \quad (7.12)$$

Where h_l is the convective heat transfer coefficient of the laminar flow section, h_t is the convective heat transfer coefficient of the turbulent flow section, and x_c is the height of the turning point when the laminar flow transforms to turbulent flow. The experimental correlation formula of the average convective heat transfer coefficient of the entire wall can be arranged as follows:

式中，h_l 是层流段的表面传热系数；h_t 是湍流段的表面传热系数；x_c 是层流转变为湍流时转折点的高度。整个壁面的平均表面传热系数实验关联式可整理为：

$$Nu = Ga^{1/3} \frac{Re}{58 Pr_s^{-1/2} \left(\frac{Pr_w}{Pr_s}\right)^{1/4} (Re^{3/4} - 253) + 9200} \tag{7.13}$$

Except that Prandtl number Pr_w is calculated with the wall temperature t_w, the reference temperature of other physical quantities is t_s.

除普朗特数 Pr_w 用壁温 t_w 计算外，其余物理量的定性温度均为 t_s。

Example 7.1 The water vapor with the pressure of 1.013×10^5Pa condenses on the square vertical wall. The size of the wall is 30cm×30cm and the wall temperature is maintained at 98℃. Try to calculate the heat transfer quantity and condensing vapor quantity per hour.

例 7.1 压力为 1.013×10^5Pa 的水蒸气在方形竖壁上凝结，壁的尺寸为 30cm × 30cm，壁温保持 98℃。试计算每小时的传热量及凝结蒸汽量。

Analysis: Firstly, calculate Reynolds number and judge whether the liquid film flow is laminar or turbulent, and then select the corresponding formula for calculation. It can be known from Eq. (7.11) that Reynolds number depends on the average convective heat transfer coefficient h. So, it cannot be solved directly. We can first assume the flow state of liquid film, then select the corresponding formula to calculate out h according to the assumed flow state.

分析：应首先计算 Re，判断液膜流动是层流还是湍流，然后选取相应的公式计算。由式（7.11）可知，Re 本身取决于平均表面传热系数 h，因此不能简单地直接求解，可先假设液膜的流态，根据假设的流态，选取相应的公式计算出 h。

And then recalculate Reynolds number with the obtained h until it is satisfied compared with the initial assumption. The liquid film is assumed to be laminar flow. According to $t_s = 100$℃, we consult the latent heat of vaporization of water vapor $r = 2257$kJ/kg. The other physical parameters can be consulted according to the average temperature of the liquid film $t_m = 99$℃.

然后，用求得的 h 重新核算 Re，直到与初始假设相比认为满意为止。假设液膜为层流，根据 $t_s = 100$℃，可查得水蒸气的汽化潜热 $r = 2257$kJ/kg。其他物性参数按照液膜平均温度 $t_m = 99$℃，可查取。

$$\rho = 958.4\text{kg/m}^3, \eta = 2.825\times10^{-4}\text{Pa}\cdot\text{s}, \lambda = 0.68\text{W/(m}\cdot\text{K)}$$

At this time, we use Eq. (7.6) of the laminar liquid film to calculate the average convective heat transfer coefficient h. Substitute the obtained parameters into the expression of h to calculate out the convective heat transfer coefficient h.

这时，选用层流液膜平均表面传热系数的计算式（7.6）计算 h。把查得的参数代入 h 的表达式，即可算出表面传热系数 h。

$$\begin{aligned} h &= 1.13\left[\frac{g\rho^2\lambda^3 r}{\eta l(t_s - t_w)}\right]^{\frac{1}{4}} \\ &= 1.13\times\left\{\frac{9.8\text{m/s}^2\times(958.4\text{kg/m}^3)^2\times[0.68\text{W/(m}\cdot\text{K)}]^3\times2257\text{kJ/kg}}{2.825\times10^{-4}\text{Pa}\cdot\text{s}\times0.3\text{m}\times2\text{K}}\right\}^{\frac{1}{4}} \\ &= 1.57\times10^4\text{W/(m}^2\cdot\text{K)} \end{aligned}$$

Recalculate Reynolds number, $Re = 59.1$ is calculated according to Eq. (7.11). This shows that it is correct to assume that the liquid film flow is laminar, then the heat transfer rate can be calculated according to Newton cooling formula.

核算 Re，按照式（7.11）计算出 $Re = 59.1$。这说明原来假设液膜为层流是正确的，传热量可按照牛顿冷却公式计算。

$$\Phi = hA(t_s - t_w) = 1.57\times10^4\text{W/(m}^2\cdot\text{K)}\times(0.3\text{m})^2\times2\text{K} = 2.83\times10^3\text{W}$$

The amount of condensed vapor is

凝结蒸汽量为

$$q_m = \frac{\Phi}{r} = \frac{2.83\times10^3\text{W}}{2.257\times10^6\text{J/kg}} = 1.25\times10^{-3}\text{kg/s} = 4.50\text{kg/h}$$

Summary This section illustrates the calculation correlation formulas of film-wise condensation, and applies it through an example.

小结 本节阐述了膜状凝结的计算关联式，并通过例题进行应用。

Question What are the main calculation steps of film-wise condensation?

思考题 膜状凝结计算的主要步骤有哪些？

Self-tests

自测题

1. How to judge the flow state of condensed liquid on the object surface?

1. 如何判断物体表面凝结液体的流动状态？

2. What is the main thermal resistance of pure vapor laminar flow film condensation heat transfer process?

2. 纯净蒸气层流膜状凝结传热过程的主要热阻是什么？

7.4 The Influencing Factors of Film-wise Condensation

7.4 膜状凝结的影响因素

In practical engineering, the film-wise condensation process is often complicated, and there are six main influencing factors.

在实际工程中，膜状凝结过程往往比较复杂，其主要影响因素有6个。

The first influencing factor is non-condensable gas. Vapor often contains non-condensable gas such as air, even if the content is small, it will also have a very harmful impact on condensation heat transfer. For example, the presence of 1% air in water vapor can reduce convective heat transfer coefficient by 60%, and the consequences are very serious.

第一个影响因素是不凝结气体。蒸气中常含有不可凝结的气体，比如空气，即使含量很少，也会对凝结传热产生十分有害的影响。例如，水蒸气中含有1%的空气能使表面传热系数降低60%，后果很严重。

On the vapor side close to the surface of liquid film, as the vapor condenses constantly and the vapor partial pressure decreases, then the partial pressure of non-condensable gas increases. Before the vapor reaches the surface of liquid film for condensation, it must diffuse through the non-condensable gas layer gathered near the interface.

在靠近液膜表面的蒸气侧，随着蒸气不断凝结，蒸气分压力减小，而不凝结气体的分压力增大。蒸气在抵达液膜表面进行凝结前，必须以扩散的方式穿过聚集在界面附近的不凝结气体层。

The non-condensable gas layer increases the resistance of the heat transfer process, and at the same time, the decrease of vapor partial pressure reduces the corresponding saturation temperature, which reduces the driving force of condensation, Δt, thus weakens the condensation process as well. Therefore, the elimination of non-condensable gas becomes the key to ensuring the design capability during the operation of the condenser.

不凝结气体层的存在增加了传热过程的阻力，同时蒸气分压力的下降使相应的饱和温度下降，减小了凝结的驱动力 Δt，也使凝结过程被削弱了。因此，在冷凝器工作中，排除不凝结气体成为保证设计能力的关键。

The second influencing factor is the number of tube rows. The geometrical arrangement of tube bundle and the physical properties of the fluid can affect the condensation heat transfer. The previously derived formula for condensation

第二个影响因素是管子排数。管束的几何布置和流体物性都会影响凝结传热。前面推导的水平管凝结传热的公式，只适用于单根水平管。对

heat transfer of a horizontal tube is only applicable to a single horizontal tube. For the heat transfer of the tube bundle consisting of n rows of horizontal tubes along the liquid flow direction, theoretically, it can be calculated by replacing the characteristic length D in the formula with nD.

于沿液流方向有 n 排水平管组成的管束的传热，理论上只要将式中的特征长度 D 换成 nD 就可以计算了。

In fact, this is an overly conservative calculation. Because the condensate on the upper tubes does not fall evenly on the lower tubes, and it will splash and impact the liquid film when falling. The degree of splashing and disturbance depends on the geometrical arrangement of tube bundle and fluid physical property, etc. The situation is complicated, so it is best to refer to the experimental data suitable for the design conditions when designing.

实际上这是过分保守的计算。因为上排管的凝结液并不是平均地落到下排管上，在落下时会产生飞溅以及对液膜的冲击扰动。飞溅和扰动的程度取决于管束的几何布置和流体物性等，情况比较复杂，设计时最好参考适合设计条件的实验资料。

The third influencing factor is condensation inside a tube. In many industrial condensers, vapor flows through inside of the tubes under the pressure difference and condenses. At this time, the heat transfer process is closely related to the vapor flow rate. Take the condensation in a horizontal tube as an example. When the vapor flow rate is relatively low, the condensate mainly accumulates at the bottom of the tube, while the vapor is located in the upper part of the tube.

第三个影响因素是管内冷凝。在许多工业冷凝器中，蒸气在压差的作用下流经管子内部并产生凝结。此时的传热过程与蒸气的流速有很大关系。以水平管中的凝结为例，当蒸气流速比较低时，凝结液主要积聚在管子的底部，而蒸气位于管子上部。

If the vapor flow rate is relatively high, then it will form annular flow. The condensate is evenly distributed around inner wall of the tube, while the center is the vapor core. As the gas and liquid flow in the tube, the thickness of liquid film increases constantly until it occupies the entire section after condensing completely. Its cross-sectional shape is shown in Fig. 7. 7.

如果蒸气流速比较高，则形成环状流动。凝结液较均匀地分布在管内壁周围，而中心为蒸气核。随着气液在管内流动，液膜厚度不断增厚，直至凝结完成时占据了整个截面。其截面形状如图 7. 7 所示。

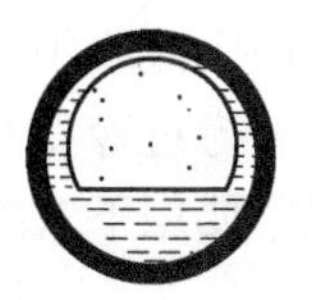

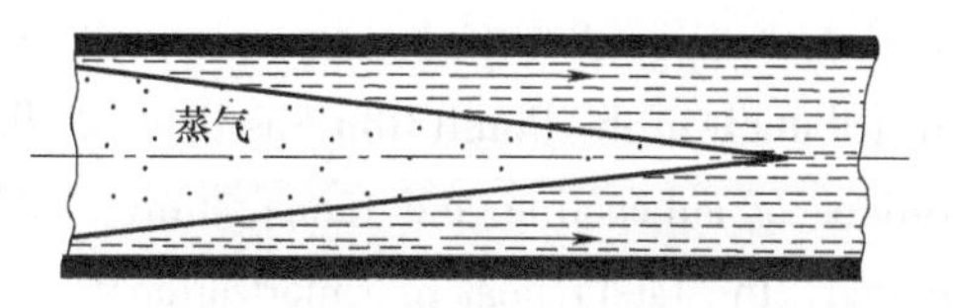

Fig. 7. 7 Schematic diagram of liquid film and vapor core during condensation inside a tube

图 7. 7 管内凝结时液膜和蒸气核示意图

The fourth influencing factor is the vapor flow rate. Nusselt's theoretical analysis ignored the influence of vapor flow rate, so it is only suitable for occasions with low flow rate, such as the condensers in power plant.

第四个影响因素是蒸气流速。努塞尔的理论分析忽略了蒸气流速的影响，因此只适用于流速较低的场合，比如电站冷凝器。

When the vapor flow rate is relatively high, the vapor can produce obvious viscous stress on the surface of the liquid film. The degree of influence varies with the flow direction of vapor in the same or different direction with the gravitational field, the flow velocity, and whether the liquid film is torn, etc.

当蒸气流速较高时，蒸气对液膜表面会产生明显的黏滞应力。其影响程度随蒸气流向与重力场同向或异向、流速大小，以及是否撕破液膜等而有所不同。

Generally speaking, when the vapor flow direction is the same as the downward flow direction of the liquid film, the liquid film will be thinned and the convective heat transfer coefficient h is increased. When the direction is opposite, the flow of liquid film will be blocked and the liquid film will be thickened, thereby reducing the convective heat transfer coefficient h.

一般来说，当蒸气流动方向与液膜向下的流动方向相同时，使液膜拉薄，表面传热系数 h 增大；方向相反时，则会阻滞液膜的流动并使其增厚，从而使表面传热系数 h 减小。

The fifth influencing factor is the superheat of vapor. The previous discussions are entirely in terms of the condensation of saturated vapor. For superheated vapor, it is proved by experiments that as long as the latent heat in the calculation formula is changed into the enthalpy difference between superheated vapor and saturated liquid, the experimental correlation of saturated vapor can also be used to calculate the condensation heat transfer coefficient of superheated vapor.

第五个影响因素是蒸气过热度。前面的讨论都是针对饱和蒸气的凝结而言的。对于过热蒸气，实验证实，只要把计算式中的潜热改为过热蒸气与饱和液的焓差，饱和蒸气的实验关联式同样可用来计算过热蒸气的凝结传热系数。

The sixth influencing factor is the subcooling degree of the liquid film and the nonlinear temperature distribution inside the liquid film. Nusselt's theoretical analysis ignored the influence of subcooling of the liquid film, and assumed that the temperature distribution in the liquid film was linear. If the actual conditions of subcooling and temperature distribution are considered, the latent heat of vaporization r shall be replaced by r' of the following formula.

第六个影响因素是液膜过冷度及液膜内温度分布非线性。努塞尔的理论分析忽略了液膜过冷度的影响，并假定液膜中温度呈线性分布。如果考虑过冷度及温度分布的实际情况，要用下式的 r' 代替汽化潜热 r。

$$r' = r + 0.68 c_p (t_s - t_w)$$

Summary　This section illustrates the six main influencing factors of film-wise condensation.

小结　本节阐述了膜状凝结的六个主要影响因素。

Question　What are the main influencing factors of the film-wise condensation process in practical engineering?

思考题　实际工程中膜状凝结过程的主要影响因素有哪些?

Self-tests

自测题

1. Try to explain the reason why non-condensable gas affects the heat transfer of film-wise condensation.

1. 试述不凝结气体影响膜状凝结传热的原因。

2. For a single horizontal tube, what factors affect the laminar film-wise condensation heat transfer?

2. 对于单根水平管，哪些因素影响层流膜状凝结传热?

授课视频

7.5 The Heat Transfer Enhancement of Film-wise Condensation

7.5 膜状凝结的传热强化

When the vapor condenses in film-wise, the thermal resistance depends on heat conduction through the liquid film layer. Therefore, reducing the thickness of the liquid film layer as much as possible is the basic idea to enhance film-wise condensation. So, two measures can be taken: ① Directly thin the liquid film sticking on the solid surface when the vapor condenses. ② Timely discharge the condensed liquid generated on the heat transfer surface to prevent the liquid film from accumulating on the heat transfer surface.

蒸气膜状凝结时，热阻取决于通过液膜层的导热。因此尽量减小液膜层的厚度，是强化膜状凝结的基本思想。为此可采取两种措施：① 蒸气凝结时直接减薄黏滞在固体表面上的液膜；② 及时将传热表面产生的凝结液体排掉，不让液膜积聚在传热表面上。

The simplest way to reduce the thickness of liquid film is to reduce the height of the heat transfer surface as much as possible for the vertical wall or the vertical tube if the process permits, or to change a vertical tube to a horizontal tube. The following focuses on the method of reducing the thickness of liquid film by surface tension. The solid surface with peaks is shown in Fig. 7. 8. The force analysis of the liquid film on the peak is carried out.

减薄液膜厚度最简单的方法是，对于竖壁或竖管，在工艺允许的情况下尽量降低传热面的高度，或者将竖管改置为横管。下面重点讨论利用表面张力减薄液膜厚度的方法。带尖峰的固体表面如图 7.8 所示，对位于尖峰上的液膜进行受力分析。

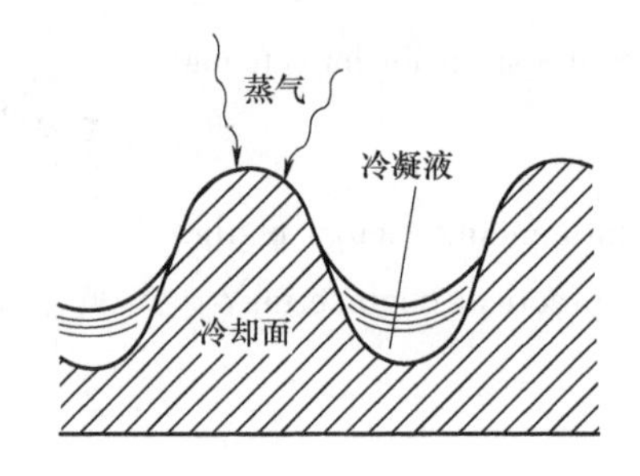

Fig. 7. 8 Effect of surface tension on the peaks

图 7. 8 尖峰上表面张力的作用

It can be seen that the surface tension of the liquid film can greatly reduce the thickness of the liquid film on the peak. According to this idea, a variety of enhanced condensation surfaces have been developed, such as the integral low rib tube, which is the earliest one, as shown in Fig. 7. 9.

可以看出，液膜的表面张力可以使尖峰上的液膜厚度大大减薄。根据这一思想，开发出了多种强化凝结表面，比如整体式低肋管就是最早的一种，如图 7. 9 所示。

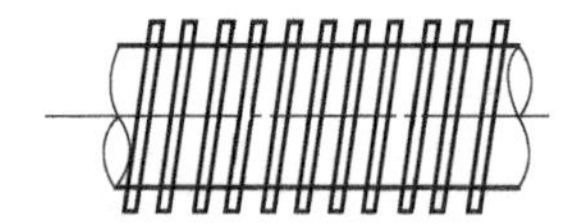
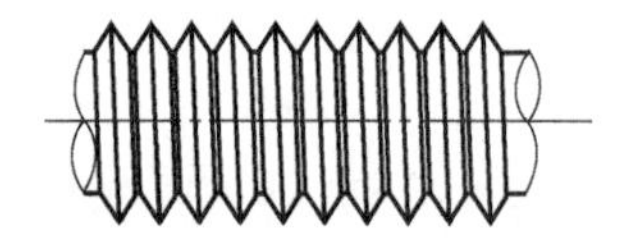

Fig. 7. 9 Integral low rib tube

图 7. 9 整体式低肋管

It was initially thought that the ribs only increased the condensation area, but the actual enhancing effect was much larger than the increase in area. This is because the liquid film on the ribs is thinned by surface tension. Subsequently, a variety of serrated surface tubes suitable for enhancing vapor condensation outside tube were developed, as shown in Fig. 7. 10.

最初人们认为肋片只是增加了凝结面积，但实际强化效果要比面积增加带来的效果大得多，这是因为肋片上的液膜受表面张力的作用而变薄。随后开发出了适用于强化蒸气管外凝结的各种锯齿表面管，如图 7. 10所示。

Fig. 7. 10 Schematic diagram of serrated surface tube

图 7. 10 锯齿表面管示意图

In the condenser of a household air conditioner, the

在家用空调的冷凝器中，制冷剂蒸

refrigerant vapor condenses inside the tube. Two and three-dimensional micro-fin tubes have been successfully developed. The thread of the two-dimensional micro-fin tube is continuous, while the thread of the three-dimensional micro-fin tube is discontinuous, as shown in Fig. 7. 11.

气在管内凝结。现已成功开发出二维与三维的微肋管。二维微肋管中螺纹是连续的，三维微肋管的螺纹是间断的，如图 7. 11 所示。

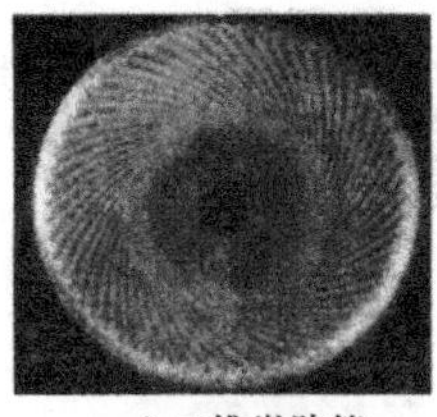
a) 二维微肋管

b) 三维微肋管

Fig. 7. 11 Photos of two and three-dimensional micro-rib tubes
图 7. 11 二维、三维微肋管照片

The above-mentioned enhanced tubes have been widely used in refrigeration and air conditioning equipment at home and abroad. The concave-convex configuration on the inner surface of the tube is the cross-section of the spiral line on the inner surface.

上述强化管已经广泛应用于国内外制冷空调设备中。管子内表面的凹凸构型，是内表面螺旋线的剖面。

When the refrigerant vapor condenses outside the tube, its condensation heat transfer coefficient is much smaller than the heat transfer coefficient of the cooling water in the tube. So, the thermal resistance of the heat transfer process is mainly on the vapor condensation side. But, after effective enhancement outside the tube, the outer thermal resistance significantly decreases, and the inner thermal resistance of the tube is highlighted. Whereupon, an enhanced tube with a spiral line on the inner surface appears, and called a double-sided enhanced tube, which can enhance the entire heat transfer process more effectively.

当制冷剂蒸气在管外凝结时，其凝结传热系数较管内冷却水的传热系数小很多。因此，传热过程的热阻主要在蒸气凝结侧。但是，管外得到有效强化后，外侧热阻明显减小，管内侧的热阻就会突显出来。于是，出现了在内表面采用螺旋线结构的强化管，称为双侧强化管，它使整个传热过程得到更有效的强化。

In engineering technology, the area of the parison tube used to manufacture the enhanced heat transfer tube is often served as the accordance for comparing the convective heat transfer coefficient. The convective heat transfer coefficient of condensation heat transfer of low-rib tubes can be increased by 2 to 4 times than that of smooth tubes. The

在工程技术中，常以制造强化传热管的坯管的面积作为比较表面传热系数的依据。低肋管凝结传热的表面传热系数可比光管提高 2~4 倍，锯齿表面管可以提高一个数量级，微肋管则一般可提高 2~3 倍。

serrated surface tube can be increased by an order of magnitude, and the micro-rib tubes can generally be increased by 2 to 3 times.

There is another way to reduce the thickness of liquid film and enhance heat transfer, which is called timely drainage. Fig. 7. 12 shows two common methods for accelerating the excretion of condensate.

减薄液膜厚度强化传热还有一种方法，叫作及时排液。图 7. 12 就是两种常见的加速排泄凝结液体的方法。

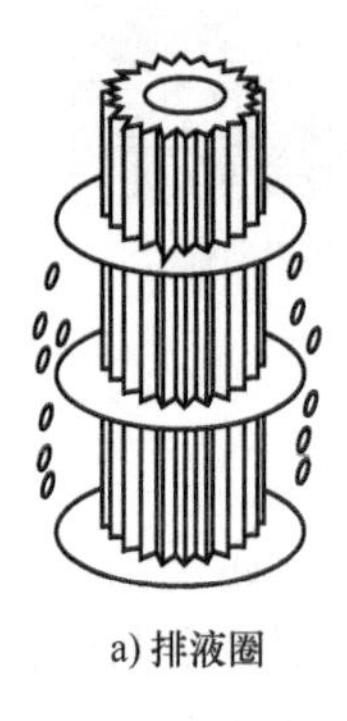

a) 排液圈

b) 泄流板

Fig. 7. 12 Measures for timely drainage

图 7. 12 及时排液的措施

The drain ring in Fig. 7. 12a is used for vertical condenser. The liquid is drained in stages during the process of liquid film flowing down, which effectively controls the thickness of liquid film. The grooves on the surface of the tube can reduce the thickness of the liquid film. The drain plate in Fig. 7. 12b is used in a horizontal type condenser, such as a condenser in a large power plant. The drain plate can prevent the condensate on the horizontal tube bundle arranged above the plate from accumulating on the tube bundle below.

图 7. 12a 中排液圈用于立式冷凝器。凝结液在液膜下流的过程中被分段排出，有效地控制了液膜的厚度。管子表面的沟槽具有减薄液膜厚度的作用。图 7. 12b 中的泄流板用于卧式冷凝器中，如大型电站的凝汽器。泄流板可使布置在该板上面水平管束上的冷凝液不会集聚到下面的管束上。

Summary This section illustrates the common measures to enhance film-wise condensation heat transfer in industry.

小结 本节阐述了工业上强化膜状凝结传热的常用措施。

Question What is the mechanism of enhanced film-wise condensation process?

思考题 强化膜状凝结过程的机理是什么？

Self-tests

自测题

1. What is the basic idea of enhanced condensation heat transfer in terms of the structure of the heat transfer surface?

1. 从传热表面的结构而言，强化凝结传热的基本思想是什么？

2. For the actual condensation heat exchanger, what are the main methods to improve the film-wise condensation heat transfer coefficient?

2. 对于实际凝结换热器，提高膜状凝结传热系数方法有哪些？

授课视频

7.6 The Modes of Boiling Heat Transfer

7.6 沸腾传热的模式

There are two forms of liquid vaporization, i. e. evaporation and boiling. Evaporation is a vaporization process that occurs on the surface of a liquid. Boiling is a violent vaporization process that takes place inside a liquid in the form of bubbles. At this time, the wall temperature exceeds the boiling point of the liquid.

液体汽化有两种形式：蒸发和沸腾。蒸发是发生在液体表面上的汽化过程。沸腾是在液体内部以产生气泡的形式进行的剧烈汽化过程。此时，壁面温度超过液体沸点。

With regard to liquid vaporization, there are many examples in daily life and industrial production. Such as the steam boiler, the evaporator in the refrigeration system, daily kettle, etc. The definition of boiling and boiling heat transfer are given below.

对于液体汽化，日常生活和工业生产中有很多的例子。比如蒸汽锅炉、制冷系统中的蒸发器、日常用的烧水壶等。下面给出沸腾及沸腾传热的定义。

Boiling is a violent vaporization process in which working medium changes from liquid to gas after absorbing heat and forms a large number of bubbles inside. Boiling heat transfer refers to a heat transfer mode in which the bubble movement generated by the vaporization of working medium takes away heat and cools it.

沸腾是指工质吸收热量后由液态转变成气态并在其内部形成大量气泡的一种剧烈的汽化过程。沸腾传热是指工质汽化产生的气泡运动带走热量并使其冷却的一种传热方式。

There are many types of boiling, such as common large vessel boiling and in-tube boiling. Large vessel boiling is also called pool boiling, while in-tube boiling is also called forced convection boiling. Each kind of boiling is also divided into subcooled boiling and saturated boiling.

沸腾的类型很多，常见的有大容器沸腾和管内沸腾。大容器沸腾也称为池内沸腾，管内沸腾也称为强制对流沸腾。每种沸腾又分为过冷沸腾和饱和沸腾。

Large vessel boiling is defined as the boiling of liquids with a free surface when a heated wall is immersed in the liquids. At this time, the bubbles can float freely and enter

大容器沸腾的定义是：加热壁面沉浸在液体中时，具有自由表面的液体所发生的沸腾。此时产生的气泡

the vessel space through the free surface of the liquid, as shown in Fig. 7. 13.

能自由浮升，穿过液体自由表面进入容器空间，如图 7. 13 所示。

Fig. 7. 13 Large vessel boiling
图 7. 13 大容器沸腾

In-tube boiling is essentially forced convection coupled with boiling. When forced convection boiling occurs in the tube, the steam will mix with the liquid flow, and various two-phase flow structures will appear, so the heat transfer mechanism is very complex.

管内沸腾本质上就是强制对流+沸腾。管内强制对流沸腾时，产生的蒸气混入液流，出现多种不同形式的两相流结构，因此传热机理很复杂。

Subcooled boiling refers to the boiling phenomenon that the main stream of liquid has not reached the saturation temperature, that is, it is in a subcooled state, but bubbles begin to appear on the wall. Saturated boiling refers to the boiling phenomenon that occurs when the main stream of liquid reaches the saturation temperature and the wall temperature is higher than the saturation temperature. There is a new concept called superheat degree, which is the difference between the wall temperature and the liquid saturation temperature.

过冷沸腾，是指液体主流尚未达到饱和温度，即处于过冷状态，但壁面上开始产生气泡的沸腾现象。饱和沸腾，是指液体主流温度达到了饱和温度而壁面温度高于饱和温度所发生的沸腾现象。这里涉及一个新概念叫作过热度，是指壁面温度与液体饱和温度的差值。

Here is a brief introduction of bubble dynamics. First, recognize the growth process of bubbles. Experiments show that bubbles usually occur only at some points of the heating surface, but not on the whole heating surface. These bubble points are called vaporization cores.

下面简单介绍一下气泡动力学。首先了解气泡的成长过程。实验表明，通常情况下沸腾时气泡只发生在加热面的某些点，而不是整个加热面上。这些产生气泡的点被称为汽化核心。

It is generally believed that the slits, cavities and pits on the wall tend to retain gas, and the liquid in the slits is much more affected by heating than the liquid on the flat surface. So, they are the best vaporization cores, as shown in Fig. 7. 14.

普遍认为，壁面上的狭缝、空穴和凹坑内容易残留气体，且狭缝中的液体受到的加热影响比平直面上的液体要多得多，因此是最好的汽化核心，如图 7. 14 所示 .

Fig. 7. 14 Vaporization core
图 7. 14 汽化核心

The bubble radius R must meet the following conditions to survive.

气泡半径 R 必须满足下列条件才能存活：

$$R \geqslant R_{\min} = \frac{2\sigma T_s}{r\rho_v (t_w - t_s)}$$

Where σ is the surface tension, r is the latent heat of vaporization, ρ_v is the vapor density, t_w is the wall temperature, and t_s is the saturation temperature at the corresponding pressure.

式中，σ 是表面张力；r 是汽化潜热；ρ_v 是蒸气密度；t_w 是壁面温度；t_s 是对应压力下的饱和温度。

With the increase of the wall superheat degree, the minimum bubble radius decreases, which leads to the increase of the number of vaporization cores on the same heating surface and enhancement of boiling heat transfer. Fig. 7. 15 shows the bubble formation process during boiling on flat bottom cup surface.

随着壁面过热度升高，最小气泡半径 $R_{\min}$减小，导致同一加热面上汽化核心数增加，使沸腾传热增强。图 7. 15 就是平底杯表面沸腾过程中气泡产生的过程。

Fig. 7. 15 Boiling picture on flat bottom cup surface
图 7. 15 平底杯表面沸腾

Then we will learn about boiling inside standpipe. The main feature of boiling inside standpipe is gas-liquid two-phase flow. The main influencing parameters are vapor content, mass flow and pressure.

再了解一下竖管内的沸腾。竖管内的沸腾主要特征为气液两相流，最主要的影响参数是含气量、质量流量和压力。

Summary This section illustrates the classification and features of boiling heat transfer, and the basic knowledge of bubble dynamics.

小结 本节阐述了沸腾传热的分类和特点，以及气泡动力学的基础知识。

Question What is the difference of heat transfer mechanisms between evaporation and boiling?

思考题 蒸发与沸腾的传热机理有什么不同?

Self-tests

自测题

1. What are large vessel boiling, in-tube boiling, subcooled boiling and saturated boiling?

1. 什么是大容器沸腾、管内沸腾、过冷沸腾和饱和沸腾?

2. Why does boiling of liquid occur only at the vaporization core of the wall?

2. 为什么液体的沸腾只发生在壁面的汽化核心处?

授课视频

7.7 The Saturated Boiling Curve in Large Vessels

7.7 大容器饱和沸腾曲线

The saturated boiling curve in large vessels characterizes the whole process of saturated boiling in a large vessel, which includes four stages with different heat transfer laws, namely natural convection region, nucleate boiling region, transition boiling region and stable film boiling region, as shown in Fig. 7.16.

大容器饱和沸腾曲线表征了大容器饱和沸腾的全部过程，包括传热规律不同的四个阶段，分别是自然对流区、核态沸腾区、过渡沸腾区和稳定膜态沸腾区，如图 7.16 所示。

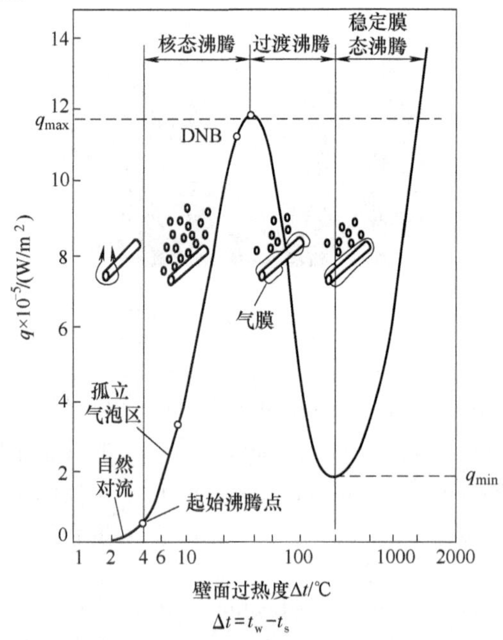

Fig. 7.16 Saturated boiling in large vessels

图 7.16 大容器饱和沸腾

The first stage is the single phase natural convection region, also known as liquid level vaporization region. The wall superheat degree is small at this stage. The saturated boiling superheat degree Δt is less than 4℃ at one atmospheric pressure. Boiling has not yet started, and heat transfer obeys the laws of single-phase natural convection. At this time, the convective heat transfer coefficient h is less than 1000W/($m^2 \cdot K$). With the increase of superheat degree Δt, the heat flux increases.

第一个阶段是单相自然对流区，也叫作液面汽化区。该阶段壁面过热度小，一个大气压下饱和沸腾过热度 $\Delta t<4$℃。沸腾尚未开始，传热服从单相自然对流规律。这时，表面传热系数 $h<1000$W/($m^2 \cdot K$)。随着过热度 Δt 的增加，热流密度 q 随之上升。

The second stage is the nucleate boiling region, also known as the saturated boiling region. With the increase of superheat degree, vaporization cores begin to appear at some points on the heating surface, and then vapor bubbles are formed.

第二个阶段叫作核态沸腾区，也叫作饱和沸腾区。随着过热度 Δt 的上升，在加热面的一些点上开始出现汽化核心，并随之形成气泡。

In the initial stage, the bubbles generated by the vaporization cores do not interfere with each other. This stage is known as isolated bubble region. The superheat degree $\Delta t \approx 4-10$℃ at this time. With the increase of superheat degree Δt, and then the heat flux q increases, the number of vaporization cores increases, and the number of bubbles also increases. The bubbles interact with each other and synthesize vapor blocks and columns. This stage is known as the interaction region.

在起始阶段，汽化核心产生的气泡互不干扰，称为孤立气泡区。此时过热度 $\Delta t \approx 4 \sim 10$℃。随着过热度 Δt 的增加，热流密度 q 随之上升，汽化核心数增加，生成的气泡数量也增加，气泡互相影响并合成气块及气柱，这一阶段称为相互影响区。

When Δt increases to a certain value, heat flux q increases to the maximum value and the bubble violently disturbs. Vaporization cores play a decisive role in the heat transfer. This stage is called nucleate boiling region, also known as bubble boiling region. It is characterized by small temperature difference, high heat transfer intensity, and maximum heat flux q at the end point. This stage is basically applied in industrial design when the superheat degree $\Delta t \approx 4-20$℃. As the superheat degree increases, the heat flux increases rapidly and eventually increases to the maximum.

当 Δt 增大到一定值时，q 增加到最大值，气泡扰动剧烈，汽化核心对传热起决定作用，该段称为核态沸腾区，也叫作泡状沸腾区。其特点为温差小，传热强度大，其终点的热流密度 q 达到最大值。工业设计中基本都应用该阶段，此时过热度 $\Delta t \approx 4 \sim 20$℃。随着过热度的增加，热流密度快速增加，并最终增加到最大值。

The third stage is the transition boiling region. Further improve the superheat degree Δt from the peak point, the

第三个阶段是过渡沸腾区。从峰值点进一步提高过热度 Δt，热流密

heat flux q will decrease. With the increase of superheat degree, the heat flux decreases rapidly and reaches the minimum q_{min}. This region is an unstable process.

度 q 减小。随着过热度 Δt 增加，热流密度 q 快速减小，并将降到最低值 q_{min}。该区段是不稳定过程。

The reason is that the growth rate of the bubble is faster than that of the bubble jumping away from the heating surface, which makes the bubbles gather and cover the heating surface and form a layer of vapor film. And the vapor removal process tends to deteriorate, which results in the decrease of heat flux when the superheat degree $\Delta t \approx$ 40–200℃.

其原因是气泡的生长速度大于气泡跃离加热面的速度，使气泡聚集覆盖在加热面上形成一层蒸气膜，而蒸气排除过程趋于恶化，致使热流密度下降，此时过热度 $\Delta t \approx$40~200℃。

The forth stage is the stable film boiling region. Starting from the minimum heat flux q_{min}, with the increase of superheat degree Δt, the bubble growth rate and the jump-off rate tend to balance. At this time, a stable vapor film layer is formed on the heating surface, and the vapor breaks away from the film layer regularly. As a result, the heat flux q increases when superheat degree Δt increases. The superheat degree Δt is greater than 200℃.

第四阶段是稳定膜态沸腾区。从最小热流密度 q_{min} 开始，随着过热度 Δt 的上升，气泡生长速度与跃离速度趋于平衡。此时在加热面上形成稳定的蒸气膜层，产生的蒸气有规律地脱离膜层，致使过热度 Δt 上升时热流密度 q 随之上升。过热温度 Δt 大于 200℃。

The characteristics of the film boiling region are as follows: ① The heat transfer in the vapor film has not only heat conduction but also heat convection. ② The radiation heat increases sharply with the increase of temperature difference, which makes the heat flux increase greatly. ③ There is something in common with film condensation in physics: The former heat must pass through the vapor film with high thermal resistance, while the latter heat must pass through the liquid film with relatively small thermal resistance.

膜态沸腾区的特点如下：① 气膜中的热量传递不仅有导热，而且有热对流。② 辐射热量随着温差的加大而剧增，使热流密度大大增加。③ 在物理上与膜状凝结具有共同点：前者热量必须穿过热阻大的气膜，后者热量必须穿过热阻相对较小的液膜。

From Fig. 7.16, we can see the change of heat flux in the boiling process. There is a peak value of the heat flux q_{max} in this process, which is called the critical heat flux, also known as burn-out point. It means that exceeding this value may cause the equipment to burn down.

从图 7.16 我们看到了这个沸腾过程中热流密度的变化。这个过程中热流密度有个峰值 q_{max}，这个峰值被称为临界热流密度，也叫作烧毁点。意味着超过该值可能会导致设备烧毁。

Generally, nucleate boiling turning point DNB is used to monitor the heat flux close to the warning value q_{max} (burn-out point), which is very important for the control of both heat flux and temperature. For the stable film boiling, because the heat must pass through the vapor film with higher thermal resistance, so its heat transfer rate is much smaller than that of nucleate boiling, and its heat transfer coefficient is much smaller than that of condensation.

一般用核态沸腾转折点 DNB 作为监视接近 q_{max}（烧毁点）的警戒，这一点对热流密度可控和温度可控都非常重要。对于稳定膜态沸腾，因为热量必须穿过热阻较大的气膜，所以传热量比核态沸腾大大减小，传热系数比凝结小得多。

Why is the heat transfer intensity in nucleate boiling region much higher than that in transition boiling region and stable film boiling region? The reason is that the formation, growth and separation of bubbles from the heating wall will cause various disturbances. Therefore, the effective way to enhance boiling heat transfer is to increase vaporization cores on the heating wall. The specific method is to add slits, holes and pits on the wall.

为什么核态沸腾区的传热强度远大于过渡沸腾区和稳定膜态沸腾区？其原因是气泡的形成、成长及脱离加热壁面引起了各种扰动。因此，强化沸腾传热的有效途径是增加加热壁面上的汽化核心。具体的方法是在壁面上增加狭缝、空穴和凹坑。

Summary This section illustrates the variation laws of the saturated boiling curve in large vessels, and analyzes the reasons for the higher heat transfer intensity of nucleate boiling.

小结 本节阐述了大容器饱和沸腾曲线的变化规律，并分析了核态沸腾传热强度较大的原因。

Question Try to compare the similarities and differences between the film condensation outside the horizontal tube and the film boiling heat transfer process outside the horizontal tube.

思考题 试对比水平管外膜状凝结及水平管外膜态沸腾传热过程的异同。

Self-tests

自测题

1. In the saturated boiling curve in large vessels, why does the curve of the stable film boiling region rise rapidly with the increase of Δt?

1. 在大容器饱和沸腾曲线中，为什么稳定膜态沸腾区的曲线会随 Δt 的增加而迅速上升？

2. What is the critical heat flux? What is the burn-out point? If it is the heating condition with constant wall temperature, will there still be burn-out phenomenon?

2. 什么是临界热流密度？什么是烧毁点？如果是定壁温加热条件，还会有烧毁现象出现吗？

3. What are the characteristics of heat transfer in each region of saturated boiling in large vessels?

3. 大容器饱和沸腾在各个区域的传热有何特点？

7.8 The Experimental Correlations of Pool Boiling Heat Transfer

7.8 大容器沸腾传热实验关联式

Boiling heat transfer is a kind of heat convection. Therefore, the Newton cooling formula is still applicable. But, there are many different calculating formulas for the boiling convective heat transfer coefficient h.

沸腾传热是对流传热的一种，因此牛顿冷却公式是仍然适用的，但沸腾传热的表面传热系数 h 有许多不同的计算公式。

$$q=h(t_w-t_s)=h\Delta t$$

First, the saturated nucleate pool boiling is studied. The main factors affecting nucleate boiling are wall superheat degree and the number of vaporization cores. The number of vaporization cores is affected by wall material, surface condition, pressure, physical properties and other factors, so the situation of boiling heat transfer is very complicated, resulting in large differences in calculation formulas.

首先研究大容器饱和核态沸腾。影响核态沸腾的因素主要是壁面过热度和汽化核心数。汽化核心数受壁面材料、表面状况、压力和物性等因素影响，所以沸腾传热的情况比较复杂，导致计算公式差异较大。

At present, there are two kinds of calculation formulas. One is for a certain liquid, the other is widely used for various liquids. The specific calculation formula is more accurate. Thus, two formulas are recommended. For saturated nucleate pool boiling of water, Mihaev formula is recommended. The application range of pressure is $10^5-4\times10^6$Pa. In this case, the calculation formula of the convective heat transfer coefficient is:

目前有两种计算式：一种是针对某一种液体，另一种是广泛适用于各种液体。针对性强的计算式精确度比较高，为此推荐两个计算式。对于水的大容器饱和核态沸腾，推荐使用米海耶夫公式，压力适用范围是 $10^5\sim4\times10^6$Pa。这时表面传热系数计算式为：

$$h=0.1224\Delta t^{2.33}p^{0.5} \tag{7.14a}$$

$$h=0.5223q^{0.7}p^{0.15} \tag{7.14b}$$

Where h is the convective heat transfer coefficient of boiling heat transfer, p is the absolute pressure of the boiling water, Δt is the wall superheat degree, and q is the heat flux.

式中，h 是沸腾传热的表面传热系数；p 是沸腾水的绝对压力；Δt 是壁面过热度；q 是热流密度。

Rohsenow obtained the following experimental correlation

罗森诺通过大量实验得出以下实验

through a large number of experiments, which is widely applicable to forced convection heat transfer.

关联式，该式是一个广泛适用的强制对流传热公式。

$$\frac{c_{pl}\Delta t}{r}=C_{wl}\left[\frac{q}{\eta_l r}\sqrt{\frac{\sigma}{g(\rho_l-\rho_v)}}\right]^{0.33}Pr_l^s \tag{7.15}$$

Eq. (7.15) can be rewritten as:

式（7.15）可以改写为：

$$q=\eta_l r\left[\frac{g(\rho_l-\rho_v)}{\sigma}\right]^{1/2}\left(\frac{c_{pl}\Delta t}{C_{wl}rPr_l^s}\right)^3 \tag{7.16}$$

Where c_{pl} is specific constant pressure heat capacity of saturated liquid; C_{wl} is empirical constants for heating surface-liquid combinations; σ is surface tension of liquid-vapor interface; s is experience index, $s=1$ for water, $s=1.7$ for other liquid.

式中，c_{pl}是饱和液体的比定压热容；C_{wl}是加热表面-液体组合情况的经验常数；σ 是液体-蒸气界面的表面张力；s 是经验指数，对于水 $s=1$，对于其他液体 $s=1.7$。

From the formula, it can be seen that the heat flux of convection heat transfer is directly proportional to the third power of temperature difference. Although the deviation between q obtained from the formula and the experimental value is up to 100%, the calculation deviation of Δt can be reduced to 33 percent when q is known. This situation is more obvious in radiation heat transfer. The heat flux must be treated carefully in the calculation.

从该公式可见，对流传热热流密度与温差的 3 次方成正比。尽管有时计算公式得到的 q 与实验值的偏差高达 100%，但已知 q 时，则可以将 Δt 计算偏差缩小到 33%。这种情况在辐射传热中更为明显，计算时必须谨慎处理热流密度。

For refrigeration media, Cooper's formula is used to calculate the boiling convective heat transfer coefficient:

对于制冷介质，沸腾传热的表面传热系数计算用库珀公式：

$$h=Cq^{0.67}M_r^{-0.5}p_r^m(-\lg p_r)^{-0.55} \tag{7.17}$$

Where the empirical constant C is:

式中，经验常数 C 为：

$$C=90\mathrm{W}^{0.33}/(\mathrm{m}^{0.66}\cdot\mathrm{K})$$

The exponent m in the formula:

公式中指数 m：

$$m=0.12-0.2\lg\{R_p\}_{\mu m}$$

Where M_r is relative molecular mass of the liquid; p_r is contrast pressure (The ratio of liquid pressure to the critical pressure of the fluid); R_p is average surface roughness ($R_p=0.3-0.4\mu m$ for the surface of general industrial tube).

式中，M_r 是液体的相对分子质量；p_r 是对比压力（液体压力与该流体的临界压力之比）；R_p 为表面平均粗糙度值（对一般工业用管材表面 $R_p=0.3\sim0.4\mu m$）。

For the critical heat flux of pool boiling, the following semi-empirical formula is recommended:

对于大容器沸腾的临界热流密度，推荐使用以下半经验公式：

$$q_{max}=\frac{\pi}{24}r\rho_v\left[\frac{g\sigma(\rho_l-\rho_v)}{\rho_v^2}\right]^{1/4}\left(\frac{\rho_l+\rho_v}{\rho_v^2}\right)^{1/2} \tag{7.18}$$

When the pressure is far away from the critical pressure, the last item at the right of Eq. (7.18) is taken as 1. While the coefficient $\pi/24$ obtained by flow analysis is replaced with the experimental value of 0.149. Then the following formula is recommended:

当压力离临界压力较远时，式（7.18）右端最后一项取为 1。同时，将流量分析得出的系数 $\pi/24$ 用实验值 0.149 代替，得到以下推荐公式：

$$q_{max}=0.149r\rho_v^{1/2}[g\sigma(\rho_l-\rho_v)]^{1/4} \tag{7.19}$$

All physical properties in the formula are obtained according to the saturation temperature of the liquid. Theoretically, this formula is only applicable to the case that the heating surface is an infinite horizontal wall. There is no characteristic length in the formula. Practically, Eq. (7.19) can be used when the characteristic length of the heating surface is more than three times of the average diameter of the bubble.

公式中所有物性均按液体的饱和温度查取。该式理论上只适用于加热面为无限大的水平壁的情形，式中没有特征长度。实际上，当加热面的特征长度大于气泡平均直径的 3 倍时，式（7.19）即可使用。

In the pool film boiling, the flow and heat transfer of vapor film are similar to that of liquid film in film-wise condensation in many aspects. For the film boiling on a horizontal tube, the convective heat transfer coefficient can be calculated as follows:

在大容器膜态沸腾中，气膜流动与传热在许多方面都类似于膜状冷凝中液膜的流动与传热。对于水平管的膜态沸腾，其表面传热系数可用以下公式计算：

$$h=0.62\left[\frac{gr\rho_v(\rho_l-\rho_v)\lambda_v^3}{\eta_v d(t_w-t_s)}\right]^{1/4} \tag{7.20}$$

In the above formula, the average temperature t_m is taken as the reference temperature for other physical properties except that the values of r and ρ_l are determined by the saturation temperature t_s. The characteristic length is the outer diameter d of the tube. If the heating surface is spherical, the coefficient in Eq. (7.20) will change from 0.62 to 0.67.

式中，除了 r 和 ρ_l 的值由饱和温度 t_s 决定外，其余物性均以平均温度 t_m 为定性温度，特征长度为管子外径 d。如果加热表面为球面，则式（7.20）中的系数 0.62 改为 0.67。

The wall temperature is generally high in film heat transfer, so it is necessary to consider the effect of thermal radiation. This effect includes two aspects: One is to directly

膜态传热时，壁面温度一般较高，因此有必要考虑热辐射的影响。该影响有两方面：一方面是直接增加

increase the heat transfer quantity, the other is to increase the film thickness thereby reducing the heat transfer quantity. So, the thermal radiation effect must be considered comprehensively. Considering the mutual influence of convection heat transfer and radiation heat transfer, Blomley suggested that the following transcendental equation can be used to calculate the convective heat transfer coefficient of composite heat transfer:

了传热量；另一方面是增大了气膜厚度，从而减少了传热量。因此，必须综合考虑热辐射效应。考虑到对流传热与辐射传热相互影响，勃洛姆来建议采用以下超越方程来计算复合传热表面传热系数：

$$h^{4/3}=h_c^{4/3}+h_r^{4/3} \tag{7.21}$$

Where h_c and h_r are the convective heat transfer coefficients calculated by the convection heat transfer and radiation heat transfer, respectively. The radiative heat transfer coefficient h_r is calculated according to the following formula:

式中，h_c 和 h_r 分别是按对流传热及辐射传热计算所得的表面传热系数，其中，辐射传热系数 h_r 根据以下公式计算：

$$h_r=\frac{\varepsilon\sigma(T_w^4-T_s^4)}{T_w-T_s} \tag{7.22}$$

Summary This section illustrates the affecting factors of nucleate boiling and the calculation formula of convection heat transfer, as well as the calculation correlation formulas for the critical heat flux of pool boiling and boiling heat transfer.

小结 本节阐述了影响核态沸腾的因素和对流传热计算公式，以及大容器沸腾的临界热流密度和沸腾传热计算关联式。

Question For different surface roughness, why is the heat transfer coefficient of nucleate boiling very different?

思考题 对于不同的表面粗糙度，为什么核态沸腾传热系数有很大的不同呢？

Self-tests

自测题

1. Briefly describe the concept of vaporization cores and the physical conditions for the production of boiling bubbles.

1. 简述汽化核心的概念、沸腾气泡产生的物理条件。

2. What is the difference between the calculation of convective heat transfer coefficient of boiling heat transfer and non phase change convective heat transfer?

2. 沸腾传热与无相变对流传热的表面传热系数计算有什么不同？

7.9 The Influencing Factors and Enhancement of Boiling Heat Transfer

7.9 沸腾传热的影响因素及强化

Boiling heat transfer is the most complicated heat transfer process with the most influencing factors in all heat transfer phenomena. The following focuses on five factors that affect pool boiling heat transfer.

沸腾传热是所有传热现象中影响因素最多、最复杂的传热过程。下面重点讨论影响大容器沸腾传热的五个因素。

The first factor is the non-condensable gas. Different from the film-wise condensation, the non-condensable gas in liquid can enhance boiling heat transfer to some extent. The reason is that the non-condensable gas will overflow from the liquid as the temperature of the working liquid rises, the tiny pits near the wall surface are activated and become the germ of bubbles. As a result, the q-Δt boiling curve will move towards the direction of decreasing Δt. Under the same superheat degree Δt, higher heat flux will be generated and heat transfer will be enhanced.

第一个因素，不凝结气体。与膜状凝结传热不同，液体中的不凝结气体会使沸腾传热得到某种程度的强化。这是因为随着工作液体温度的升高，不凝结气体会从液体中逸出，使壁面附近的微小凹坑得到活化，成为气泡的胚芽，从而使 q-Δt 沸腾曲线向着 Δt 减小的方向移动，即在相同的过热度 Δt 下产生更高的热流密度，强化了传热。

For the boiling heat transfer equipment under stable operation, unless non-condensable gas is continuously injected into the working liquid. Otherwise, they will not play an enhancing role once they escape.

对于处于稳定运行下的沸腾传热设备来说，除非不断地向工作液体注入不凝结气体，否则它们一经逸出，也就起不到强化作用了。

The second factor is the subcooling degree. In the pool boiling, the temperature of most fluid is lower than the saturation temperature under the corresponding pressure, the temperature difference is called subcooling degree, and the boiling is called subcooled boiling. The subcooling degree only affects subcooled boiling and doesn't affect saturated boiling. In the initial stage of nucleate boiling, the natural convection heat transfer mode still accounts for a considerable proportion. While in the natural convection heat transfer, $h \propto (t_w - t_f)^{1/4}$, therefore subcooling can enhance heat transfer in this region.

第二个因素，过冷度。在大容器沸腾中，大部分流体的温度低于相应压力下的饱和温度，其温差称为过冷度，这种沸腾称为过冷沸腾。过冷度只影响过冷沸腾，不影响饱和沸腾。在核态沸腾起始段，自然对流传热模式还占相当大的比例。而自然对流传热时，$h \propto (t_w - t_f)^{1/4}$，因此，过冷会使该区域的传热有所增强。

The third factor is the height of liquid level. In the pool boiling, when the liquid level on the heat transfer surface is high enough, the boiling heat transfer coefficient is independent of the height of liquid level. However, when the liquid level drops to a certain value, which is called the critical liquid level, the convective heat transfer coefficient will increase obviously with the decrease of liquid level. For water at atmospheric pressure, the critical liquid level is about 5mm.

第三个因素，液位高度。在大容器沸腾中，当传热表面上的液位足够高时，沸腾传热表面传热系数与液位高度无关。但当液位降低到一定值，这个值称为临界液位，表面传热系数会随液位的降低而明显升高。对于常压下的水，其临界液位约为 5mm。

The fourth factor is the acceleration of gravity g. With the development of aerospace technology, the study of heat transfer laws under high gravity and micro-gravity has been booming but still far from mature. The existing results show that in the range of $0.1-100\times9.8\text{m/s}^2$, the acceleration of gravity g has no effect on nucleate boiling heat transfer, but has an effect on natural convection heat transfer. Based on Eq. (7.23), with the increase of the acceleration of gravity g, Nusselt number increases and then heat transfer is enhanced.

第四个因素，重力加速度 g。随着航空航天技术的发展，超重力和微重力条件下传热规律的研究得到蓬勃发展，但还远不成熟。现有成果表明，从 $0.1\sim100\times9.8\text{m/s}^2$ 的范围内，g 对核态沸腾传热规律没有影响，但对自然对流传热有影响。由式（7.23）可知，随着 g 增加，Nu 升高，传热得到加强。

$$Gr=\frac{g\alpha\Delta t l^3}{\nu^2} \qquad Nu=C(Gr\ Pr)^n \tag{7.23}$$

The fifth factor is the structure of boiling surface. The tiny pits on boiling surface are most likely to be vaporization cores. So, the more pits, the more vaporization cores, and heat transfer will be enhanced. The study of enhancing boiling heat transfer is mainly to increase surface pits. The basic principle of enhancing boiling heat transfer is to increase the vaporization cores on the heating surface as much as possible.

第五个因素，沸腾表面的结构。沸腾表面上的微小凹坑最容易成为汽化核心，因此，凹坑越多，汽化核心就越多，传热就会得到强化。强化沸腾传热的研究主要是增加表面凹坑。强化沸腾传热的基本原则是，尽量增加加热面上的汽化核心。

For pool boiling, there are two commonly used methods of processing pits now. ① Physical and chemical means such as sintering, brazing, flame spraying and ionization deposition are used to form the porous structure on the heat exchange surface. ② Mechanical processing methods are used to create porous structures on the heat transfer surfaces.

对于大容器沸腾，目前有两种常用的凹坑加工手段。① 采用烧结、钎焊、火焰喷涂和电离沉积等物理与化学手段，在传热表面上形成多孔结构。② 采用机械加工方法在传热表面上制成多孔结构。

For in-tube boiling, at present there are two enhanced

对于管内沸腾，目前有如下两种强

surface structures as follows. The first is the internal thread steel tube; The second is two-dimensional and three-dimensional micro rib tube.

化表面结构：第一种是内螺纹钢管；第二种是二维、三维微肋管。

Summary This section illustrates the influencing factors of nucleate boiling heat transfer and the methods for enhancing boiling heat transfer.

小结 本节阐述了核态沸腾传热的影响因素和强化沸腾传热的方法。

Question What is the basic idea of enhancing boiling heat transfer in terms of the structure of heat transfer surface?

思考题 从传热表面的结构而言，强化沸腾传热的基本思想是什么？

Self-tests

自测题

1. What are the influencing factors of nucleate boiling heat transfer?

2. What are the methods of enhancing boiling heat transfer?

1. 核态沸腾传热的影响因素有哪些？

2. 强化沸腾传热的方法有哪些？

Exercises

习题

7.1 When pouring a glass of water on a hot iron plate, the surface of the plate immediately produces a number of small jumping water droplets, and the droplets can remain for quite a while without being vaporized. Try to explain this phenomenon, often called the Leidenfrost phenomenon, from the viewpoint of Heat Transfer, and find the point on the boiling heat transfer curve where the state begins to form.

7.1 当把一杯水倒在一块炽热的铁板上时，板面立即会产生许多跳动着的小水滴，而且可以维持相当一段时间而不被汽化掉。试从传热学的观点来解释这一现象［常称为莱顿弗罗斯特（Leidenfrost）现象］，并从沸腾传热曲线上找出开始形成这一状态的点。

7.2 The cooling curve of steel element with a diameter of 6mm is shown in Fig. 7.17 when it is quenched in the water at 98℃. The initial temperature of the steel element is 800℃. Try to analyze the properties of the heat transfer process represented by each section of the curve.

7.2 直径为6mm的钢元件在98℃水中淬火时的冷却曲线如图7.17所示，钢元件初温为800℃，试分析曲线各段所代表的传热过程的性质。

7.3 The diameter of the heating surface at the bottom of a copper pan is 30cm, which is required to produce 2.3kg saturated water vapor per hour when boiling at the atmospheric pressure of 1.013×10^5Pa. Try to determine the temperature of the water contact surface when the bottom of the pot is clean.

7.3 一铜制平底锅底部的受热面直径为30cm，要求其在1.013×10^5Pa的大气压下沸腾时每小时能产生2.3kg饱和水蒸气。试确定锅底干净时其与水接触面的温度。

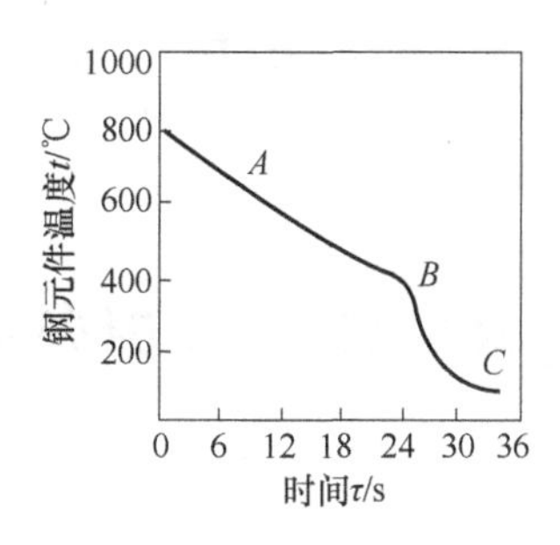

Fig. 7. 17 The cooling curve of steel element

图 7. 17 钢元件的冷却曲线

7. 4 A mechanically polished thin-walled stainless steel tube with a diameter of 3. 5mm and a length of 100mm is placed in a water container at a pressure of 1.013×10^5Pa. And the water temperature is close to the saturation temperature. Electrify both ends of the stainless steel tube as a heating surface. Try to calculate the value of convective heat transfer coefficient between the water and the steel tube surface when the heating power is 1. 9W and 100W, respectively.

7. 4 一直径为 3. 5mm、长为 100mm 的机械抛光的薄壁不锈钢管，被置于压力为 1.013×10^5Pa 的水容器中，水温已接近饱和温度，对该不锈钢管两端通电以作为加热表面。试计算当加热功率为 1. 9W 及 100W 时，水与钢管表面间的表面传热系数。

7. 5 When the boiling experiment in a large vessel in the laboratory is conducted at the pressure of 1.013×10^5Pa, the method of heating with high current through small diameter stainless steel tube is adopted. In order to demonstrate the entire nucleate boiling region at a voltage not higher than 220V, try to estimate the required resistance per meter length of the stainless steel tube when the diameter of the selected stainless steel tube is set as 3mm in diameter and 100mm in length.

7. 5 在实验室内进行压力为 1.013×10^5Pa 的大容器沸腾实验时，采用大电流通过小直径不锈钢管的方法加热。为了能在电压不高于 220V 的情形下演示整个核态沸腾区域，试估算所需的不锈钢管的每米长电阻应为多少？设选定的不锈钢管的直径为 3mm，长为 100mm。

7. 6 The cooling water with an average temperature of 15℃ and a flow rate of 1. 5m/s flows through a horizontally placed copper tube with an outer diameter of 32mm and an inner diameter of 28mm. The water vapor with the saturation pressure of 0.024×10^5Pa condenses outside the copper tube, which is 1. 5m long. Try to calculate the amount of condensed water per hour (the thermal resistance of the copper tube is not considered).

7. 6 平均温度为 15℃、流速为 1. 5m/s 的冷却水，流经外径为 32mm、内径为 28mm 的水平放置的铜管。饱和压力为 0.024×10^5Pa 的水蒸气在铜管外凝结，管长 1. 5m，试计算每小时的凝结水量（铜管的热阻可不考虑）。

Chapter 8 Basic Laws of Thermal Radiation and Radiation Properties

第 8 章 热辐射基本定律和辐射特性

8.1 Thermal Radiation Phenomenon

8.1 热辐射现象

As we know, radiation is a phenomenon of electromagnetic waves transferring energy. According to the different ways of generating electromagnetic waves, different frequencies of electromagnetic waves can be obtained. Common electromagnetic waves include radio waves, infrared ray, visible light, ultraviolet ray, X-ray and γ-ray etc.

我们知道，辐射是电磁波传递能量的现象。按照产生电磁波的不同方式，可以得到不同频率的电磁波。常见的电磁波有无线电波、红外线、可见光、紫外线、X 射线及 γ 射线等。

Thermal radiation is a way of transferring energy in the form of electromagnetic waves generated by thermal motion. It belongs to the basic mode of heat transfer, along with heat conduction and heat convection. Thermal radiation has the following five characteristics.

热辐射是通过热运动产生的电磁波来传递能量的方式。它与热传导、热对流都属于热量传递的基本方式。热辐射具有以下五个特点：

First, as long as the temperature of any object is higher than 0K, that is, absolute zero, it will constantly emit thermal radiation to the surrounding space, and absorb thermal radiation at the same time.

第一，对于任何物体，只要温度高于 0K，即绝对零度，就会不停地向周围空间发出热辐射，同时也在吸收热辐射。

Second, thermal radiation can spread in vacuum with the highest efficiency. In other words, the energy transfer of thermal radiation does not require the presence of other media.

第二，热辐射可以在真空中传播，且效率最高。也就是说，热辐射的能量传递不需要其他介质的存在。

Third, the radiation heat transfer process is accompanied by the conversion of energy form. When an object emits thermal radiation, thermal energy is converted into radiant energy. When absorbing thermal radiation, the radiant

第三，辐射传热过程伴随能量形式的转变。物体发射热辐射时热能转变为辐射能，吸收热辐射时辐射能又转变为热能。

energy is converted into thermal energy.

Fourth, thermal radiation has strong directionality. When calculating thermal radiation, positive and negative signs are often used to indicate its direction.

第四，热辐射具有强烈的方向性。进行热辐射计算时，常用正负号表示其方向。

Fifth, the intensity of radiant energy depends on the fourth power of thermodynamic temperature. Following the basic law of thermal radiation: Stefan-Boltzmann law.

第五，辐射能强度取决于热力学温度的 4 次方。遵循热辐射的基本定律：斯特藩-玻尔兹曼定律。

From the above five characteristics, it can be seen that thermal radiation is significantly different from heat conduction and thermal convection previously learned.

从上述五个特点可以看出，热辐射和前面学习的热传导、热对流具有明显的不同。

Radiation heat transfer refers to the total effect of mutual radiation and absorption between and among objects. When the object is in thermal equilibrium with the environment, the thermal radiation on its surface is still going on, but its net radiation heat transfer is equal to zero.

辐射传热，是指物体之间相互辐射和吸收的总效果。当物体与环境处于热平衡时，其表面的热辐射仍在不停地进行，但其净的辐射传热量等于零。

Thermal radiation has the commonness of general radiation phenomenon. Now let's describe the characteristics of thermal radiation from the perspective of electromagnetic wave.

热辐射具有一般辐射现象的共性。下面从电磁波的角度描述热辐射的特性。

The propagation speed of electromagnetic waves, i. e. speed of light: $c=f\lambda$。Where c represents the speed of light, f represents frequency, and λ represents the wavelength. According to the above equation, the wavelength distribution figure of electromagnetic wave spectrum at different frequencies can be obtained, as shown in Fig. 8. 1.

电磁波的传播速度（即光速）：$c=f\lambda$。其中，c 表示光速，$c=3\times10^8$ m/s，f 表示频率；λ 表示波长。根据上式，可以得到不同频率的电磁波谱的波长分布图，如图 8. 1 所示。

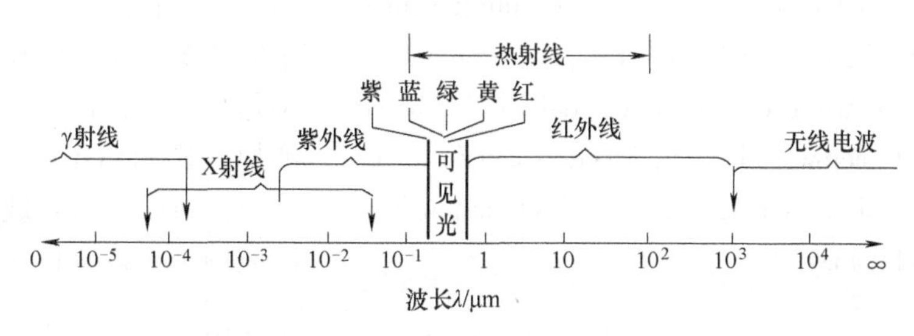

Fig. 8. 1 Electromagnetic wave spectrum

图 8. 1 电磁波谱图

As shown in Fig. 8. 1, the wavelength range of electromagnetic wave spectrum is $\lambda = 0 - \infty$. We can also see the wavelength range of several common electromagnetic waves. In industry, the wavelength range of significant thermal radiation (below 2000K) is $\lambda = 0.1 - 100\mu m$. Among them, the wavelength range of sunlight is $\lambda = 0.2 - 2\mu m$, the wavelength range of visible light is $\lambda = 0.38 - 0.76\mu m$, the wavelength range of infrared ray $\lambda = 0.76 - 1000\mu m$.

图 8.1 中显示，电磁波谱的波长范围 $\lambda = 0 \sim \infty$。从中还能看出常见的几种电磁波的波长范围。工业上有实际意义的热辐射（2000K 以下）波长区段 $\lambda = 0.1 \sim 100\mu m$。其中，太阳光的波长范围 $\lambda = 0.2 \sim 2\mu m$，可见光的波长范围 $\lambda = 0.38 \sim 0.76\mu m$，红外线的波长范围 $\lambda = 0.76 \sim 1000\mu m$。

Electromagnetic waves of various wavelengths have different applications. Among them, ultraviolet ray has the following advantages: ① It can be used for sterilization in hospitals, indoors and so on; ② It promotes bone development. But it also has some disadvantages: ① It causes skin aging and wrinkle; ② It produces spots; ③ It causes dermatitis. The sunscreen we use in summer is to reduce the damage of ultraviolet to the skin.

各种波长的电磁波有不同的应用。其中，紫外线具有以下优点：① 可用于医院及室内等杀菌消毒；② 促进骨骼发育。但是它也有缺点：① 使皮肤老化产生皱纹；② 产生斑点；③ 造成皮肤炎。夏天人们使用的防晒产品，就是为了减少紫外线对皮肤的伤害。

X-rays are also called Roentgen rays. It can penetrate objects and be used in medical perspective. X-rays are used for chest radiography in hospitals, and can also be used for flaw detection in industry. Infrared rays can be divided into two parts: When $\lambda < 25\mu m$, called near infrared; when $\lambda > 25\mu m$, called far infrared. For example, far infrared heating technology and infrared imaging technology used in industry.

X 射线，又称为伦琴射线。它可以穿透物体，用于医学透视；医院拍胸片就是用的 X 射线，工业上也可用来探伤。红外线可以分为两部分：$\lambda < 25\mu m$ 时，称为近红外线；$\lambda > 25\mu m$，称为远红外线。比如工业上用的远红外加热技术、红外线成像技术等。

The electromagnetic wave with $\lambda = 1mm - 1m$ is called microwave. Microwave can pass through plastic, glass and ceramic products. But it can be absorbed by things like water with polar molecules and produces internal heat sources inside the things, so that the things can be heated uniformly. The microwave oven used in our daily life is based on the principle of microwave heating. Electromagnetic waves with $\lambda > 1m$ are called as long waves, which are widely used in radio technology.

将波长 $\lambda = 1mm \sim 1m$ 的电磁波称为微波。微波可以穿过塑料、玻璃、陶瓷制品，却会被像水那样具有极性分子的物体吸收，并在物体内产生内热源，从而使物体均匀地得到加热。人们日常生活中所使用的微波炉就是利用了微波加热的原理。波长 $\lambda > 1m$ 的电磁波称为长波，广泛应用于无线电技术。

When thermal radiation is projected on the surface of an object, generally, three phenomena will occur, namely absorption, reflection and penetration, as shown in Fig. 8. 2.

当热辐射投射到物体表面上时，一般会发生三种现象，即吸收、反射和穿透，如图 8.2 所示。

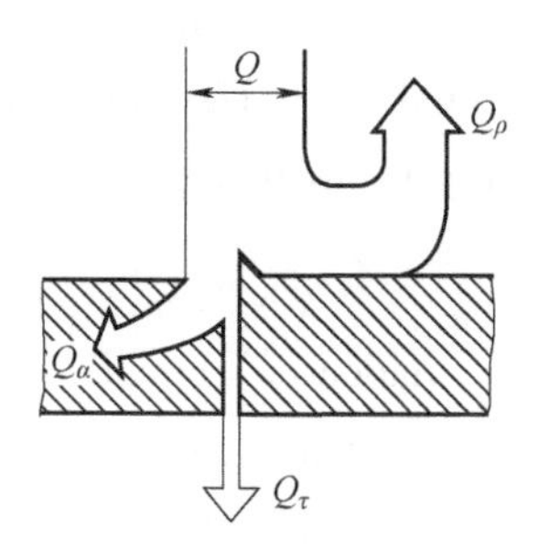

Fig. 8. 2 Absorption, reflection and penetration
图 8. 2 吸收、反射和穿透

If Q is used to represent the total radiant energy, Q_α represents the part of energy absorbed by the object, Q_ρ represents the part of energy reflected by the object, and Q_τ represents the part of energy that penetrates the object. Then, Eq. (8. 1) can be obtained:

如果用 Q 来表示总的辐射能量，Q_α 表示被物体吸收的那一部分能量，Q_ρ 表示被物体反射的那一部分能量，Q_τ 表示穿透物体的那一部分能量。那么，可以得到式（8. 1）：

$$Q=Q_\alpha+Q_\rho+Q_\tau \tag{8.1}$$

We divide both sides of the formula by Q to get

将式子两边同除以 Q，可以得到

$$\frac{Q_\alpha}{Q}+\frac{Q_\rho}{Q}+\frac{Q_\tau}{Q}=1$$

$$\alpha+\rho+\tau=1 \tag{8.2}$$

Where α means absorptivity, ρ means reflectivity, and τ represents transmissivity.

式中，α 表示吸收比（率）；ρ 表示反射比（率）；τ 表示穿透比（率）。

For different objects, their absorptivity, reflectivity and transmissivity are different. For most solids and liquids, the transmissivity τ is equal to 0, the absorptivity α plus the reflectivity ρ equals to 1. When radiation enters the surface of solid or liquid, it can be absorbed within a very short distance. For metal conductors, this distance is only 1μm; and for nonconductive materials, the distance is less than 1mm. Therefore, it can be considered that solids and liquids do not allow thermal radiation to penetrate, namely the transmissivity is equal to 0.

对于不同物体，它们的吸收比、反射比和穿透比不尽相同。对于大多数的固体和液体，穿透比 τ 等于 0，吸收比 α 加上反射比 ρ 等于 1。当辐射能进入固体或液体表面后，在极短的距离内就能被吸收完。对于金属导体，这一距离仅为 1μm；对于非导电体材料，距离小于 1mm。因此，可以认为固体和液体不允许热辐射穿透，即穿透比等于 0。

For gas without particles, the reflectivity ρ is equal to 0, the absorptivity α plus the transmissivity τ equals 1. For black bodies, the absorptivity is equal to 1; For mirrors or

对于不含颗粒的气体：反射比 ρ 等于 0，吸收比 α 加上穿透比 τ 等于 1。对于黑体：吸收比 $\alpha=1$；对于镜体

white bodies, the reflectivity is equal to 1; and for transparent bodies, the transmissivity is equal to 1. The so-called a black body, a white body and a transparent body are all assumed to be ideal models.

或白体：反射比 $\rho=1$；对于透明体：穿透比 $\tau=1$。所谓的黑体、镜体（白体）和透明体都是假定的理想模型。

Summary This section illustrates the definition of thermal radiation, five characteristics of thermal radiation, electromagnetic spectrum, the application of electromagnetic waves of various wavelengths, and the absorptivity, reflectivity and transmissivity of thermal radiation by objects.

小结 本节阐述了热辐射的定义、热辐射的五个特点、电磁波谱、各种波长电磁波的应用，以及物体对热辐射的吸收比、反射比和穿透比。

Question What are the main differences between thermal radiation and heat conduction, thermal convection?

思考题 热辐射与热传导、热对流有什么主要区别?

Self-tests

自测题

1. Try to describe the wavelength range of thermal radiation that has practical industrial significance, and the concepts of near infrared and far infrared radiation.

1. 试述工业上有实际意义的热辐射波长范围，以及近红外、远红外辐射概念。

2. What are the differences between the absorptivity, reflectivity and transmissivity of different objects?

2. 不同物体的吸收比、反射比和穿透比有什么区别?

授课视频

8.2 A Black Body Model

8.2 黑体模型

The reflection phenomenon of radiant energy projected on the surface of an object is also the same as visible light, and includes specular reflection and diffuse reflection, which depends on the uneven size of the surface, that is, the roughness of the surface.

辐射能投射到物体表面后的反射现象也和可见光一样，有镜面反射和漫反射之分。这取决于表面不平整尺寸的大小，即表面的粗糙程度。

The roughness mentioned here is relative to the wavelength of thermal radiation. When the uneven size of the surface is smaller than the wavelength of the irradiation, specular reflection is formed and the incident angle is equal to the reflection angle at the time. As shown in Fig. 8.3, a highly polished metal plate is an example of specular reflection.

这里所说的粗糙程度，是相对于热辐射的波长而言的。当表面的不平整尺寸小于投入辐射的波长时，形成镜面反射，此时入射角等于反射角。如图 8.3 所示，高度磨光的金属板就是镜面反射的实例。

When the uneven size of the surface is greater than the wavelength of the irradiation, diffuse reflection is formed. At this time, the radiation projected on the surface of the object from a certain direction is reflected in all directions of space, as shown in Fig. 8. 4. Diffuse reflection is relatively common. Generally, the surface of engineering materials always forms diffuse reflection.

当表面的不平整尺寸大于投射辐射的波长时，形成漫反射。这时，从某一方向投射到物体表面上的辐射向空间各个方向反射出去，如图 8.4 所示。漫反射比较常见，一般工程材料表面形成的都是漫反射。

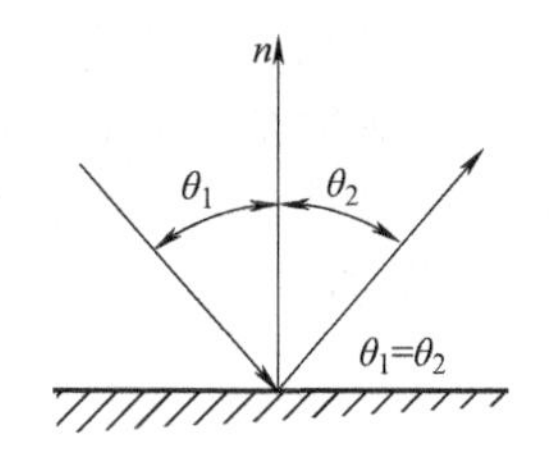

Fig. 8. 3 Specular reflection

图 8. 3 镜面反射

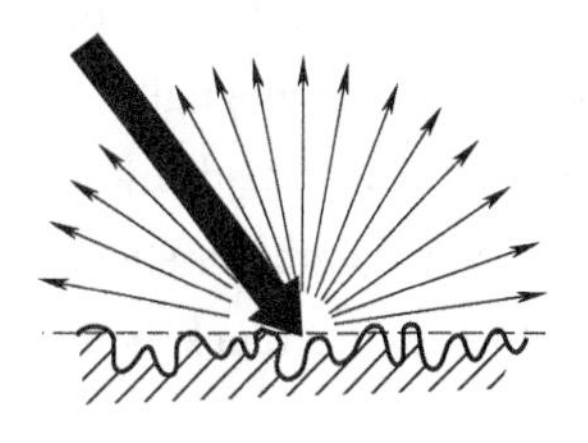

Fig. 8. 4 Diffuse reflection

图 8. 4 漫反射

The absorptivity α, reflectivity ρ, and transmissivity τ of different objects in nature are very different due to different specific conditions, which brings great difficulties to the research of thermal radiation. For convenience, the research should start with the ideal object.

自然界不同物体的吸收比 α、反射比 ρ 和穿透比 τ，因具体条件不同而千差万别，这给热辐射研究带来很大困难。为了方便，从理想物体入手进行研究。

The object that can absorb all the thermal radiation energy projected on its surface is called a black body, or an absolute black body. Its absorptivity $\alpha=1$. The black body is a kind of scientific hypothetical model, which does not exist in reality, but an approximate black body can be artificially created, as shown in Fig. 8. 5.

能吸收投射到其面上的所有热辐射能的物体叫作黑体，或者叫作绝对黑体。其吸收比 $\alpha=1$。黑体是一种科学的假想模型，现实生活中并不存在，但可人工制造出近似的黑体，如图 8. 5 所示。

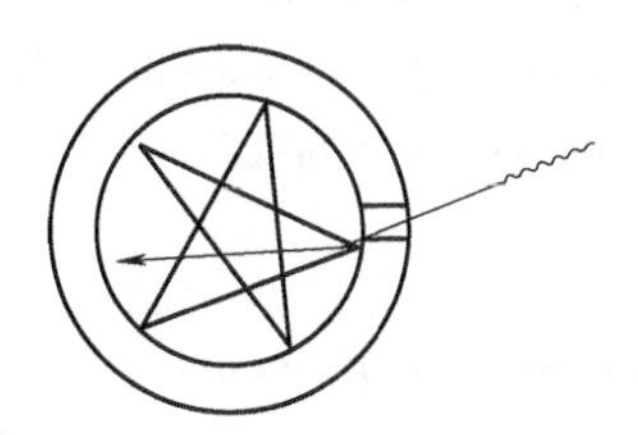

Fig. 8. 5 A black body model

图 8. 5 黑体模型

Choose a material with relatively large absorptivity to make a cavity, and open a small hole on the side of the cavity, then try to keep the cavity wall temperature uniform. At

选用吸收比较大的材料制造一个空腔，并在空腔边上开一个小孔，再设法使空腔壁面保持均匀的温度，

this time, the small hole on the cavity has the characteristics of the black body. This kind of cavity with a small hole and uniform temperature is a black body model.

这时空腔上的小孔就具有黑体的特性。这种带有小孔的温度均匀的空腔就是一个黑体模型。

This is because when the radiation energy enters the cavity through the small hole, it will undergo multiple absorption and reflection in the cavity. And after every absorption, the radiation energy is reduced in proportion to the absorption rate of the inner wall. The energy that can finally escape the small hole is negligible, and can be considered to be completely absorbed by the cavity. Therefore, in terms of radiation characteristics, the small hole has the same properties as the black body surface.

这是因为当辐射能经小孔射入空腔时，在空腔内要经历多次吸收和反射，而每经过一次吸收，辐射能会按内壁吸收比的份额减少。最终能离开小孔的能量已微乎其微，可以认为完全被空腔吸收。所以，就辐射特性而言，小孔具有黑体表面一样的性质。

It's worth noting that the absorptivity of the cavity material itself has no effect on the black body model in principle. At a certain ratio of the small hole area to the total area of the cavity, the greater the absorptivity of the material itself, the greater the effective absorptivity of the black body model. And the smaller the proportion of the small hole area to the total area of the cavity, the higher the absorptivity of the small hole. The black body is of special importance in thermal radiation analysis. The black body has the strongest radiation ability among the objects at the same temperature.

值得注意的是，空腔材料本身的吸收比原则上对黑体模型没有影响。在一定的小孔面积与空腔总面积之比下，材料本身的吸收比越大，黑体模型的有效吸收比越大；小孔面积占空腔总面积的比例越小，小孔的吸收比就越高。黑体在热辐射分析中有其特殊的重要性，在相同温度的物体中黑体的辐射能力最强。

The object with reflectivity $\rho = 1$ is called a mirror or a white body. When it is diffuse reflection, it is called an absolute white body. The object with transmissivity $\tau = 1$ is called an absolute transparent body. The black body, the white body and the transparent body mentioned here are all ideal models. Next, let's learn the expression of thermal radiation energy.

反射比 $\rho = 1$ 的物体叫作镜面或白体。当为漫反射时称作绝对白体。穿透比 $\tau = 1$ 的物体叫作绝对透明体。这里所说的黑体、白体和透明体，都是理想模型。接下来学习热辐射能量的表示方法。

(1) The emissive power, usually expressed by E. Its definition is that the energy within the entire wavelength range that an object radiates in all directions of the hemispherical space above it in unit time and in unit surface area. The unit is W/m^2.

(1) 辐射力，通常用 E 表示。它定义为物体在单位时间内，单位表面面积向其上的半球空间的所有方向辐射出去的全部波长范围内的能量，单位为 W/m^2。

(2) The spectral emissive power E_λ, also known as monochromatic emissive power. Its definition is that the energy

(2) 光谱辐射力 E_λ，又称作单色辐射力。它定义为单位时间内，单

within a unit wavelength including the wavelength λ that an object radiates to all directions of the hemispherical space above it in unit time and in unit surface area. The unit is W/(m^2 · m) or W/(m^2 · μm). Note that m in the denominator of unit represents the width of unit wavelength. The unit of m is too large for the wavelength width of thermal radiation, so μm is often used instead.

位表面面积向其上的半球空间所有方向辐射出去的包含波长 λ 在内的单位波长内的能量，单位为 W/(m^2 · m) 或 W/(m^2 · μm)。注意：单位分母中的 m 表示单位波长的宽度。m 这个单位对于热辐射的波长宽度而言太大，因此常用 μm 来代替。

From the definitions of the emissive power E and the spectral emissive power E_λ, it is not difficult to see that the emissive power E is equal to the integral of the spectral emissive power E_λ over the whole wavelength range, as shown in Eq. (8.3).

从辐射力 E 与光谱辐射力 E_λ 的定义不难看出，辐射力就是光谱辐射力 E_λ 在整个波长范围内的积分，见式（8.3）。

$$E=\int_0^\infty E_\lambda \mathrm{d}\lambda \tag{8.3}$$

Note that physical quantities related to the black body are represented by subscript b. For example, the emissive power of the black body is E_b, and the spectral emissive power of the black body is $E_{b\lambda}$.

注意：与黑体相关的物理量均采用下标 b 表示，如黑体的辐射力为 E_b，黑体的光谱辐射力为 $E_{b\lambda}$。

Summary This section illustrates a black body model and the two reflections of solid surfaces, namely specular reflection and diffuse reflection, and two expressions of thermal radiation energy, namely emissive power and spectral emissive power.

小结 本节阐述了黑体模型和固体表面的两种反射，即镜面反射和漫反射；以及热辐射能量的两种表示方法，即辐射力和光谱辐射力。

Question Why should the instruction of hemispherical space and all wavelengths be added when defining the emissive power of an object?

思考题 为什么在定义物体的辐射力时要加上“半球空间”及“全部波长”的说明？

Self-tests

自测题

1. Why is the concept of the black body introduced into the theory of thermal radiation? What are its main characteristics?

1. 在热辐射理论中为什么要引入黑体的概念？其有什么主要特征？

2. What is the relationship between the emissive power E and the spectral emissive power E_λ?

2. 辐射力 E 与光谱辐射力 E_λ 之间具有怎样的关系？

8.3 Planck's Law

Planck's law: The variation law of the spectral emissive power $E_{b\lambda}$ of a black body with wavelength λ (and temperature T). The expression is as follows:

8.3 普朗克定律

普朗克定律：黑体的光谱辐射力 $E_{b\lambda}$ 随波长 λ（及温度 T）的变化规律。其表达式为：

$$E_{b\lambda}=\frac{c_1\lambda^{-5}}{e^{c_2/(\lambda T)}-1} \tag{8.4}$$

Where λ denotes wavelength with the unit of m, T is the temperature of a black body with the unit of K, c_1 is the first radiation constant with the value of 3.7419×10^{-16} W · m^2, and c_2 is the second radiation constant with the value of 1.4388×10^{-2} W · K. Fig. 8.6 depicts the variation relationship of the spectral emissive power of a black body with wavelength λ and temperature T.

式中，λ 是波长（m）；T 是黑体温度（K）；c_1 是第一辐射常数，其值为 3.7419×10^{-16} W · m^2；c_2 是第二辐射常数，其值为 1.4388×10^{-2} W · K。图 8.6 描绘了黑体光谱辐射力随波长 λ 和温度 T 的依变关系。

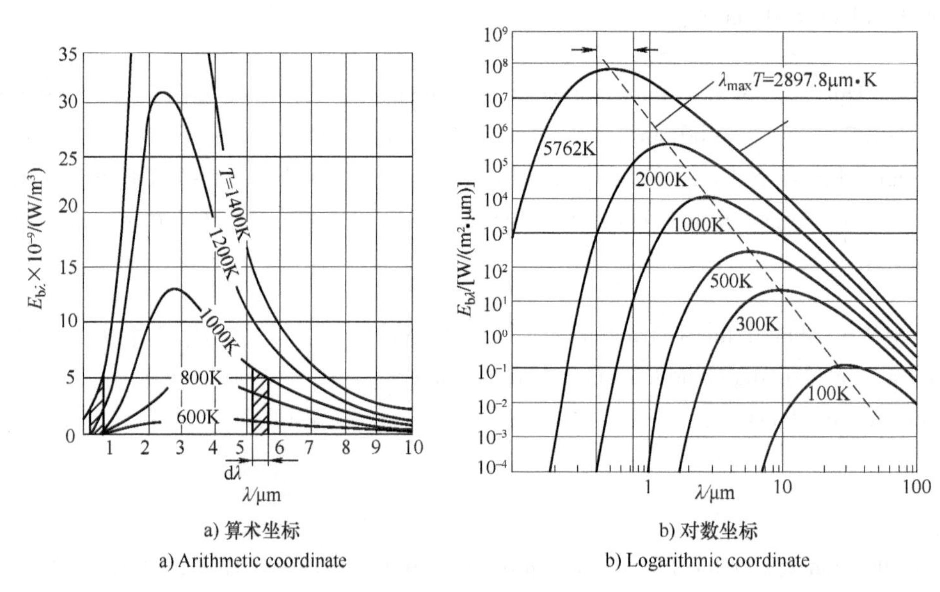

Fig. 8.6 Planck's law illustration
图 8.6 普朗克定律图示

In this figure, the abscissa denotes the wavelength, while the ordinate denotes the spectral emissive power of a black body. The curves in this figure indicate that the spectral

图中，横坐标表示波长，纵坐标表示黑体的光谱辐射力。从图中曲线可以看出，不同温度下的黑体光谱

emissive power of a black body at different temperatures increases first and then decreases with the increase of wavelength. At the same wavelength, its spectral emissive power increases as the temperature rises.

辐射力均随着波长的增加先增大后减小。在同一波长下，随着温度升高，其光谱辐射力增强。

Enlarging the figure shows that the maximum spectral radiation power of a black body at different temperatures lies in a straight line. The highest point of different curves, that is, the peak of different curves, moves to the left, that is, to a shorter wavelength.

将图放大可以看出，不同温度下黑体的最大光谱辐射力位于一条直线上。不同曲线的最高点即不同曲线的波峰向左移动，即移向较短的波长。

From this phenomenon, we can conclude that the wavelength λ_m corresponding to the maximum spectral emissive power $E_{b\lambda}$ is inversely proportional to the temperature T, which is called Wien's displacement law. This law is also called the fourth basic law of black body radiation.

从这一现象可以得出：对应于最大光谱辐射力 $E_{b\lambda}$ 的波长 λ_m 与 T 成反比，该关系称为维恩位移定律。该定律称为黑体辐射第四个基本定律。

$$\lambda_m T = 2.9\times10^{-3}\,\mathrm{m\cdot K} \tag{8.5}$$

Example 8.1 Try to calculate the wavelength λ_m corresponding to the maximum monochromatic emissive power of the black body at 2000K and 5800K, respectively.

例 8.1 试分别计算温度为 2000K 和 5800K 的黑体的最大单色辐射力所对应的波长 λ_m。

Solution: This problem can be directly calculated by the expression of Wien's displacement law.

解： 此题可以直接应用维恩位移定律表示式来计算。

$$T=2000\mathrm{K}\text{ 时，}\lambda_m = \frac{2.9\times10^{-3}\,\mathrm{m\cdot K}}{2000\mathrm{K}} = 1.45\times10^{-6}\,\mathrm{m} = 1.45\,\mu\mathrm{m}$$

$$T=5800\mathrm{K}\text{ 时，}\lambda_m = \frac{2.9\times10^{-3}\,\mathrm{m\cdot K}}{5800\mathrm{K}} = 0.50\times10^{-6}\,\mathrm{m} = 0.50\,\mu\mathrm{m}$$

The calculation results show that the wavelength is located in the infrared region (0.76−1000μm) when the temperature is 2000K; It is located in visible light region (0.38−0.76μm) when the temperature is 5800K.

从计算结果可以看出，温度为 2000K 时，波长位于红外线波段（0.76~1000μm）；温度为 5800K 时，波长位于可见光波段（0.38~0.76μm）。

Summary This section illustrates Planck's law of black body radiation and Wien's displacement law, and their respective expressions and physical meanings.

小结 本节阐述了黑体辐射的普朗克定律和维恩位移定律，以及它们各自的表达式及物理意义。

Question What is the variation law of spectral emissive power of the black body at different temperatures with wavelength?

思考题 不同温度下黑体的光谱辐射力随着波长的变化规律是什么？

Self-tests

自测题

1. Describe the expression of Wien's displacement law, and consider its action in natural science and engineering applications.

2. How is the radiation energy of the black body distributed according to the wavelength?

1. 描述维恩位移定律的表达式，并考虑它在自然科学及工程应用中的作用。

2. 黑体的辐射能按波长是怎样分布的？

授课视频

8.4 Stefan-Boltzmann's Law

8.4 斯特藩-玻尔兹曼定律

Stefan-Boltzmann's law indicates that the emissive power of a black body is directly proportional to the fourth power of its thermodynamic temperature, also known as the law of the fourth power. The expression is as follows:

斯特藩-玻尔兹曼定律：黑体辐射力正比于其热力学温度的四次方，也叫作四次方定律。其表达式为：

$$E_b = \sigma T^4 \tag{8.6}$$

Where $\sigma = 5.67\times10^{-8}\mathrm{W/(m^2 \cdot K^4)}$ is known as Stefan-Boltzmann constant. In order to calculate the high temperature conveniently, divide the thermodynamic temperature T by 100. Eq. (8.6) can be rewritten as follows:

式中，$\sigma = 5.67\times10^{-8}\mathrm{W/(m^2 \cdot K^4)}$称作斯特藩-玻尔兹曼常数。为了计算高温方便，将热力学温度T除以100，式（8.6）可改写为：

$$E_b = C_0\left(\frac{T}{100}\right)^4 \tag{8.7}$$

Where $C_0 = 5.67\mathrm{W/(m^2 \cdot K^4)}$ is the black body radiation coefficient. The variation relationship of the spectral emissive power of a black body with wavelength λ and temperature T in Fig. 8.6 shows that the area under the spectral emissive power curve is the emissive power of the black body at the temperature. So, the following equation is obtained:

式中，黑体辐射系数$C_0 = 5.67\mathrm{W/(m^2 \cdot K^4)}$。图8.6黑体光谱辐射力随波长$\lambda$和温度$T$的依变关系表明，光谱辐射力曲线下的面积就是该温度下黑体的辐射力。因而有以下式子：

$$E_b = \int_0^{\infty} E_{b\lambda} d\lambda = \int_0^{\infty} \frac{c_1 \lambda^{-5}}{e^{c_2/(\lambda T)} - 1} d\lambda = \sigma T^4 \tag{8.8}$$

Eq. (8.8) represents the relationship between the emissive power E_b and the spectral emissive power $E_{b\lambda}$ of a black body, and also represents the relationship between Planck's law and Stefan-Boltzmann's law.

式（8.8）表示了黑体的辐射力 E_b 和光谱辐射力 $E_{b\lambda}$ 的关系，也表示了普朗克定律和斯特藩-玻尔兹曼定律之间的关系。

Example 8.2 A black body is placed in a factory building at room temperature of 27℃. Ty to find the emissive power of the black body surface under the condition of heat balance. If the black body is heated to 327℃, how much is its emissive power?

例 8.2 一黑体置于室温为27℃的厂房中，试求在热平衡条件下黑体表面的辐射力。如将黑体加热到327℃，它的辐射力又是多少？

Solution: For the first question, the black body temperature is the same as room temperature under the condition of heat balance, that is, 27℃. It can be converted to the thermodynamic temperature of (27+273)K, and then substitute it into the formula of Steffan-Boltzmann law to calculate the emissive power E_{b1}.

解：对于此题第一问，在热平衡条件下黑体温度与室温相同，即等于27℃。换算为热力学温度为（27+273）K，代入斯特藩-玻尔兹曼定律的公式，可以计算出它的辐射力 E_{b1}。

$$E_{b1} = C_0 \left(\frac{T}{100}\right)^4 = 5.67\text{W/(m}^2 \cdot \text{K}^4) \times \left(\frac{27+273\text{K}}{100}\right)^4 = 459\text{W/m}^2$$

Similarly, when the temperature is 327℃, its emissive power E_{b2} can be calculated with the equation.

同样，当温度为327℃时，代入公式计算出它的辐射力 E_{b2}。

$$E_{b2} = C_0 \left(\frac{T}{100}\right)^4 = 5.67\text{W/(m}^2 \cdot \text{K}^4) \times \left(\frac{327+273\text{K}}{100}\right)^4 = 7348\text{W/m}^2$$

The calculation results show that the emissive power is directly proportional to the fourth power of thermodynamic temperature, therefore, the emissive power increases sharply with the increase of temperature. Although the temperature T_2 is only twice of temperature T_1, the ratio of the emissive power is as high as 16 times.

计算结果表明，辐射力与热力学温度的四次方成正比，所以随着温度的升高辐射力急剧增大。虽然温度 T_2 仅为 T_1 的两倍，而辐射力之比却高达16倍。

Stefan-Boltzmann's law can calculate the emissive power of a black body over all wavelengths. How to calculate the emissive power of a black body in a certain wavelength range? Therefore, the concept of "black body radiation function" is introduced. If the radiation energy emitted of

斯特藩-玻尔兹曼定律可以求出整个波长范围内的黑体辐射力。某波段内的黑体辐射力如何计算呢？为此，引入"黑体辐射函数"这个概念。如果计算黑体在波长 λ_1 至

the black body in the wavelengths λ_1 and λ_2 range is calculated, as shown in the shadow area of Fig. 8.7, the calculation formula is as follows:

λ_2 区段内所发射的辐射能，如图 8.7 中阴影面积所示，计算式为：

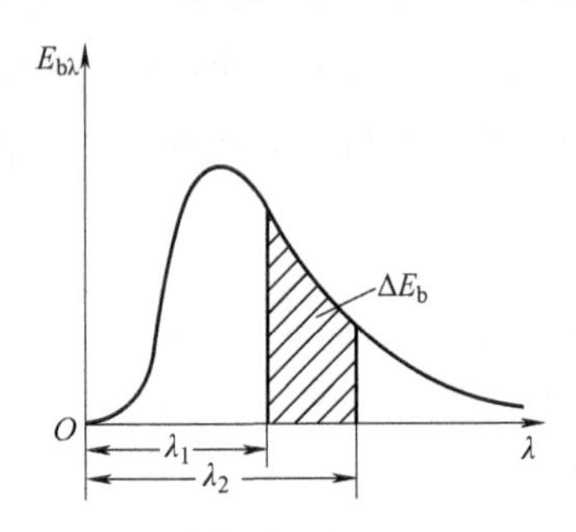

Fig. 8.7 Radiation energy in a specific wavelength range

图 8.7 特定波长区段内的黑体辐射能

$$\Delta E_b = \int_{\lambda_1}^{\lambda_2} E_{b\lambda} d\lambda \tag{8.9}$$

The radiation energy of this wavelength range is expressed as a percentage of the emissive power of the black body at the same temperature, as shown in the following equation:

将该波段的辐射能表示成同温度下黑体辐射力的百分数，如下式：

$$F_{b(\lambda_1-\lambda_2)} = \frac{\int_{\lambda_1}^{\lambda_2} E_{b\lambda} d\lambda}{\int_0^{\infty} E_{b\lambda} d\lambda} = \frac{1}{\sigma T^4}\int_{\lambda_1}^{\lambda_2} E_{b\lambda} d\lambda = \frac{1}{\sigma T^4}\left(\int_0^{\lambda_2} E_{b\lambda} d\lambda - \int_0^{\lambda_1} E_{b\lambda} d\lambda\right)$$
$$= F_{b(0-\lambda_2)} - F_{b(0-\lambda_1)} \tag{8.10}$$

Where $F_{b(0-\lambda_1)}$ and $F_{b(0-\lambda_2)}$ are called black body radiation function (percentage). If the wavelength and thermodynamic temperature are known, the values of the black body radiation function can be found from the black body radiation function table. By conversion of Eq. (8.10), the black body emissive power of certain wavelength range $\lambda_1-\lambda_2$ can be obtained, written as:

式中，$F_{b(0-\lambda_1)}$ 和 $F_{b(0-\lambda_2)}$ 称为黑体辐射函数（百分数）。如果已知波长和热力学温度，可以从黑体辐射函数表中查出黑体辐射函数的数值。将式（8.10）进行变换，可以求出某波段 λ_1 到 λ_2 内的黑体辐射力，记作：

$$E_{b(\lambda_1-\lambda_2)} = F_{b(\lambda_1-\lambda_2)} E_b = (F_{b(0-\lambda_1)} - F_{b(0-\lambda_2)}) E_b \tag{8.11}$$

Example 8.3 When the temperature is 1000K, 1400K, 3000K, and 6000K, respectively, try to calculate the share of visible light and infrared radiation in the total black body radiation.

例 8.3 试分别计算温度为 1000K、1400K、3000K、6000K 时，可见光和红外线辐射在黑体总辐射中所占的份额。

Solution: The wavelength λ of the spectral emissive power of any black body ranges from 0 to ∞. For visible light, its wavelength ranges from 0. 38μm to 0. 76μm. For infrared ray, its wavelength ranges from 0. 76μm to 1000μm. Multiply the given temperature in the project by 0. 38μm, 0. 76μm, 1000μm respectively and get each λT value.

解： 任何黑体发出光谱辐射力的波长 λ 范围都是 0~∞，对于可见光，其波长范围为 0. 38 ~ 0. 76μm。对于红外线，其波长范围为 0. 76 ~ 1000μm。将题中给定的温度分别乘以 0. 38μm、0. 76μm、1000μm，从而得到各 λT 值。

Then, according to these λT values, find the respective energy share value $F_b(0-\lambda)$ in the black body radiation function table. According to Eq. (8. 11), calculate the respective share of visible light and infrared radiation. The calculated results are listed in Table 8. 1.

然后，根据这些 λT 值，在黑体辐射函数表中查得各自的能量份额值 $F_b(0-\lambda)$。再根据式（8. 11），计算出可见光和红外线辐射各自占的份额，计算得到的结果列于表 8. 1 中。

Table 8. 1 Respective shares of visible light and infrared radiation

表 8. 1 可见光和红外线辐射各自占的份额

温度/K	$\lambda_1=0.38\mu m$		$\lambda_2=0.76\mu m$		$\lambda_3=1000\mu m$	
	$\lambda T/(\mu m\cdot K)$	$F_{b(0-\lambda_1)}(\%)$	$\lambda T/(\mu m\cdot K)$	$F_{b(0-\lambda_2)}(\%)$	$\lambda T/(\mu m\cdot K)$	$F_{b(0-\lambda_3)}(\%)$
1000	380	<0. 1	760	<0. 1	1×10^6	100
1400	532	<0. 1	1064	0. 12	1.4×10^6	100
3000	1140	0. 14	2280	11. 5	3×10^6	100
6000	2280	11. 5	4560	57. 0	6×10^6	100

温度/K	所占份额（%）	
	可见光 $F_{b(\lambda_2-\lambda_1)}=F_{b(0-\lambda_2)}-F_{b(0-\lambda_1)}$	红外线 $F_{b(\lambda_3-\lambda_2)}=F_{b(0-\lambda_3)}-F_{b(0-\lambda_2)}$
1000	<0. 1	>99. 9
1400	0. 12	99. 88
3000	11. 4	88. 5
6000	45. 5	43. 0

Discussions: When T is less than 1000K, the proportion of visible light in black body radiation is much less than 1/1000. As the temperature rises to about 3000K, the proportion of visible light can reach more than 10%. Radiation of most real objects is mainly infrared radiation. When an incandescent lamp gives off light, the temperature of tungsten wire is no more than 3000K. So, an incandescent lamp emits less than 10% of visible light, and more than 90% of infrared ray.

讨论： 在 $T<1000K$ 时，黑体辐射中可见光的比例远不到 1/1000。当温度上升到 3000K 左右时，可见光的比例才可达到 10%以上。大多数实际物体的辐射以红外辐射为主。白炽灯发光时钨丝的温度不超过 3000K。所以，白炽灯发出的可见光比例不到 10%，90%以上是红外线。

Summary This section illustrates Stefan-Boltzmann's law of black body radiation and its relationship with Planck's law, as well as the definition, expression and physical meaning of black body radiation function.

小结 本节阐述了黑体辐射的斯特藩-玻尔兹曼定律及其与普朗克定律的关系，以及黑体辐射函数的定义、表达式及物理意义。

Question How is the radiation energy of a black body distributed with wavelength and temperature?

思考题 黑体的辐射能随波长 λ 和温度 T 是怎样分布的？

Self-tests

自测题

1. How to calculate the emissive power of the black body at this temperature from the spectral emissive power?

1. 如何由光谱辐射力计算该温度下黑体的辐射力？

2. What is the black body radiation function? How to calculate the emissive power of the black body in a certain wavelength range from the black body radiation function?

2. 什么是黑体辐射函数？如何由黑体辐射函数计算某波段内的黑体辐射力？

授课视频

8.5 Lambert's Law

8.5 兰贝特定律

The emissive power indicates the ability of an object to emit radiation energy in general, but it does not explain the distribution regularity of radiation energy in different directions of space. Therefore, the concept of solid angle is introduced. Solid angle or micro-element solid angle indicates the size of the space in a certain direction.

辐射力从总体上说明了物体发射辐射能的能力，但未说明辐射能量在空间不同方向上的分布规律，为此引入立体角。立体角或微元立体角表示某一方向的空间所占的大小。

The solid angle is defined as the area of the sphere divided by the square of its radius. Its calculation formula is:

立体角的定义：球面面积除以球半径的平方。计算公式为：

$$\Omega=\frac{A_c}{r^2} \tag{8.12}$$

Its unit is sr (i. e. spatial degree or steradian). Then, the micro-element solid angle is:

单位：sr（空间度或球面度）。那么，微元立体角为：

$$d\Omega=\frac{dA_c}{r^2} \tag{8.13}$$

As shown in Fig. 8. 8, in the spherical coordinate system, φ is called the longitude angle, and θ is called the latitude angle. The direction of space can be expressed by the longitude angle and latitude angle of the direction.

如图 8.8 所示，在球坐标系中，φ 称为经度角，θ 称为纬度角，空间的方向可以用该方向的经度角和纬度角来表示。

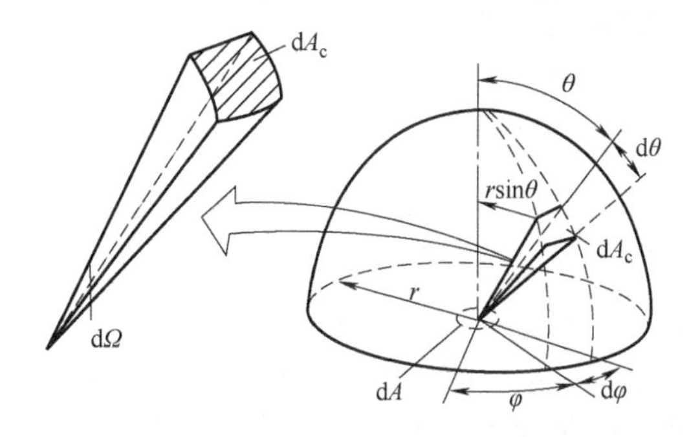

Fig. 8. 8 Schematic diagram of solid angle

图 8.8 立体角示意图

According to the definition of the micro-element solid angle, we can get:

根据微元立体角的定义，可得：

$$d\Omega=\frac{dA_c}{r^2}=\frac{rd\theta \cdot r\sin\theta d\varphi}{r^2}=\sin\theta d\theta d\varphi \tag{8.14}$$

Suppose that the energy emitted by micro-element area dA of a black body to the direction of spatial latitude angle θ in micro-element solid angle $d\Omega$ is $d\Phi(\theta)$, then the experimental measurements show that:

设面积为 dA 的黑体微元面积，向围绕空间纬度角 θ 方向的微元立体角 $d\Omega$ 内辐射出去的能量为 $d\Phi(\theta)$，则实验测定表明：

$$\frac{d\Phi(\theta)}{dAd\Omega}=I\cos\theta \tag{8.15}$$

Where I is a constant and independent of the θ direction. This equation can be expressed in another form:

式中，I 是常数，与 θ 方向无关。此式还可以表示为另一种形式：

$$\frac{d\Phi(\theta)}{dA\cos\theta d\Omega}=I \tag{8.16}$$

Where $dA\cos\theta$ can be regarded as the area seen from the θ direction, called visible area, as shown in Fig. 8. 9. The physical quantity at the left of Eq. (8. 16) is the energy emitted by unit visible area of a black body to unit solid angle in any direction of space, and is called directional radiation intensity.

式中，$dA\cos\theta$ 可以视为从 θ 方向看过去的面积，称为可见面积，如图 8.9 所示。式（8.16）左端的物理量，是从黑体单位可见面积发射出去的落到空间任意方向的单位立体角中的能量，称为定向辐射强度。

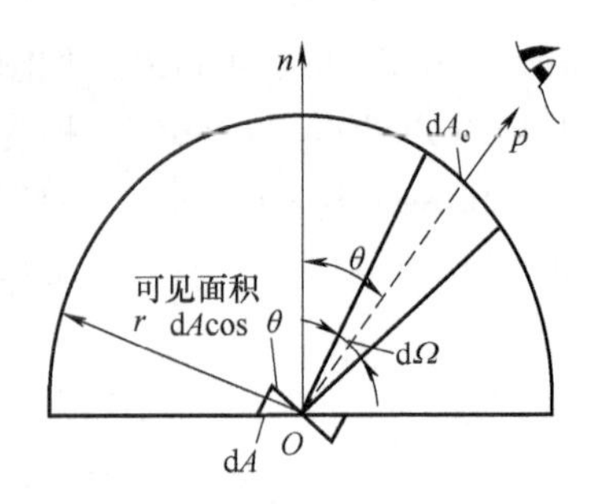

Fig. 8.9 Definition diagram of directional radiation intensity
图 8.9 定向辐射强度的定义图

The definition of **directional radiation intensity I** can also be expressed as the energy of all wavelengths emitted by an object in the unit solid angle in a unit time on the unit visible area in the vertical emission direction.

定向辐射强度 I 的定义也可表述为：单位时间内，物体在垂直发射方向的可见单位面积上，在单位立体角内发射的一切波长的能量。

The directional radiation intensity of a black body is a constant and independent of direction, which is Lambert's law. Lambert's law gives the distribution regularity of radiation energy of a black body in the spatial direction. Namely, the directional radiation intensities in all directions of hemispherical space are equal, also known as the third basic law of black body radiation.

黑体的定向辐射强度 I 是个常量，与方向无关，这就是**兰贝特（朗伯）定律**。兰贝特定律给出了黑体辐射能按空间方向的分布规律，即在半球空间的各个方向上的定向辐射强度相等，也称为黑体辐射的第三个基本定律。

Note that the directional radiation intensity is measured by the unit visible area, if the unit actual radiation area is used as the measurement basis, then it is the result showed in Eq. (8.15). This equation shows that the energy emitted by unit area of a black body is unevenly distributed in different directions of space, and changes according to the cosine law of the space latitude angle θ. Its value is the largest in the direction perpendicular to the surface, and is 0 in the direction parallel to the surface. This is another expression of Lambert's law, called cosine law.

注意：定向辐射强度是以单位可见面积作为度量依据的，如果以单位实际辐射面积为度量依据，则是式（8.15）所示的结果。该式表明，黑体单位面积辐射出去的能量在空间的不同方向分布是不均匀的，按空间纬度角 θ 的余弦规律变化，在垂直于该表面的方向最大，而与表面平行的方向为 0。这是兰贝特定律的另一种表达方式，称为**余弦定律**。

Multiply the two sides of Eq. (8.15) by $d\Omega$ separately, then integrate the whole hemisphere space to obtain the energy emitted by the unit area of a black body and falling into the whole hemisphere space, and that is the emissive power of the black body:

将式（8.15）两端各乘以 $d\Omega$，然后对整个半球空间做积分，就得到从单位黑体表面发射出去落到整个半球空间的能量，即黑体的辐射力：

$$E_b = \int_{\Omega=2\pi} I \cos \theta \mathrm{d}\Omega = I_b \pi \quad (8.17)$$

According to Eq. (8.17), for the radiation that obeys Lambert's law, its emissive power E is equal to π times the directional radiation intensity I.

从式（8.17）可以得到，遵守兰贝特定律的辐射，数值上其辐射力 E 等于定向辐射强度 I 的 π 倍。

The four basic laws of a black body radiation are summarized as follows:

黑体辐射的四个基本定律总结如下：

(1) The emissive power of black body E_b is determined by Stefan-Boltzmann's law, which is directly proportional to its thermodynamic temperature to the fourth power.

（1）黑体的辐射力 E_b 由斯特藩-玻尔兹曼定律确定，辐射力正比于其热力学温度的四次方。

(2) The distribution regularity of radiation energy of a black body by wavelength obeys Planck's law.

（2）黑体辐射能量按波长的分布规律服从普朗克定律。

(3) The distribution regularity of radiation energy of a black body by space direction obeys Lambert's law.

（3）黑体辐射能量按空间方向的分布规律服从兰贝特定律。

(4) The spectral emissive power $E_{b\lambda}$ of a black body has a peak, and the wavelength λ_m corresponding to this peak is determined by Wien's displacement law. With the increase of temperature, the λ_m moves to the direction of short wavelength.

（4）黑体的光谱辐射力 $E_{b\lambda}$ 有个峰值，与此峰值对应的波长 λ_m 由维恩位移定律确定。随着温度的升高，λ_m 向波长短的方向移动。

Summary This section illustrates the definitions, expressions and physical meanings of the solid angle, directional radiation intensity, and Lambert's law.

小结 本节阐述了立体角和定向辐射强度的定义、表达式及物理意义，以及兰贝特定律。

Question The directional radiation intensity is independent of space. Does it mean that the radiation energy of a black body is uniformly distributed in all directions of hemispheric space?

思考题 定向辐射强度与空间无关，是否意味着黑体的辐射能在半球空间各方向上是均匀分布的？

Self-tests

自测题

1. Briefly describe the contents of the four basic laws of black body radiation and their relationship

1. 简述黑体辐射四个基本定律的内容及其相互关系。

2. Briefly describe the physical significance of emissive power and directional radiation intensity. What's the relationship between them?

2. 简述辐射力和定向辐射强度的物理意义。它们之间有什么关系？

3. How is the radiation of a black body distributed in the spatial direction?

3. 黑体的辐射按空间方向是怎样分布的?

授课视频

8.6 The Thermal Radiation Characteristics of Actual Objects

8.6 实际物体的辐射特性

Fig. 8.10 is a schematic diagram of spectral emissive power of actual objects. It can be seen from the figure that: ① At the same temperature, a black body has the strongest ability to emit thermal radiation, including all directions and all wavelengths. ② The emission capacity of an actual object surface is lower than that of the black body at the same temperature. ③ The spectral emissive power of an actual object often varies irregularly with wavelength.

图8.10所示为实际物体的光谱辐射力示意图，从图中可以看出：① 同温度下，黑体发射热辐射的能力最强，包括所有方向和所有波长。② 实际物体表面的发射能力低于同温度下的黑体。③ 实际物体的光谱辐射力往往随波长做不规则变化。

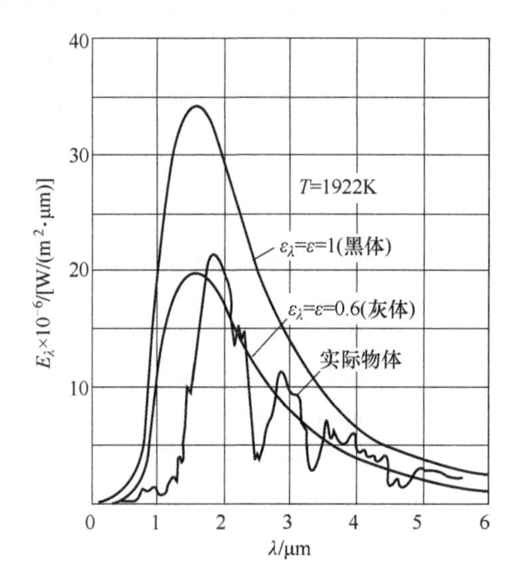

Fig. 8.10 Schematic diagram of spectral emissive power of actual objects

图8.10 实际物体的光谱辐射力示意图

Therefore, for the calculation of thermal radiation of actual objects, the concept of emissivity needs to be defined. Emissivity, also known as blackness, refers to the ratio of total hemispherical emissive power of an actual object to

因此，对于实际物体的热辐射计算，需定义发射率的概念。发射率，也称为黑度，是指相同温度下实际物体的半球总辐射力与黑体半

that of the black body at the same temperature, which is usually expressed by the letter ε.

球总辐射力之比，通常用字母 ε 表示。

$$\varepsilon=\frac{E}{E_b}=\frac{\int_0^{\infty}E_{\lambda}\mathrm{d}\lambda}{\sigma T^4}=\frac{\int_0^{\infty}\varepsilon(\lambda)E_{b\lambda}\mathrm{d}\lambda}{\sigma T^4} \tag{8.18}$$

When the emissivity ε is known, the emissive power E of an actual object can be determined by the fourth power law:

已知发射率 ε，实际物体的辐射力 E 可用四次方定律确定：

$$E=\varepsilon E_b=\varepsilon\sigma T^4=\varepsilon C_0\left(\frac{T}{100}\right)^4 \tag{8.19}$$

The spectral emissivity $\varepsilon(\lambda)$ is the ratio of the spectral emissive power of an actual object to that of the black body at the same temperature. The spectral emissivity of actual objects is also known as monochromatic blackness.

光谱发射率 $\varepsilon(\lambda)$：实际物体的光谱辐射力 E_{λ} 与同温度下黑体的光谱辐射力 $E_{b\lambda}$ 之比。实际物体的光谱发射率又称为单色黑度。

$$\varepsilon(\lambda)=E_{\lambda}/E_{b\lambda} \tag{8.20}$$

What needs to be explained is that Eq. (8.20) is only for the average situation of direction and spectrum. In fact, the emission capacity of actual surface varies with direction and spectrum. For a specified direction θ and wavelength λ, the directional emissivity $\varepsilon(\theta)$ needs to be defined, and the expression is as follows:

需要说明：式（8.20）只是针对方向和光谱的平均情况，实际上真实表面的发射能力随方向和光谱是变化的。对于某一指定的方向 θ 和波长 λ，需要定义定向发射率 $\varepsilon(\theta)$，其表达式如下：

$$\varepsilon(\theta)=\frac{I(\theta)}{I_b(\theta)} \tag{8.21}$$

That is the ratio of the directional radiation intensity of an actual object to that of a black body. A black body, a white body, a gray body and so on are ideal models, but the thermal radiation characteristics of actual objects are not exactly the same as that of these ideal models, which can be explained as follows:

即实际物体的定向辐射强度与黑体的定向辐射强度之比。黑体、白体、灰体等都是理想模型，而实际物体的辐射特性并不完全与这些理想物体相同，说明如下：

① The difference of the emissive powers between an actual object and a black body or a gray body can be clearly seen from Fig. 8.10; ② The emissive power of an actual object is not exactly proportional to its thermodynamic temperature to the fourth power; ③ The directional radiation intensity of an actual object also does not strictly comply with Lambert's law.

① 实际物体与黑体或灰体的辐射力的差别从图 8.10 中可以明显看出；② 实际物体的辐射力并不完全与热力学温度的四次方成正比；③ 实际物体的定向辐射强度也不严格遵守兰贝特定律。

Corresponding to the emissive power E_b, spectral emissive power $E_{b\lambda}$, and directional radiation intensity $I(\theta)$ of a black body, when calculating the radiation energy of an actual object, the above differences are attributed to three correction coefficients, namely emissivity ε, spectral emissivity $\varepsilon(\lambda)$ and directional emissivity $\varepsilon(\theta)$. The emissivity of common material surface is shown in Table 8.2.

对应于黑体的辐射力 E_b、光谱辐射力 $E_{b\lambda}$和定向辐射强度 $I(\theta)$，计算实际物体的辐射能时，上述差别都归于三个修正系数，即发射率 ε、光谱发射率 $\varepsilon(\lambda)$ 和定向发射率 $\varepsilon(\theta)$。常用材料表面的发射率见表 8.2。

Table 8.2 The emissivity of common material surface

表 8.2 常用材料表面的发射率

材料类别和表面状况	温度/℃	法向发射率 ε_n
磨光的铬	150	0.058
铬镍合金	52~1034	0.64~0.76
灰色、氧化的铅	38	0.28
镀锌的铁皮	38	0.23
具有光滑的氧化层表皮的钢板	20	0.82
氧化的钢	200~600	0.8
磨光的铁	400~1000	0.14~0.38
氧化的铁	125~525	0.78~0.82
磨光的铜	20	0.03
氧化的铜	50	0.6~0.7
磨光的黄铜	38	0.05
无光泽的黄铜	38	0.22
磨光的铝	50~500	0.04~0.06

The following factors affecting the emissivity of objects can be summarized from Table 8.2:

从表 8.2 中可以总结出影响物体发射率 ε 的因素：

(1) The emissivity of an object surface depends on the kind of substance, surface temperature and surface situation, which indicates that the emissivity is only related to the radiant object itself and does not involve external conditions.

(1) 物体表面的发射率取决于物质种类、表面温度和表面状况。说明发射率只与发射辐射的物体本身有关，而不涉及外界条件。

(2) For the same metal material, the emissivity of the highly polished surface is very small, while the emissivity of the rough surface and the oxidized surface is usually several times of that of the polished surface.

(2) 同一金属材料，高度磨光表面的发射率很小，而粗糙表面和受氧化表面的发射率常是磨光表面的数倍。

(3) The emissivity of most nonmetallic materials is very

(3) 大部分非金属材料的发射率值

high, generally between 0.85-0.95, and has little to do with the surface situation (including color). In the absence of data, 0.90 can be approximately taken.

都很高，一般在0.85~0.95范围内，且与表面状况（包括颜色在内）的关系不大，在缺少资料时，可近似取0.90。

The directional radiation intensity is independent of direction, in other words, the surface complying with Lambert's law, is called diffuse surface. The directional emissivity of an actual object does not exactly comply with Lambert's law, but we still approximately think that the most engineering materials comply with Lambert's law. That is, the actual surface is generally assumed to be diffuse surface in engineering.

定向辐射强度与方向无关，或者说，满足兰贝特定律的表面，称为漫反射表面。虽然实际物体的定向发射率并不完全符合兰贝特定律，但仍然近似地认为大多数工程材料服从兰贝特定律：即工程上一般都将真实表面假设为漫反射表面。

Summary　This section illustrates the thermal radiation characteristics of actual objects, and the definitions, the physical meanings and expressions of emissivity, spectral emissivity, directional emissivity.

小结　本节阐述了实际物体的辐射特性，以及发射率、光谱发射率、定向发射率的定义、物理意义及表达式。

Question　How to understand that the radiation energy from the black-body model (i.e. the small hole in the wall of the cavity with uniform temperature) has diffuse properties?

思考题　如何理解从黑体模型（即温度均匀的空腔器壁上的小孔）发出的辐射能具有漫反射特性？

Self-tests

自测题

1. What are the characteristics of spectral emissivity of actual objects compared with black and gray bodies?

1. 实际物体的光谱发射率与黑体、灰体相比有什么特点？

2. What are the main factors affecting the emissivity of the actual object surface?

2. 影响实际物体表面发射率的主要因素是什么？

授课视频

8.7 The Absorptivity of Actual Objects

8.7 实际物体的吸收比

When the external radiation energy is projected onto the surface of an object, how does the object absorb the irradiation?

当外界的辐射能投射到物体表面时，该物体对投入辐射的吸收情况如何呢？

Irradiation refers to the total radiation energy projected from the outside (which may be several objccts) onto the unit surface area of an object per unit time. The percentage of irradiation (including all wavelengths) absorbed by an object is called the absorptivity of the object, expressed as the letter α.

投入辐射是指单位时间内从外界（可能是几个物体）投射到物体单位表面面积上的总辐射能。物体对投入辐射（含各种波长）所吸收的百分数，叫作该物体的吸收比，用字母 α 表示。

$$\alpha=\frac{Q_{\alpha}}{Q} \tag{8.22}$$

The percentage of the radiation energy with specific wavelength absorbed by an object is called spectral absorptivity $\alpha(\lambda)$, also called monochromatic absorptivity.

把物体吸收某一特定波长辐射能的百分数，叫作光谱吸收比 $\alpha(\lambda)$，也叫作单色吸收比。

$$\alpha(\lambda,T_1)=\frac{\text{吸收的某一特定波长的能量}}{\text{投入的某一特定波长的能量}} \tag{8.23}$$

Generally speaking, the spectral absorptivity of the object is related to wavelength. The spectral absorptivity of the object varies with wavelength, which indicates the selective absorption characteristic of an actual object. Fig. 8.11 shows that the spectral absorptivity of the metal conductors varies with wavelength at room temperature. It can be seen from the figure that for some materials such as polished aluminum and copper, their spectral absorptivity varies slightly with wavelength.

一般地说，物体的光谱吸收比与波长有关。物体的光谱吸收比随波长变化，体现了实际物体具有选择性吸收的特性。图 8.11 所示为金属导电体在室温下光谱吸收比随波长的变化。从图中可以看出，有些材料比如磨光的铝和磨光的铜，光谱吸收比随波长的变化不大。

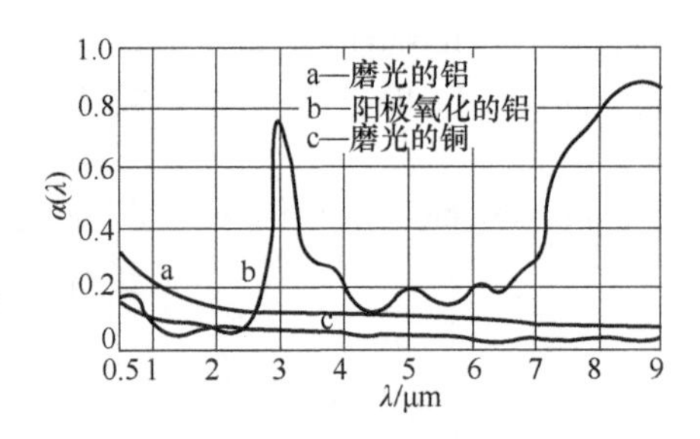

Fig. 8.11 Spectral absorptivity of the different metal materials

图 8.11 不同金属材料的光谱吸收比

Fig. 8.12 shows the relationship between the spectral absorptivity of non-conductive materials and wavelength. We can see from the figure that the spectral absorptivity $\alpha(\lambda)$ of material such as white ceramic tile is less than 0.2 in the range of wavelength less than 2μm. But in the range of

图 8.12 所示为非导电体材料的光谱吸收比同波长的关系。从图中可以看出，比如白瓷砖，在波长小于 2μm 的范围内光谱吸收比 $\alpha(\lambda)$ 小于 0.2；而在波长大于 5μm 的范

wavelength more than 5μm, the spectral absorptivity $\alpha(\lambda)$ is more than 0.9. $\alpha(\lambda)$ varies greatly with wavelength.

围内光谱吸收比 $\alpha(\lambda)$ 高于 0.9。$\alpha(\lambda)$ 随波长的变化很大。

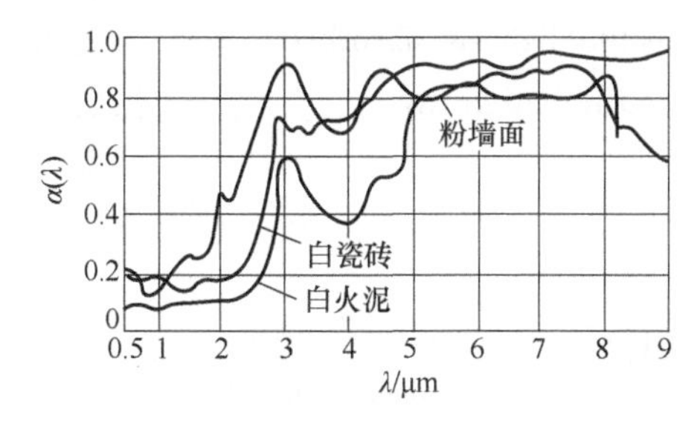

Fig. 8.12 Spectral absorptivity of non-conductive materials

图 8.12 非导电体材料的光谱吸收比

The property that the absorption capacity of the actual object to the irradiation varies with its wavelength is called the selective absorption of the object. In other words, the spectral absorptivity of the object varies with wavelength. There are many examples of the selective absorption properties of the objects (That is, the absorption of the irradiation at some wavelengths is more, while the absorption of the irradiation at other wavelengths is less) in practical production.

实际物体对投入辐射的吸收能力随其波长不同而变化的特性，称为物体的吸收具有选择性。或者说，物体的光谱吸收比随波长而异。物体的选择性吸收特性（即对有些波长的投入辐射吸收多，而对另一些波长的投入辐射吸收少），在实际生产中的例子很多。

For example, the use process of vegetable greenhouse makes use of the selectivity of glass to absorb radiation energy. When the sunlight strikes the glass, most of the solar energy can enter the greenhouse because the transmissivity of glass to radiation energy with wavelength less than 3μm is very large; However, because of the low temperature of objects in the greenhouse, most of the radiation energy is in the infrared range with a wavelength greater than 3μm. Since the transmissivity of glass to radiation energy with wavelength greater than 3μm is very small, the loss of radiation energy to the outside of the greenhouse is prevented. This is called the greenhouse effect.

如蔬菜温室的使用过程就是利用了玻璃对辐射能吸收的选择性。当太阳光照射到玻璃上时，由于玻璃对波长小于 3μm 的辐射能的穿透比很大，可使大部分太阳能进入温室；但温室中的物体由于温度较低，其绝大部分辐射能位于波长大于 3μm 的红外范围内，而玻璃对波长大于 3μm 的辐射能的穿透比很小，从而阻止了辐射能向温室外的散失，这就是所谓的“温室效应”。

Everything in the world presents different colors is also because of the selectivity of absorption and radiation. When the sunlight strikes the surface of an object, it will be black if the object absorbs almost all of visible light; It

世上万物呈现不同颜色的主要原因，也在于选择性的吸收和辐射。当阳光照射到一个物体表面上时，如果该物体几乎全部吸收可见光，

will be white if it reflects almost all of visible light; And it will be gray if the object absorbs and reflects all colors of visible light. If only one wavelength of visible light is reflected and almost all other visible light is absorbed, it presents the color of the reflected radiation ray.

它就呈黑色；如果几乎全部反射可见光，它就呈白色；如果几乎均匀地吸收各色可见光，并均匀地反射各色可见光，它就呈灰色。如果只反射了一种波长的可见光，而几乎全部吸收了其他可见光，则它就呈现被反射的这种辐射线的颜色。

But everything often has two aspects, people take advantage of selective absorption, and are also puzzled at the same time. Because the absorptivity of an object is related to the irradiation wavelength, which brings great trouble to the calculation of radiation heat transfer in engineering.

但所有事情往往都有双面性，人们在利用选择性吸收的同时，也为其伤透了脑筋。因为物体的吸收比与投入辐射波长有关的特性，给工程中辐射传热计算带来了巨大麻烦。

The absorptivity α of actual objects depends on two factors: the absorption object itself and the characteristics of the irradiation. Between them, the situation of the absorption object itself includes the type of material, surface temperature and surface condition.

实际物体的吸收比 α 取决于两方面因素：吸收物体本身情况和投入辐射的特性。其中，吸收物体本身的情况包括物质的种类、表面温度和表面状况。

The absorptivity α of actual objects is related to not only their surface properties and temperature (T_1), but also the energy distribution of irradiation by wavelength. The energy distribution of irradiation by wavelength also depends on the property and temperature (T_2) of the object emitting the irradiation. If subscripts 1 and 2 are used to represent the object under study and the object generating the irradiation respectively, then the absorptivity α_1 of object 1 can be expressed as follows:

实际物体的吸收比 α 不仅与自身表面性质和温度（T_1）有关，还与投入辐射按波长的能量分布有关。投入辐射按波长的能量分布又取决于发出投入辐射的物体的性质和温度（T_2）。如果用下标 1、2 分别代表所研究的物体和产生投入辐射的物体，则物体 1 的吸收比 α_1 可表示为：

$$\alpha_1=\frac{\text{吸收的总能量}}{\text{投入的总能量}}=\frac{\int_0^{\infty}\alpha(\lambda,T_1)\varepsilon(\lambda,T_2)E_{b\lambda}(T_2)\mathrm{d}\lambda}{\int_0^{\infty}\varepsilon(\lambda,T_2)E_{b\lambda}(T_2)\mathrm{d}\lambda}$$
$$=f(T_1,T_2,\text{表面 1 的性质},\text{表面 2 的性质}) \tag{8.24}$$

Eq. (8.24) shows that the absorptivity α_1 of object 1 is related to the surface property and temperature (T_1) of object 1, and the property and temperature (T_2) of object 2 emitting irradiation. If the irradiation comes from

式（8.24）表明，物体 1 的吸收比 α_1 与物体 1 自身表面性质和温度（T_1）以及发出投入辐射的物体 2 的性质和温度（T_2）有关。如果投入辐射来自

a black body, since its emissivity is equal to 1, then the absorptivity α_1 of object 1 can be expressed as:

黑体，由于其发射率为 1，则物体 1 的吸收比 α_1 可表示为：

$$\alpha_1 = \frac{\int_0^\infty \alpha(\lambda, T_1) E_{b\lambda}(T_2) \mathrm{d}\lambda}{\int_0^\infty E_{b\lambda}(T_2) \mathrm{d}\lambda} = \frac{\int_0^\infty \alpha(\lambda, T_1) E_{b\lambda}(T_2) \mathrm{d}\lambda}{\sigma T_2^{\ 4}} \tag{8.25}$$

$$= f(T_1, T_2, \text{表面 1 的性质})$$

Eq. (8.25) shows that the absorptivity α_1 of object 1 is only related to its surface property and temperature (T_1) and the temperature (T_2) of the black body emitting irradiation in this case. Fig. 8.13 shows the relationship between the absorptivity of some object surfaces to black body radiation and temperature. And the temperature of material T_1 is 294K in the figure.

式（8.25）表明，在这种情况下物体 1 的吸收比 α_1 只与其自身表面性质和温度（T_1）以及发出投入辐射的黑体的温度（T_2）有关。图 8.13 所示为一些物体表面对黑体辐射的吸收比与温度的关系。图中材料的自身温度 T_1 为 294K。

As can be seen from Fig. 8.13, for the irradiation emitted by the black body, the absorptivity of the listed objects is strongly related to the temperature of the irradiation. For the irradiation emitted by the actual object, the range of the absorptivity of an object will be larger, so it is very difficult to take such a complex situation into account in the practical project calculation.

由图 8.13 可以看出，对于黑体发出的投入辐射，所列物体的吸收比与投入辐射的温度有很大关系。对于实际物体发出的投入辐射，物体的吸收比变化范围会更大，在实际工程计算中要顾及如此复杂的情况很困难。

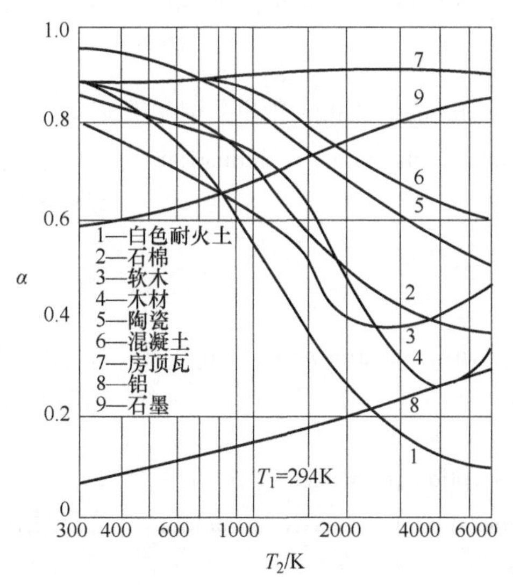

Fig. 8.13 Relationship between the absorptivity of object surfaces to black body radiation and temperature

图 8.13 物体表面对黑体辐射的吸收比与温度的关系

Therefore, the concept of a gray body is introduced, and the object whose spectral absorptivity is independent of wavelength is defined as a gray body. For the gray body, at a certain temperature:

因此，引入灰体的概念，将光谱吸收比与波长无关的物体定义为灰体。对于灰体，在自身一定温度下有：

$$\alpha=\alpha(\lambda)=\text{const} \tag{8.26}$$

At this time, the absorptivity α is a constant regardless of the irradiation distribution. A gray body is also an ideal model. For the calculation of radiation heat transfer, the actual body is generally treated as gray body in industry. Since actual objects are more or less selective in absorbing radiation energy, then why can they be assumed to be gray bodies in engineering calculations?

此时，不管投入辐射分布如何，吸收比 α 皆为常数。灰体也是一种理想模型。工业上的辐射传热计算，一般都按灰体来处理。既然实际物体或多或少都对辐射能的吸收具有选择性，为什么工程计算又可假定灰体呢?

For engineering calculations, as long as the wavelength is in the studied range, the spectral absorptivity is basically independent of wavelength, then the assumption of a gray body can be established and without the requirement that the spectral absorptivity is constant in the whole wavelength range. In the common temperature range in engineering, many engineering materials have this characteristic.

对于工程计算而言，只要在所研究的波长范围内，光谱吸收比基本与波长无关，则灰体的假定即可成立，而不必要求光谱吸收比在全波段范围内为常数。在工程常见的温度范围内，许多工程材料都具有这一特点。

Summary This section illustrates the absorptivity of actual objects, the concept of a gray body and its engineering applications.

小结 本节阐述了实际物体的吸收比、灰体的概念及其工程应用。

Question As glass can pass through visible light, why can it be treated as a gray body in the range of industrial thermal radiation?

思考题 既然玻璃可以透过可见光，为什么在工业热辐射范围内可以作为灰体处理?

Self-tests

自测题

1. What is irradiation, absorptivity, spectral absorptivity?

1. 什么是投入辐射、吸收比、光谱吸收比?

2. What are the selective absorption properties of the objects? What factors determine the absorptivity of actual objects?

2. 什么是物体的选择性吸收特性?实际物体的吸收比取决于哪些因素?

3. What is a gray body? Under what conditions can an actual object be regarded as a gray body?

3. 什么是灰体?实际物体在什么样的条件下可以看成是灰体?

8.8 Kirchhoff's Law

8.8 基尔霍夫定律

What is the intrinsic connection between radiation and absorption of actual objects? In 1859, Kirchhoff answered the question with a thermodynamic method, namely Kirchhoff's law. This law can be derived by studying the radiation heat transfer between two surfaces. As shown in Fig. 8.14, two parallel plane plates are very close to each other, and all the radiation energy from one plate falls on the other.

实际物体的辐射与吸收之间有什么内在联系呢？1859 年，基尔霍夫用热力学方法回答了该问题，即基尔霍夫定律。这个定律可以通过研究两个表面之间的辐射传热导出。如图 8.14 所示，两块平行平板，相距很近，从一块板发出的辐射能全部落到另一块板上。

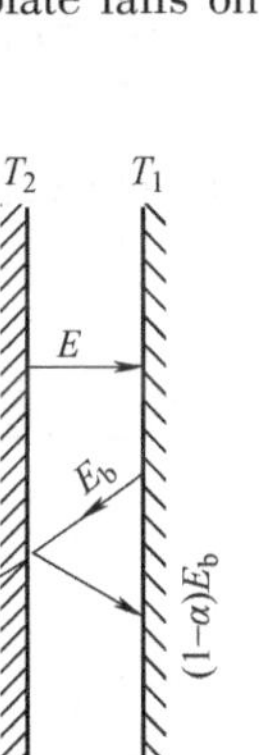

Fig. 8.14 Radiation heat transfer between two parallel plane plates

图 8.14 平行平板间的辐射传热

Suppose that plate 1 is a black body, its surface temperature is T_1, the emissive power is E_b, and the absorptivity $\alpha_b = 1$. Plate 2 is an arbitrary object, its surface temperature is T_2, the emissive power is E, and the absorptivity is α. Now let's investigate the difference between the energy emission and absorption of plate 2.

假设板 1 是黑体，其表面温度为 T_1，辐射力为 E_b，吸收比 $\alpha_b = 1$；板 2 是任意物体，其表面温度为 T_2，辐射力为 E，吸收比为 α。现在来考察板 2 的能量收支差额。

For plate 2 itself, the energy emitted by plate 2 per unit area in unit time is E, which can be fully absorbed when it is projected on the black body surface 1. At the same time, the energy emitted by the black body surface 1 is E_b, which can only be partly absorbed when it falls on the

对于板 2 自身，其单位面积在单位时间内发射出的能量为 E，这份能量投射在黑体表面 1 上时被全部吸收。同时，黑体表面 1 发出的能量为 E_b，这份能量落到板 2 上时只能

plate 2. The part of energy is αE_b, and the rest of energy $(1-\alpha)E_b$ is reflected to plate 1, and is fully absorbed by the black body surface 1.

被吸收一部分，这部分能量为 αE_b，其余部分能量 $(1-\alpha)E_b$ 被反射回板 1，并被黑体表面 1 全部吸收。

The difference between the energy emission and absorption of plate 2 is the heat flux q of radiation heat transfer between the two plates, which can be expressed as:

板 2 的能量支出与收入的差额，即两板间辐射传热的热流密度 q，可表示为：

$$q = E - \alpha E_b$$

When the system is at the same temperature, that is, under thermal equilibrium condition, then $q=0$. Thus, the above equation becomes:

当体系处于温度相等的状态，即处于热平衡条件下时，有 $q=0$，于是上式变为：

$$\frac{E}{\alpha} = E_b$$

When this relationship is extended to any object, the above equation can be written as:

把这种关系推广到任意物体时，上式可写为：

$$\frac{E_1}{\alpha_1} = \frac{E_2}{\alpha_2} = \cdots = \frac{E}{\alpha} = E_b \tag{8.27}$$

The equation can also be rewritten as:

该式也可以改写为：

$$\alpha = \frac{E}{E_b} = \varepsilon \tag{8.28}$$

The above two equations are two mathematical expressions of Kirchhoff's law. Eq. (8.27) can be described as that under thermal equilibrium condition, the ratio of the self emissive power of any object to its absorptivity from black body radiation identically equals to the emissive power of the black body at the same temperature. Eq. (8.28) can be briefly described as that under thermal equilibrium, the absorptivity of any object to the black body irradiation is equal to the emissivity of the object at the same temperature.

上述两式即基尔霍夫定律的两种数学表达式。式（8.27）可以表述为，在热平衡条件下，任何物体的自身辐射和它对来自黑体辐射的吸收比的比值，恒等于同温度下黑体的辐射力。式（8.28）可简述为：在热平衡时，任意物体对黑体投入辐射的吸收比，等于同温度下该物体的发射率。

Kirchhoff's law tells us that the absorptivity of an object is equal to the emissivity. However, the equation has strict restrictions: ① The whole system is in thermal equilibrium; ② The projected radiation source must be the black body at the same temperature.

基尔霍夫定律告诉人们，物体的吸收比等于发射率。但该式成立有严格的限制条件：① 整个系统处于热平衡状态；② 投射辐射源必须是同温度下的黑体。

Actually, in the engineering calculation of radiation heat transfer, the irradiation is neither black body radiation nor in thermal equilibrium. Then, under what premise can these two conditions be removed? Let's study the case of a diffuse gray body.

实际上，在进行工程辐射传热计算时，投入辐射既非黑体辐射，更不会处于热平衡。那么在什么前提下，这两个条件可以去掉呢？下面研究漫反射灰体的情形。

Firstly, according to the definition of a gray body, its absorptivity is independent of wavelength and is a constant at a certain temperature. Secondly, the emissivity of an object is a physical parameter and independent of environmental conditions. Suppose that a gray body is in thermal equilibrium with a black body at a certain temperature T. According to Kirchhoff's law, the absorptivity of the gray body is equal to its emissivity. We can think that $\alpha(T)=\varepsilon(T)$.

首先，按灰体的定义，其吸收比与波长无关，在一定温度下是一个常数。其次，物体的发射率是物性参数，与环境条件无关。假设在某一温度 T 下一个灰体与黑体处于热平衡。按基尔霍夫定律可以得到，灰体的吸收比等于其发射率，可以认为：$\alpha(T)=\varepsilon(T)$。

Then, considering the change of the environment, the received radiation of the gray body does not come from the black body at the same temperature, but keeps itself temperature unchanged. At this time, in consideration of the above characteristics of the emissivity and the absorptivity of a gray body, obviously there should still be $\alpha(T)=\varepsilon(T)$. As a result, its absorptivity must be equal to its emissivity for the diffuse gray surface, namely $\alpha=\varepsilon$.

然后，考虑改变该灰体的环境，使其所受到的辐射不是来自同温度下的黑体，但保持其自身温度不变。此时考虑到发射率及灰体吸收比的上述性质，显然仍应有 $\alpha(T)=\varepsilon(T)$。所以，对于漫（反射）灰表面，一定有吸收比等于发射率，即 $\alpha=\varepsilon$。

For the diffuse gray body, regardless of whether the irradiation comes from a black body or is in thermal equilibrium conditions, the absorptivity identically equals to the emissivity at the same temperature. This conclusion brings substantial simplification to radiation heat transfer calculations and is widely used in engineering calculations.

对于漫（反射）灰体，不论投入辐射是否来自黑体，也不论是否处于热平衡条件，其吸收比恒等于同温度下的发射率。这个结论对辐射传热计算带来实质性的简化，广泛应用于工程计算。

In most cases, an object can be treated as a gray body. According to Kirchhoff's law, the greater the emissive power of an object, the greater its absorption capacity. In other words, the objects that are good at radiation must be good at absorption, and vice versa. Therefore, the emissive power of the black body is the strongest at the same temperature.

在大多数情况下，物体可作为灰体，由基尔霍夫定律可知，物体的辐射力越大，其吸收能力也越大。换句话说，善于辐射的物体必善于吸收，反之亦然。所以，同温度下黑体的辐射力最大。

Summary This section illustrates the relationship between the absorptivity and emissivity of actual objects, and focu-

小结 本节阐述了实际物体的吸收比与发射率的关系，重点学习了基

ses on Kirchhoff's law and the concept of a diffuse gray body.

尔霍夫定律和漫反射灰体的概念。

Question The interior of a building is painted with white ash, even in a sunny day, why is it always dark inside when looking at the window of the building from a distance?

思考题 某楼房室内用白灰粉刷，即使在晴朗的白天，远眺该楼房的窗口时，为什么总觉得里面黑洞洞的？

Self-tests

自测题

1. What is Kirchhoff's law and what are the restrictions on its establishment?

1. 什么是基尔霍夫定律？其成立的限制条件是什么？

2. What is a diffuse gray body? What is the significance of its characteristics to the engineering calculation of radiation heat transfer?

2. 什么是漫灰体？其特性对辐射传热的工程计算有何意义？

Exercises

习题

8.1 Approximating the surface of the sun as a black body at $T = 5800\text{K}$, try to determine the percentage of visible light in the radiation energy emitted by the sun.

8.1 把太阳表面近似地看成是 $T=5800\text{K}$ 的黑体，试确定太阳发出的辐射能中可见光所占的百分数。

8.2 The average temperature of the flame in a furnace is 1500K, and there is a fire hole in the furnace wall. Try to calculate the power radiated from the fire hole when it is open. What is the spectral emissive power of wavelength 2μm in this radiant energy? At which wavelength is the spectral emissive power the greatest?

8.2 一炉膛内火焰的平均温度为1500K，炉墙上有一着火孔，试计算当着火孔打开时从孔向外辐射的功率。该辐射能中波长为2μm的光谱辐射力是多少？哪种波长下的光谱辐射力最大？

8.3 A space craft has a sunny diffuse panel on its outer shell. The back of the plate can be considered as adiabatic, and the sunny side obtains the irradiation $G = 1300\text{W/m}^2$ from the sun. The spectral emissivity of the surface: $\varepsilon(\lambda) = 0.5$ when $0 \leqslant \lambda \leqslant 2\mu\text{m}$; $\varepsilon(\lambda) = 0.2$, when $\lambda \geqslant 2\mu\text{m}$. Try to determine the temperature value when the surface temperature of the plate is in steady-state. To simplify the calculation, assume that the solar radiation energy is concentrated within 0-2μm.

8.3 在一空间飞行物的外壳上有一块向阳的漫反射面板。板背面可以认为是绝热的，向阳面得到的太阳投入辐射 $G=1300\text{W/m}^2$。该表面的光谱发射率：$0 \leqslant \lambda \leqslant 2\mu\text{m}$ 时，$\varepsilon(\lambda) = 0.5$；$\lambda \geqslant 2\mu\text{m}$ 时，$\varepsilon(\lambda) = 0.2$。试确定当该板表面温度处于稳态时的温度值。为简化计算，设太阳的辐射能均集中在0~2μm内。

8.4 Measured by a specific instrument, a black body furnace emits radiation (within a hemisphere) at a wavelength of 0.7μm with an energy of 10^8W/m^3. What temperature does the black body furnace work at? What is the heating power of radiant black body furnace under this working condition? The area of the radiation hole is 4×10^{-4}m^2.

8.4 用特定的仪器测得，一黑体炉发出的波长为 0.7μm 的辐射能（在半球范围内）为 10^8W/m^3，试问该黑体炉工作在多高的温度下？该工况下辐射黑体炉的加热功率为多大？辐射小孔的面积为 4×10^{-4}m^2。

8.5 Try to determine the luminous efficiency of an electric bulb with an electrical power of 100W. Assume that the tungsten filament of the bulb can be viewed as a 2900K black body with a 2mm×5mm rectangular sheet geometry.

8.5 试确定一个电功率为 100W 的电灯泡发光效率。假设该灯泡的钨丝可看成是 2900K 的黑体，其几何形状为 2mm×5mm 的矩形薄片。

8.6 The spectral absorption ratio of a selective absorbing surface varies with λ is as shown in Fig. 8.15. Try to calculate the solar energy absorbed per unit area of the surface and the total absorptivity of solar radiation when the solar irradiation is G=800W/m^2.

8.6 选择性吸收表面的光谱吸收比随 λ 变化的特性如图 8.15 所示，试计算当太阳投入辐射 G=800W/m^2 时，该表面单位面积上所吸收的太阳能量及对太阳辐射的总吸收比。

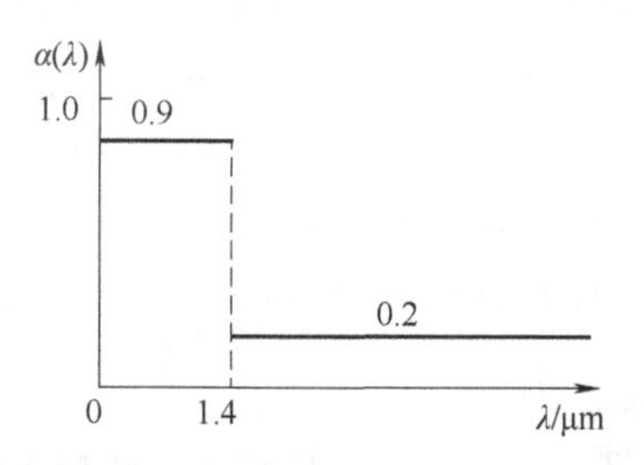

Fig. 8.15 Illustration of exercise 8.6

图 8.15 习题 8.6 图

Chapter 9 Calculation of Radiation Heat Transfer

第 9 章 辐射传热计算

授课视频

9.1 Angle Factor of Radiation Heat Transfer

9.1 辐射传热的角系数

Both the emission and absorption of thermal radiation have spatial directivity. For two surfaces in the space, the radiation heat transfer quantity between them has a great relationship with their relative positions. Fig. 9.1 shows two isothermal surfaces with two extreme arrangements: In Fig. 9.1a, the two surfaces are infinitely close to each other, in which case the mutual radiation heat transfer quantity between them is the largest; In Fig. 9.1b, the two surfaces are on the same plane, in which case the mutual radiation heat transfer quantity between them is zero.

热辐射的发射和吸收均具有空间方向性。对于空间中的两个表面，它们之间的辐射传热量与它们的相对位置有很大关系。图 9.1 所示两个等温表面，有两种极端布置情况：图 9.1a 中两表面无限接近，此时它们之间相互的辐射传热量最大；图 9.1b 中两表面位于同一平面上，此时它们相互间的辐射传热量为零。

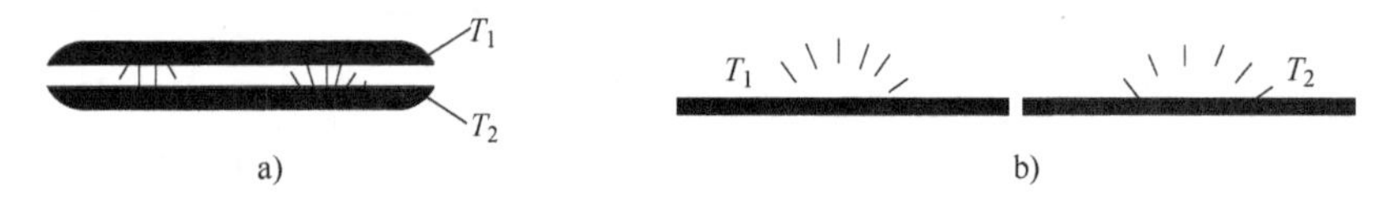

Fig. 9.1 The relative position of two surfaces in the space

图 9.1 两平面在空间中的相对位置

From Fig. 9.1, we can see that when the relative positions of the two surfaces are different, the percentage of radiation energy emitted from one surface and falling on the other surface will vary accordingly, thus affecting the heat transfer quantity.

由图 9.1 可以看出，两个表面间的相对位置不同时，一个表面发出而落到另一个表面上的辐射能的百分数随之不同，从而影响传热量。

When calculating the radiation heat transfer, we need to consider the surface geometric factors that affect radiation heat transfer, including the surface geometry, the area size and the relative position of each surface. These factors are often dealt with by angle factor. The "angle factor"

在计算辐射传热时，需考虑影响辐射传热的表面几何因素，包括表面几何形状、面积大小和各表面的相对位置。这些因素常用"角系数"来处理。"角系数"是 20 世纪 20

was put forward with the development of radiation heat transfer calculation on solid surfaces in the 1920s, also known as shape factor or visual factor, etc.

年代随着固体表面辐射传热计算的发展而提出的，又称为形状因子、可视因子等。

Suppose there are two arbitrarily placed surfaces 1 and 2 filled with transparent media. The percentage of the radiation energy emitted by surface 1 and falling on surface 2, is called the angle factor of surface 1 to surface 2, denoted as $X_{1,2}$. $X_{1,2}$ can be expressed as:

假设有两个任意放置的表面 1 和表面 2，其间充满透明介质。表面 1 发出的辐射能落到表面 2 上的百分数，称为表面 1 对表面 2 的角系数，记作 $X_{1,2}$。$X_{1,2}$可表示为：

$$X_{1,2}=\frac{\text{The thermal radiation received by surface 2 from surface 1}}{\text{Effective radiation from surface 1(total radiant energy)}} \tag{9.1}$$

Similarly, the angle factor of surface 2 to surface 1 is defined, denoted as $X_{2,1}$. Please note that $X_{1,2}$ and $X_{2,1}$ have different meanings and different values.

同理，可定义表面 2 对表面 1 的角系数，记作 $X_{2,1}$。请注意，$X_{1,2}$和$X_{2,1}$表示的意义不同，其值也不相等。

For the two extreme arrangements between two isothermal surfaces in Fig. 9.1: When the two surfaces are infinitely close in Fig. 9.1a, it is not difficult to get $X_{1,2}=X_{2,1}=1$ according to the definition of the angle factor; For the two surfaces on the same plane in Fig. 9.1b, $X_{1,2}=X_{2,1}=0$ according to the definition of the angle factor.

对于图 9.1 中两个等温表面间的两种极端布置情况：对于图 9.1a 中当两表面无限接近时，根据角系数的定义，不难得出 $X_{1,2}=X_{2,1}=1$；对于图 9.1b 中两表面位于同一平面上，根据角系数的定义 $X_{1,2}=X_{2,1}=0$。

When discussing the angle factor, it is assumed that: ① The studied surface is diffuse; ② The radiant heat flux emitted from different positions on the studied surface is uniform. Under these two assumptions, the changes in the surface temperature and emissivity of an object only affect the amount of radiant energy emitted by the object, without affecting the relative distribution in space. Therefore, the percentage of radiation energy falling on other surfaces is not affected.

在讨论角系数时，假定：① 所研究的表面是漫射的；② 在所研究表面的不同位置上向外发射的辐射热流密度是均匀的。在这两个假定下，物体的表面温度及发射率的改变，只影响该物体向外发射的辐射能多少，而不影响在空间的相对分布，因而不影响辐射能落到其他表面上的百分数。

It can be seen that the angle factor is a pure geometric factor and has nothing to do with the temperature and emissivity of the two surfaces, which brings great convenience to its calculation. For practical engineering problems, although

由此可以得知，角系数是一个纯几何因子，与两个表面的温度及发射率没有关系。这给角系数的计算带来很大方便。对于实际工程问题，

these assumptions may not be satisfied, the resulting deviations are generally within the allowable range of engineering calculations. So, this treatment method is widely used in engineering.

虽然不一定满足这些假定，但由此造成的偏差一般均在工程计算允许的范围内。因此，这种处理方法在工程中广为采用。

For the convenience of discussion, an object is treated as a black body when studying the angle factor. All the conclusions are suitable for the diffuse gray surfaces. Next, according to the definition of the angle factor, calculate the angle factor of micro-element surface to micro-element surface, the angle factor of micro-element surface to surface, and the angle factor of surface to surface, respectively.

为了方便讨论，在研究角系数时把物体作为黑体来处理，所得到的结论对于漫灰表面均适合。接下来，根据角系数的定义，分别计算微元面对微元面的角系数、微元面对面的角系数及面对面的角系数。

(1) The angle factor of micro-element surface to micro-element surface

(1) 微元面对微元面的角系数

Fig. 9. 2 shows the radiation heat transfer between two black body micro-element surfaces. The angle factor of the micro-element surface dA_1 to the micro-element surface dA_2 is denoted as $X_{d1,d2}$. According to the definition equation of the angle factor, we can get:

图 9. 2 所示为两个黑体微元面之间的辐射传热。把微元面 dA_1 对微元面 dA_2 的角系数记为 $X_{d1,d2}$。根据角系数的定义式，可以得到：

$$X_{d1,d2}=\frac{\text{落到 } dA_2 \text{ 上由 } dA_1 \text{ 发出的辐射能}}{dA_1 \text{ 向外发出的总辐射能}}=\frac{I_{b1}\cos\theta_1 dA_1 d\Omega_1}{E_{b1}dA_1} \tag{9.2}$$

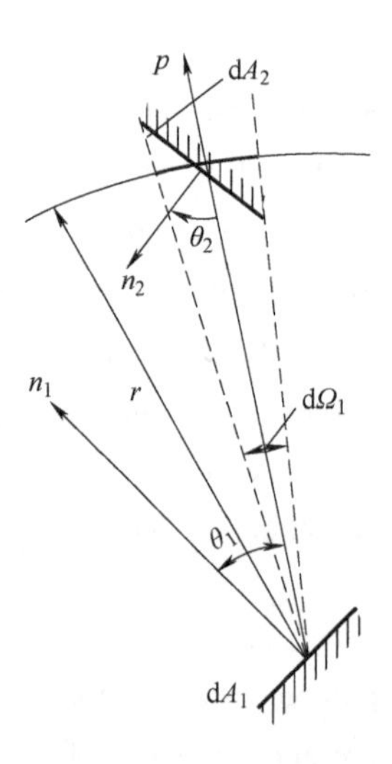

Fig. 9. 2 The radiation heat transfer between two black body micro-element surfaces

图 9. 2 两个黑体微元面之间的辐射传热

Where I_{b1} is the directional radiation intensity of the micro-element surface dA_1, then the emissive power

其中，I_{b1}为微元面 dA_1 的定向辐射强度，那么辐射力

$$E_{b1} = \pi I_{b1}$$

$d\Omega_1$ is the micro-element solid angle

$d\Omega_1$ 为微元立体角

$$d\Omega_1 = \frac{dA_2 \cos\theta_2}{r^2}$$

Substitute E_{b1} and $d\Omega_1$ into Eq. (9.2) to get:

将E_{b1}和$d\Omega_1$ 代入式 (9.2)，可得到：

$$X_{d1,d2} = \frac{dA_2 \cos\theta_1 \cos\theta_2}{\pi r^2} \tag{9.3}$$

Similarly, we can also get:

类似地，也可以得到：

$$X_{d2,d1} = \frac{dA_1 \cos\theta_1 \cos\theta_2}{\pi r^2} \tag{9.4}$$

From Eq. (9.3) and Eq. (9.4), it can be seen that the angle factor is only related to the area of the receiving surface and not to the area of the emitting surface.

从式 (9.3) 和式 (9.4) 可以看出，角系数仅与接收面的面积有关，与发射面的面积无关。

(2) The angle factor of micro-element surface to surface.

(2) 微元面对面的角系数

From the definition of the angle factor, the angle factor of the micro-element surface dA_1 to the receiving surface A_2, is equal to the integral of $X_{d1,d2}$ over the surface A_2. Similarly, the angle factor $X_{d2,1}$ of the micro-element surface dA_2 to the receiving surface A_1 equals to $X_{d2,d1}$ integrated on the A_1 surface.

由角系数的定义可知，微元面 dA_1 对面 A_2 的角系数，等于角系数 $X_{d1,d2}$在接收面 A_2 上的积分。同理，微元面 dA_2 对面 A_1 的角系数 $X_{d2,1}$，等于 $X_{d2,d1}$在接收面 A_1 上的积分。

$$X_{d1,2} = \int_{A_2} X_{d1,d2} \tag{9.5}$$

$$X_{d2,1} = \int_{A_1} X_{d2,d1} \tag{9.6}$$

(3) The angle factor of surface to surface.

(3) 面对面的角系数

According to Eq. (9.3) the angle factor of the micro-element surface dA_1 to the micro-element surface dA_2, and Eq. (9.5) the angle factor of micro-element surface to surface, we can get the angle factor $X_{1,2}$ of the surface A_1 to the

由微元面 dA_1 对微元面 dA_2 的角系数的计算式 (9.3) 和微元面对面的角系数的计算式 (9.5)，可以得到面 A_1 对面 A_2 的角系数 $X_{1,2}$， surface A_2, and $X_{2,1}$ of the surface A_2 to the surface A_1:

以及面 A_2 对面 A_1 的角系数 $X_{2,1}$：

$$X_{1,2}=\frac{1}{A_1}\int_{A_1}\int_{A_2}\left(\frac{\cos\theta_1\cos\theta_2\mathrm{d}A_2}{\pi r^2}\right)\mathrm{d}A_1=\frac{1}{A_1}\int_{A_1}\int_{A_2}X_{\mathrm{d}1,\mathrm{d}2}\mathrm{d}A_1 \tag{9.7}$$

$$X_{2,1}=\frac{1}{A_2}\int_{A_1}\int_{A_2}\left(\frac{\cos\theta_1\cos\theta_2\mathrm{d}A_1}{\pi r^2}\right)\mathrm{d}A_2=\frac{1}{A_2}\int_{A_1}\int_{A_2}X_{\mathrm{d}2,\mathrm{d}1}\mathrm{d}A_2 \tag{9.8}$$

Summary This section illustrates the definition of the angle factor, the angle factor of micro-element surface to micro-element surface, the angle factor of micro-element surface to surface, and the angle factor of surface to surface.

小结 本节阐述了角系数的定义、微元面对微元面的角系数、微元面对面的角系数和面对面的角系数。

Question Why is the angle factor introduced and what is its physical meaning?

思考题 为什么要引入角系数？其物理意义是什么？

Self-tests

自测题

1. When calculating the radiation heat transfer, what surface geometric factors should be considered that affect the radiation heat transfer?

1. 在计算辐射传热时，需考虑影响辐射传热的哪些表面几何因素？

2. What is the angle factor? Under what premises can it be concluded that the angle factor is a pure geometric factor?

2. 什么是角系数？角系数是一个纯几何因子的结论在什么前提下得出的？

授课视频

9.2 The Properties of the Angle Factor

9.2 角系数的性质

According to the definition and analytical expressions of angle factor learned earlier, the algebraic properties of the angle factor can be deduced.

根据前面学习的角系数的定义和各解析式，可推导出角系数的代数性质。

(1) The first property of the angle factor: relativity.

(1) 角系数第一个性质：相对性。

From the calculation equations (9.3) and (9.4) of the angle factors $X_{\mathrm{d}1,\mathrm{d}2}$ and $X_{\mathrm{d}2,\mathrm{d}1}$ of micro-element surface to micro-element surface, we can get:

由微元面对微元面的角系数 $X_{\mathrm{d}1,\mathrm{d}2}$ 和 $X_{\mathrm{d}2,\mathrm{d}1}$ 的计算式（9.3）和式（9.4），可以得到：

$$dA_1X_{d1,d2}=dA_2X_{d2,d1} \tag{9.9}$$

From the calculation equations (9.7) and (9.8) of the angle factor $X_{1,2}$ and $X_{2,1}$ of surface to surface, we can also get:

由面对面的角系数 $X_{1,2}$ 和 $X_{2,1}$ 的计算式（9.7）和式（9.8），也可以得出：

$$A_1X_{1,2}=A_2X_{2,1} \tag{9.10}$$

This is the expression of the relativity of the angle factor between two surfaces of finite size. For the relativity of the angle factor between two finite-size surfaces A_1 and A_2, it can be obtained by analyzing the radiation heat transfer quantity between two black body surfaces shown in Fig. 9. 3. We record the radiation heat transfer quantity between the two surfaces as $\Phi_{1,2}$, then $\Phi_{1,2}=A_1E_{b1}X_{1,2}-A_2E_{b2}X_{2,1}$. When the temperatures T_1 and T_2 of the two surfaces are equal, the net radiation heat transfer quantity is zero.

这是两个有限大小表面之间角系数的相对性表达式。对于两个有限大小表面 A_1 和 A_2 之间角系数的相对性，可以通过分析图 9. 3 所示两个黑体表面间的辐射传热量而获得。将两个表面间的辐射传热量记为 $\Phi_{1,2}$，则有 $\Phi_{1,2}=A_1E_{b1}X_{1,2}-A_2E_{b2}X_{2,1}$。当两个表面的温度 $T_1=T_2$ 时，净辐射传热量为零。

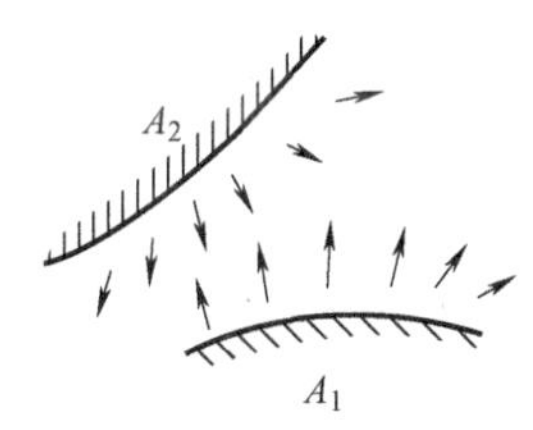

Fig. 9. 3 Schematic diagram of radiation heat transfer between two black body surfaces

图 9. 3 两黑体表面间的辐射传热示意图

(2) The second property of the angle factor: integrity.

（2）角系数的第二个性质：完整性。

Fig. 9. 4 shows a closed system composed of n surfaces. According to the law of energy conservation, radiation energy emitted from any surface must all fall on each surface of the closed system. Therefore, the angle factors between any surface and each surface in the closed cavity has the following relationship:

图 9. 4 所示为由 n 个表面组成的封闭系统。据能量守恒定律，从任何一个表面发射出的辐射能，必全部落到封闭系统的各表面上。因此，封闭腔内任何一个表面与各表面之间的角系数存在下列关系：

$$X_{1,1}+X_{1,2}+X_{1,3}+\cdots+X_{1,n}=\sum_{i=1}^{n}X_{1,i}=1 \tag{9.11}$$

This expression is called integrity of the angle factor. When surface 1 is a non-concave surface, $X_{1,1}=0$. If surface 1 is a concave surface shown by the dotted line in the figure,

此表达式称为角系数的完整性。当表面 1 为非凹表面时，$X_{1,1}=0$。若表面 1 为图中虚线所示的凹表面时，

then the angle factor $X_{1,1}$ is not zero.

则表面 1 对自己本身的角系数 $X_{1,1}$ 不为零。

(3) **The third property of the angle factor: additive property.**

(3) **角系数的第三个性质：可加性。**

As shown in Fig. 9. 5, calculate the angle factor of surface 1 to surface 2, the total energy emitted from surface 1 and falling on surface 2 is equal to the sum of radiation energy falling on all parts of surface 2. If surface 2 is divided into two surfaces, which are recorded as 2A and 2B respectively, it can be proved by the law of energy conservation:

如图 9. 5 所示，计算表面 1 对表面 2 的角系数。从表面 1 发出而落到表面 2 上的总能量等于落到表面 2 上各部分的辐射能之和。如果将表面 2 分为两个面，分别记作 2A 和 2B，通过能量守恒定律可证明：

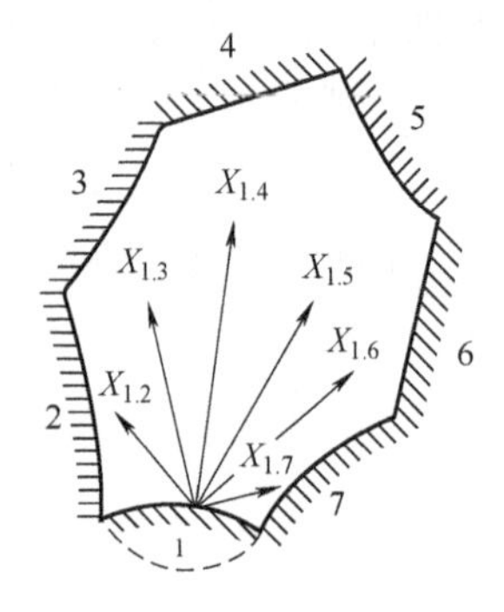

Fig. 9. 4 Diagram of closed surface radiation
图 9. 4 封闭表面辐射图

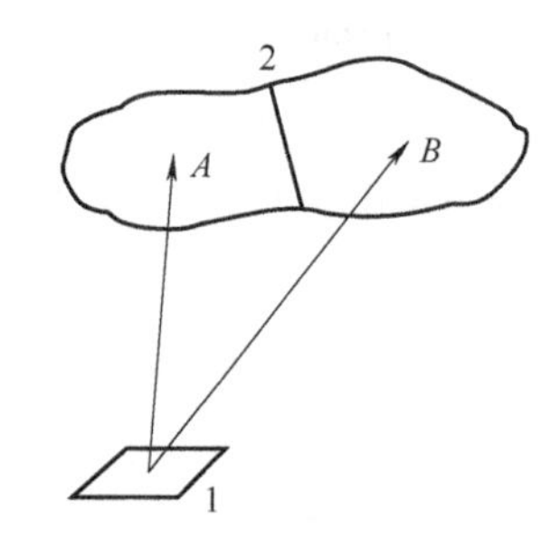

Fig. 9. 5 Radiation relation diagram
图 9. 5 辐射关系图

$$\Phi_{1,2}=\Phi_{1,2A}+\Phi_{1,2B}$$
$$\Rightarrow A_1E_{b1}X_{1,2}=A_1E_{b1}X_{1,2A}+A_1E_{b1}X_{1,2B}$$

cancel out A_1E_{b1} and get

约去 A_1E_{b1}，得到

$$X_{1,2}=X_{1,2A}+X_{1,2B} \tag{9.12}$$

If surface 2 is further divided into n surfaces, we can get:

如果把表面 2 进一步分为 n 个面，则有：

$$X_{1,2}=\sum_{i=1}^{n}X_{1,2i} \tag{9.13}$$

Let's look at the energy conservation situation of surface 2 to surface 1. The total radiation energy emitted from surface 2 and falling on surface 1 is equal to the sum of the radiation energy emitted from all parts of surface 2 and falling on surface 1.

再看面 2 对面 1 的能量守恒情况。从表面 2 发出落到表面 1 上的总辐射能，等于从表面 2 的各个组成部分发出而落到表面 1 上的辐射能之和。

$$\Phi_{2,1}=\Phi_{2A,1}+\Phi_{2B,1}$$
$$\Rightarrow A_2E_{b2}X_{2,1}=A_{2A}E_{b2}X_{2A,1}+A_{2B}E_{b2}X_{2B,1}$$

From the above equation, there is no additive property to the angle factor of surface 2 to surface 1 in Fig. 9. 5. That is, when additive property of the angle factor is used, only the second subscript in the sign of the angle factor is additive, and the first subscript in the sign of the angle factor does not have such a relationship.

由上式可知，图 9. 5 中表面 2 对表面 1 的角系数不存在可加性。即利用角系数可加性时，只有对角系数符号中的第二个角码是可加的，对角系数符号中的第一个角码则不存在这样的关系。

Summary　This section illustrates the three properties of the angle factor: relativity, integrity, and additive property.

小结　本节阐述了角系数的三个性质：相对性、完整性、可加性。

Question　Why is the angle factor $X_{1,1}$ for itself not zero when the surface 1 in Fig. 9. 4 is a concave surface?

思考题　为什么图 9. 4 中表面 1 为凹表面时对自身的角系数 $X_{1,1}$ 不为零？

Self-tests

自测题

1. What are the characteristics of the angle factor? What is the physical background to these characteristics?

1. 角系数有哪些特性？这些特性的物理背景是什么？

2. What is the essence of the relativity of the angle factor between two surfaces of finite size?

2. 两个有限大小表面之间角系数的相对性实质是什么？

授课视频

9. 3　The Calculation Method of the Angle Factor

9. 3　角系数的计算方法

The angle factor is the basic parameter to calculate the radiation heat transfer between objects. Direct integration method and algebraic analysis method are often used to solve the angle factor.

角系数是计算物体间辐射传热所需的基本参数。求解角系数常用直接积分法和代数分析法。

1. The direct integration method

1. 直接积分法

The direct integration method is directly based on the basic definition of the angle factor and obtains the angle factor

直接积分法是直接按照角系数的基本定义，通过求解多重积分而获得

by solving multiple integrals. Fig. 9.6 shows two finite surfaces A_1 and A_2. The angle factor $X_{1,2}$ of the surface A_1 to the surface A_2 is denoted as:

角系数的方法。如图 9.6 所示，两个有限大小的面 A_1、A_2，面 A_1 对面 A_2 的角系数 $X_{1,2}$ 计算式为：

$$X_{1,2}=\frac{1}{A_1}\int_{A_1}\int_{A_2}\frac{\cos\theta_1\cos\theta_2 \mathrm{d}A_2\mathrm{d}A_1}{\pi r^2}$$

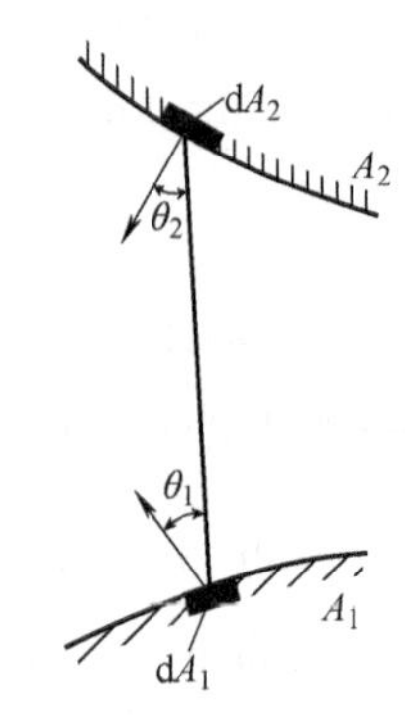

Fig. 9.6 Diagram of direct integration method

图 9.6 直接积分法的图示

This is the integral expression for solving the angle factor between any two surfaces, which is a quadruple integral. When calculating, we often encounter some mathematical difficulties and need to use some special skills.

这是求解任意两表面之间角系数的积分表达式，是一个四重积分。计算时，常会遇到一些数学上的困难，需采用专门的技巧。

In engineering, a large number of the angle factors of geometric structure have been drawn into diagrams. Some reference books give some calculation equations of the angle factors of two-dimensional geometric structures, and the calculation equations of the angle factors of several typical three-dimensional geometric structures and some engineering calculation diagrams. We can directly find the angle factor by charting.

工程上已将大量几何结构角系数的求解结果绘制成图线。一些参考书给出了一些二维几何结构角系数的计算公式，以及几种典型三维几何结构角系数的计算式和工程计算图线，可以通过查图直接查出角系数。

2. The algebraic analysis method

2. 代数分析法

The algebraic analysis method is to obtain a set of algebraic equations by using the relativity, integrity, and additive property of angle factors, then the angle factor is obtained by solving the equations. Note that the premise of using this method is that the system must be closed. If the system is not closed, we can make an imaginary surface to let it closed.

代数分析法是利用角系数的相对性、完整性和可加性获得一组代数方程，然后通过求解方程获得角系数。注意：利用该方法的前提是系统必须是封闭的。如果系统不封闭，可以做假想面令其封闭。

Fig. 9. 7 shows a closed system consisting of three non-concave surfaces with areas A_1, A_2, and A_3, respectively. According to the integrity and relativity of the angle factor, the following 6 algebraic equations can be obtained:

图 9. 7 所示是由三个非凹表面组成的封闭系统，面积分别为 A_1、A_2 和 A_3。根据角系数的完整性和相对性，可以得到以下六个代数方程：

$$X_{1,2}+X_{1,3}=1 \qquad A_1X_{1,2}=A_2X_{2,1}$$
$$X_{2,1}+X_{2,3}=1 \qquad A_1X_{1,3}=A_3X_{3,1}$$
$$X_{3,1}+X_{3,2}=1 \qquad A_2X_{2,3}=A_3X_{3,2}$$

All angle factors can be obtained by solving this closed system of equations. For example, $X_{1,2}$ is:

求解这个封闭系统的方程组，可得所有角系数。比如 $X_{1,2}$ 为：

$$X_{1,2}=\frac{A_1+A_2-A_3}{2A_1} \tag{9.14}$$

If the lengths of the three surfaces on the cross section of the system are l_1, l_2 and l_3 respectively, and their widths are equal, Eq. (9. 14) can be written as:

如果系统横截面上三个表面的长度分别为 l_1、l_2 和 l_3，且宽度相等，则式（9. 14）可写为：

$$X_{1,2}=\frac{l_1+l_2-l_3}{2l_1} \tag{9.15}$$

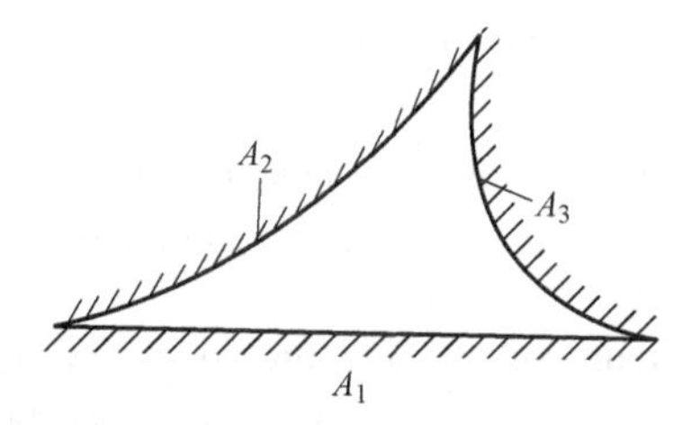

Fig. 9. 7 A closed system consisting of three non-concave surfaces

图 9. 7 由三个非凹表面组成的封闭系统

In a similar way, other angle factors can be obtained. Fig. 9. 8 shows an unclosed system composed of two surfaces, How to calculate the angle factor between surfaces A_1 and A_2 in the diagram?

同理，可求出其他角系数。图 9. 8 所示是由两个表面组成的不封闭系统，如何求解图中表面 A_1 和 A_2 之间的角系数？

First, make two imaginary surfaces *ac* and *bd*, which represent two surfaces extending infinitely in the direction perpendicular to the paper surface, and form a closed system together with surfaces A_1 and A_2. In this system, according to the integrity of the angle factor, the angle factor of the surface A_1 to A_2 is:

首先，做两个假想面 *ac*、*bd*，代表垂直于纸面方向上无限延伸的两个表面，连同表面 A_1、A_2 构成一个封闭系统。在此系统内，根据角系数的完整性，表面 A_1 对 A_2 的角系数为：

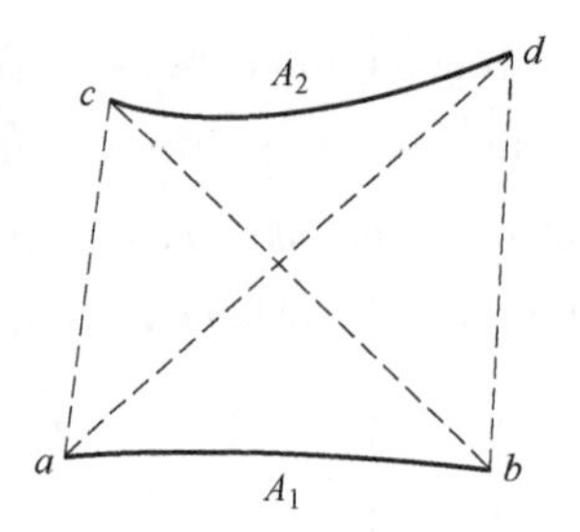

Fig. 9. 8 Schematic diagram of an unclosed system composed of two surfaces

图 9. 8 由两个表面组成的不封闭系统示意图

$$X_{ab,cd}=1-X_{ab,ac}-X_{ab,bd} \tag{9.16}$$

Then, make two imaginary surfaces *ad* and *bc*, and figures *abc* and *abd* are regarded as two closed systems composed of three surfaces respectively. Apply Eq. (9. 15) to get:

然后，再做两个假想面 *ad*、*bc*，把图形 *abc* 和 *abd* 看成各由三个表面组成的两个封闭系统。直接应用式（9. 15）可以得到：

$$X_{ab,ac}=\frac{ab+ac-bc}{2ab}$$

$$X_{ab,bd}=\frac{ab+bd-ad}{2ab}$$

Substitute $X_{ab,ac}$ and $X_{ab,bd}$ into the above equation, and obtain:

将 $X_{ab,ac}$和 $X_{ab,bd}$代入上式，可得到：

$$X_{ab,cd}=\frac{(bc+ad)-(ac+bd)}{2ab}=\frac{\text{Sum of intersecting lines}-\text{Sum of uncrossed lines}}{2\times\text{The section length of surface } A_1} \tag{9.17}$$

This method is called cross line method. Note that the so-called cross line and non-cross line here refers to the section line of the virtual surface or the auxiliary line.

该方法称为交叉线法。注意：这里所谓的交叉线和不交叉线都是指虚拟面的断面线，或者说是辅助线。

Summary This section illustrates the two methods for solving the angle factor, namely the direct integration method and the algebraic analysis method.

小结 本节阐述了求解角系数的两种方法，即直接积分法和代数分析法。

Question Why must a closed system be used to calculate the angle factor between one surface and other surfaces?

思考题 为什么计算一个表面与其他表面之间的角系数时必须采用封闭系统？

Self-tests

自测题

1. What properties of the angle factor are used when solving the angle factor by algebraic analysis method?

1. 用代数分析法求解角系数时利用了角系数的哪些性质？

2. How to solve the angle factor between surfaces A_1 and A_2 in an unclosed system composed of two surfaces?

2. 如何求解由两个表面构成的不封闭系统中表面 A_1 和 A_2 之间的角系数?

9.4 Radiation Heat Transfer in a Closed Cavity with Two Surfaces

9.4 两表面封闭腔的辐射传热

Among the modes of heat transfer, both heat conduction and convection heat transfer occur between objects in direct contact, while radiation heat transfer can occur between two surfaces separated by vacuum or heat permeable medium. The heat permeable medium here refers to the medium that does not participate in thermal radiation, such as air. Radiation heat transfer between solid surfaces refers to the absence of heat permeable medium between solid surfaces.

在热量传递的方式中，导热和对流传热都发生在直接接触的物体之间，而辐射传热可发生在两个被真空或透热介质隔开的表面之间。这里的透热介质指的是不参与热辐射的介质，如空气。本节所讨论的固体表面间的辐射传热是指表面之间不存在透热介质的情形。

When calculating the net radiation heat between a surface and the outside, the computation model must be a closed cavity containing the surfaces under study. The surfaces of the closed cavity for radiation heat transfer can be physically all real or partially imaginary.

当计算一个表面与外界的净辐射热量时，计算模型必须是包含所研究表面在内的一个封闭腔。辐射传热封闭腔的表面可以是物理上全部真实的，也可以是部分虚构的。

This section discusses the closed system composed of two surfaces. The focus is the calculation method of radiation heat transfer between gray body surfaces. Radiation heat transfer of a closed system with two black body surfaces is shown in Fig. 9.9. If the black body surfaces 1 and 2 are infinitely long in the direction perpendicular to the paper surface, then the net radiation heat transfer between surfaces 1 and 2 is as follows:

本节讨论由两个表面组成的封闭系统，重点在于灰体表面间辐射传热的计算方法。两黑体表面封闭系统的辐射传热如图 9.9 所示。黑体表面 1、2 在垂直纸面方向上为无限长，则表面 1、2 间的净辐射传热量为：

$$\begin{aligned}\Phi_{1,2} &= A_1 E_{b1} X_{1,2} - A_2 E_{b2} X_{2,1} \\ &= A_1 X_{1,2}(E_{b1} - E_{b2}) = A_2 X_{2,1}(E_{b1} - E_{b2})\end{aligned} \tag{9.18}$$

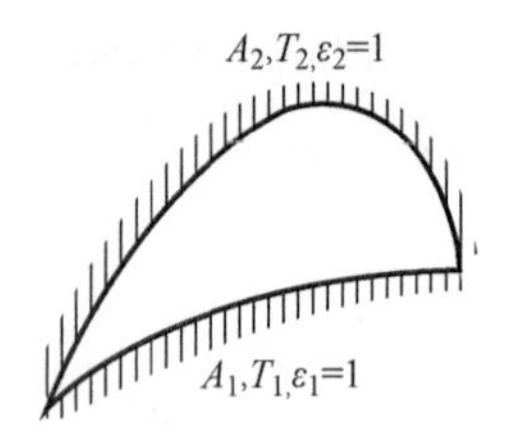

Fig. 9.9　Radiation heat transfer of a closed system with two black body surfaces

图 9.9　两黑体表面封闭系统的辐射传热

Where $A_1 E_{b1} X_{1,2}$ represents the part of heat radiation emitted by surface 1 reaching surface 2, and $A_2 E_{b2} X_{2,1}$ represents the part of heat radiation emitted by surface 2 reaching surface 1. The above equation shows that the key to the calculation of radiation heat transfer in a black body system is to find the angle factor.

式中，$A_1 E_{b1} X_{1,2}$表示表面 1 发出的热辐射到达表面 2 的部分；$A_2 E_{b2} X_{2,1}$表示表面 2 发出的热辐射到达表面 1 的部分。式（9.18）表明，黑体系统辐射传热计算的关键在于求角系数。

For the radiation heat transfer of a closed cavity composed of two diffuse gray surfaces, the situation is much more complicated. Because: ① The absorptivity of the gray body surface is less than 1, the radiation energy input to the gray body surface is not absorbed at one time, but is reflected and absorbed several times. ② The radiation energy emitted from a gray body surface includes not only its own emissive power, but also the reflected radiation energy, which adds complexity to the calculation of radiation heat transfer.

对于两漫灰表面组成的封闭腔的辐射传热，情况就要复杂得多。因为：① 灰体表面的吸收比小于 1，投入到灰体表面上的辐射能不是一次完成吸收，而是经过多次反射和吸收；② 由一个灰体表面向外发射的辐射能，除了其自身的辐射力，还包括被反射的辐射能，这增加了辐射传热计算的复杂性。

In order to reduce the trouble caused by multiple reflections between gray bodies to the calculation of radiation heat transfer, the concepts of irradiation G and radiosity J need to be introduced. The total radiation energy per unit area of a surface projected from the outside in unit time is called the irradiation of the surface, denoted as G, and the unit is W/m^2.

为了减少灰体间的多次反射给辐射传热计算带来的麻烦，需要引入投入辐射 G 和有效辐射 J 的概念。单位时间内外界投射到某一表面的单位面积上的总辐射能，称为该表面的投入辐射，记为 G，单位为 W/m^2。

The total radiation energy per unit area leaving the surface in unit time is called radiosity, and denoted as J. Radiosity J include not only the emitted radiation E of the surface itself, but also the part ρG reflected by the surface in the irradiation G. It can be expressed as:

单位时间内离开该表面单位面积的总辐射能，称为有效辐射，记作 J。有效辐射 J 不仅包括表面自身的发射辐射 E，还包括投入辐射 G 中被表面反射的部分 ρG。可表示为：

$$J=E+\rho G=\varepsilon E_b+(1-\alpha)G$$

Fig. 9. 10 shows the self-emission and absorption of external radiation by solid surface 1.

图 9. 10 所示为固体表面 1 的自身发射与吸收外界辐射的情形。

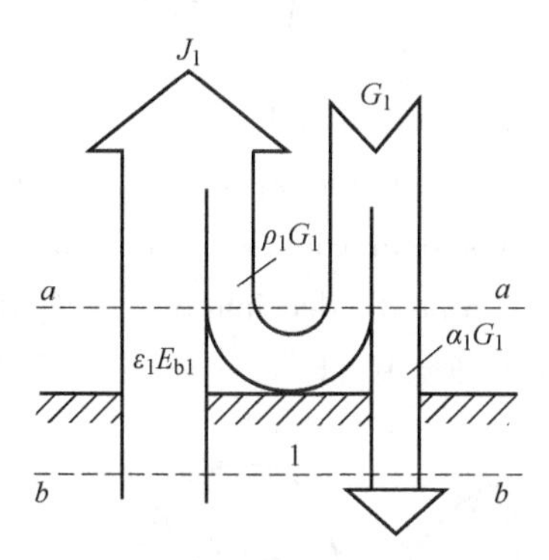

Fig. 9. 10 Self-emission and absorption of external radiation by solid surface 1

图 9. 10 固体表面 1 的自身发射与吸收外界辐射

Write the energy budget of surface 1 at the outer *a-a* and the inner *b-b* very close to the surface, respectively. At the outer *a-a* of surface 1, the calculation formula of the net radiation heat transfer quantity q is:

分别对离表面非常近的外部 *a-a* 处与内部 *b-b* 处两个位置写出表面 1 的能量收支。从表面 1 外部 *a-a* 处来观察，净辐射传热量 q 计算式为：

$$q=J_1-G_1$$

At the inner *b-b* of surface 1, the net radiation heat transfer quantity q between the surface and the outside is:

从表面 1 内部 *b-b* 处来观察，该表面与外界的净辐射传热量 q 为：

$$q=E_1-\alpha_1 G_1=\varepsilon_1 E_{b1}-\alpha_1 G_1$$

Combine the above two formulas and eliminate G_1 in the formula to get:

将以上两式联立，消去式中 G_1 得到：

$$q=\varepsilon_1 E_{b1}-\alpha_1(J_1-q)$$

The expression of J_1 is obtained after sorting:

整理后得到 J_1 的表达式为：

$$J_1=\frac{\varepsilon_1 E_{b1}}{\alpha_1}-\frac{(1-\alpha_1)q}{\alpha_1}$$

Taking into account $\alpha_1=\varepsilon_1$, the following equation can be obtained:

考虑到 $\alpha_1=\varepsilon_1$，可得下式：

$$J_1=E_{b1}-\left(\frac{1}{\varepsilon_1}-1\right)q$$

Remove the subscript in the above formula to get the general expression formula:

将上式中的下标去掉，即得一般表达式：

$$J=E_b-\left(\frac{1}{\varepsilon}-1\right)q \tag{9.19}$$

Note that the quantities in this formula are all relative to the same surface and the net heat released to the outside is positive. The two-dimensional closed system composed of two isothermal diffuse gray surfaces can be divided into the following four situations, as shown in Fig. 9. 11.

注意：该式中各个量均是对同一表面而言的，而且以向外界的净放热量为正值。对于两个等温的漫灰表面组成的二维封闭系统，可分为以下四种情形，如图 9. 11 所示。

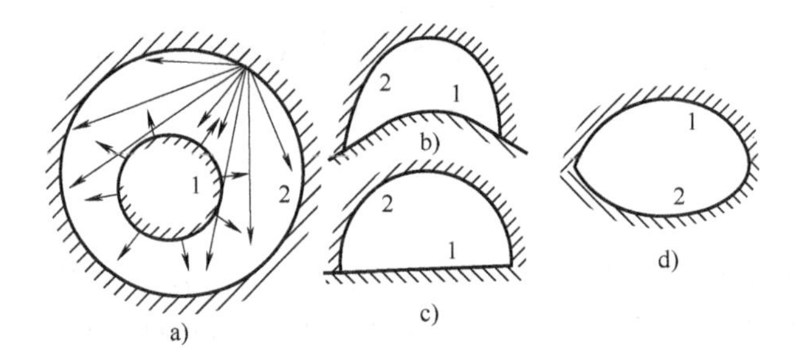

Fig. 9. 11　Two-dimensional closed system composed of two isothermal diffuse gray surfaces

图 9. 11　两个等温的漫灰表面组成的二维封闭系统

Fig. 9. 11a represents a two-dimensional closed system, where 1 and 2 are cylindrical surfaces, or a three-dimensional closed system, where 1 and 2 are spherical surfaces. The systems represented by Fig. 9. 11b, c and d are infinitely long in the direction perpendicular to the paper surface and are two-dimensional systems. In either case, the calculation formula of the radiation heat transfer between surfaces 1 and 2 is:

图 9. 11a 所示情形可代表二维封闭系统，其中，1、2 为圆柱面；也可代表三维封闭系统，其中，1、2 为球面。图 9. 11b、c、d 所代表的系统，在垂直于纸面方向无限长，为二维系统。无论对于哪种情形，表面 1、2 间辐射传热量的计算式为：

$$\Phi_{1,2}=A_1J_1X_{1,2}-A_2J_2X_{2,1} \tag{a}$$

Where $A_1J_1X_{1,2}$ represents the part of the radiosity from surface 1 reaching surface 2, and $A_2J_2X_{2,1}$ represents the part of the radiosity from surface 2 reaching surface 1. Apply Eq. (9. 19) to get:

式中，$A_1J_1X_{1,2}$表示表面 1 发出的有效辐射到达表面 2 的部分；$A_2J_2X_{2,1}$表示表面 2 发出的有效辐射到达表面 1 的部分。应用式（9. 19），可得到：

$$J_1A_1=A_1E_{b1}-\left(\frac{1}{\varepsilon_1}-1\right)\Phi_{1,2} \tag{b}$$

$$J_2A_2=A_2E_{b2}-\left(\frac{1}{\varepsilon_2}-1\right)\Phi_{2,1} \tag{c}$$

According to the law of energy conservation:

根据能量守恒定律有:

$$\Phi_{1,2}=-\Phi_{2,1} \tag{d}$$

Substitute the above three equations (b), (c) and (d) into equation (a) to get:

将上面三式(b)、(c)、(d)代入式(a),得到:

$$\Phi_{1,2}=\frac{E_{b1}-E_{b2}}{\dfrac{1-\varepsilon_1}{\varepsilon_1 A_1}+\dfrac{1}{A_1 X_{1,2}}+\dfrac{1-\varepsilon_2}{\varepsilon_2 A_2}} \tag{9.20}$$

If A_1 is used as the calculation area, Eq. (9.20) can be rewritten as:

若用 A_1 作为计算面积,式(9.20)可改写为:

$$\Phi_{1,2}=\frac{A_1 X_{1,2}(E_{b1}-E_{b2})}{X_{1,2}\left(\dfrac{1}{\varepsilon_1}-1\right)+1+\dfrac{A_1}{A_2}X_{1,2}\left(\dfrac{1}{\varepsilon_2}-1\right)} \tag{9.21}$$

Define system blackness (or system emissivity) ε_s as:

定义系统黑度(或称为系统发射率)ε_s 为:

$$\varepsilon_s=\frac{1}{1+X_{1,2}\left(\dfrac{1}{\varepsilon_1}-1\right)+X_{2,1}\left(\dfrac{1}{\varepsilon_2}-1\right)} \tag{9.22}$$

Substitute the system blackness into $\Phi_{1,2}$ to get the calculation equation of gray body radiation heat transfer:

将系统黑度代入 $\Phi_{1,2}$,可以得到灰体辐射传热计算式为:

$$\Phi_{1,2}=\varepsilon_s A_1 X_{1,2}(E_{b1}-E_{b2}) \tag{9.23}$$

Compared with the calculation equation of black body radiation heat transfer, the calculation equation of gray body system adds a correction factor ε_s. The value of ε_s is less than 1, and it is the factor of the influence of multiple absorption and reflection on the heat transfer caused by the emissivity value of the gray body system being less than 1.

与黑体辐射传热计算式相比,灰体系统的计算式多了一个修正因子 ε_s。ε_s 的值小于 1,它是由于灰体系统发射率小于 1 引起的多次吸收与反射对传热量影响的因子。

For the following three special cases, the calculation equation of radiation heat transfer $\Phi_{1,2}$ can be further simplified.

对于下列三种特殊情形,辐射传热量 $\Phi_{1,2}$的计算式可进一步简化。

The first special case: Surface 1 is a plane or convex surface (Fig. 9.11a, b, c), at this time $X_{1,2}$ equals 1, so the system emissivity can be simplified as:

第一种特殊情形: 表面 1 为平面或凸表面(图 9.11a、b、c),此时 $X_{1,2}=1$,于是系统发射率可简化为:

$$\varepsilon_s = \frac{1}{\dfrac{1}{\varepsilon_1} + \dfrac{A_1}{A_2}\left(\dfrac{1}{\varepsilon_2} - 1\right)}$$

At this time, the radiation heat transfer $\Phi_{1,2}$ is:

此时，辐射传热量 $\Phi_{1,2}$ 为：

$$\Phi_{1,2} = \frac{A_1(E_{b1} - E_{b2})}{\dfrac{1}{\varepsilon_1} + \dfrac{A_1}{A_2}\left(\dfrac{1}{\varepsilon_2} - 1\right)} = \varepsilon_s A_1 \times 5.67\mathrm{W/(m^2 \cdot K^4)}\left[\left(\frac{T_1}{100}\right)^4 - \left(\frac{T_2}{100}\right)^4\right] \tag{9.24}$$

The second special case: The difference between surface areas A_1 and A_2 is very small, that is, A_1/A_2 is close to 1, so the system emissivity can be simplified as:

第二种特殊情形：表面面积 A_1 和 A_2 相差很小，即 A_1/A_2 趋近于 1，于是系统发射率可以简化为：

$$\varepsilon_s = \frac{1}{\dfrac{1}{\varepsilon_1} + \dfrac{1}{\varepsilon_2} - 1}$$

At this time, the radiation heat transfer $\Phi_{1,2}$ is:

此时，辐射传热量 $\Phi_{1,2}$ 为：

$$\Phi_{1,2} = \frac{A_1(E_{b1} - E_{b2})}{\dfrac{1}{\varepsilon_1} + \dfrac{1}{\varepsilon_2} - 1} = \frac{A_1 \times 5.67\mathrm{W/(m^2 \cdot K^4)}\left[\left(\dfrac{T_1}{100}\right)^4 - \left(\dfrac{T_2}{100}\right)^4\right]}{\dfrac{1}{\varepsilon_1} + \dfrac{1}{\varepsilon_2} - 1} \tag{9.25}$$

Practically radiation heat transfer between infinite parallel plates belongs to this special case.

实际上无限大平行平板间的辐射传热就属于此种特例。

The third special case: Surface area A_2 is much larger than A_1, namely A_1/A_2 approaches zero, so the system emissivity can be simplified as:

第三种特殊情形：表面面积 A_2 比 A_1 大得多，即 $A_1/A_2 \to 0$，于是系统发射率可简化为：

$$\varepsilon_s = \varepsilon_1$$

At this time, the radiation heat transfer $\Phi_{1,2}$ is:

此时，辐射传热量 $\Phi_{1,2}$ 为：

$$\Phi_{1,2} = \varepsilon_1 A_1(E_{b1} - E_{b2}) = \varepsilon_1 A_1 \times 5.67\mathrm{W/(m^2 \cdot K^4)}\left[\left(\frac{T_1}{100}\right)^4 - \left(\frac{T_2}{100}\right)^4\right] \tag{9.26}$$

For the case that the system emissivity ε_s is equal to ε_1, it is not necessary to know the area A_2 of the shelled object 2 and its emissivity ε_2 for radiation heat transfer calculation. This is the case for small objects in large rooms, such as the radiation heat dissipation of high temperature

对于系统发射率 ε_s 等于 ε_1 这种情况，辐射传热计算不需要知道包壳物体 2 的面积 A_2 及其发射率 ε_2。对于大房间内的小物体，比如高温管道的辐射散热及气体容器或管道

pipelines, the radiation error of thermocouple temperature measurement in gas containers or pipelines, and other practical problems.

内热电偶测温的辐射误差等实际问题，都属于此种情况。

Summary This section illustrates the radiation heat transfer calculation of the closed system with two black body surfaces and the closed system with two diffuse gray surfaces, and the simplified calculation of the radiation heat transfer between the diffuse gray surfaces in three special cases.

小结 本节阐述了两黑体表面封闭系统和两漫灰表面封闭系统的辐射传热计算，以及三种特殊情形下漫灰表面间的辐射传热简化计算。

Question Why is a closed cavity model used to calculate the net radiation heat transfer between a surface and the outside?

思考题 为什么计算一个表面与外界之间的净辐射传热量时要采用封闭腔模型？

Self-tests

自测题

1. What are self radiation, irradiation and radiosity of a surface? What is the effect of radiosity on the calculation of radiation heat transfer in gray body surface system?

1. 什么是表面的自身辐射、投入辐射及有效辐射？有效辐射对于灰体表面系统辐射传热计算有什么作用？

2. What complexity is added to the radiation heat transfer calculation for the actual surface closed system compared with the black body surface closed system?

2. 实际表面封闭系统与黑体表面封闭系统相比，辐射传热计算增加了哪些复杂性？

授课视频

9.5 Radiation Heat Transfer Calculation of a Closed Cavity with Two Surfaces

9.5 两表面封闭腔的辐射传热计算

Example 9.1 There is a liquid oxygen storage container with a double-wall silver-plated interlayer structure as shown in Fig. 9.12. The temperature of the inner surface of the outer wall is $t_{w1} = 20℃$, and the temperature of the outer surface of the inner wall is $t_{w2} = -183℃$. The emissivity of the silver-plated wall is $\varepsilon = 0.02$. Try to calculate

例 9.1 液氧储存容器为双壁镀银的夹层结构，如图 9.12 所示。外壁内表面温度 $t_{w1} = 20℃$，内壁外表面温度 $t_{w2} = -183℃$，镀银壁的发射率 $\varepsilon = 0.02$。试计算由于辐射传热每单位面积容器壁的散热量。

the heat dissipation per unit area of the container wall due to radiation heat transfer.

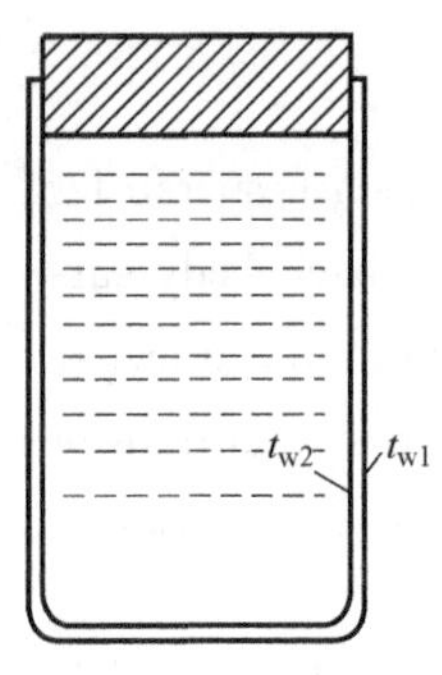

Fig. 9. 12 Schematic diagram of liquid oxygen storage container

图 9. 12 液氧储存容器示意图

Solution: Because the gap between the container interlayers is very small, it can be considered as radiation heat transfer between infinite parallel planes. The area ratio of the two planes A_1/A_2 approaches 1. Therefore, the radiation heat dissipation per unit area of the container wall can be calculated according to the second special case of the radiation heat transfer calculation of the closed cavity with two surfaces. First, convert t_{w1} and t_{w2} into thermodynamic temperatures T_{w1} and T_{w2}.

解：因为容器夹层的间隙很小，可认为属于无限大平行表面间的辐射传热问题。两平面的面积比 $A_1/A_2 \to 1$，因此，容器壁单位面积的辐射散热量可按照两表面封闭腔的辐射传热计算的第二种特殊情形计算。首先将 t_{w1} 和 t_{w2} 换算为热力学温度 T_{w1} 和 T_{w2}。

$$T_{w1} = t_{w1} + 273\text{K} = (20+273)\text{K} = 293\text{K}$$

$$T_{w2} = t_{w2} + 273\text{K} = (-183+273)\text{K} = 90\text{K}$$

Next, substitute T_{w1} and T_{w2} into the formula of the second special case to calculate the heat transfer quantity per unit area, $q_{1,2}$.

然后，将 T_{w1} 和 T_{w2} 代入第二种特殊情形计算公式，计算得到单位面积的传热量 $q_{1,2}$。

$$q_{1,2} = \frac{C_0\left[\left(\frac{T_1}{100}\right)^4 - \left(\frac{T_2}{100}\right)^4\right]}{\frac{1}{\varepsilon_1} + \frac{1}{\varepsilon_2} - 1} = \frac{5.67\text{W}/(\text{m}^2 \cdot \text{K}^4) \times [(2.93\text{K})^4 - (0.9\text{K})^4]}{\frac{1}{0.02} + \frac{1}{0.02} - 1} = 4.18\text{W}/\text{m}^2$$

Discussion: The silver-plated walls have a great effect on reducing radiation heat dissipation. As a comparison, if the emissivity of the inner and outer walls $\varepsilon_1 = \varepsilon_2 = 0.8$, then $q_{1,2} = 276\text{W}/\text{m}^2$ by calculation, indicating that the heat dissipation is increased by 66 times.

讨论：镀银壁对降低辐射散热量作用极大。作为比较，如果设内外壁面的发射率 $\varepsilon_1 = \varepsilon_2 = 0.8$，计算得到 $q_{1,2} = 276\text{W}/\text{m}^2$，即散热量增加 66 倍。

If a vacuum interlayer is not used, but the insulating materials are laid on the outside of the container for thermal insulation. Take the thermal conductivity of the insulating material as 0.05W/(m·K), which is already a very good thermal insulating material. The calculation is based on the heat conduction of a one-dimensional plane plate. The required wall thickness δ of the thermal insulating material should satisfy the following formula. It shows the insulation effectiveness of low-emissivity vacuum interlayer.

如果不采用抽真空的夹层，而采用在容器外敷设保温材料的方法来绝热，取保温材料的导热系数为0.05W/(m·K)(这已经是相当好的保温材料了)，则按一维平板导热问题来估算，所需的保温材料壁厚δ应满足下式。由此可见，抽真空的低发射率夹层保温的有效性。

$$4.18\,\mathrm{W/m^2} = 0.05\,\mathrm{W/(m\cdot K)} \times \frac{[20-(-183)]\,\mathrm{K}}{\delta}$$

$$\delta = 2.43\,\mathrm{m}$$

Example 9.2 A steel pipe with a diameter of $d=50$mm and a length of $l=8$m, is placed in a brick channel with a cross section of 0.2m×0.2m. If the temperature and emissivity of the steel pipe are $t_1=250$℃ and $\varepsilon_1=0.79$ respectively. And the temperature and emissivity of the brick channel wall are $t_2=27$℃, $\varepsilon_2=0.93$ respectively. Try to calculate the radiation heat loss of the steel pipe.

例9.2 一根直径$d=50$mm、长度$l=8$m的钢管，被置于横断面为0.2m×0.2m的砖槽道内。若钢管的温度和发射率分别为$t_1=250$℃、$\varepsilon_1=0.79$，砖槽壁面温度和发射率分别为$t_2=27$℃，$\varepsilon_2=0.93$，试计算该钢管的辐射热损失。

Solution: This is a three-dimensional question. Because the length-to-diameter ratio of the steel pipe l/d is much bigger than 1, it can be treated approximately as a two-dimensional problem. The surface of the steel pipe is convex, and the radiation heat loss of the steel pipe can be calculated directly according to the first special case of the radiation heat transfer calculation of the two-surface closed cavity. Substitute the specific values into the formula to get the radiation heat.

解：这是一个三维问题。因为钢管的长径比$l/d\gg1$，可近似地按二维问题处理。钢管的表面为凸面，可直接按照两表面封闭腔的辐射传热计算的第一种特殊情形计算钢管的辐射散热损失。把具体数值代入公式，可得到辐射热量为：

$$\Phi = \frac{A_1 C_0\left[\left(\frac{T_1}{100}\right)^4-\left(\frac{T_2}{100}\right)^4\right]}{\frac{1}{\varepsilon_1}+\frac{A_1}{A_2}\left(\frac{1}{\varepsilon_2}-1\right)} = \frac{3.14\times0.05\mathrm{m}\times8\mathrm{m}\times5.67\mathrm{W/(m^2\cdot K^4)}\times[(5.23\mathrm{K})^4-(3.00\mathrm{K})^4]}{\frac{1}{0.79}+\frac{3.14\times0.05\mathrm{m}\times8\mathrm{m}}{4\times0.2\mathrm{m}\times8\mathrm{m}}\times\left(\frac{1}{0.93}-1\right)}$$

$$= 3.71\,\mathrm{kW}$$

Discussion: This problem can also approximately adopt the model of $A_1/A_2\to0$. In this case, the calculation result is as follows. It can be seen that the calculated results of the

讨论：这一问题也可近似采用$A_1/A_2\to0$的模型，此时计算结果如下，可见两种模型的计算结果仅相差1%。

two models are only 1% different.

$$\Phi_{1,2}=\varepsilon_1 A_1(E_{b1}-E_{b2})=\varepsilon_1 A_1\times 5.67\text{W}/(\text{m}^2\cdot\text{K}^4)\left[\left(\frac{T_1}{100}\right)^4-\left(\frac{T_2}{100}\right)^4\right]=3.754\text{kW}$$

Example 9.3 A cylindrical buried heating furnace with a diameter of $d=0.75$m, is heated by the electric heating method as shown in Fig. 9.13. During the operation, the furnace top cover needs to be removed for a period of time. Suppose that the temperature of the cylinder body is 500K at this time, the cylinder bottom is 650K, and the ambient temperature is 300K. Try to calculate the heat loss per unit time during the removal of the top cover. Both the cylinder body and the bottom can be used as black body.

例 9.3 一个直径 $d=0.75$m 的圆筒形埋地式加热炉，采用电加热方法加热，如图 9.13 所示。在操作过程中，需要将炉子顶盖移去一段时间。设此时筒身温度为 500K，筒底温度为 650K，环境温度为 300K。试计算顶盖移去期间单位时间内的热损失，设筒身及底面均可作为黑体。

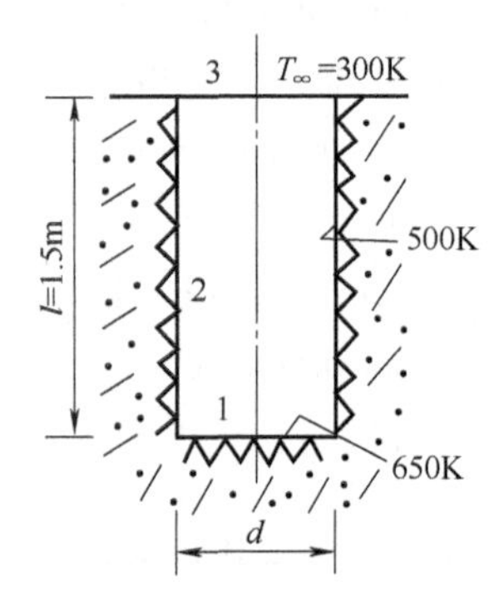

Fig. 9.13 Schematic diagram of the cylindrical buried heating furnace

图 9.13 圆筒形埋地式加热炉示意图

Solution: The radiation heat loss from the side wall and bottom of the heating furnace to the plant through the top cover opening is almost absorbed by the objects in the plant, and the proportion returned to the heating furnace is almost zero. Therefore, the top cover opening can be regarded as an imaginary black body surface, and its temperature is equal to the ambient temperature.

解： 从加热炉的侧壁与底面通过顶盖开口散失到厂房中的辐射热量几乎全部被厂房中的物体吸收，返回到加热炉内的比例几乎为零。因此，顶盖开口处可当作一个假想的黑体表面，其温度等于环境温度。

In this way, a black body closed cavity composed of three isothermal surfaces is formed. Mark the bottom of the cylinder as surface 1, the cylinder body as surface 2, and the top cover as surface 3. Then the radiation energy Φ dissipated by the heating furnace into the plant is equal to $\Phi_{2,3}$ plus $\Phi_{1,3}$.

这样就形成了由三个等温表面组成的黑体封闭腔，将筒底记作面 1，筒身记作面 2，顶盖记作面 3。那么，加热炉散失到厂房中的辐射能 Φ 等于 $\Phi_{2,3}$加上 $\Phi_{1,3}$。

$$\Phi=\Phi_{2,3}+\Phi_{1,3}=A_2X_{2,3}(E_{b2}-E_{b3})+A_1X_{1,3}(E_{b1}-E_{b3})$$

According to $r_2/l=0.375/1.5=0.25$, $l/r_1=1.5/0.375=4$, look up the angle factor diagram between two coaxial parallel disks, we can get $X_{1,3}=0.06$. According to the integrity of the angle factor, we can get:

根据$r_2/l=0.375/1.5=0.25$，$l/r_1=1.5/0.375=4$，查两同轴平行圆盘间的角系数图，可以得到$X_{1,3}=0.06$。根据角系数的完整性，可得：

$$X_{1,2}=1-X_{1,3}=1-0.06=0.94$$

According to the relativity of the angle factor, we can get:

根据角系数的相对性，可得：

$$X_{2,1}=\frac{A_1}{A_2}X_{1,2}=\frac{3.14\times0.75^2/4}{3.14\times0.75\times1.5}\times0.94=0.118$$

According to the symmetry, we can get $X_{2,1}=X_{2,3}$, finally the specific values are substituted into the calculation formula of Φ, the total radiation energy is equal to 1542W.

再根据对称性可得到$X_{2,1}=X_{2,3}$，最后将具体数值代入Φ的计算公式，可得到总的辐射能等于1542W。

Summary This section illustrates the calculation steps and methods of radiation heat transfer of the closed cavity system composed of two surfaces through examples.

小结 本节通过例题阐述了由两表面组成的封闭腔系统的辐射传热计算步骤及方法。

Question Why can the silver-plated wall greatly reduce the radiation heat dissipation?

思考题 为什么镀银壁面能大大降低辐射散热量？

Self-tests

自测题

1. What properties of the angle factor need to be used to solve the radiation heat transfer between two surfaces?

1. 求解两表面之间的辐射传热时需要采用角系数的哪些性质？

2. Under what circumstances should the first, second and third special cases be used for the calculation of radiation heat transfer in two-surface closed cavities?

2. 两表面封闭腔的辐射传热计算，在什么情况下采用第一、二、三种特殊情形？

授课视频

9.6 Radiation Heat Transfer in Multi-surface Closed Systems

9.6 多表面封闭系统的辐射传热

In the multi-surface closed system, the net radiation heat transfer quantity of one surface is equal to the sum of heat transfer quantity with other surfaces. The network method

在多表面系统中，一个表面的净辐射传热量等于与其余表面的传热量之和。求解一个表面的净辐射传热

is often used to solve the net radiation heat transfer quantity of a surface.

量常采用网络法。

The principle of the network method is to liken the current, potential difference and resistance in electricity to the heat flux, thermal potential difference and thermal resistance in thermal radiation, and to liken circuit to the heat flux path of radiation heat transfer. According to the calculation formula of the radiosity of a diffuse gray surface:

网络法的原理是：用电学中的电流、电位差和电阻比拟热辐射中的热流、热势差与热阻，用电路来比拟辐射传热的热流路。根据一个漫灰表面的有效辐射的计算式：

$$J=E_b-\left(\frac{1}{\varepsilon}-1\right)q \tag{9.27}$$

By transforming this formula, it can be written as

变换后可写为

$$q=\frac{E_b-J}{\frac{1-\varepsilon}{\varepsilon}} \quad 或 \quad \Phi=\frac{E_b-J}{\frac{1-\varepsilon}{A\varepsilon}} \tag{9.28}$$

Where E_b-J is called as surface thermal potential difference; $\frac{1-\varepsilon}{\varepsilon}$ or $\frac{1-\varepsilon}{A\varepsilon}$ is called as surface radiation thermal resistance. The meaning of surface radiation thermal resistance is that it is equivalent to the presence of thermal resistance on the surface when the surface of an object cannot absorb all the projected energy. Each actual object surface has a surface radiation thermal resistance, as shown in Fig. 9. 14.

式中，E_b-J 称为表面热势差；$\frac{1-\varepsilon}{\varepsilon}$ 或$\frac{1-\varepsilon}{A\varepsilon}$称为表面辐射热阻。表面辐射热阻的含义是：当物体表面不能全部吸收投射能量时，相当于表面存在热阻。每一个实际物体表面都有一个表面辐射热阻，如图 9. 14 所示。

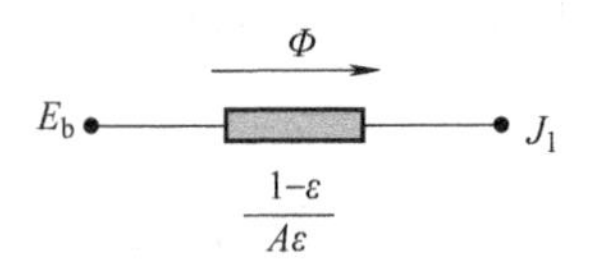

Fig. 9. 14 Surface radiation thermal resistance

图 9. 14 表面辐射热阻

For the black body surface, its emissivity $\varepsilon=1 \Rightarrow R_r=0$, that is, the surface thermal resistance of a black body is equal to zero. For radiation heat transfer quantity between diffuse gray surfaces 1 and 2 is:

对黑体表面，其发射率 $\varepsilon=1 \Rightarrow R_r=0$，即黑体的表面热阻等于零。对于两漫灰表面 1、2 间的辐射传热量：

$$\Phi_{1,2}=A_1J_1X_{1,2}-A_2J_2X_{2,1}$$

According to the relativity of angle factor $A_1X_{1,2}=A_2X_{2,1}$, we can get:

根据角系数相对性 $A_1X_{1,2}=A_2X_{2,1}$，可得：

$$\Phi_{1,2}=A_1X_{1,2}(J_1-J_2) \tag{9.29}$$

By transforming the formula, it can be written as:

将式子变换后，可写为：

$$\Phi_{1,2}=A_1X_{1,2}(J_1-J_2)=\frac{J_1-J_2}{\dfrac{1}{A_1X_{1,2}}} \tag{9.30}$$

Where J_1-J_2 is the spatial heat potential difference; $1/(A_1X_{1,2})$ is spatial radiation thermal resistance; E_b is equivalent to the power supply potential; and J is equivalent to the node voltage, as shown in Fig. 9. 14 and Fig. 9. 15. There is a spatial radiation thermal resistance between each pair of surfaces.

式中，J_1-J_2 是空间热势差；$1/(A_1X_{1,2})$ 是空间辐射热阻；E_b 相当于电源电动势；J 相当于节点电压，如图 9. 14 和图 9. 15 所示。每一对表面之间都有一个空间辐射热阻。

For the closed system composed of two diffuse gray surfaces, according to the fact that each surface has a surface radiation thermal resistance and each pair of surfaces has a spatial radiation thermal resistance, its equivalent network diagram can be drawn as shown in Fig. 9. 16.

对于两漫灰表面组成的封闭系统，根据每一个表面都有一个表面辐射热阻，每一对表面间都有一个空间辐射热阻，可画出其等效网络图，如图 9. 16 所示。

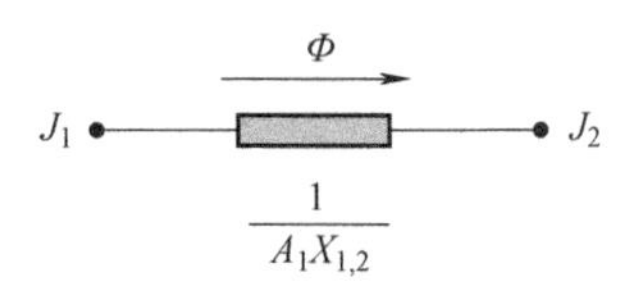

Fig. 9. 15　Spatial radiation thermal resistance

图 9. 15　空间辐射热阻

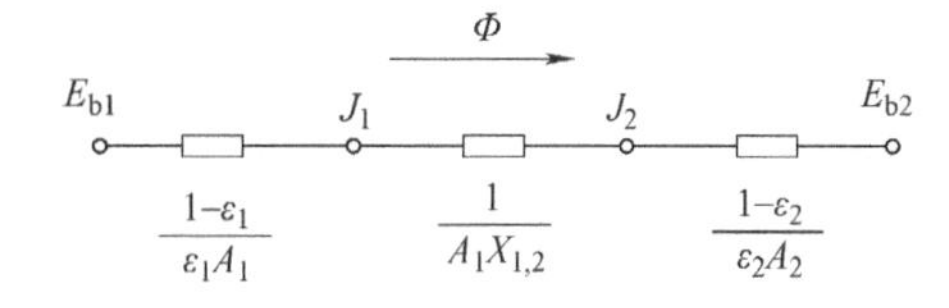

Fig. 9. 16　The equivalent network diagram

图 9. 16　等效网络图

With reference to the calculation of series circuit, the heat transfer calculation formula can be written. This is the same as the previous calculation formula obtained by the net heat method.

参照串联电路的计算，可写出传热量计算式。这与前面采用净热量法得到的计算公式相同。

$$\Phi_{1,2}=\frac{E_{b1}-E_{b2}}{\dfrac{1-\varepsilon_1}{\varepsilon_1A_1}+\dfrac{1}{A_1X_{1,2}}+\dfrac{1-\varepsilon_2}{\varepsilon_2A_2}} \tag{9.31}$$

A closed system consisting of three surfaces is shown in Fig. 9. 17. The steps to solve the radiation heat transfer by using the network method are as follows:

由三个表面组成的封闭系统如图 9. 17 所示。应用网络法求解其辐射传热的步骤如下：

(1) Draw the equivalent network diagram, as shown in Fig. 9. 18.

(1) 画出等效网络图，如图 9. 18 所示。

(2) List the heat flux equations of each node; The heat flux equations of nodes J_1, J_2, J_3 are as follows:

(2) 列出各节点的热流方程，节点 J_1、J_2、J_3 的热流方程分别为：

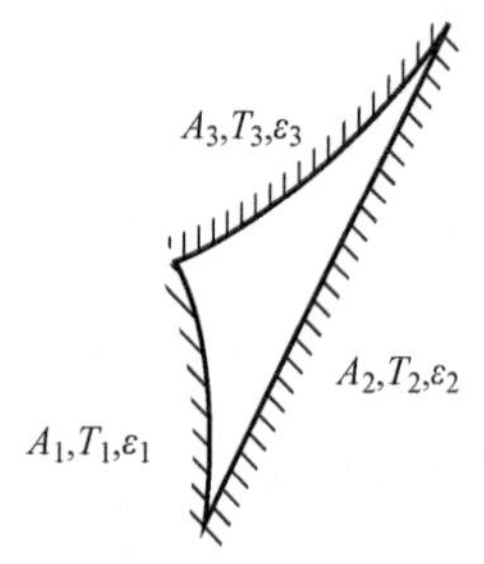

Fig. 9. 17 A closed system consisting of three surfaces

图 9. 17 由三个表面组成的封闭系统

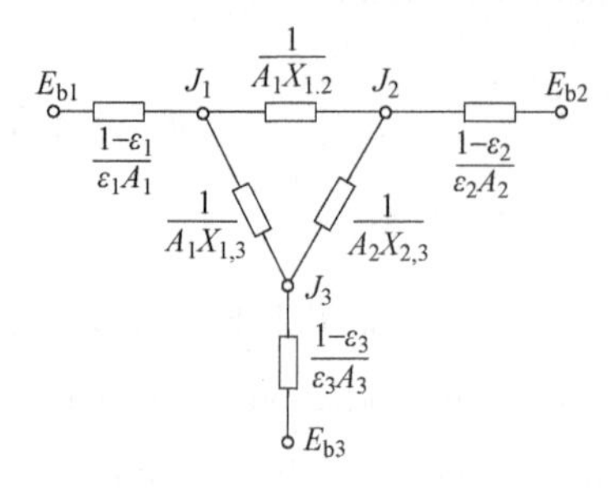

Fig. 9. 18 The equivalent network diagram of a closed system consisting of three surfaces

图 9. 18 由三个表面组成的封闭系统的等效网络图

$$J_1:\ \frac{E_{b1}-J_1}{\dfrac{1-\varepsilon_1}{\varepsilon_1 A_1}}+\frac{J_2-J_1}{\dfrac{1}{A_1X_{1,2}}}+\frac{J_3-J_1}{\dfrac{1}{A_1X_{1,3}}}=0$$

$$J_2:\ \frac{E_{b2}-J_2}{\dfrac{1-\varepsilon_2}{\varepsilon_2 A_2}}+\frac{J_1-J_2}{\dfrac{1}{A_1X_{1,2}}}+\frac{J_3-J_2}{\dfrac{1}{A_2X_{2,3}}}=0$$

$$J_3:\ \frac{E_{b3}-J_3}{\dfrac{1-\varepsilon_3}{\varepsilon_3 A_3}}+\frac{J_1-J_3}{\dfrac{1}{A_1X_{1,3}}}+\frac{J_2-J_3}{\dfrac{1}{A_2X_{2,3}}}=0$$

(3) Solve the above equation group to obtain the surface radiosity J_i of the nodes.

(3) 求解上面的方程组，得到节点的表面有效辐射 J_i。

(4) Calculate the net radiation heat transfer quantity of each surface according to the equation $\Phi_i=\dfrac{E_{bi}-J_i}{\dfrac{1-\varepsilon_i}{A_i\varepsilon_i}}$.

(4) 按照公式 $\Phi_i=\dfrac{E_{bi}-J_i}{\dfrac{1-\varepsilon_i}{A_i\varepsilon_i}}$，计算每个表面的净辐射传热量。

There are two important special cases of the three-surface closed system:

三表面封闭系统有两个重要特例：

(1) There is a black body surface with zero surface thermal resistance. Its network diagram is shown in Fig. 9. 19.

(1) 有一个表面为黑体，即表面热阻为零，其网络图如图 9. 19 所示。

At this time, the temperature of the surface is generally known, which means that J_3 is known, $J_3=E_{b3}$. There are only two unknown nodes J_1 and J_2, and the heat flux

此时该表面的温度一般是已知的，意味着 J_3 已知，$J_3=E_{b3}$。只有两个未知节点 J_1、J_2，列出这两个节

equations of nodes J_1 and J_2 are listed. Solve the equation group to obtain J_1 and J_2, then calculate the net heat transfer quantity on each surface.

点的热流方程，求解方程组可得到 J_1、J_2，再计算各表面的净传热量。

(2) There is an adiabatic surface whose net heat transfer quantity of the surface is zero. Its network diagram is shown in Fig. 9. 20. The difference with a black body surface is that the temperature of the surface 3 is unknown. But it still absorbs and emits thermal radiation, and both are equal.

(2) 有一个表面绝热，即该表面的净传热量为零，其网络图如图 9.20 所示。与黑体表面不同的是，该表面 3 的温度未知，但是其仍吸收和发射热辐射，且两者相等。

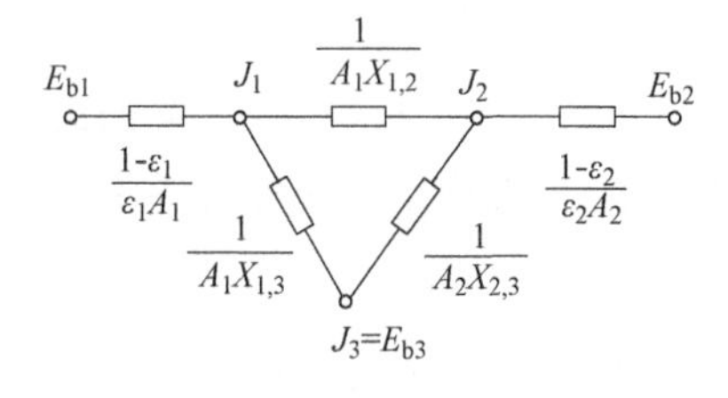

Fig. 9. 19 The system network diagram with surface 3 as a black body

图 9. 19 表面 3 为黑体的系统网络图

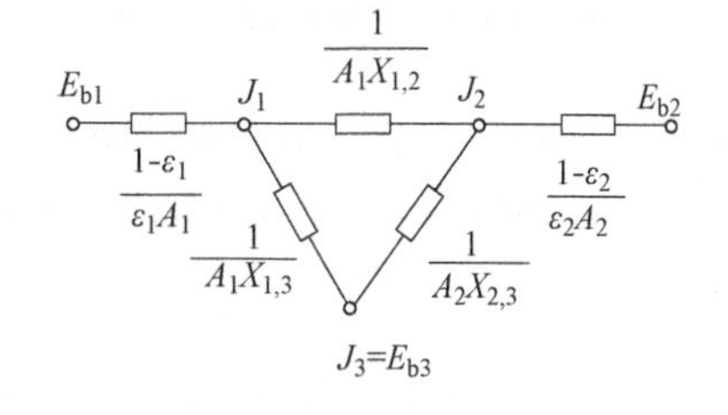

Fig. 9. 20 The system network diagram with surface 3 as an adiabatic surface

图 9. 20 表面 3 为绝热表面的系统网络图

It will affect the radiation heat transfer of other surfaces due to the directivity of heat radiation. The surface with undetermined temperature and zero net radiation heat transfer is called as the re-radiating surface. Another way of the network diagram is shown in Fig. 9. 21 at this time.

由于热辐射具有方向性，因此它会影响其他表面的辐射传热。这种表面温度未定而净辐射传热量为零的表面，称为重辐射面。此时网络图的另一种表示方法如图 9. 21 所示。

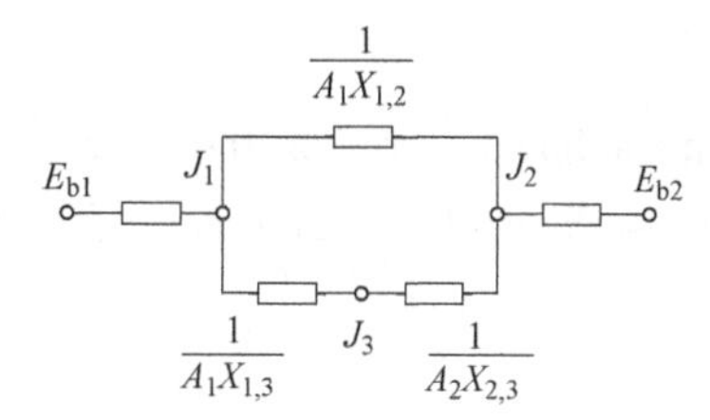

Fig. 9. 21 Another way of the equivalent network diagram

图 9. 21 等效网络图的另一种表示形式

In this case, the net radiation heat transfer quantity between surfaces 1 and 2 is calculated as Eq. (9. 32):

此种情况下，表面 1、2 之间净辐射传热量按式 (9. 32) 计算：

$$\Phi_{1,2}=\frac{E_{b1}-E_{b2}}{\Sigma R_t} \tag{9.32}$$

The total thermal resistance can be calculated similarly to series and parallel resistors by Eq. (9. 33):

类似于串、并联电阻，此时总热阻可用式 (9. 33) 进行计算：

$$\Sigma R_t=\frac{1-\varepsilon_1}{\varepsilon_1A_1}+\frac{1-\varepsilon_2}{\varepsilon_2A_2}+R_{eq} \tag{9.33}$$

According to the electrical principle, the equivalent resistance R_{eq} of the parallel circuit is calculated as follows:

按电学原理，并联电路的等效电阻R_{eq}计算公式为：

$$\frac{1}{R_{eq}}=\frac{1}{\frac{1}{A_1X_{1,2}}}+\frac{1}{\frac{1}{A_1X_{1,3}}+\frac{1}{A_2X_{2,3}}} \tag{9.34}$$

There are often the re-radiating surfaces in engineering radiation calculations. For example, the refractory furnace wall with good insulation in electric furnaces and heating furnaces is this kind of thermal insulation surface.

在工程辐射计算中，常会遇到有重辐射面的情形，比如电炉及加热炉中，保温很好的耐火炉墙就是这种绝热表面。

Summary This section illustrates the network method for calculating the radiation heat transfer of three-surface closed systems.

小结 本节阐述了三表面封闭系统的辐射传热计算的网络法。

Question Compared with the black body surface, the re-radiating surface also has $J=E_b$. Does this mean that the re-radiating surface has the same properties as the black body?

思考题 重辐射面与黑体表面相比，均有$J=E_b$。这是否意味着重辐射面与黑体表面具有相同的性质？

Self-tests

自测题

1. What is the re-radiating surface? What is surface radiation thermal resistance? What is spatial radiation thermal resistance?

1. 什么是重辐射面？什么是表面辐射热阻？什么是空间辐射热阻？

2. Briefly describe the calculation method for solving the radiation heat exchange of a three-surface closed system using the network method.

2. 简述采用网络法求解三表面封闭系统辐射传热的计算方法。

授课视频

9.7 Radiation Heat Transfer Calculation for Multi-surface Closed Systems

9.7 多表面封闭系统的辐射传热计算

Example 9.4 Two parallel plates with a size of 1m×2m and a spacing of 1m are placed in a large factory building at room temperature $t_3=27$℃. The back of the plates does

例 9.4 两块尺寸均为1m×2m，间距为1m的平行平板，置于室温$t_3=27$℃的大厂房内。平板背面不参

not transfer heat. The temperature and emissivity of the two plates are known to be $t_1 = 827℃$, $t_2 = 327℃$, $\varepsilon_1 = 0.2$ and $\varepsilon_2 = 0.5$, respectively. Try to calculate the net radiation heat dissipation of each plate and the radiation heat obtained by the factory wall.

与传热。已知两板的温度和发射率分别为$t_1 = 827℃$、$t_2 = 327℃$和$\varepsilon_1 = 0.2$、$\varepsilon_2 = 0.5$。试计算每块板的净辐射散热量及厂房墙壁所得到的辐射热量。

Solution: This question is about the radiation heat transfer among three gray surfaces. Because the surface area, A_3 of the factory wall is very large and the surface thermal resistance can be taken as zero, $J_3 = E_{b3}$ is a known quantity and its equivalent network diagram is shown in Fig. 9. 22

解：本题是三个灰表面间的辐射传热问题。因厂房墙壁表面面积A_3很大，其表面热阻可取为零，因此$J_3 = E_{b3}$是个已知量，其等效网络图如图9.22所示。

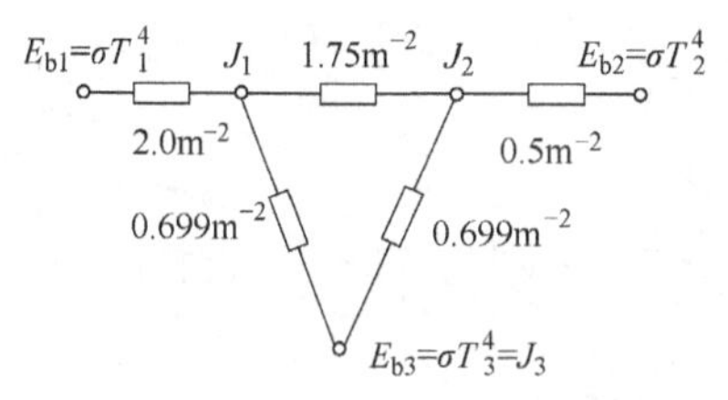

Fig. 9. 22 Equivalent network diagram

图9.22 等效网络图

According to the geometric characteristics such as the size and spacing of the two parallel plates 1, 2 given in the question, we get $X/H = 2$ and $Y/H = 1$. By the angle factor diagram between two parallel rectangular surfaces, we can find $X_{1,2} = X_{2,1} = 0.285$, and $X_{1,3} = X_{2,3} = 1 - X_{1,2} = 1 - 0.285 = 0.715$.

根据题中给定的两平行平板1、2的尺寸及间距等几何特性，有$X/H = 2$、$Y/H = 1$。由两平行长方形表面间的角系数图9.23，可以查出$X_{1,2} = X_{2,1} = 0.285$，而$X_{1,3} = X_{2,3} = 1 - X_{1,2} = 1 - 0.285 = 0.715$。

Where surface thermal resistance is:

其中，表面热阻为：

$$\frac{1-\varepsilon_1}{\varepsilon_1 A_1} = \frac{1-0.2}{0.2\times 2m^2} = 2.0m^{-2}$$

$$\frac{1-\varepsilon_2}{\varepsilon_2 A_2} = \frac{1-0.5}{0.5\times 2m^2} = 0.5m^{-2}$$

Spatial thermal resistance is:

空间热阻为：

$$\frac{1}{A_1 X_{1,2}} = \frac{1}{2m^2\times 0.285} = 1.75m^{-2}$$

$$\frac{1}{A_1 X_{1,3}} = \frac{1}{2m^2\times 0.715} = 0.699m^{-2}$$

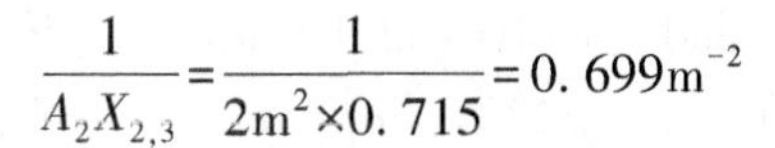

$$\frac{1}{A_2X_{2,3}}=\frac{1}{2\text{m}^2\times0.715}=0.699\text{m}^{-2}$$

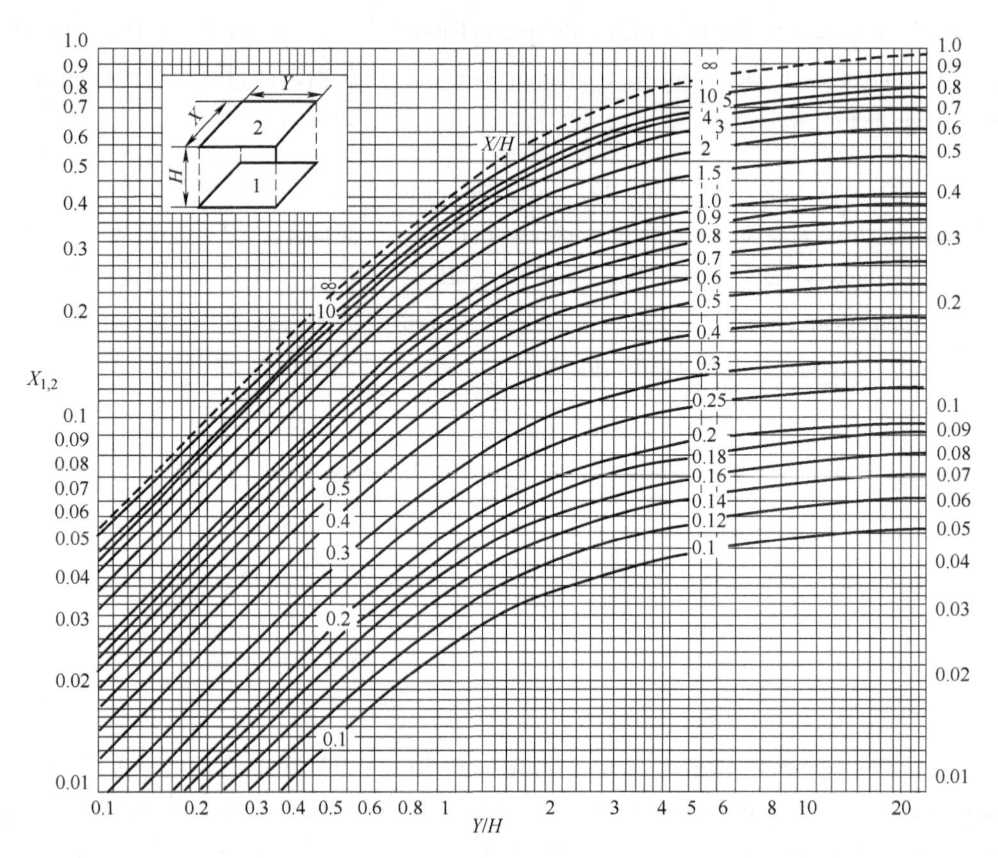

Fig. 9.23 The angle factor between two parallel rectangular surfaces

图 9.23 两平行长方形表面间的角系数

The above thermal resistance values have been marked in the network diagram. Then Kirchhoff's law of DC circuits is applied to nodes to get:

以上各热阻的数值已标出在网络图中，然后对节点应用直流电路的基尔霍夫定律，得：

$$J_1: \quad \frac{E_{b1}-J_1}{2\text{m}^{-2}}+\frac{J_2-J_1}{1.75\text{m}^{-2}}+\frac{E_{b3}-J_1}{0.699\text{m}^{-2}}=0$$

$$J_2: \quad \frac{J_1-J_2}{1.75\text{m}^{-2}}+\frac{E_{b3}-J_2}{0.699\text{m}^{-2}}+\frac{E_{b2}-J_2}{0.5\text{m}^{-2}}=0$$

In the above two equations, there are only two unknown quantities J_1 and J_2. Next, calculate the black body emissive power E_{b1}, E_{b2} and E_{b3} of three surfaces:

在上述两个方程式中，只有两个未知量 J_1、J_2。接下来，计算出三个表面的黑体辐射力 E_{b1}、E_{b2}、E_{b3}为：

$$E_{b1}=C_0\left(\frac{T_1}{100}\right)^4=5.67\text{W}/(\text{m}^2\cdot\text{K}^4)\times\left(\frac{1100}{100}\text{K}\right)^4=83.01\times10^3\text{W}/\text{m}^2=83.01\text{kW}/\text{m}^2$$

$$E_{b2}=C_0\left(\frac{T_2}{100}\right)^4=5.67\text{W}/(\text{m}^2\cdot\text{K}^4)\times\left(\frac{600}{100}\text{K}\right)^4=7.348\times10^3\text{W}/\text{m}^2=7.348\text{kW}/\text{m}^2$$

$$E_{b3}=C_0\left(\frac{T_3}{100}\right)^4=5.67\text{W}/(\text{m}^2\cdot\text{K}^4)\times\left(\frac{300}{100}\text{K}\right)^4=459\text{W}/\text{m}^2=0.459\text{kW}/\text{m}^2$$

Substitute into the above equations about J_1 and J_2, and solving them simultaneously to get:

代入上述关于 J_1、J_2 的方程式中，联立求解得：

$$J_1=18.33\text{kW}/\text{m}^2,\quad J_2=6.437\text{kW}/\text{m}^2$$

Thus, the radiative heat transfer quantity of plate 1 is:

于是，板 1 的辐射传热量为：

$$\Phi_1=\frac{E_{b1}-J_1}{\dfrac{1-\varepsilon_1}{\varepsilon_1 A_1}}=\frac{83.01\times10^3-18.33\times10^3}{2}\text{W}=32.34\times10^3\text{W}=32.34\text{kW}$$

The radiative heat transfer of plate 2 is:

板 2 的辐射传热量为：

$$\Phi_2=\frac{E_{b2}-J_2}{\dfrac{1-\varepsilon_2}{\varepsilon_2 A_2}}=\frac{7.348\times10^3-6.437\times10^3}{0.5}\text{W}=1.822\times10^3\text{W}=1.822\text{kW}$$

The radiative heat transfer quantity of the factory wall is:

厂房墙壁的辐射传热量为：

$$\Phi_3=\frac{E_{b3}-J_1}{0.699\text{m}^{-2}}+\frac{E_{b3}-J_2}{0.699\text{m}^{-2}}=-\left(\frac{E_{b1}-J_1}{2\text{m}^{-2}}+\frac{E_{b2}-J_2}{0.5\text{m}^{-2}}\right)=-(\Phi_1+\Phi_2)$$
$$=-(32.34\times10^3\text{W}+1.822\times10^3\text{W})=-34.16\text{kW}$$

Discussion: In this example, the net heat transfer of surfaces 1 and 2 are both positive, indicating that both surfaces release heat to the environment. According to the law of energy conservation, this energy must be absorbed by the wall. The negative sign in the above result indicates this physical meaning.

讨论： 本题中表面 1 和表面 2 的净传热量均为正值，说明两个表面都向环境放出了热量。按照能量守恒定律，这份能量一定被墙壁所吸收。上述结果中的负号就表示了这一物理意义。

Example 9.5 In Example 9.4, assume that the wall of the large factory is a re-radiating surface. When other conditions remain unchanged, try to calculate the net radiation heat dissipation of the higher temperature surface.

例 9.5 假定例 9.4 中大房间的墙壁为重辐射表面，在其他条件不变时，试计算温度较高表面的净辐射散热量。

Solution: The difference between this example and Example 9.4 is that the wall of the factory is regarded as an insulating surface. Therefore, the wall of the room cannot

解： 本例与例 9.4 的区别在于把房间墙壁看成是绝热表面。于是，房间墙壁不能把热量传向外界，其辐

transfer heat to the outside. The radiation network diagram is as shown in Fig. 9. 21. As other conditions remain unchanged, each thermal resistance value, and values of E_{b1} and E_{b2} in the above example are still valid in this example.

射网络如图 9. 21 所示。因其他条件不变，上例中各热阻值及 E_{b1} 和 E_{b2} 值在本例中仍然有效。

$$R_1=\frac{1-\varepsilon_1}{\varepsilon_1 A_1}=2\text{m}^{-2}, R_2=\frac{1-\varepsilon_2}{\varepsilon_2 A_2}=0.5\text{m}^{-2}$$

$$R_{1,2}=\frac{1}{A_1 X_{1,2}}=1.75\times10^{-2}\text{m}^{-2}, R_{1,3}=\frac{1}{A_1 X_{1,2}}=0.699\text{m}^{-2}, R_{2,3}=R_{1,3}=0.699\text{m}^{-2}$$

$$E_{b1}=83.01\text{kW/m}^2, E_{b2}=7.348\text{kW/m}^2$$

The equivalent resistance of the series and parallel circuits is:

串、并联电路部分的等效电阻为：

$$\frac{1}{R_{eq}}=\frac{1}{R_{1,2}}+\frac{1}{R_{2,3}+R_{1,3}}=\frac{1}{1.75\text{m}^{-2}}+\frac{1}{0.699\text{m}^{-2}+0.699\text{m}^{-2}}=1.29\text{m}^2$$

$$R_{eq}=\frac{1}{1.29\text{m}^2}=0.78\text{m}^{-2}$$

The total resistance between E_{b1} and E_{b2} is:

在 E_{b1} 和 E_{b2} 之间的总阻值为：

$$\sum R=R_1+R_{eq}+R_2=(2+0.78+0.5)\text{m}^{-2}=3.28\text{m}^{-2}$$

Finally, we can get that the net radiation heat dissipation of the higher temperature surface is:

最后可以得到，温度较高的表面的净辐射散热量为：

$$\Phi_{1,2}=\frac{E_{b1}-E_{b2}}{\sum R}=\frac{83.01\times10^3\text{W/m}^2-7.348\times10^3\text{W/m}^2}{3.28\text{m}^{-2}}$$
$$=23.07\times10^3\text{W}=23.07\text{kW}$$

Discussion: After surface 3 is changed to a re-radiating surface, the heat transfer situation has important changes. First, the net heat transfer of the high-temperature surface 1 is reduced by 29%. Second, surface 2 is a net heat-releasing surface in the above example, but now it becomes a net heat-absorbing surface. When calculating the radiation heat transfer of a multi-surface system, it is necessary to be careful to determine whether one of the surfaces is a re-radiating surface. From the point of view of mathematical and physical modeling, this is equivalent to giving the thermal boundary conditions correctly.

讨论： 表面 3 改为重辐射面后，传热情况发生了重要变化。首先高温表面 1 的净传热量减少了 29%。其次表面 2 在上例中也是一个净放热表面，而此时变成了一个净吸热表面。在进行多表面系统辐射传热计算时，确定其中某个表面是否为重辐射面必须慎重。从数学和物理建模的角度看，这相当于要正确地给出热边界条件。

Summary This section illustrates the methods and steps of calculating the radiation heat transfer of a multi-surface system by using the network method in two special cases.

小结 本节阐述了两种特殊情况下采用网络法计算多表面系统的辐射传热的方法和步骤。

Question What practical role does the network method have in the calculation of radiation heat transfer in multi-surface systems?

思考题 网络法对于多表面系统的辐射传热计算有什么实际作用?

Self-tests

自测题

1. For a multi-surface system with known temperature, try to summarize the basic steps for solving the net radiation heat transfer of each surface.

1. 对于温度已知的多表面系统,试总结求解每一表面净辐射传热量的基本步骤。

2. What are the conditions for determining a surface as a re-radiating surface in the calculation of radiation heat transfer in a multi-surface system?

2. 在多表面系统辐射传热计算时,确定某个表面为重辐射面的条件是什么?

授课视频

9.8 The Characteristics of Gas Radiation

9.8 气体辐射的特点

In the common temperature range in industry, the diatomic gases with symmetrical molecular structure, such as air, oxygen, hydrogen, and nitrogen etc., have no ability to emit and absorb radiation energy and can be considered as a transparent body of thermal radiation. But the triatomic or polyatomic gases, such as CO_2, water vapor, ozone, sulfur dioxide, methane, chlorofluorocarbons and CO with asymmetric molecular structure, have a strong ability to absorb and emit heat radiation.

在工业上常见的温度范围内,分子结构对称的双原子气体(如空气、氧气、氢气、氮气等)并无发射和吸收辐射能的能力,可认为是热辐射的透明体。但是对于 CO_2(二氧化碳)、水蒸气、臭氧、二氧化硫、甲烷、氯氟烃等三原子、多原子气体,以及分子结构不对称的 CO(一氧化碳),却具有很强的吸收和发射热辐射的本领。

Gas radiation is different from solid and liquid radiation, and it has the following characteristics:

气体辐射不同于固体和液体辐射,它具有以下特点:

The first characteristic is that gas radiation has a strong selectivity to wavelength. Gases are radiative only in certain

第一个特点,气体辐射对波长有强烈的选择性,气体只在某些波长区

wavelength ranges and correspondingly they have absorption ability only in the same wavelength ranges. Usually, this radiant wavelength range is called optical band. Outside the optical band, the gas neither radiates nor absorbs and shows the properties of a transparent body to thermal radiation.

段内具有辐射能力，相应地也只在同样的波长区段内才具有吸收能力，通常把这种有辐射能力的波长区段称为光带。在光带以外，气体既不辐射，也不吸收，对热辐射呈现透明体的性质。

For example, ozone can absorb almost all ultraviolet rays with a wavelength less than 0. 3μm and also has strong absorption effect for rays with a wavelength between 0. 3μm and 0. 4μm. Therefore, the ozone in the atmosphere can protect humans from the damage of ultraviolet rays.

例如，臭氧几乎能全部吸收波长小于 0. 3μm 的紫外线，对波长在 0. 3~0. 4μm 范围内的射线也有较强的吸收作用。因而，大气层中的臭氧能保护人类不受紫外线伤害。

Carbon dioxide has three main optical bands, while water vapor has three main optical bands, as shown in Fig. 9. 24. It can be seen that these optical bands are located in the infrared wavelength range and the optical bands of carbon dioxide and water vapor overlap in two places. Because gases have the characteristic of selective absorption of radiation wavelength, gases are not gray bodies.

如图 9. 24 所示，二氧化碳的主要光带有三段，水蒸气的主要光带也有三段。可以看出，这些光带均位于红外线的波长范围，而且二氧化碳和水蒸气的光带有两处是重叠的。由于气体对辐射波长具有选择性吸收的特点，所以气体不是灰体。

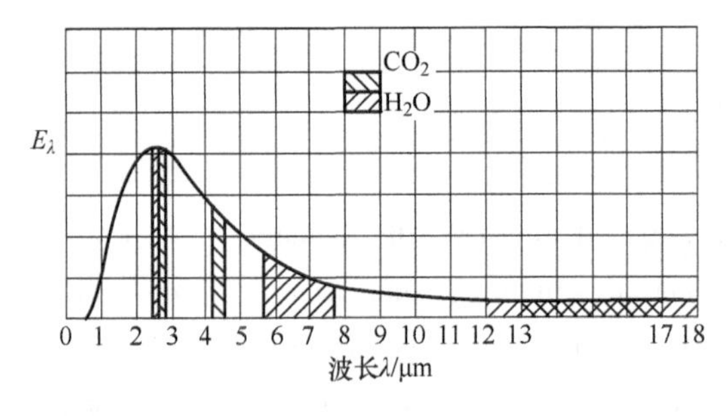

Fig. 9. 24 The main optical bands of carbon dioxide and water vapor

图 9. 24 二氧化碳和水蒸气的主要光带

The second characteristic is that the radiation and absorption of gases take place in the entire volume of space, while the radiation and absorption of solids and liquids take place on the surface. In terms of absorption, the radiation energy projected on the interface of the gas layer is absorbed and decayed during the radiation stroke. In terms of radiation, the radiation received on the interface of the gas layer is gas radiation of the entire volume.

第二个特点，气体的辐射和吸收是在整个容积空间中进行的，而固体和液体的辐射及吸收都在表面进行。就吸收而言，投射到气体层界面上的辐射能要在辐射行程中被吸收而衰减；就辐射而言，气体层界面上所接收到的辐射为整个容积的气体辐射。

All this shows that the radiation and absorption of gas take

这都说明，气体的辐射和吸收是在

place in the entire volume, which is related to the shape and volume of the container. Therefore, referring to the emissivity and absorptivity of gas, we also have to explain the shape and size of the container where the gas resides besides other conditions. When the thermal radiation enters the absorptive gas layer, it is decayed by gas absorption along the way. The extent of attenuation depends on the radiation intensity and the number of gas molecules encountered along the way. The number of gas molecules is related to the ray stroke length and gas density ρ.

整个容积中进行的，与容器的形状和容积有关。因此，论及气体的发射率和吸收比，除其他条件外，还必须说明气体所处容器形状和容积大小。当热辐射进入吸收性气体层时，因沿途被气体吸收而削弱，削弱的程度取决于辐射强度及途中所碰到的气体分子数目。气体分子数目与射线行程长度及气体密度 ρ 有关。

As shown in Fig. 9. 25, examine the decay law of the spectral radiation with wavelength λ in the gas. Assuming that the intensity of the spectral radiation projected to the gas interface at $x=0$ is $I_{\lambda,0}$, and the radiation intensity becomes $I_{\lambda,x}$ after a distance of x. After passing through the micro-element gas layer dx, its attenuation is $\mathrm{d}I_{\lambda,x}$. It has been proved theoretically that $\frac{\mathrm{d}I_{\lambda,x}}{I_{\lambda,x}}$ is proportional to the stroke dx. Set the proportional coefficient is k_λ, so we can get Eq. (9. 35), where the negative sign indicates absorption.

如图 9. 25 所示，考察波长为 λ 的光谱辐射在气体内的削弱规律。假设投射到气体界面 $x=0$ 处的光谱辐射强度为 $I_{\lambda,0}$，通过一段距离 x 后，该辐射强度变为 $I_{\lambda,x}$。再通过微元气体层 dx 后，其衰减量为 $\mathrm{d}I_{\lambda,x}$。理论上已经证明$\frac{\mathrm{d}I_{\lambda,x}}{I_{\lambda,x}}$与行程 d$x$ 正比。设比例系数为 k_λ，则有式（9. 35）。式中负号表示吸收。

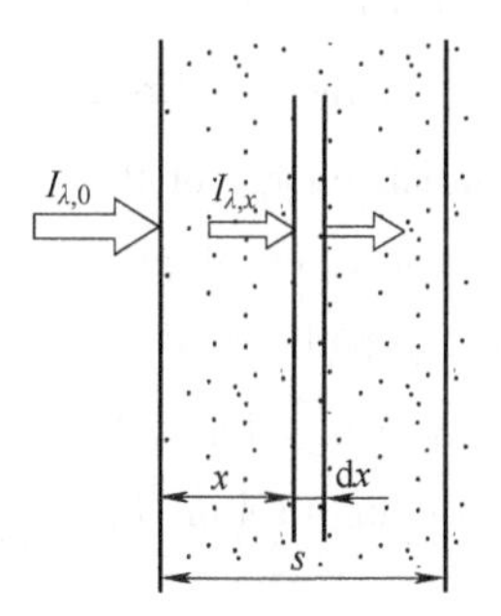

Fig. 9. 25 Attenuation of spectral radiation as it passes through the gas layer

图 9. 25 光谱辐射通过气体层时的衰减情况

$$\frac{\mathrm{d}I_{\lambda,x}}{I_{\lambda,x}}=-k_\lambda\,\mathrm{d}x \tag{9. 35}$$

Where k_λ is the spectral attenuation coefficient, and its unit is m^{-1}. k_λ depends on the gas type, density and heat radiation wavelength. Integrate Eq. (9. 35) to get the spectral radiant intensity:

式中，k_λ 是光谱减弱系数（m^{-1}），k_λ 取决于气体的种类、密度及热辐射波长。对式（9. 35）积分可得光谱辐射强度为：

$$\int_{I_{\lambda,0}}^{I_{\lambda,s}} \frac{dI_{\lambda,x}}{I_{\lambda,x}} = -\int_0^s k_\lambda dx$$

$$I_{\lambda,s} = I_{\lambda,0} e^{-k_\lambda s} \tag{9.36}$$

This formula is called Beer's law, where s is the path length of the radiation ray often called the ray path length, which depends on the shape and size of the container. This formula shows that its intensity decays exponentially when the spectral radiation travels in the absorptive gas. The Beer formula can be rewritten as:

该式称为贝尔定律，式中，s 是辐射射线通过的路程长度，常称为射线程长，取决于容器的形状和大小。该式表明光谱辐射在吸收性气体中传播时，强度按指数规律衰减。贝尔公式可改写为：

$$\frac{I_{\lambda,s}}{I_{\lambda,0}} = e^{-k_\lambda s} = \tau(\lambda, s) \tag{9.37}$$

This formula is the monochromatic transmissivity of the gas layer with a thickness of s. For pure gas, if the reflectivity is zero, then the absorptivity is:

此式是厚度为 s 的气体层的单色穿透比。对于纯净气体，反射比为零，则吸收比为：

$$\alpha(\lambda, s) = 1 - \tau(\lambda, s) = 1 - e^{-k_\lambda s} \tag{9.38}$$

According to Kirchhoff's law, the spectral emissivity of the gas layer is:

根据基尔霍夫定律，气体层的光谱发射率为：

$$\varepsilon(\lambda, s) = \alpha(\lambda, s) = 1 - e^{-k_\lambda s} \tag{9.39}$$

The most concern in engineering is to determine the sum of the radiation energy of the gas in all optical bands. Then, it is necessary to determine the emissivity ε_g of the gas first, then use the formula to calculate the radiation energy of the gas. Due to the volumetric radiation characteristics of gas, the emissivity ε_g is closely related to the ray path length s, which depends on the shape and size of the gas volume. As shown in Fig. 9.26, the ray path length radiating to A or B is different from different directions.

工程中最关心的是确定气体在所有光带内辐射能的总和。于是需要首先确定气体的发射率 ε_g，然后用公式计算气体的辐射能。由于气体的容积辐射特性，发射率 ε_g 与射线程长 s 关系密切，而 s 取决于气体容积的形状和尺寸。如图 9.26 所示，从不同方向辐射到 A 或 B 处的射线程长各不相同。

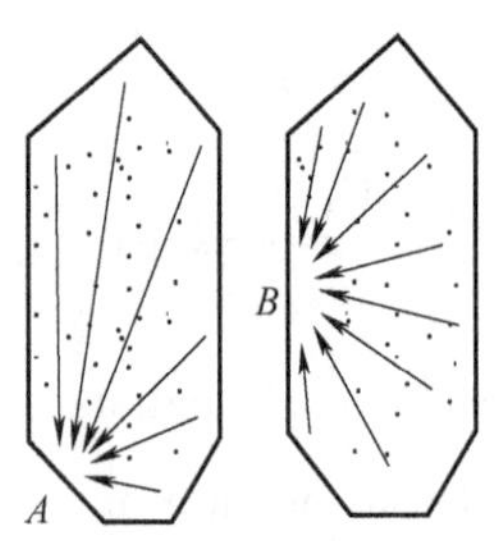

Fig. 9.26 Radiation of gases to different regions

图 9.26 气体对不同区域的辐射

In order to make the ray path length uniform, the concept of equivalent hemisphere is introduced. And the volume that is not spherical is equivalent to hemisphere, then its radius R is the equivalent ray path length, as shown in Fig. 9. 27. The average ray path length of gas of some typical geometric volume to the entire envelope wall is listed in the following Table 9. 1. In the absence of data, the average ray path length of gas of any geometric shape to the entire envelope wall can be calculated as follows.

为了使射线程长均匀，引入当量半球的概念，将不是球形的容积等效为半球，则其半径 R 就是等效的射线程长，如图 9.27 所示。一些典型几何容积的气体，对整个包壁的平均射线程长列于表 9.1 中。在缺少资料的情况下，任意几何形状气体对整个包壁辐射的平均射线程长可按下式计算。

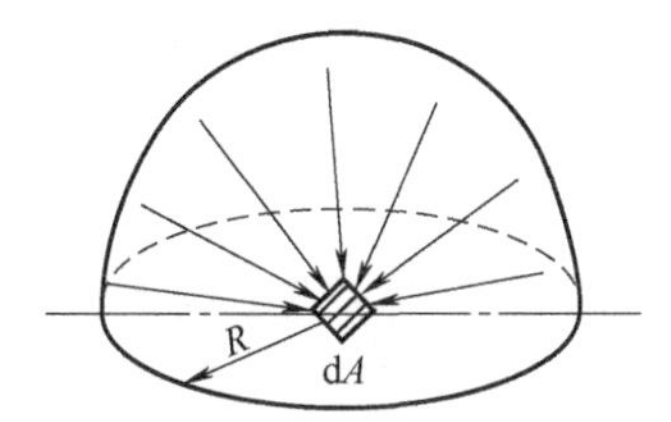

Fig. 9. 27 Radiation of the gas in the hemisphere to the center of the sphere

图 9. 27 半球体中的气体向球体中心的辐射

Table 9. 1 Average ray path length of gas radiation

表 9. 1 气体辐射平均射线程长

气体容积的形状	特性尺度	受到气体辐射的位置	平均射线程长
球	直径 d	整个包壁或壁上的任何地方	$0.6d$
立方体	边长 b	整个包壁	$0.6b$
高度等于直径的圆柱体	直径 d	底面圆心 整个包壁	$0.77d$ $0.6d$
两无限大平行平板之间	平板间距 H	平板	$1.8H$
无限长圆柱体	直径 d	整个包壁	$0.9d$

$$s=3.6\frac{V}{A} \tag{9.40}$$

Where V is the gas volume with the unit of m^3, and A is the envelope wall area with the unit of m^2.

式中，V 是气体容积（m^3）；A 是包壁面积（m^2）。

Summary This section illustrates the characteristics and the calculation of gas radiation.

小结 本节阐述了气体辐射的特点和计算。

Question Can the gas be treated as a gray body? Please explain the reason.

思考题 气体能当灰体来处理吗？请说明原因。

Self-tests

自测题

1. What are the main differences in the radiation characteristics between gases and general solids?

2. Is it meaningful to talk about the emissivity and absorptivity of a gas apart from the geometric space in which the gas is located?

1. 气体与一般固体比较，其辐射特性有什么主要差别?

2. 离开了气体所处的几何空间而谈论气体的发射率与吸收比有没有实际意义?

授课视频

9.9 Control of Radiation Heat Transfer

9.9 辐射传热的控制

The enhancing and weakening of heat transfer are an important task in the study of heat transfer. From the radiation heat transfer in a closed system composed of two surfaces, we can know that there are two main ways to enhance radiation heat transfer: ① Increase the emissivity; ② Increase the angle factor. There are three main ways to weaken radiation heat transfer: ① Reduce the emissivity; ② Reduce the angle factor; ③ Add the heat shield.

传热的强化和削弱是传热学研究的重要问题。通过两表面组成的封闭系统辐射传热可知，强化辐射传热的主要途径有两种：① 提高发射率；② 增大角系数。削弱辐射传热的主要途径有三种：① 降低发射率；② 减小角系数；③ 加入遮热板。

This section mainly discusses the ways to weaken radiation heat transfer. Among them, inserting a heat shield is a common means to weaken radiation heat transfer. The so-called heat shield refers to a thin plate inserted between two radiation heat exchange surfaces to weaken the radiation heat exchange. Inserting a heat shield is equivalent to reducing the surface emissivity.

本节主要讨论削弱辐射传热的方式。其中，插入遮热板是削弱辐射传热常用的一种手段。所谓遮热板，是指插入两个辐射传热表面之间，以削弱辐射传热的薄板。插入遮热板相当于降低了表面发射率。

In order to explain the working principle of the heat shield, the following analysis focuses on the changes in radiation heat transfer caused by inserting a metal sheet between two parallel plates. Assuming that the flat plates and the metal sheet are gray bodies and their absorptivity is equal to the emissivity. As studied before, we know that for the closed system composed of two infinite plates 1 and 2, the heat transfer quantity is:

为了说明遮热板的工作原理，下面分析在两平行平板之间插入一块金属薄板所引起的辐射传热的变化。假设平板和金属薄板都是灰体，并且它们的吸收比等于发射率。由前面的学习知道，对于两个无限大平板 1、2 组成的封闭系统，其传热量为：

$$\Phi_{1,2}=\frac{E_{b1}-E_{b2}}{\frac{1-\varepsilon_1}{A_1\varepsilon_1}+\frac{1}{A_1X_{1,2}}+\frac{1-\varepsilon_2}{A_2\varepsilon_2}}$$

Assuming that the area of the two plates is equal, the emissivity is equal to ε, and the angle coefficient is 1, then the radiation heat flux and system emissivity between the plates are as follows:

假设两平板的面积相等、发射率均为 ε、角系数均为 1，则平板间的辐射传热流量和系统发射率如下：

$$q_{1,2}=\frac{E_{b1}-E_{b2}}{\frac{1}{\varepsilon_1}+\frac{1}{\varepsilon_2}-1}=\varepsilon_s(E_{b1}-E_{b2}) \qquad \varepsilon_s=\frac{1}{\frac{1}{\varepsilon_1}+\frac{1}{\varepsilon_2}-1}$$

Next, we insert a heat shield 3 with emissivity still equal to ε between the two flat plates, as shown in Fig. 9. 28.

接下来在两块平板之间插入一块发射率仍为 ε 的遮热板3，如图9. 28 所示。

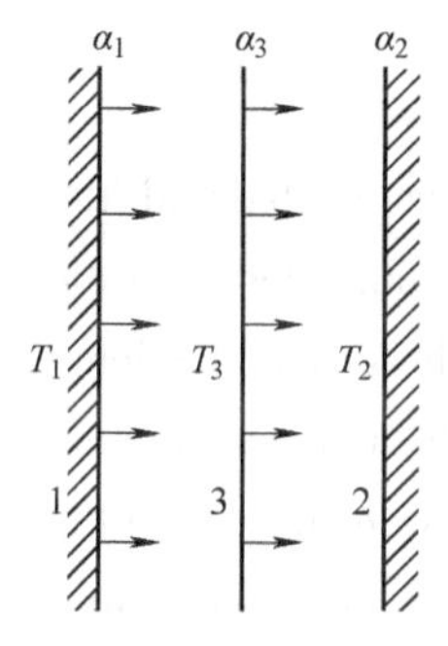

Fig. 9. 28　Schematic diagram of two steady-state heat exchange systems

图 9. 28　两个稳态热交换系统的示意图

In this way, the three plates form two steady-state heat exchange systems. The system emissivity of both systems consisting of surface 1 and 3, and surfaces 3 and 2 is ε_s. The heat fluxes of radiation heat transfer of the two systems consisting of surface 1 and 3, and surfaces 3 and 2 are $q_{1,3}$ and $q_{3,2}$.

这样，三个平板就组成了两个稳态传热系统，表面 1、3 和表面 3、2 两个系统的系统发射率相同，都是 ε_s。表面 1、3 和表面 3、2 两个系统的辐射传热热流量为 $q_{1,3}$和 $q_{3,2}$：

$$q_{1,3}=\varepsilon_s(E_{b1}-E_{b3})$$
$$q_{3,2}=\varepsilon_s(E_{b3}-E_{b2})$$

Under thermal steady-state conditions, $q_{1,2}=q_{1,3}=q_{3,2}$, combining the three equations above we can get:

在热稳态条件下有，$q_{1,2}=q_{1,3}=q_{3,2}$，将上面的三个方程联立可以得到：

$$q_{1,2}=\frac{1}{2}\varepsilon_s(E_{b1}-E_{b2})$$

As a result, compared with the radiation heat transfer without metal sheet, the insertion of the heat shield reduces the

由此得到，与未加金属薄板时的辐射传热相比，插入遮热板使辐射传

radiation heat transfer quantity by half.

热量减少了一半。

In order to make the effect of weakening radiation heat transfer more significant, the metal sheet with low emissivity is used as the heat shield actually. When a heat shield fails to meet the requirements of weakening heat tranfer, a multi-layer heat shield can be used. The heat shield is widely used in engineering. Here are four application instances.

为使削弱辐射传热的效果更为显著，实际上都采用发射率低的金属薄板作为遮热板。当一块遮热板达不到削弱传热的要求时，可以采用多层遮热板。遮热板的工程应用广泛，以下是四个应用实例：

(1) In the steam turbine, it is used to reduce the radiation heat transfer between the inner and outer sleeves. The general structure of the steam inlet pipes of domestically made 300,000kW steam turbine cylinder used at high and medium pressures is shown in Fig. 9. 29.

(1) 在汽轮机中用于减少内外套管间辐射传热。国产 30 万 kW 汽轮机高中压气缸进汽连接管的大致结构如图 9. 29 所示。

The inner sleeve of the steam turbine is connected with the inner cylinder, and the outer sleeve is connected with the outer cylinder. High-temperature steam flows into the inner cylinder through the inner sleeve, and the wall temperature of the inner sleeve is relatively high. In order to reduce the radiation heat transfer between the inner and outer sleeves, a cylindrical heat shield made of stainless steel is placed between them.

汽轮机的内套管与内缸连接，外套管与外缸连接。高温蒸汽经内套管流入内缸，内套管的壁温较高。为减少内外套管间的辐射传热，在其间安置了一个用不锈钢制成的圆筒形遮热罩。

(2) The heat shield is used in cryogenic containers that store liquid gas. The schematic diagram of the container storing liquid nitrogen and liquid oxygen is shown in Fig. 9. 30.

(2) 遮热板应用于储存液态气体的低温容器。储存液氮、液氧的容器示意图如图 9. 30 所示。

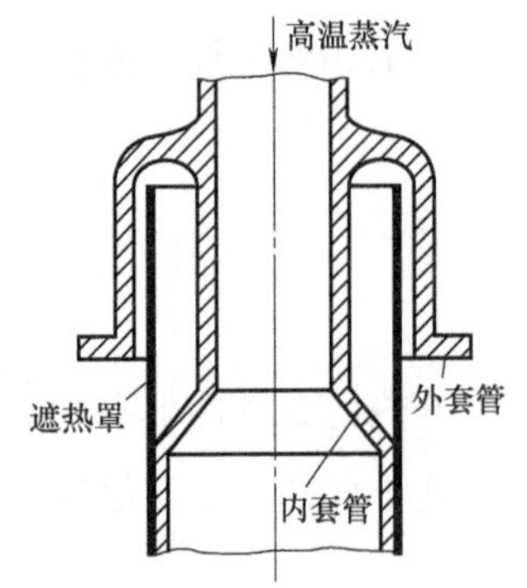

Fig. 9. 29 The heat shield at the steam inlet connection pipe

图 9. 29 进汽连接管处的遮热罩

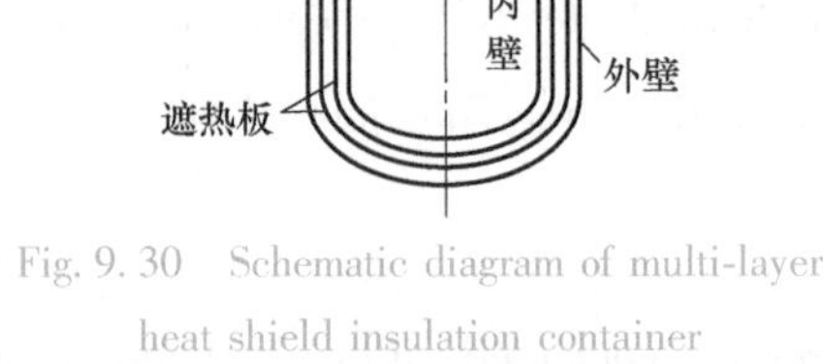

Fig. 9. 30 Schematic diagram of multi-layer heat shield insulation container

图 9. 30 多层遮热板保温容器示意图

In order to improve the insulation effect, the method of using a heat shield and vacuuming is adopted. The heat shield is made of plastic film and is coated with a metal foil layer with a large reflectivity. The thickness of the foil layer is about 0. 01-0. 05mm, and a light weight and low thermal conductivity material is embedded between the foil layers as a separation layer. A high vacuum is drawn from the insulation layer.

为了提高保温效果，采用遮热板并抽真空的方法。遮热板用塑料薄膜制成，其上涂以反射比很大的金属箔层。箔层厚约为 0. 01～0. 05mm，箔间嵌以质轻且导热系数小的材料作为分隔层，绝热层中抽成高度真空。

(3) The heat shield is used for super heat insulation tubing. A large amount of oil in the world is buried kilometer or even thousands of meters below the stratum, and its viscosity is very high. It is necessary to inject high-temperature and high-pressure steam to dilute the oil during mining.

（3）遮热板用于超级隔热油管。世界上有大量石油埋藏于地层下千米乃至数千米处，黏度很大。开采时需注射高温高压蒸汽，以使石油稀释。

In the process of delivering steam to several kilometers below the ground, reducing heat loss is a very important task. The super heat insulation tubing is manufactured by a multi-layer heat shield similar to a low-temperature heat preservation container, and is pumped vacuum. The cross section is shown in Fig. 9. 31.

在将蒸汽输送到地面下数千米处的过程中，减少散热损失是件很重要的工作。超级隔热油管是采用了类似低温保温容器的多层遮热板，并用抽真空的方式制造而成的，截面图如图 9. 31 所示。

(4) The heat shield is used to improve the accuracy of temperature measurement. Fig. 9. 32 shows the temperature measurement of a single-layer heat shield extraction thermocouple.

（4）遮热板用于提高温度测量的准确度。图 9. 32 所示为单层遮热罩抽气式热电偶测温图。

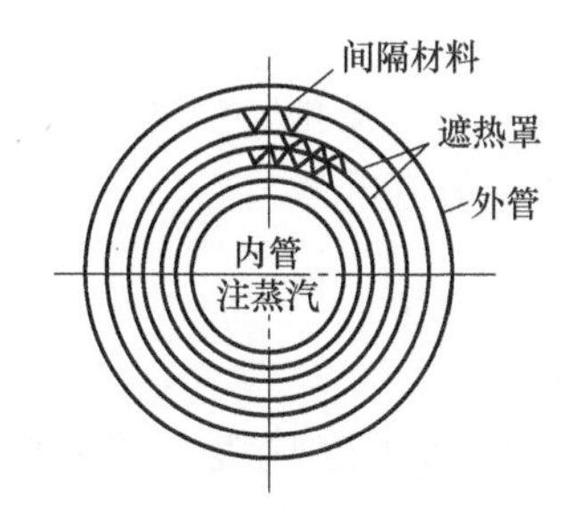

Fig. 9. 31 Multi-layer heat shield super insulation oil pipe

图 9. 31 多层热屏蔽超级绝缘油管

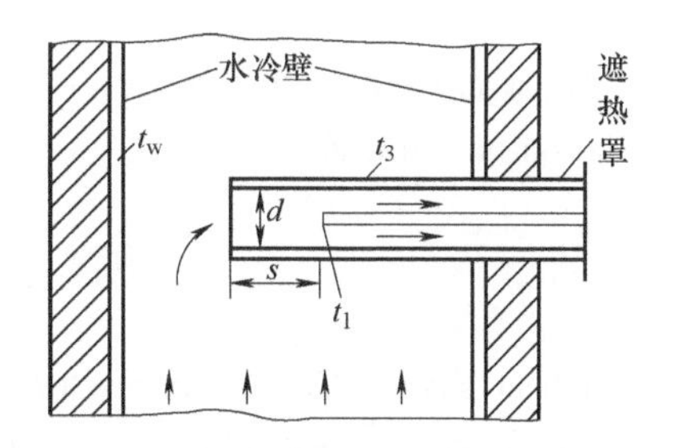

Fig. 9. 32 Single layer heat shield extraction thermocouple temperature measurement schematic

图 9. 32 单层遮热罩抽气式热电偶测温图

If a bare thermocouple is used to measure the temperature of a high-temperature airflow, the high-temperature airflow transfers heat to the thermocouple through heat convection.

如果使用裸露热电偶测量高温气流的温度，高温气流以对流方式把热量传给热电偶，同时热电偶又以辐

At the same time, the thermocouple transfers heat to the lower temperature container wall by thermal radiation. When the convective heat transfer quantity of the thermocouple is equal to its radiation heat dissipation, the temperature of the thermocouple no longer changes. This temperature is the indicating temperature of the thermocouple. The indicating temperature must be lower than the true temperature of the gas, so it will cause temperature measurement errors.

射的方式把热量传给温度较低的容器壁。当热电偶的对流传热量等于其辐射散热量时，热电偶的温度就不再变化。此温度即为热电偶的指示温度。指示温度必低于气体的真实温度，造成测温误差。

When using a heat shield extraction thermocouple, the thermocouple is protected by the heat shield. Radiation heat dissipation is reduced and the extraction effect enhances the convective heat transfer between the gas and the thermocouple. At this time, the indicating temperature of the thermocouple can be closer to the true temperature of the gas, which reduces the temperature measurement error. The effect is more obvious when using a multi-layer heat shield. It is worth pointing out that in order to make the heat shield effectively shield the thermocouple, the distance s between the thermocouple and the port of the heat shield should be greater than $(2-2.2)d$.

使用遮热罩抽气式热电偶时，热电偶在遮热罩保护下辐射散热减少，而抽气作用又增强了气体与热电偶间的对流传热。此时热电偶的指示温度可更接近于气体的真实温度，使测温误差减小。采用多层遮热罩时效果更加明显。值得指出的是，为使遮热罩能对热电偶有效地起到屏蔽作用，热电偶离开遮热罩端口的距离 s 应大于 $(2\sim2.2)d$。

Summary This section illustrates the ways to enhance and weaken radiation heat transfer, and the principle and application of heat shield.

小结 本节阐述了强化和削弱辐射传热的途径，以及遮热板的原理及其应用。

Question What are the main methods to enhance and weaken radiation heat transfer?

思考题 强化和削弱辐射传热的方法主要有哪些？

Self-tests

自测题

1. What is a heat shield? Try to analyze the principle of heat shield and its effect in weakening radiation heat transfer.

1. 什么是遮热板？试分析遮热板的原理及其在削弱辐射传热中的作用。

2. What are the causes of measurement error when using a thermocouple to measure the temperature of high temperature gas? What can be done to reduce the measurement error?

2. 用热电偶测量高温气体温度时，产生测量误差的原因是什么？可以采取什么措施来减小测量误差？

Exercises

习题

9.1 There are two tiny areas, A_1 and A_2, as shown in Fig. 9.33, $A_1 = 2\times10^{-4}\,m^2$, and $A_2 = 3\times10^{-4}\,m^2$. A_1 is a diffuse surface with the radiation force $E_1 = 5\times10^4\,W/m^2$. Try to calculate the radiation energy emitted by A_1 and falling on A_2.

9.1 设有如图 9.33 所示的两个微小面积 A_1、A_2，$A_1 = 2\times10^{-4}\,m^2$，$A_2 = 3\times10^{-4}\,m^2$。$A_1$ 为漫反射表面，辐射力 $E_1 = 5\times10^4\,W/m^2$。试计算由 A_1 发出而落到 A_2 上的辐射能。

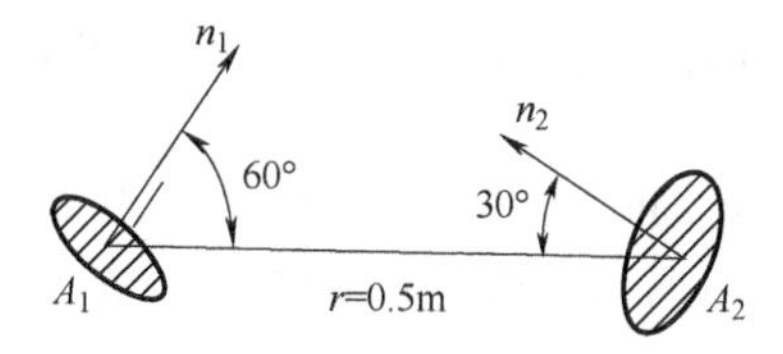

Fig. 9.33 Figure for exercise 9.1

图 9.33 习题 9.1 图

9.2 Try to determine the angle factor $X_{1,2}$ in Fig. 9.34 by a simple method.

9.2 试用简捷方法确定图 9.34 中的角系数 $X_{1,2}$。

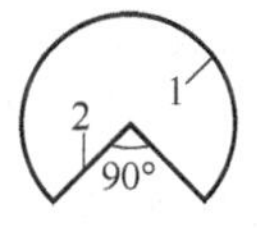

a) 在垂直于纸面方向无限长

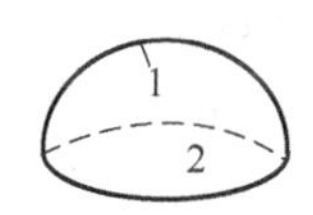

b) 半球内表面与底面

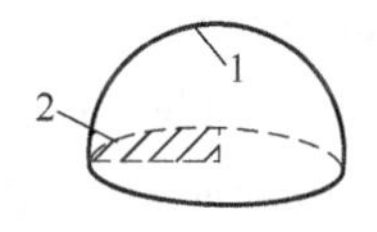

c) 半球内表面与1/4底面

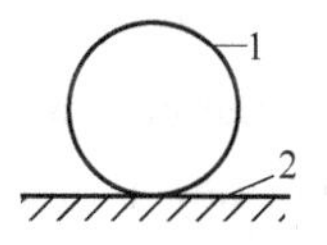

d) 球与无限大平面

Fig. 9.34 Figure for exercise 9.2

图 9.34 习题 9.2 图

9.3 Try to determine the angle factor $X_{1,2}$ of the set structure in Fig. 9.35a and b.

9.3 试确定图 9.35a、b 中几何结构的角系数 $X_{1,2}$。

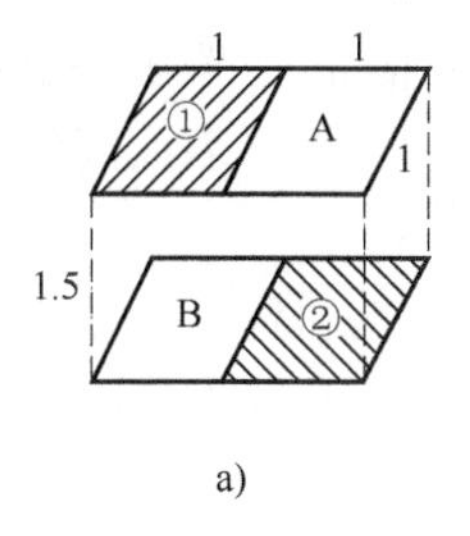

a)

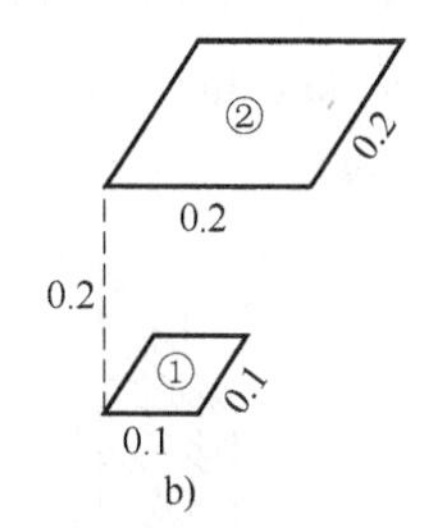

b)

Fig. 9.35 Figures for exercise 9.3

图 9.35 习题 9.3 图

9.4 Two black bodies of equal area are closed in an adiabatic envelope. Assume that the temperatures of the two black bodies are T_1 and T_2, respectively, and their relative positions are arbitrary. Try to draw the network

9.4 两个面积相等的黑体被置于一绝热的包壳中，假定两黑体的温度分别为 T_1 与 T_2，且相对位置是任意的。试画出该辐射传热系统

diagram of the radiation heat transfer system, and derive the expression of the surface temperature T_3 of the adiabatic envelope.

的网络图，并导出绝热包壳表面温度 T_3 的表达式。

9.5 The surface temperatures of two infinite plates are t_1 and t_2, and the emissivity is ε_1 and ε_2, respectively. The emissivity of the heat shield between the two plates is ε_2. Try to draw the network diagram of radiation heat transfer between the three plates under steady state.

9.5 两块无限大平板的表面温度分别为 t_1 及 t_2，发射率分别为 ε_1 及 ε_2，其间遮热板的发射率为 ε_2。试画出稳态时三板之间辐射传热的网络图。

9.6 There is a horizontal square solar collector with a side length of 1.1m, the emissivity of the absorbing surface is $\varepsilon=0.2$, and the absorptivity of solar energy is $\alpha_s=0.9$. The solar irradiation $G=800\text{W/m}^2$, the temperature of the absorbing surface of the collector is 90℃. The ambient temperature is 30℃, and the sky can be regarded as a black body of 23K. Try to determine the efficiency of this collector. The absorbent surface is directly exposed to the air and there is no sandwich on it (the collector efficiency is defined as the ratio of solar radiation absorbed by the collector to solar irradiation).

9.6 有一水平放置的正方形太阳能集热器，边长为 1.1m，吸热表面发射率 $\varepsilon=0.2$，对太阳能的吸收比 $\alpha_s=0.9$，当太阳的投入辐射 $G=800\text{W/m}^2$ 时，测得集热器吸热表面的温度为 90℃。此时环境温度为 30℃，天空可视为 23K 的黑体，试确定此集热器的效率。设吸热表面直接暴露于空气中，其上无夹层（集热器效率定义为集热器所吸收的太阳辐射能与太阳投入辐射之比）。

9.7 It is known that the temperatures of the two perpendicular square surfaces (Fig. 9.36) are $T_1=1000\text{K}$ and $T_2=500\text{K}$, respectively, and their emissivity is $\varepsilon_1=0.6$ and $\varepsilon_2=0.8$ respectively. The two surfaces are located in an adiabatic room. Try to calculate the net radiation heat transfer between the two surfaces.

9.7 已知两个相互垂直的正方形表面（图 9.36）的温度分别是 $T_1=1000\text{K}$、$T_2=500\text{K}$，发射率分别为 $\varepsilon_1=0.6$、$\varepsilon_2=0.8$，该两表面置于一绝热的房间内。试计算该两表面间的净辐射传热量。

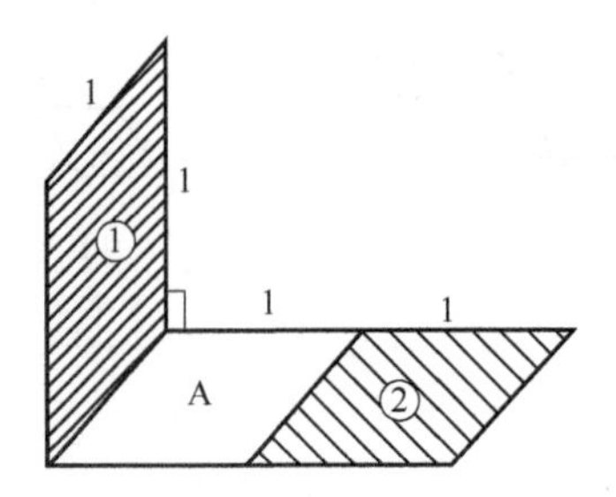

Fig. 9.36 Figure for exercise 9.7

图 9.36 习题 9.7 图

Chapter 10 Analysis of the Heat Transfer Process and Thermal Calculation of Heat Exchanger

第 10 章 传热过程分析与换热器热计算

10.1 Analysis and Calculation Methods of Heat Transfer Process

10.1 传热过程分析与计算方法

We have learned that the heat transfer process refers to the process of heat transfer between two fluids through a solid wall in Chapter 1, which is a specific concept. The basic heat transfer modes of the heat transfer process contain heat conduction, heat convection and heat radiation. In the heat transfer process, the basic equation of heat transfer calculation is:

在第 1 章中已经学过，传热过程是指两流体间通过固体壁面进行的换热过程，是一个特指的概念。传热过程的基本传热方式包含热传导、热对流和热辐射。在传热过程中，传热量计算的基本方程为：

$$\Phi = kA(t_{f1} - t_{f2}) \tag{10.1}$$

Where k is the heat transfer coefficient, also known as the overall heat transfer coefficient. For the heat transfer process of the heat transfer area with different structures, the calculation formula of k is different. Eq. (10.1) assumes that the temperature of two heat transfer fluids is constant, or two temperatures represent the average temperature of two fluids. How to calculate the overall heat transfer coefficient k for the heat transfer process of the heat transfer area with different structures?

式中，k 是传热系数，也称为总传热系数。对于不同结构传热面的传热过程，k 的计算公式不同。式（10.1）假定两种换热流体的温度恒定，或者两个温度代表两种流体的平均温度。对于不同结构传热面的传热过程，传热系数 k 如何计算呢？

For plane walls, the calculation formula of k in the heat transfer process is:

对于平壁，传热过程 k 的计算公式为：

$$k = \frac{1}{\frac{1}{h_1} + \frac{\delta}{\lambda} + \frac{1}{h_2}} \tag{10.2}$$

Two points need to be explained: ① The appropriate calculation correlations of h_1 and h_2 should be selected according to the specific working condition. ② The convective heat transfer coefficient h_t should adopt the equivalent heat transfer coefficient or composite convective heat transfer coefficient when considering radiation heat transfer. h_c represents the convective heat transfer coefficient, and h_r represents the heat transfer coefficient converted from radiation heat transfer. The equivalent heat transfer coefficient h_t is calculated as follows:

需要说明两点：① 式中，h_1 和 h_2 需根据具体工况选择相应的计算关联式；② 考虑辐射传热时，表面传热系数 h_t 应采用等效传热系数，或者复合表面传热系数。用 h_c 表示表面传热系数，h_r 表示辐射传热折算的传热系数。等效传热系数 h_t 计算方法如下：

For single phase convection heat transfer:

对于单相对流传热：

$$h_t = h_c + h_r \tag{10.3}$$

For film boiling heat transfer:

对于膜态沸腾传热：

$$h_t^{4/3} = h_c^{4/3} + h_r^{4/3} \tag{10.4}$$

The conversion formula of radiant heat transfer:

辐射传热的折算公式为：

$$h_r(T_1 - T_2) = \varepsilon\sigma(T_1^4 - T_2^4) \tag{10.5}$$

The radiant heat transfer is converted into convection heat transfer, and the converted heat transfer coefficient h_r can be obtained. The heat transfer calculation of cylindrical walls (i. e. circular tubes) is as shown in Fig. 10. 1. The convective heat transfer quantity inside the tube:

也就是，把辐射传热量折算成对流传热量，可求出折算的传热系数 h_r。对于圆筒壁（即圆管）的传热计算如图 10.1 所示。管内对流传热量为：

$$\Phi = h_i \pi d_i l(t_{f1} - t_{w1}) \tag{10.6}$$

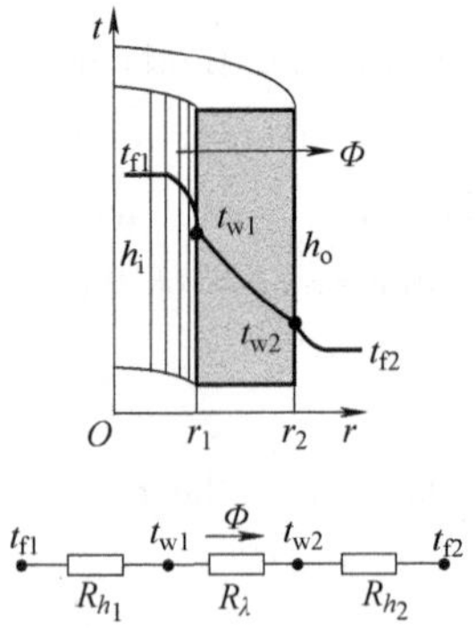

Fig. 10. 1 Heat transfer process through the wall of a circular tube
图 10.1 通过圆管壁的传热过程

The thermal conduction quantity of the tube wall:

管壁的导热量为:

$$\Phi = \frac{t_{w1} - t_{w2}}{\frac{1}{2\pi l \lambda}\ln\left(\frac{d_o}{d_i}\right)} \tag{10.7}$$

The convective heat transfer quantity outside the tube:

管外对流传热量为:

$$\Phi = h_o \pi d_o l (t_{w2} - t_{f2}) \tag{10.8}$$

Then, the partial thermal resistances of the three heat transfer links are:

那么，三个传热环节的分热阻分别是:

$$R_{h_i} = \frac{1}{h_i \pi d_i l}, \quad R_\lambda = \frac{\ln(d_o/d_i)}{2\pi l \lambda}, \quad R_{h_o} = \frac{1}{h_o \pi d_o l}$$

The total heat transfer equation of the heat transfer process of a circular tube is that Φ is equal to the total temperature difference divided by the total thermal resistance of the heat transfer process. Put the area in the numerator and the reciprocal of the denominator is written as k_o, which is the overall heat transfer coefficient.

圆管传热过程的总传热方程是: Φ 等于总温差除以传热过程的总热阻。把面积提到分子上，分母的倒数记作 k_o，也就是总传热系数。

$$\Phi = \frac{\pi d_o l (t_{fi} - t_{fo})}{\frac{d_o}{h_i d_i} + \frac{d_o}{2\lambda}\ln\left(\frac{d_o}{d_i}\right) + \frac{1}{h_o}} = k_o \pi d_o l (t_{fi} - t_{fo}) \tag{10.9}$$

It is usually based on the outside surface area of the tube in engineering calculation of k_o. The overall heat transfer coefficient:

工程计算时，k_o 常以管外侧表面面积为基准。总传热系数为:

$$k_o = \frac{1}{\frac{d_o}{h_i d_i} + \frac{d_o}{2\lambda}\ln\left(\frac{d_o}{d_i}\right) + \frac{1}{h_o}} \tag{10.10}$$

For the heat transfer calculation of finned wall, taking a flat wall as an example, as shown in Fig. 10. 2. The overall area of the finned wall:

对于肋壁的传热计算，以平壁为例，如图 10. 2 所示。肋壁面的总面积为:

$$A_o = A_1 + A_2 \tag{10.11a}$$

The convective heat transfer quantity on the left side of the wall:

壁面左侧的对流传热量为:

$$\Phi = h_i A_i (t_{fi} - t_{wi}) \tag{10.11b}$$

The heat conduction quantity of the flat wall:

平壁面的导热量为:

$$\Phi = \frac{\lambda}{\delta} A_i (t_{wi} - t_{wo}) \tag{10.11c}$$

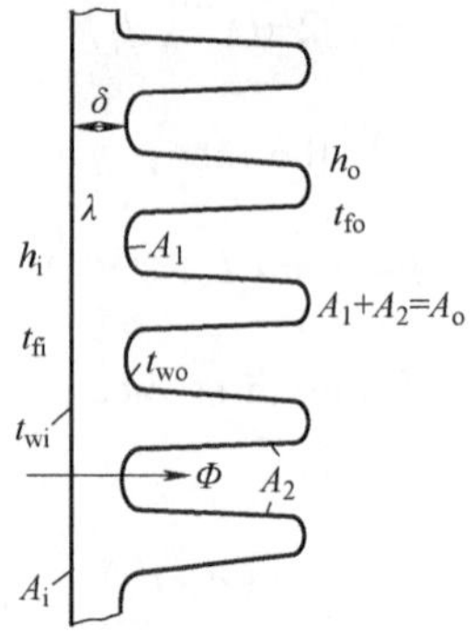

Fig. 10.2 Heat transfer process through the finned wall

图 10.2 通过肋壁的传热过程

The convective heat transfer quantity on the right side of the wall is divided into two parts: base tube heat transfer quantity and fin heat transfer quantity.

壁面右侧的对流传热量分为两部分:平壁传热量和翅片传热量。

$$\begin{aligned}\Phi &= h_o A_1 (t_{wo} - t_{f2}) + h_o \eta_f A_2 (t_{wo} - t_{f2}) \\ &= h_o \eta_o A_o (t_{wo} - t_{f2})\end{aligned} \tag{10.12}$$

The overall efficiency of finned surface:

肋面总效率为:

$$\eta_o = \frac{A_1 + \eta_f A_2}{A_o} \tag{10.13}$$

Where η_f is fin efficiency.

式中,η_f 是肋效率。

The calculation of overall heat transfer quantity for finned wall is often based on the surface area of the non-finned side. If there is no fin inside the tube, the inner surface area A_i is used as a base.

对于肋壁的总传热量计算,常以无肋片侧的表面面积为基准。如果左侧无肋片,以左侧表面面积 A_i 为基准。

$$\Phi = \frac{A_i (t_{f1} - t_{f2})}{\dfrac{1}{h_i} + \dfrac{\delta}{\lambda} + \dfrac{A_i}{h_o \eta_o A_o}} = k A_i (t_{f1} - t_{f2}) \tag{10.14}$$

Define $\beta = A_o / A_i$ as the finned coefficient. The overall heat transfer coefficient in Eq. (10.14) is:

定义$\beta = A_o / A_i$ 为肋化系数。式(10.14)中总传热系数为:

$$k = \frac{1}{\dfrac{1}{h_i} + \dfrac{\delta}{\lambda} + \dfrac{1}{h_o \eta_o \beta}} \tag{10.15}$$

Therefore, as long as $\eta_o\beta>1$, the fin has the effect of enhancing heat transfer.

因此，只要 $\eta_o\beta>1$，肋片就可以起到强化传热的效果。

Question The tube with external fins and the tube with outside insulation layers, both of them increase the thermal resistance of heat conduction and decrease the convective heat transfer resistance at the same time. Why can the fins enhance heat transfer but the insulation layers reduce heat transfer?

思考题 圆管外加装肋片，与圆管外敷设保温层，它们都会使导热热阻增大，同时对流传热热阻减小。为什么肋片能增强传热，而保温层却削弱传热呢？

This question involves a new concept, i. e., critical thermal insulation diameter. The circular tube becomes a multilayer cylindrical wall after it is laid with the insulation layer, and the calculation formula of the heat transfer process is as follows:

这个问题涉及一个新概念——临界热绝缘直径。圆管外敷设保温层后，变成了多层圆筒壁，传热过程计算式为：

$$\Phi=\frac{\pi l(t_{fi}-t_{fo})}{\dfrac{1}{h_i d_i}+\dfrac{1}{2\lambda_1}\ln\left(\dfrac{d_{o1}}{d_i}\right)+\dfrac{1}{2\lambda_2}\ln\left(\dfrac{d_{o2}}{d_{o1}}\right)+\dfrac{1}{h_o d_{o2}}} \tag{10.16}$$

The numerator is the overall temperature difference, and the denominator is the total thermal resistance, including two convective heat transfer resistances and two thermal conduction resistances. It can be seen that Eq. (10.16) has an additional thermal conduction resistance $\dfrac{1}{2\lambda_2}\ln\left(\dfrac{d_{o2}}{d_{o1}}\right)$ compared with the calculation formula for the heat transfer process of a circular tube, and the overall thermal conduction resistance must have increased.

分子是总温差，分母是总热阻，包括两个对流传热热阻和两个导热热阻。可以看出，式（10.16）比圆管的传热过程计算式多了一项导热热阻$\dfrac{1}{2\lambda_2}\ln\left(\dfrac{d_{o2}}{d_{o1}}\right)$，总导热热阻肯定增大了。

At the same time, the outer diameter of the tube increases from d_{o1} to d_{o2}, so, the convective heat transfer resistance on the outside $\dfrac{1}{h_o d_{o2}}$ is reduced. It can be seen that the insulation layer increases the thermal resistance of heat conduction and weakens heat transfer; and at the same time decreases the convective heat transfer resistance and enhances heat transfer. Then, is the comprehensive effect weakened or enhanced?

同时，管外径由 d_{o1} 增大到了 d_{o2}，所以外侧的对流传热热阻$\dfrac{1}{h_o d_{o2}}$减小了。由此可见，保温层使导热热阻增大，传热削弱；同时，使对流传热热阻减小，传热增强。那么，综合效果到底是削弱还是增强呢？

Analysis: The formula for the above heat transfer process

分析：把上述传热过程计算式可看

can be regarded as a function of Φ for d_{o2}, then take the first derivative of Φ with respect to d_{o2}, and set it equal to zero, then we can find $d_{o2}=\frac{2\lambda_2}{h_o}$, written as d_{cr} and called the critical thermal insulation diameter. This shows that there does exist an extreme value of Φ for d_{o2}, and it should be the maximum according to the basic laws of heat transfer.

作 Φ 为 d_{o2} 的函数，然后 Φ 对 d_{o2} 求一阶导数，并令其等于 0，于是可求出 $d_{o2}=\frac{2\lambda_2}{h_o}$，记作 d_{cr}，称为临界热绝缘直径。这说明 Φ 对于 d_{o2} 确实存在一个极值，根据热量传递的基本规律可知，应该是极大值。

That is to say the Φ value increases gradually when d_{o2} varies between d_{o1} and d_{cr}, and the Φ value decreases gradually when d_{o2} is greater than d_{cr}. It can be seen that critical thermal insulation diameter is the outer diameter of the insulation layer corresponding to the minimum total thermal resistance or maximum heat transfer quantity. The above formula can be rewritten as:

也就是说，当 d_{o2} 在 $d_{o1}\sim d_{cr}$ 范围内变化时，Φ 值逐渐增大；当 d_{o2} 大于 d_{cr} 时，Φ 值逐渐减小。可见，临界热绝缘直径是对应于最小总热阻（或最大传热量）的保温层外径。上式可改写为：

$$\frac{h_o d_o}{\lambda}=Bi=2 \tag{10.17}$$

Eq. (10.17) shows that the heat loss can be further reduced by increasing the thickness of insulation layer when the Biot number of the outer surface of the tube insulation layer is greater than 2. Conversely, heat transfer can be enhanced by increasing the thickness of insulation layer when Biot number is less than 2.

式（10.17）说明，当管道保温层外表面的 Bi 大于 2 时，增大保温层厚度可进一步减少热损失。相反，当 Bi 小于 2 时，增大保温层厚度反而起到增强传热的作用。

Take the thermal conductivity of a representative thermal insulation material is 0.1W/(m·K), and the convective heat transfer coefficient of air is 9W/(m^2·K), then the critical thermal insulation diameter can be calculated as 22mm. Generally the outside diameter of a power pipe is greater than 22mm, so, the critical thermal insulation diameter is usually not considered.

取代表性的保温材料导热系数为 0.1W/(m·K)，空气的表面传热系数为 9W/(m^2·K)，可计算出临界热绝缘直径为 22mm。一般动力管道外径都大于 22mm，所以通常不必考虑临界热绝缘直径。

Summary This section illustrates the calculation methods of heat transfer quantity and heat transfer coefficient in heat transfer process of the heat transfer area with different structures, and the physical meaning and calculation method of critical thermal insulation diameter.

小结 本节阐述了不同结构传热面的传热过程传热量和传热系数的计算方法，以及临界热绝缘直径的物理意义和计算方法。

Question Why is the overall heat dissipation capacity corresponding to the critical thermal insulation diameter a maximum?

思考题 为什么说临界热绝缘直径对应的总散热量是极大值?

Self-tests

自测题

1. Briefly describe the physical meaning and calculation method of critical thermal insulation diameter.

1. 简述临界热绝缘直径的物理意义及计算方法。

2. Why does the critical thermal insulation diameter need to be considered when adding thermal insulation material to the circular tube, but the plane wall need not?

2. 为什么在给圆管加保温材料时需要考虑临界热绝缘直径的问题,而平壁不需要考虑?

授课视频

10.2 Application of Calculation Method for the Heat Transfer Process

10.2 传热过程的计算方法应用

Example 10.1 The outer diameter of the steam pipe is 80mm, and the wall thickness is 3mm. The outside is coated with a 40mm thick cement perlite insulation layer, and its thermal conductivity is a function of average temperature $\bar{t}$ of insulation layer $\bar{\lambda}_2 = 0.0651 + 0.000105\bar{t}$. The steam temperature $t_{fi} = 150℃$ in the pipe, the ambient temperature $t_{\infty} = 20℃$, the composite convective heat transfer coefficient of the outer surface of the insulation layer to the environment $h_o = 7.6\text{W}/(\text{m}^2 \cdot \text{K})$, the convective heat transfer coefficient of the steam in the pipe $h_i = 16\text{W}/(\text{m}^2 \cdot \text{K})$, and the thermal conductivity of the steel pipe wall is $46.2\text{W}/(\text{m} \cdot \text{K})$. Try to find the heat loss per pipe length.

例 10.1 蒸汽管道的外径为 80mm,壁厚为 3mm,外侧包有厚 40mm 的水泥珍珠岩保温层,其导热系数是保温层平均温度 $\bar{t}$ 的函数 $\bar{\lambda}_2 = 0.0651 + 0.000105\bar{t}$。管内蒸汽温度 $t_{fi} = 150℃$,环境温度 $t_{\infty} = 20℃$,保温层外表面对环境的复合表面传热系数 $h_o = 7.6\text{W}/(\text{m}^2 \cdot \text{K})$,管内蒸汽的表面传热系数 $h_i = 116\text{W}/(\text{m}^2 \cdot \text{K})$,钢管壁的导热系数为 $46.2\text{W}/(\text{m} \cdot \text{K})$。试求每米管长的热损失。

Analysis: This is a heat transfer process of double-layer cylinder wall, the calculation formula for the heat transfer process of the multilayer cylindrical wall can be used to calculate the heat loss per pipe length, that is, the total

分析: 这是一个双层圆筒壁的传热过程,要计算每米管长的热损失,可采用多层圆筒壁的传热过程计算公式,也就是总温差除以总热阻,

temperature difference divided by the total thermal resistance, thc total thcrmal rcsistancc includcs two convective heat transfer resistances, plus thermal conduction resistances of two layers of the cylindrical wall.

这里总热阻包括两个对流传热热阻，加上两层圆筒壁的导热热阻。

The difficulty of the question is to calculate the average thermal conductivity λ_2 of cement perlite. The thermal conductivity is related to the average wall temperature, but the temperature on both sides of the wall is unknown, and it is also difficult to measure in practical engineering. To calculate the thermal conductivity of cement perlite, it is necessary to assume its wall temperature.

本题的难点是计算水泥珍珠岩的导热系数 λ_2 的平均值。该导热系数与壁面的平均温度有关，但是壁面两侧的温度都不知道，实际工程中测量也很困难。要计算水泥珍珠岩的导热系数，需要先假定其壁面温度。

The inside temperature of the insulation laycr can bc regarded as the steam temperature in the pipe, because the convective heat transfer resistance in the pipe and the thermal conduction resistance of the pipe wall are very small, it can be seen clearly in the following number comparison. The outside temperature of the insulation layer can be assumed to be 30℃, and it will be corrected later. We can get results that meet certain requirements after several iterations of calculation.

保温层的内侧温度可看成与管内的蒸汽温度相同，因为管内对流传热热阻和管壁的导热热阻都很小，从下面的数值对比中可以清楚地看到这一点。保温层外侧温度可以先假设为 30℃，以后再进行修正。经过数次迭代计算，即可得到满足一定要求的结果。

This kind of question "what to solve, what to assume" is a typical nonlinear problem. The iterative method is an effective solution method for nonlinear problems. According to known conditions, we can first find: The outer diameter of insulation layer $d_o = 0.16\text{m}$, the internal diameter of the pipe $d_i = 0.074\text{m}$, heat loss per pipe length:

这种"要求解什么，需假设什么"的问题是典型的非线性问题。对于非线性问题，"迭代法"是一种行之有效的求解方法。根据已知条件，首先可求出：保温层外径 $d_o = 0.16\text{m}$，管道内径 $d_i = 0.074\text{m}$，每米管长的热损失：

$$\Phi = k\pi d_o l(t_{fi} - t_\infty) = k\pi \times 0.16\text{m} \times 1\text{m} \times (150℃ - 20℃) = 65.3k\text{m}^2 \cdot ℃$$

Where k is the undetermined quantity. The heat transfer process of multi-layer cylindrical wall based on the outer diameter is calculated as follows:

式中，k 是待求量。以外径为基准的多层圆筒壁传热过程计算式为：

$$\Phi = \frac{\pi d_{o2} l(t_{fi} - t_{fo})}{\dfrac{d_{o2}}{h_i d_i} + \dfrac{d_{o2}}{2\lambda_1}\ln\left(\dfrac{d_{o1}}{d_i}\right) + \dfrac{d_{o2}}{2\lambda_2}\ln\left(\dfrac{d_{o2}}{d_{o1}}\right) + \dfrac{1}{h_o}}$$

From this formula, the calculation formula of the overall heat transfer coefficient k is as follows:

由该式可得总传热系数 k 的计算式为:

$$k=\frac{1}{\frac{d_{o2}}{h_i d_i}+\frac{d_{o2}}{2\lambda_1}\ln\left(\frac{d_{o1}}{d_i}\right)+\frac{d_{o2}}{2\lambda_2}\ln\left(\frac{d_{o2}}{d_{o1}}\right)+\frac{1}{h_o}}$$

Substitute the known parameters, the formula contains an undetermined average thermal conductivity λ_2 of the insulation layer:

代入已知参数，式中含有一个待求的保温层导热系数 λ_2 的平均值:

$$k=1/\left[\frac{1}{16\text{W}/(\text{m}^2\cdot\text{K})}\times\frac{0.16\text{m}}{0.074\text{m}}+\frac{0.16\text{m}}{2\times46.2\text{W}/(\text{m}\cdot\text{K})}\times\ln\frac{0.08}{0.074}+\frac{0.16\text{m}}{2\bar{\lambda}_2}\times\ln\frac{0.16\text{m}}{0.08\text{m}}+\frac{1}{7.6\text{W}/(\text{m}^2\cdot\text{K})}\right] \tag{a}$$

Based on the above assumptions, the average temperature of the insulation layer can be calculated:

根据上述假设条件，可计算出保温层的平均温度为:

$$\bar{t}=\frac{1}{2}\times(150℃+30℃)=90℃$$

Then, the average thermal conductivity of the insulation layer can be calculated:

于是可计算出保温层的平均导热系数为:

$$\bar{\lambda}_2=(0.0651+0.000105\times90)\text{W}/(\text{m}\cdot\text{K})=0.0746\text{W}/(\text{m}\cdot\text{K})$$

By substituting into the above formula (a) to get the overall heat transfer coefficient:

代入式 (a)，可求出总传热系数为:

$$k=\frac{1}{0.0186+0.000135+0.743+0.132}\text{W}/(\text{m}^2\cdot\text{K})=1.119\text{W}/(\text{m}^2\cdot\text{K})$$

The convective heat transfer resistance in the pipe and the thermal conduction resistance of the pipe wall are very small from the comparison of four partial thermal resistances, in particular the thermal conduction resistance of the pipe wall can be ignored. This shows that it is reasonable to take the inner surface temperature of the insulation layer as 150℃. Then, the heat loss per pipe length is:

从四个分热阻的对比来看，管内对流传热热阻和管壁的导热热阻均很小，特别是管壁热阻完全可以忽略不计。这说明，将保温层内表面的温度取作 150℃ 是合理的。于是每米管长的热损失为:

$$\Phi=65.3\times1.119\text{W}=73.1\text{W}$$

This is not the final answer, because the outside temperature of the insulation layer of 30℃ is assumed and needs to be checked. From Newton cooling formula, the outside temperature can be obtained:

这并不是最后答案，因为保温层的外表面温度 30℃ 是假设的，需要加以校核。由牛顿冷却公式可得到外表面温度为:

$$t_{wo}=\frac{\Phi}{\pi d_0 h_0}+t_{\infty}=\frac{73.1\text{W}}{\pi\times 0.16\text{m}\times 7.6\text{W/(m}^2\cdot\text{K)}}+20℃=39.1℃$$

Then, take it as the outside temperature of the insulation layer and recalculate:

再以此作为保温层外表面温度，重新计算：

$$\bar{\lambda}_2=\left(0.0651+0.000105\times\frac{150+39.1}{2}\right)\text{W/(m}\cdot\text{K)}=0.0750\text{W/(m}\cdot\text{K)}$$

$$k=1.124\text{W/(m}^2\cdot\text{K)},\Phi=73.4\text{W}$$

Discussion: The calculation process must adopt the iterative method because the thermal conductivity is a function of temperature. The relative deviation of the thermal conductivity of the insulation material is less than 1% in two adjacent calculations of this example, so, it can bc considered that the iterative computation has converged as an engineering calculation, and the last assumed value is the undetermined value.

讨论：由于导热系数是温度的函数，计算过程必须采用迭代法。本例相邻两次计算中，保温材料导热系数的相对偏差已小于1%，因此作为工程计算，可以认为迭代计算已收敛，最后一次假设值即待求值。

At the same time, we also get an empirical knowledge: The convective heat transfer resistance of the medium in the pipe is generally much smaller than the thermal conduction resistance of the insulation layer for insulated pipes that transport water or high-pressure steam, so, the pipe wall temperature can be taken as the average temperature of the medium in the pipe. This approach is particularly useful for a concise analysis of engineering heat transfer problems.

同时还得到一条经验：对于输送水或者高压蒸汽的保温管道，管内介质的对流传热热阻一般比保温层的导热热阻要小得多，因而可取管壁温度等于管内介质的平均温度。这种做法对于工程传热问题的简明分析特别有用。

Example 10.2 The outer diameter of aluminum wire is 5.1mm, and the wire is coated with polyvinyl chloride with thermal conductivity of 0.15W/(m · K) as an insulation layer. The ambient temperature is 40℃, and the surface temperature of aluminum wire is limited below 70℃. The composite convective heat transfer coefficient between the surface of the insulation layer and the environment is 10W/(m^2 · K). Calculate the heat dissipating capacity per meter of wire with different insulation thickness of δ.

例 10.2 铝电线外径为5.1mm，外包导热系数 λ = 0.15W/(m · K)的聚氯乙烯作为绝缘层。环境温度为40℃，铝线表面温度限制70℃以下。绝缘层表面与环境间的复合表面传热系数为10W/(m^2·K)。求绝缘层厚度 δ 不同时每米电线的散热量。

Analysis: Obviously, the research object of this question

分析：显然，本题的研究对象是铝

is the insulation layer of aluminum wire. The outer diameter of the wire plus twice the thickness of the insulation layer 2δ is the outer diameter d_o of the insulation layer.

电线外的绝缘层。电线外径加上两倍的绝缘层厚度 2δ，也就是绝缘层外径 d_o。

$$d_o = d_i + 2\delta$$

According to the superposition principle of thermal resistance in series in the calculation formula of steady-state heat transfer, the heat dissipating capacity per meter of wire can be calculated as long as the temperature difference in the numerator corresponds to the thermal resistance in the denominator.

根据串联热阻叠加原理，在稳态传热计算式中，只要分子上的温差与分母中的热阻相对应，即可求出每米电线的散热量为：

$$\frac{\Phi}{l} = \frac{\pi(t_{wi} - t_{fo})}{\frac{1}{2\lambda}\ln\left(\frac{d_o}{d_i}\right) + \frac{1}{h_o d_o}}$$

Substitute the known parameters into the above formula, and the heat dissipating capacity per meter of wire can be obtained:

将已知参数代入上式，可得每米电线的散热量为：

$$\frac{\Phi}{l} = \frac{\pi\times(70℃ - 40℃)}{\frac{1}{2\times0.15\mathrm{W/(m\cdot K)}}\times\ln\frac{d_o}{0.0051\mathrm{m}} + \frac{1}{10\mathrm{W/(m^2\cdot K)}d_o}}$$

Φ/l is a function of the outer diameter d_o of the insulation layer and also a function of the insulation thickness δ. Take d_o as 10 - 70mm and the calculated results are plotted in Fig. 10. 3.

Φ/l 是绝缘层外径 d_o 的函数，也是绝缘层厚度 δ 的函数。取 d_o 为 10~70mm，计算结果用图线绘于图 10. 3 中。

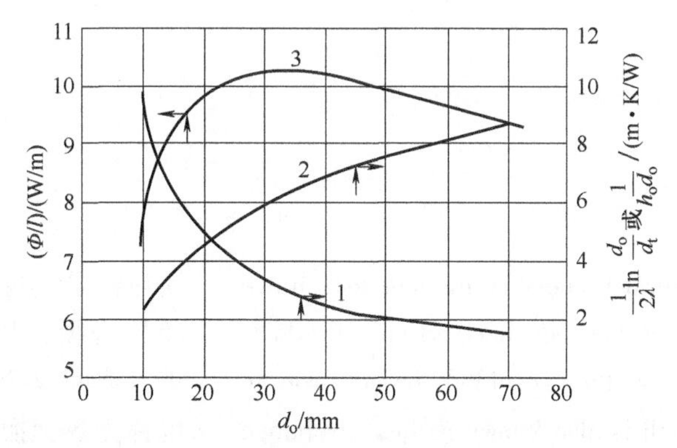

Fig. 10. 3 The relationship between wire heat dissipation capacity and outer diameter of insulation layer

图 10. 3 电线散热量与绝缘层外径的关系

In Fig. 10.3, the abscissa is the outer diameter d_o of the insulation layer, the ordinate respectively represents the thermal conduction resistance of the insulation layer (curve 2), convective heat transfer resistance on the outside of the insulation layer (curve 3) and heat dissipating capacity (curve 1). We can see that the heat dissipating capacity reaches the maximum value when the outer diameter is 30mm. When the outer diameter of the insulation layer is less than 30mm, increasing the thickness of the insulation layer will not weaken the heat dissipation, but will increase the heat dissipation.

图 10.3 中，横坐标为绝缘层外径 d_o，纵坐标分别表示绝缘层导热热阻（曲线 2）、绝缘层外侧对流传热热阻（曲线 3）和散热量（曲线 1）。可以看出，外径为 30mm 时散热量达到最大值。当绝缘层外径小于 30mm 时，增加其厚度非但不会削弱散热，反而会增强散热。

This situation is advantageous for wires because it can increase the ability of the electric current to pass through. The thickness of the insulation layer is about 1mm for the actual product corresponding to the wire in this question, which is within the favourable range for heat dissipation.

对电线来说，处于这种情况下是有利的，因为可以增加电流的通过能力。本题中电线对应的实际产品，所采用的绝缘层厚度约为 1mm，处于对散热有利的范围内。

Summary This section illustrates the calculation method of the heat conduction problem by the iterative method, and the influence of variation of insulation layer thickness on the surface heat dissipation through two examples.

小结 本节通过两个例题说明了用迭代法求解导热问题的计算方法，以及绝缘层厚度的变化对表面散热的影响。

Question What is the physical significance of critical thermal insulation diameter?

思考题 临界热绝缘直径的物理意义是什么？

授课视频

10.3 The Types of Heat Exchangers

10.3 换热器的类型

Heat exchanger is a general process equipment to achieve heat transfer and exchange between hot and cold fluids to meet the requirements of the specified process or to recycle waste heat, which is also known as heat exchange equipment, as shown in Fig. 10.4. There exist three heat transfer modes in heat exchangers, i. e. heat conduction, convection heat transfer and radiation heat transfer.

换热器是实现冷热流体之间热量传递与交换，以满足规定的工艺要求，或回收余热的通用工艺设备，也称为热交换器，如图 10.4 所示。在换热器中存在三种传热方式：热传导、对流传热和辐射传热。

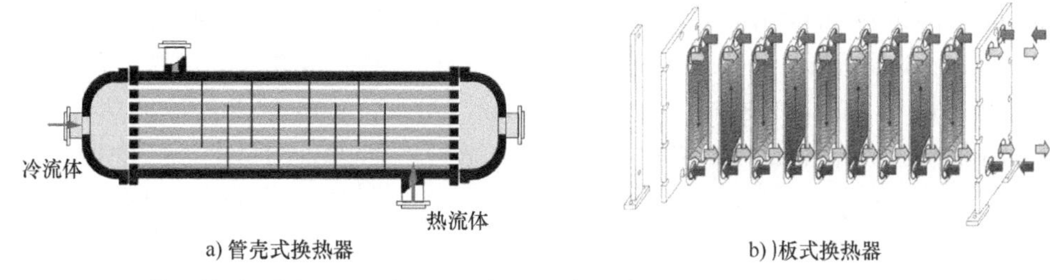

Fig. 10. 4 Schematic diagram of principle and structure of heat exchanger

图 10. 4 换热器的结构和原理示意图

Heat exchanger is a general equipment widely used in energy power system and the process industry. The investment of heat exchangers accounts for 10%-20% of total investment in a chemical plant, their mass accounts for about 40% of total equipment mass, and their maintenance workload accounts for 60% of total maintenance workload.

换热器是能源动力系统和过程工业广泛使用的一种通用设备。在化工厂中，换热设备投资占总投资的10%~20%，质量约占设备总质量的40%，检修工作量占总检修工作量的60%。

The investment of heat exchange equipment accounts for about 35%-40% of total investment in the refinery. The investment of heat exchange equipment including boiler accounts for about 70% of total investment in a thermal power system. The mass of evaporators and condensers accounts for about 70%-80% of gross mass in refrigeration air conditioning and heat pump systems.

在炼油厂中，换热设备投资约占总投资的35%~40%。在热力发电系统中，换热设备包括锅炉的投资约占总投资的70%。在制冷空调和热泵系统中，蒸发器和冷凝器的质量约占总质量的70%~80%。

There are many types of heat exchangers, as shown in Fig. 10. 5, such as shell and tube heat exchanger, plate heat exchanger, heat pipe heat exchanger, fin tube heat exchanger, cooling tower, cooling water tower; and including boiler, radiation heat transfer occurs in the furnace and convection heat transfer is carried on both side walls of smoke tubes or water tubes.

换热器的结构形式很多，如图 10.5 所示，有管壳式换热器、板式换热器、热管换热器、翅片管换热器、冷却塔、凉水塔；包括锅炉，炉膛内为辐射传热，烟管或水管壁面两侧为对流传热。

Fig. 10. 5 Structural form of heat exchanger

图 10. 5 换热器的结构形式

Heat exchangers can be divided into three types according to working principle: mixing type, regenerator type and dividing wall type. The following is a brief introduction to the principle and characteristics of each heat exchanger.

按照工作原理，换热器可分为三种：混合式、蓄热式和间壁式。下面简单介绍各种换热器的原理和特点。

(1) Mixing type heat exchanger, also known as direct contact heat exchanger.

(1) 混合式换热器，又称为直接接触式换热器。

Its working principle is that the cold and hot fluids directly contact, mixing with each other for heat transfer. Its characteristics include high heat transfer efficiency, large heat transfer area per unit volume, simple structure and cheap price etc. It is often made into towers, as shown in Fig. 10.6. This is a cooling tower or cool water tower commonly used in thermal power plants and chemical production.

其工作原理是：冷、热流体直接接触，彼此混合换热。其特点是传热效率高，单位体积提供的传热面积大，结构简单，价格便宜等。常做成塔状，如图 10.6 所示，这是火力发电厂和化工生产中常用的凉水塔或冷却塔。

Fig. 10.6 Cool water tower or cooling tower

图 10.6 凉水塔或冷却塔

(2) Regenerator type heat exchanger, also known as recover heating type heat exchanger.

(2) 蓄热式换热器，又称为回热式换热器。

Its working principle is that the regenerator consisting of solid materials such as fillers or porous grid bricks alternately contacts with the hot and cold fluids, thereby transferring heat from hot fluid to cold fluid. It is characterized by compact structure, cheap price, large heat transfer area per unit volume and is suitable for gas-gas heat transfer.

其工作原理是：由固体材料（如填料或多孔性格子砖等）构成的蓄热体，与热流体和冷流体交替接触，从而把热量从热流体传递给冷流体。其特点是结构紧凑、价格便宜、单位体积传热面大，适合于气-气换热。

(3) Dividing wall type heat exchanger, also known as surface type heat exchanger.

(3) 间壁式换热器，又称为表面式换热器。

Its working principle is that two fluids transfer heat through the solid wall. So, it is a typical heat transfer

其工作原理是：两种流体通过固体壁面进行热量传递。因此是一个典

process, as shown in Fig. 10. 7. First, hot fluid transfers heat to one side of the wall by convection heat transfer, then the wall transfers heat to the other side by heat conduction; finally cold fluid takes heat away by convection heat transfer with the other side of the wall.

型的传热过程，如图 10.7 所示。首先，热流体通过对流传热把热量传递给壁面一侧，接着，壁面通过导热把热量传递到另一侧；然后，冷流体与另一侧壁面进行对流传热把热量带走。

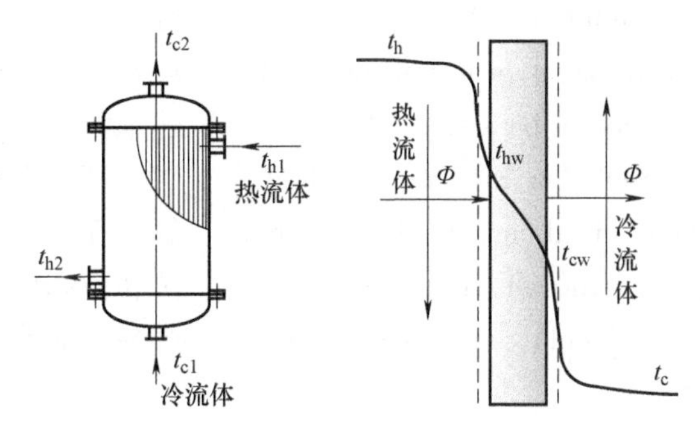

Fig. 10. 7 Heat transfer process of dividing wall type heat exchanger

图 10. 7 间壁式换热器的传热过程

The dividing wall type heat exchanger has a strong adaptability, so more than 90% of the current projects use the dividing wall type heat exchanger. According to the different structure types of heat transfer surfaces, the dividing wall type heat exchanger is divided into double-pipe type, shell and tube type, cross-flow type, plate type, spiral plate type and so on. The following is a brief introduction to the principle and characteristics of each heat exchanger.

间壁式换热器具有适应性强的特点，所以目前工程中 90%以上都采用间壁式换热器。根据换热面的结构形式不同，间壁式换热器又分为套管式、管壳式、交叉流式、板式和螺旋板式等。下面简单介绍每一种换热器的原理和特点。

1) **Double-pipe heat exchanger** is composed of two straight tubes with different diameters assembled into concentric sleeve, which is the most simple dividing wall type heat exchanger, as shown in Fig. 10. 8. The inner and outer tubes flow through different fluids respectively for convection heat transfer.

1) **套管式换热器**，将两种直径不同的直管组装成同心套管而成，是最简单的一种间壁式换热器，如图 10. 8 所示。内管和外管分别流过不同的流体，进行对流换热。

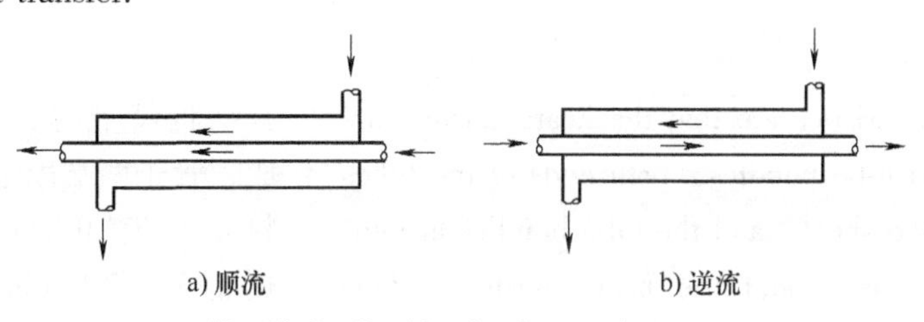

Fig. 10. 8 Double-pipe heat exchangers

图 10. 8 套管式换热器

There are two fluid flow modes in the double-pipe heat exchanger, i. e. parallel-flow and counter-flow. This heat exchanger is suitable for the case of small heat transfer or small fluid flow.

套管式换热器中两种流体流动方式有顺流和逆流两种。该换热器适用于传热量不大或流体流量不大的情形。

Its advantages include simple structure, wide working range (such as high temperature and high pressure), convenient increase and decrease of heat transfer area, and high heat transfer coefficient on both sides of the heat transfer surface due to increased flow rate of both fluids. Its disadvantages include large metal consumption per unit heat transfer area, not compact enough, troublesome maintenance, cleaning and disassemble, and easy to cause leakage at detachable joints.

其优点是：结构简单，工作适应范围广（如高温、高压），传热面积增减方便，可通过提高两种流体流速来获得传热面两侧较高的传热系数。其缺点是：单位换热面积的金属消耗量大，不够紧凑，检修、清洗和拆卸比较麻烦，在可拆连接处容易造成泄漏。

2) **Shell and tube heat exchanger, also known as tubular heat exchanger**, as shown in Fig. 10. 9. It is the most typical dividing wall type heat exchanger, which has a long history and dominates the market, and especially widely used in large installations, more than 70% of projects use shell and tube heat exchangers.

2）**管壳式换热器，又称为列管式换热器**，如图 10. 9 所示。它是最典型的间壁式换热器，使用历史悠久，占据市场主导地位，尤其在大型装置中普遍采用，工程中 70%以上都是管壳式换热器。

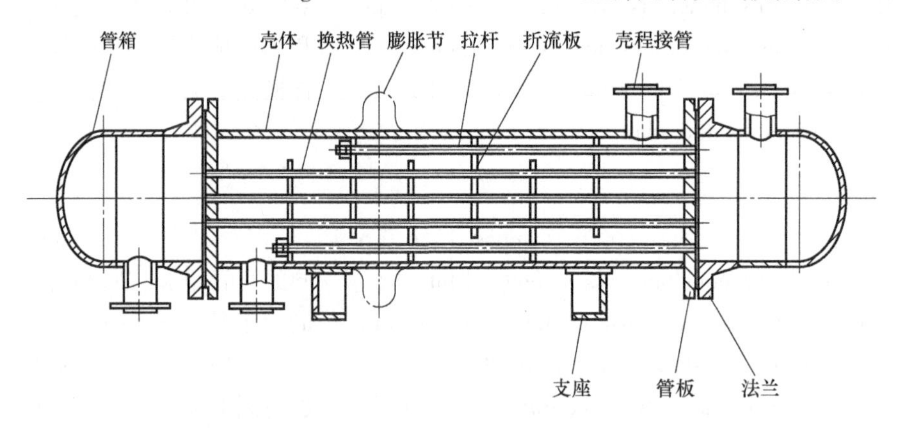

Fig. 10. 9 Shell and tube heat exchanger
图 10. 9 管壳式换热器

Its structural characteristics are that the heat transfer surface is composed of tube bundles, both ends of the tubes are fixed on the tube sheet, and the tube bundles are encapsulated in the shell, and the heat transfer surface through which the two fluids flow is divided into tube side and shell side.

其结构特点是：传热面由管束组成，管子两端固定在管板上，管束封装在壳体内，两种流体流经的传热面分为管程和壳程。

Its performance advantages include strong structure, high reliability, strong adaptability (such as high temperature and high pressure), convenient manufacture, high handling capacity, wide selection of materials, low production costs, convenient cleaning of heat transfer surface. Its disadvantages are that it is inferior to plate heat exchanger in terms of heat transfer efficiency, compactness of the structure and metal consumption per unit heat transfer area.

其性能优点是：结构坚固，可靠性高，适应性强（如高温、高压），易于制造，处理能力大，选材范围广，生产成本较低，换热表面的清洗比较方便。其缺点是：在其传热效率、结构的紧凑度，以及单位换热面积的金属消耗量等方面不如板式换热器。

In order to increase the flow rate of the fluid in the tube side, a partition plate can be arranged in the tube box at one end of the heat exchanger, which constitutes a two-tube-side structure. U-shaped tubes can be used to avoid thermal stress caused by thermal expansion of the tubes. The tube bundle is supported by the arched baffles, which can change the fluid flow state to enhance heat transfer.

为了提高管程流体的流速，在换热器的一端管箱内可设置一块隔板，而构成双管程结构。换热管束可采用 U 形管，以避免管子受热膨胀而引起热应力。管束采用弓形折流板支撑，这可以改变流体流动状态，以强化传热。

3) **Cross flow heat exchanger**, it is another form of dividing wall type heat exchanger. Its main features include that cold and hot fluids flow in a cross pattern, some can be mixed freely and others cannot be mixed. According to the structure of the heat transfer surfaces, cross flow heat exchangers can be divided into tube bundle type, fin tube type, tube belt type and plate fin type, as shown in Fig. 10. 10.

3) **交叉流换热器**，是间壁式换热器的又一种形式。其主要特点是：冷、热流体呈交叉状流动，有的可自由掺混，有的不能掺混。根据换热表面的结构不同，交叉流换热器又分为管束式、管翅式、管带式和板翅式四种，如图 10. 10 所示。

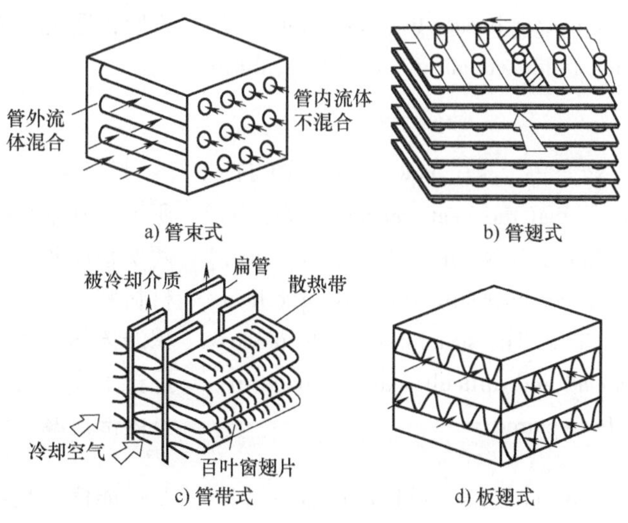

Fig. 10. 10 Cross flow heat exchangers

图 10. 10 交叉流换热器

Steam superheater, economizer and air preheater in boiler device are tube bundle type heat exchanger. The evaporators and condensers in household air conditioners mostly adopt fin tube type heat exchanger. The radiator of automobile engine is usually of tube-belt type heat exchanger, where the heat transfer tube is generally elliptical tube or flat tube, and multi-layer pleated fins are arranged outside the tube to enhance heat transfer in air side. Plate fin type heat exchanger is widely used in low temperature engineering.

锅炉装置中的蒸汽过热器、省煤器、空气预热器都采用管束式换热器。家用空调器中的蒸发器与冷凝器，多采用管翅式换热器。汽车发动机的散热器常采用管带式，其中换热管一般为椭圆管或扁管，管外布置多层百褶翅片，以强化空气侧的换热。板翅式换热器广泛应用于低温工程中。

4) **Plate (surface) heat exchanger**, it is a heat exchanger through the plate surface for heat transfer, as shown in Fig. 10. 4b. Its performance features include that the fluid reaches a turbulent state at a lower velocity, so heat transfer coefficient is high; It is made of plates, which is convenient for mass production and low in manufacturing cost; Its pressure resistance is worse than that of tubular heat exchanger.

4) **板（面）式换热器**，它是通过板面进行传热的换热器，如图 10. 4b 所示。其性能特点是：流体在较低的速度下达到湍流状态，传热系数大；采用板材制作，便于大规模生产，制造成本较低；其耐压性能比管式换热器差。

Its structural features are that it is superimposed in parallel by a group of thin plane plates with the same geometric structure, cold and hot fluids flow in each channel at intervals, and heat transfer is realized through plates. In order to enhance heat transfer and increase plate stiffness, various corrugations are produced on the plane plate. Plate type heat exchanger is easy to disassemble and clean, and suitable for fluids containing readily scaling materials.

其结构特点是：由一组几何结构相同的薄平板平行叠加而成，冷、热流体间隔地在每个通道中流动，通过板片实现换热。为强化换热并增加板片的刚度，平板上通常压制有各种波纹。板式换热器拆卸、清洗方便，适用于含有易结垢物的流体。

5) **Spiral plate heat exchanger**, as shown in Fig. 10. 11. Its structural features are that the heat transfer surface is rolled from two metal plates, it is simple to manufacture, high material utilization, compact structure and large heat transfer area per unit volume. The structural disadvantages are large welding workload and difficult sealing, once internal leakage cannot be repaired.

5) **螺旋板式换热器**，如图 10. 11 所示。其结构特点是：换热表面由两块金属板卷制而成，制造简单，材料利用率高，结构紧凑，单位体积的传热面积大。其结构缺点是：焊接工作量大，密封比较困难，一旦内部泄漏无法修复。

Its performance advantages are that cold and hot fluids flow spirally in a spiral channel, which can achieve complete counter flow, high heat transfer efficiency; Self-

其性能优点是：冷热流体在螺旋状通道内做螺旋流动，可实现完全逆流，换热效率高；流体有自冲刷的

brushing effect of the fluid, and is not easy to scale. So, the heat exchanger is suitable for liquid-liquid heat transfer, gas-liquid heat transfer, especially for heating or cooling of high viscosity fluids and heat transfer of suspensions containing solid particles.

作用，不易结垢。因此，该换热器适用于液-液换热、气-液换热，尤其适合于高黏度流体的加热或冷却，以及含有固体颗粒的悬浮液的换热。

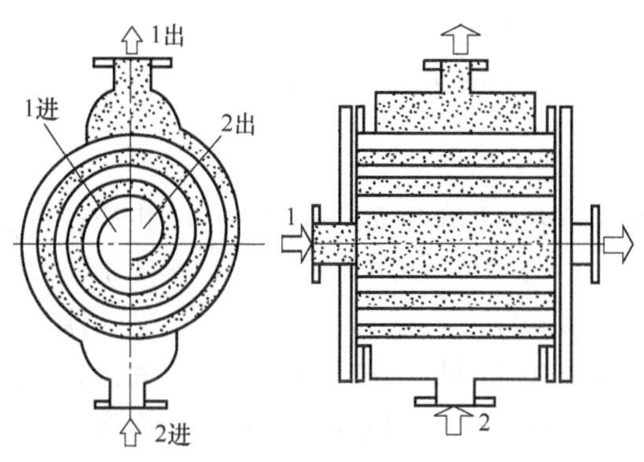

Fig. 10. 11　Spiral plate heat exchanger

图 10. 11　螺旋板式换热器

Heat exchangers can be divided into compact heat exchanger and non-compact heat exchanger according to the entire compactness. There are two measures of compactness of heat exchanger: ① The heat transfer area contained in a unit volume, called the heat transfer area density β; ② Hydraulic diameter (i. e. equivalent diameter d_h).

换热器按整体紧凑度又可分为紧凑式换热器和非紧凑式换热器。换热器的紧凑度衡量指标有两个：① 单位体积内所包含的换热面积，称为换热面积密度 β；② 水力直径(即当量直径 d_h)。

The heat exchanger with heat transfer area density $\beta \geqslant 700\mathrm{m}^2/\mathrm{m}^3$, or hydraulic diameter $d \leqslant 6\mathrm{mm}$, is generally called compact heat exchanger. For example, plate heat exchangers are mostly compact heat exchanger. There are four ways to increase the compactness of heat exchangers: ① Reduction of tube diameter; ② Adopting plate structure; ③ Adopting various finned surfaces (i. e. extended surfaces); ④ Adopting winding tube structure etc.

一般将换热面积密度大于或等于 $700\mathrm{m}^2/\mathrm{m}^3$，或水力直径小于或等于6mm的换热器，称为紧凑式换热器，如板式换热器多属于紧凑式。提高换热器紧凑度的途径有四种：① 减小管径；② 采用板式结构；③ 采用各种肋化表面（即扩展表面）；④ 采用缠绕管式结构等。

The following four factors should be considered in the selection of heat exchangers: ① Fluid properties, including pressure, temperature, corrosivity, etc.; ② Heat transfer efficiency, resistance characteristics and service life; ③ Total costs, including acquisition and operation costs; ④ Anti-scaling performance, or cleaning and maintenance requirements.

换热器选型时通常要考虑以下四种因素：① 流体性质，包括压力、温度、腐蚀性等；② 换热效率、阻力特性及使用寿命；③ 总费用，包括购置费和运行费；④ 抗结垢性能，或清洗维修要求。

The structural development of shell and tube heat exchanger is mainly reflected in the following three aspects: ① High-efficiency enhanced tube, such as cross stripe tube, spiral groove tube, zooming tube, finned tube, etc.; ② Inserts in the tube, such as torsion band, spiral torsion plate, static mixer and spiral spring, etc.; ③ Tube bundle support, such as baffle rod, spiral baffle, spiral fin, petal orifice baffle, etc.

管壳式换热器的结构发展主要体现在以下三个方面：① 高效强化管，如横纹管、螺旋槽管、缩放管、翅片管等；② 管内插入物，如扭带、螺旋扭片、静态混合器、螺旋弹簧等；③ 管束支撑，如折流杆、螺旋折流板、螺旋肋片、花瓣孔板等。

The internal structure of the baffle rod shell and tube heat exchanger is shown in Fig. 10. 12. The shell side fluid flows longitudinally, with small flow resistance, which eliminates the vibration of the tube bundle and flow dead zone, and the heat transfer efficiency is high at high Reynolds number.

折流杆管壳式换热器的内部结构如图 10. 12 所示。壳程流体呈纵向流动，流动阻力小，消除了管束振动和流动死区，高雷诺数时传热效率高。

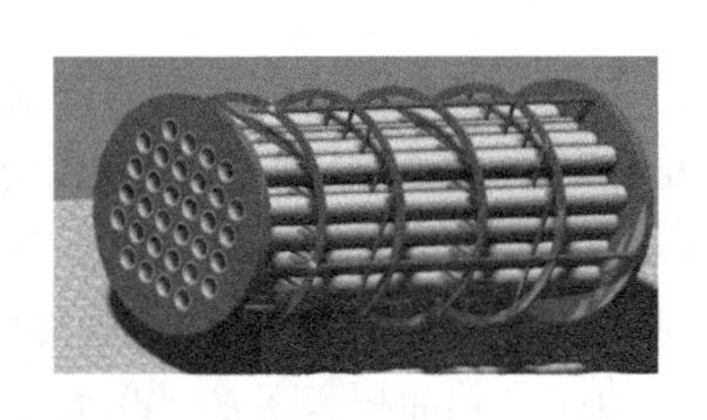

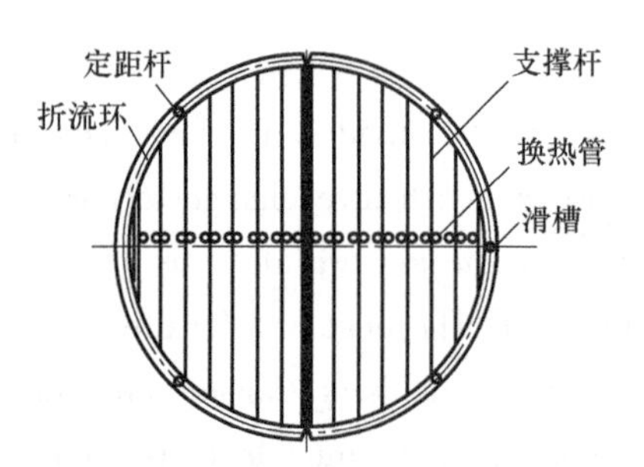

Fig. 10. 12 The internal structure of the baffle rod shell and tube heat exchanger

图 10. 12 折流杆管壳式换热器的内部结构

The spiral baffle shell and tube heat exchanger, as shown in Fig. 10. 13, makes the shell side fluid flow as a whole spiral, which eliminates the flow dead zone, and has large heat transfer coefficient and good comprehensive heat transfer performance.

螺旋折流板管壳式换热器，如图 10. 13 所示，使壳程流体呈整体螺旋流动，消除了流动死区，传热系数大，综合传热性能好。

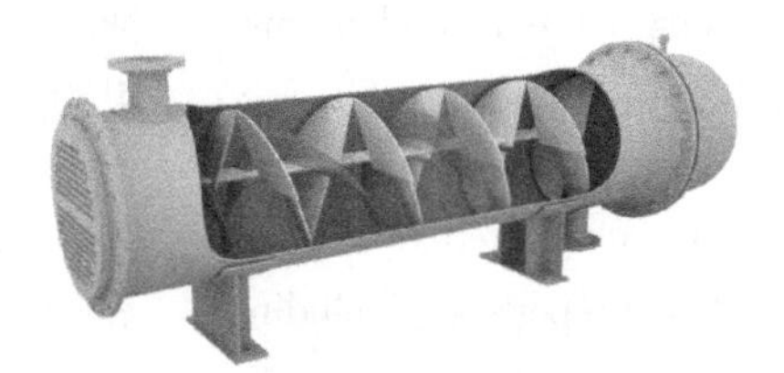

Fig. 10. 13 The spiral baffle shell and tube heat exchanger

图 10. 13 螺旋折流板管壳式换热器

The spiral finned self-support and petal orifice baffle heat exchangers are the patented technology developed by the

螺旋肋片自支撑和花瓣孔板支撑换热器是作者研发的专利技术，如

author, as shown in Fig. 10. 14. It has the advantages of simple structure, convenient processing, spiral flow of the shell side fluid around a single heat transfer tube, no flow dead zone, large heat transfer coefficient and good comprehensive heat transfer performance.

图 10. 14 所示。其结构简单，加工方便，使壳程流体围绕单根换热管做螺旋流动，无流动死区，传热系数大，综合传热性能好。

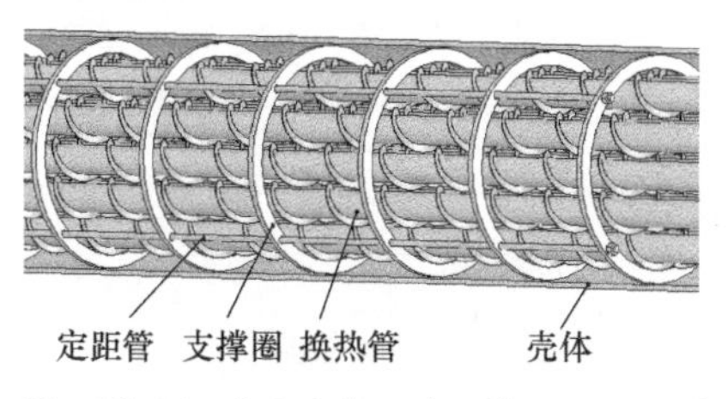

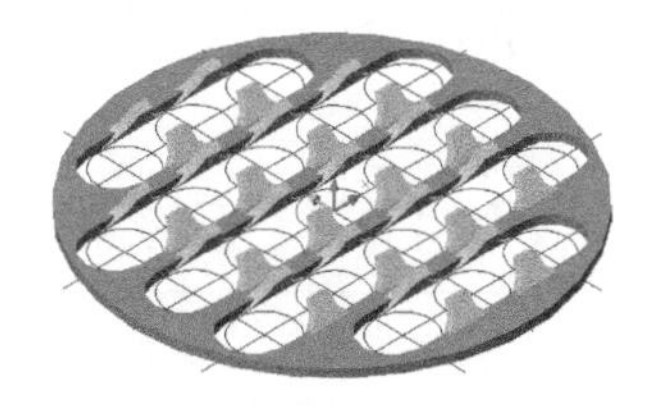

Fig. 10. 14 Spiral finned self-support and petal orifice baffle heat exchangers

图 10. 14 螺旋肋片自支撑和花瓣孔板支撑换热器

Summary This section illustrates the structure, working principle and type of heat exchangers, and introduces the general situation of the structure development of heat exchangers.

小结 本节阐述了换热器的结构形式、工作原理和类型，以及换热器的结构发展概况。

Question What are the main types of heat exchangers? What are the characteristics of each? What factors should be considered when selecting a heat exchanger?

思考题 换热器主要有哪些类型？各有什么特点？换热器的选型应考虑哪些因素？

Self-tests

1. What are the main structure types of shell and tube heat exchangers?

2. What are the characteristics of the dividing wall type heat exchanger?

自测题

1. 管壳式换热器有哪些主要结构类型？

2. 间壁式换热器的特点是什么？

授课视频

10.4 Calculation of the Mean Temperature Difference for Fully Parallel-flow or Fully Counter-flow Heat Exchangers

10.4 纯顺流或逆流换热器的平均温差计算

The equation of heat transfer process is as following:

传热过程的方程式为：

$$\Phi = kA(t_{f1} - t_{f2})$$

Where the temperature of the two fluids is averaged. While in the actual heat transfer process, the temperature of the two fluids is always changing, as shown in Fig. 10. 15.

式中，两种流体的温度取平均值。而实际的传热过程中，两种流体的温度始终在变化，如图 10. 15 所示。

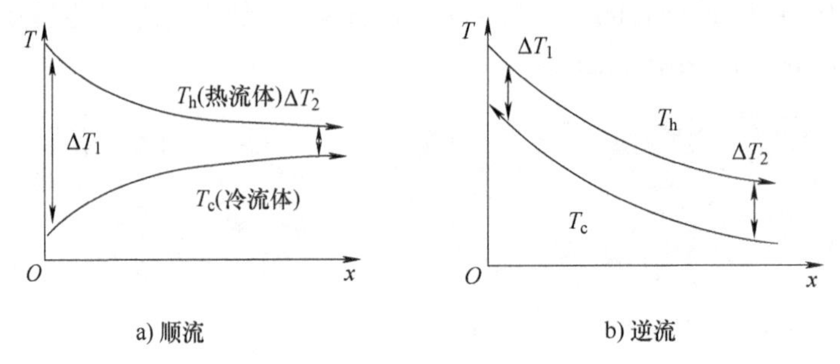

Fig. 10. 15 The temperature variation of the heat transfer process in heat exchanger

图 10. 15 换热器中传热过程的温度变化

These are the temperature variation curves of two fluids in heat exchangers under the conditions of parallel-flow and counter-flow. The inlets of two fluids are located at the same side of heat exchanger for parallel-flow. The temperature of hot fluid constantly decreases from left to right, while the temperature of cold fluid constantly rises from left to right. The temperature difference at the inlet side is large, while that at the outlet side is small.

这是换热器中两种流体在顺流和逆流状态下温度的变化曲线。顺流时两种流体的进口位于换热器的同一端，热流体从左向右温度不断降低，而冷流体从左向右温度不断升高；进口端的温差很大，而出口端的温差很小。

The inlets of two fluids are located at both sides of heat exchanger for counter-flow. The temperature of hot fluid constantly decreases from left to right, while the temperature of cold fluid constantly rises from right to left. The temperature difference is fairly uniform throughout the heat transfer process.

逆流时两种流体的进口分别位于换热器的两端，热流体从左向右温度不断降低，而冷流体从右向左温度不断升高，整个传热过程的温差比较均匀。

The mean temperature difference for fully parallel-flow and fully counter-flow heat exchanger is calculated as follows. As shown in Fig. 10. 16, if the temperature difference is not constant along the whole heat transfer surface, then the equation for calculating heat transfer rate on the whole heat transfer surface is:

下面计算纯顺流及纯逆流换热器的平均温差。如图 10. 16 所示，如果温差沿整个传热面不是常数，那么整个传热面上计算热流量的方程式为：

$$\Phi = kA\Delta t_m \tag{10.18}$$

Where Δt_m is the mean temperature difference over the whole heat transfer area, also called as the mean temperature pressure.

式中，Δt_m 是整个传热面积上的平均温差，或称为平均温压。

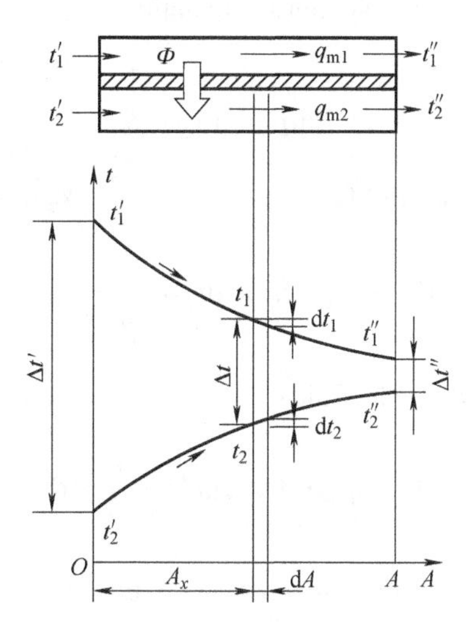

Fig. 10. 16 Derivation of the mean temperature difference for the parallel-flow

图 10. 16 顺流平均温差的推导

Now let's deduce the formula of mean temperature difference and make the following assumptions for the parallel-flow: ① The mass flow rate of cold and hot fluids and specific heat capacity are constant and no phase change; ② The heat transfer coefficient over the whole heat transfer area is constant; ③ The heat exchanger has no heat dissipation loss; ④ The thermal conduction quantity of fluid along the flow direction on heat transfer surface can be ignored.

下面来推导平均温差的计算公式，以顺流情况为例，进行以下假设：① 冷、热流体的质量流量及比热容均为常数，并且没有相变；② 传热系数在整个传热面上为常数；③ 换热器没有散热损失；④ 传热面上流体沿流动方向的导热量可忽略不计。

To calculate the mean temperature difference over the whole heat transfer area, the first thing is to know the relationship between the local temperature difference at any position and the corresponding heat transfer area. Then, integrate along the whole heat transfer area and calculate the average value.

要计算沿整个传热面的平均温差，首先要知道任意位置的局部温差随对应部分传热面积的变化关系。然后，再沿整个传热面积进行积分，求平均值。

On the basis of the above assumptions, and the inlet and outlet temperatures of hot and cold fluids are known. Now calculate the heat transfer quantity on the micro element heat transfer area of dA in the figure, the temperature difference on both sides of dA:

在上述假设的基础上，已知冷、热流体的进、出口温度，下面计算图中微元传热面 dA 上的传热量，dA 的两侧温差为：

$$\Delta t = t_1 - t_2 \tag{a}$$

On the micro element area of dA, the heat transfer quantity of two fluids is:

在微元面 dA 上，两种流体的换热量为：

$$\mathrm{d}\Phi = k\mathrm{d}A\Delta t \tag{b}$$

The heat release quantity for the hot fluid:

对于热流体的放热量：

$$\mathrm{d}\Phi = -q_{m1}c_1\mathrm{d}t_1 \Rightarrow \mathrm{d}t_1 = -\frac{1}{q_{m1}c_1}\mathrm{d}\Phi \tag{c}$$

The heat absorption quantity for the cold fluid:

对于冷流体的吸热量：

$$\mathrm{d}\Phi = q_{m2}c_2\mathrm{d}t_2 \Rightarrow \mathrm{d}t_2 = \frac{1}{q_{m2}c_2}\mathrm{d}\Phi \tag{d}$$

$$\Delta t = t_1 - t_2 \Rightarrow \mathrm{d}(\Delta t) = \mathrm{d}t_1 - \mathrm{d}t_2 \tag{e}$$

Substitute dt_1 and dt_2 into Eq. (e), the sum of the two terms in brackets is written as μ for writing convenience:

把 dt_1 和 dt_2 代入式（e），为书写方便把括号中两项之和记作 μ，可得：

$$\mathrm{d}(\Delta t) = -\left(\frac{1}{q_{m1}c_1} + \frac{1}{q_{m2}c_2}\right)\mathrm{d}\Phi = -\mu\mathrm{d}\Phi \tag{f}$$

Substitute $\mathrm{d}\Phi = k\mathrm{d}A\Delta t$ into Eq. (f) to get:

把 $\mathrm{d}\Phi = k\mathrm{d}A\Delta t$ 代入式（f），可得：

$$\mathrm{d}(\Delta t) = -\mu k\mathrm{d}A\Delta t \tag{g}$$

Separate variables, divide both sides of Eq. (g) by Δt:

分离变量，式（g）两边同时除以 Δt，得：

$$\frac{\mathrm{d}(\Delta t)}{\Delta t} = -\mu k\mathrm{d}A \tag{h}$$

Eq. (h) integrates Δt and dA respectively, the integral limits of Δt are $\Delta t'$ and Δt_x, and the integral limits of A are 0 and A_x respectively:

式（h）分别对 Δt 和 dA 进行积分，Δt 的积分限分别是 $\Delta t'$ 和 Δt_x，A 的积分限分别是 0 和 A_x：

$$\int_{\Delta t'}^{\Delta t_x}\frac{\mathrm{d}\Delta t}{\Delta t} = -\mu k\int_0^{A_x}\mathrm{d}A \tag{i}$$

$$\ln\frac{\Delta t_x}{\Delta t'} = -\mu kA_x \tag{j}$$

$$\Delta t_x = \Delta t'\exp(-\mu kA_x) \tag{k}$$

As we can see that the local temperature difference Δt_x changes exponentially with the partial heat transfer area

可见，当地温差 Δt_x 随部分换热面积 A_x 呈指数规律变化。那么，沿整

A_x. Then, the mean temperature difference over the whole heat transfer area is:

个换热面的平均温差为:

$$\Delta t_m = \frac{1}{A}\int_0^A \Delta t_x \mathrm{d}A_x = \frac{1}{A}\int_0^A \Delta t' \exp(-\mu k A_x)\mathrm{d}A_x \tag{l}$$

Integrate the above equation to get:

积分可得:

$$\Delta t_m = -\frac{\Delta t'}{\mu k A}[\exp(-\mu k A) - 1] \tag{m}$$

When $A_x = A$, corresponding to the right boundary in Fig. 10. 16, $\Delta t_x = \Delta t''$:

当 $A_x = A$ 时，对应于图 10. 16 中右边界，$\Delta t_x = \Delta t''$，得

$$\ln\frac{\Delta t''}{\Delta t'} = -\mu k A \tag{n}$$

$$\frac{\Delta t''}{\Delta t'} = \exp(-\mu k A) \tag{o}$$

Replace the other terms in Eq. (m) except the temperature difference with Eq. (n) and Eq. (o) to get the calculation formula of logarithmic mean temperature difference:

将式（m）中除温差外其他项用式（n）和式（o）代换，可得对数平均温差的计算式为:

$$\Delta t_m = \frac{\Delta t'' - \Delta t'}{\ln\frac{\Delta t''}{\Delta t'}} = \frac{\Delta t' - \Delta t''}{\ln\frac{\Delta t'}{\Delta t''}} \tag{10.19}$$

The derivation process for counter-flow is exactly the same as that for parallel-flow, and the final calculation formula of logarithmic mean temperature difference is the same as that for the parallel-flow completely. So, the logarithmic mean temperature difference can be written as a unified calculation formula whether it is parallel-flow or counter-flow:

逆流时推导过程与顺流完全一样，最终得到的对数平均温差的计算式与顺流时形式也相同。因此，不管是顺流还是逆流，对数平均温差计算式可写成统一形式:

$$\Delta t_m = \frac{\Delta t_{max} - \Delta t_{min}}{\ln\frac{\Delta t_{max}}{\Delta t_{min}}} \tag{10.20}$$

The logarithmic mean temperature difference between parallel-flow and counter-flow is calculated by the same formula, but the calculation process is different. For parallel-flow, $\Delta t'$ is the inlet temperature of hot fluid minus that of cold fluid, and $\Delta t''$ is the outlet temperature of hot fluid minus that of cold fluid, which summarize as "inlet minus inlet, outlet minus outlet".

虽然顺流和逆流的对数平均温差计算式形式相同，但是计算过程不同。顺流时，$\Delta t'$ 是热流体的进口温度减去冷流体的进口温度，$\Delta t''$ 是热流体的出口温度减去冷流体的出口温度，概括为“进口减进口（温度），出口减出口（温度）”。

For counter-flow, $\Delta t'$ is the inlet temperature of hot fluid minus the outlet temperature of cold fluid, and $\Delta t''$ is the outlet temperature of hot fluid minus the inlet temperature of cold fluid, which summarize as "inlet minus outlet, outlet minus inlet".

逆流时，$\Delta t'$是热流体的进口温度减去冷流体的出口温度，$\Delta t''$是热流体的出口温度减去冷流体的进口温度，可以概括为“进口减出口（温度），出口减进口（温度）”。

It is more convenient to calculate the logarithmic mean temperature difference according to the temperature distribution in Fig. 10. 15. There is also a simple way to calculate the mean temperature difference, namely the arithmetic mean temperature difference, that is the large temperature difference plus small temperature difference and divided by 2.

按照图 10. 15 中温度分布计算对数平均温差会更方便。平均温差还有一种简单的计算方法，即算术平均温差，也就是大温差加上小温差除以 2。

The arithmetic mean temperature difference is equivalent to a linear change in temperature, so, it's always greater than the logarithmic mean temperature difference in the same inlet and outlet temperatures. When the ratio of large to small temperature difference is less than or equal to 2, the difference between them is less than 4%. When the ratio of large to small temperature difference is less than or equal to 1. 7, the difference between them is less than 2. 3%.

算术平均温差相当于温度呈直线变化的情况，因此，总是大于相同进出口温度下的对数平均温差。当大小温差之比小于或等于 2 时，两者的差别小于 4%。当大小温差之比小于或等于 1. 7 时，两者的差别小于 2. 3%。

Summary This section illustrates the calculation methods of the logarithmic mean temperature difference for fully parallel-flow and fully counter-flow heat exchangers.

小结 本节阐述了纯顺流和逆流换热器的对数平均温差计算方法。

Question What is the effect of different fluid arrangement on the mean temperature difference?

思考题 不同的流体布置方式对平均温差有什么影响？

Self-tests

自测题

1. What is the parallel-flow and counter-flow arrangement of heat exchanger? What are the characteristics of these two arrangements?

1. 什么叫作换热器的顺流布置和逆流布置？这两种布置方式有何特点？

2. How to calculate the logarithmic mean temperature difference of a heat exchanger?

2. 换热器的对数平均温差如何计算？

10.5 The Calculation of Mean Temperature Difference for Heat Exchanger with Complex Arrangement

10.5 复杂布置换热器的平均温差计算

The logarithmic mean temperature difference discussed earlier is only for fully parallel-flow and fully counter-flow. However, the flow state in the actual heat exchanger is very complex, generally there are both parallel-flow and counter-flow, as shown in Fig. 10.17. For this complex flow state, the theoretical method can be used to analyze it, but the mathematical derivation is very complicated.

前面讨论的对数平均温差只是针对纯顺流和纯逆流情况，而实际换热器中流动状态非常复杂，一般既有顺流，又有逆流，如图 10.17 所示。对于这种复杂流动状态，可采用理论方法进行分析，但数学推导非常复杂。

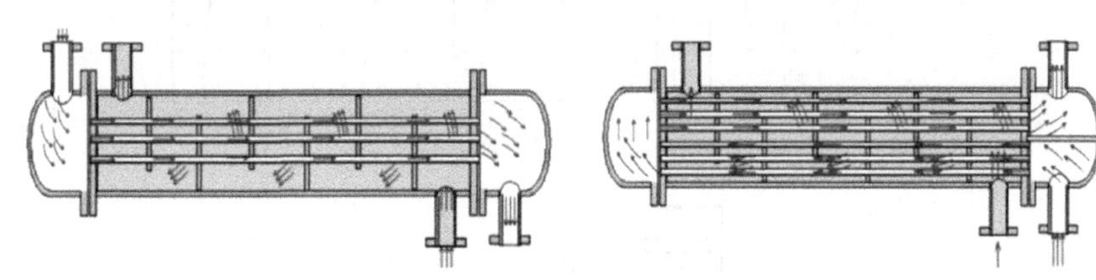

Fig. 10.17 Complex flow state in heat exhangers

图 10.17 换热器中复杂流动状态

The fact shows that the mean temperature difference of fully counter-flow is the largest. Therefore, the logarithmic mean temperature difference for the heat exchanger with complex arrangement can be calculated on the basis of fully counter-flow heat exchanger, then be corrected again. The expression is:

事实证明，纯逆流的平均温差最大。因此，复杂布置换热器的对数平均温差可在纯逆流换热器的基础上进行计算，然后再加以修正。表达式为：

$$\Delta t_m = \Psi(\Delta t_m)_{ctf} \tag{10.21}$$

Where Ψ is the correction coefficient of temperature difference less than 1, which can be obtained by consulting figures. Fig. 10.18 and Fig. 10.19 show the correction coefficient of temperature difference in shell and tube heat exchangers.

式中，Ψ 是小于 1 的温差修正系数，可查图得到。图 10.18 和图 10.19 给出了管壳式换热器的温差修正系数。

Fig. 10.20 and Fig. 10.21 show the correction coefficient of temperature difference in cross flow heat exchangers.

图 10.20 和图 10.21 给出了交叉流换热器的温差修正系数。

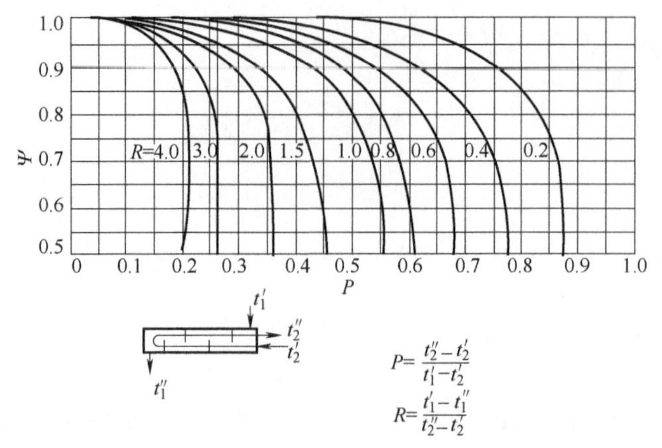

Fig. 10.18 Ψ value of 1 shell side and 2, 4, 6, 8, ⋯ tube sides

图 10.18 壳侧 1 程、管侧 2, 4, 6, 8, …程的 Ψ 值

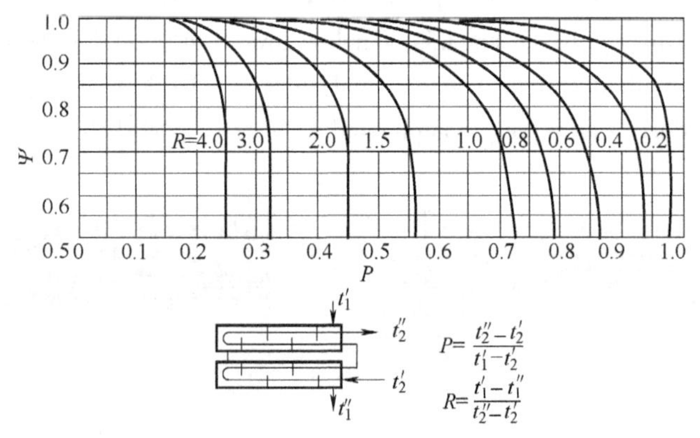

Fig. 10.19 Ψ value of 2 shell sides and 2, 4, 6, 8, ⋯ tube sides

图 10.19 壳侧 2 程、管侧 2, 4, 6, 8, …程的 Ψ 值

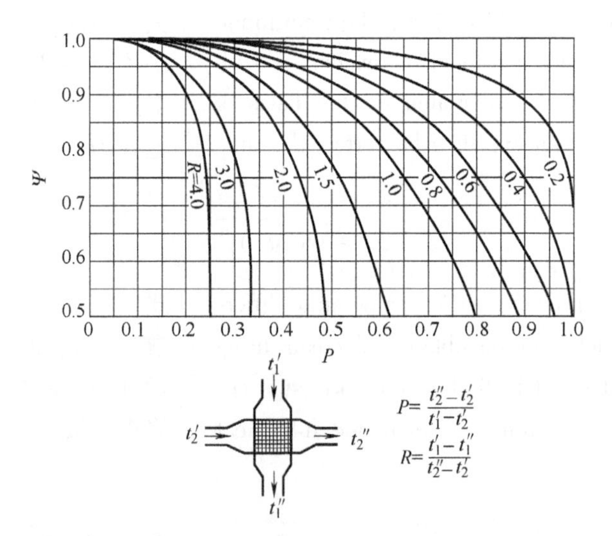

Fig. 10.20 Ψ value of a cross flow in which neither fluid mixes

图 10.20 一次交叉流，两种流体各自都不混合的 Ψ 值

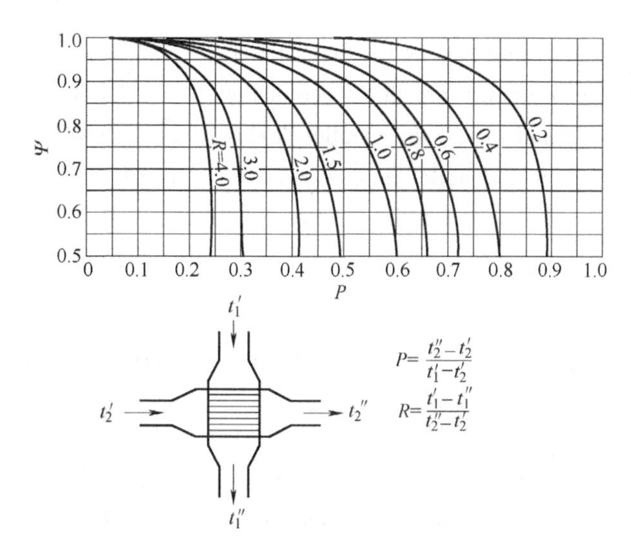

Fig. 10. 21 Ψ value of a cross flow when one fluid is mixed and the other fluid is not mixed

图 10. 21 一次交叉流，一种流体混合，另一种流体不混合时的 Ψ 值

The following four points should be noted to find the correction coefficient of temperature difference by using these curves:

利用曲线求温差修正系数需注意以下四点：

(1) P is the abscissa and R is the parameter in all the figures. The value of Ψ depends on the dimensionless parameters P and R. The subscripts 1 and 2 in the formula represent the two fluids respectively. The superscript $'$ is for inlet and $''$ for outlet.

(1) 图中，均以 P 为横坐标，R 为参量，Ψ 值取决于无量纲参数 P 和 R。式中，下标 1、2 分别表示两种流体，上标$'$表示进口，$''$表示出口。

$$P=\frac{t_2''-t_2'}{t_1'-t_2'},\qquad R=\frac{t_1'-t_1''}{t_2''-t_2'} \tag{10.22}$$

(2) The physical meaning of P is the ratio of the actual temperature rise of fluid 2 to the theoretical maximum temperature rise, so it may only be less than 1.

(2) P 的物理意义是：流体 2 的实际温升与理论上所能达到的最大温升之比，所以只能小于 1。

(3) The physical meaning of R is the ratio of temperature difference between two fluids, that is the ratio of heat capacity.

(3) R 的物理意义是：两种流体的温差之比，也就是热容量之比。

(4) Pay attention to the number of shell side and tube side when consulting the figures for shell and tube heat exchanger.

(4) 对于管壳式换热器，查图时要注意“壳程”数和“管程”数。

The comparison of various flow states in heat exchangers is as follows: Parallel-flow and counter-flow are two extreme cases, the logarithmic mean temperature difference of the counter-flow is the largest, and that of the parallel-flow is the smallest at the same inlet and outlet temperatures. In addition, the outlet temperature of hot fluid t''_h must be higher than that of cold fluid t''_c in parallel-flow. But in counter-flow, the outlet temperature of cold fluid t''_c may be higher than that of hot fluid t''_h. It can be seen that the heat exchange intensity is the highest in counter-flow.

下面对换热器各种流动状态进行对比：顺流和逆流是两种极端情况，在相同的进出口温度下，逆流的对数平均温差最大，顺流的最小。而且顺流时，热流体的出口温度 t''_h 一定高于冷流体的出口温度 t''_c。而逆流时，冷流体的出口温度 t''_c 可能会高于热流体的出口温度 t''_h。可见逆流时换热强度最大。

The temperature of the fluid undergoing phase transition remains constant for heat exchangers with phase transition, such as evaporator or condenser. So, there is no sense to discuss parallel-flow or counter-flow.

对于有相变的换热器，如蒸发器或冷凝器，发生相变的流体温度不变。所以，不存在顺流还是逆流的问题。

Since the temperature difference of heat exchanger in counter-flow is the largest and the heat exchange intensity is also the largest, then, are all the heat exchangers designed to be counter-flow? Of course not. Because many factors should be considered in the design of heat exchanger, not only heat exchange strength.

既然逆流布置时换热器的温差最大，换热强度也最大，那么是不是所有换热器都设计成逆流形式呢? 当然不是，因为换热器设计要考虑很多因素，而不仅仅是换热强度。

The highest temperatures of hot and cold fluids are located on the same side of the heat exchanger in counter-flow, which makes the wall temperature particularly high and may cause large thermal stress, then may cause damage to the heat exchanger. Therefore, it needs to be specially designed to parallel-flow for high temperature heat exchanger.

逆流时，冷热流体的最高温度都位于换热器的同侧，使该处壁温特别高，会造成较大的热应力，可能对换热器造成破坏。因此，对于高温换热器，需要专门设计成顺流。

Example 10. 3 An oil cooler was tested for heat transfer, the parameters were measured as follows: inlet oil temperature $t'_1 = 49.9$℃, inlet water temperature $t'_2 = 21.4$℃, outlet oil temperature $t''_1 = 44.6$℃, outlet water temperature $t''_2 = 24$℃, mass flow of water $q_m = 21.5 \times 10^3$kg/h, heat transfer area $A = 2.85\text{m}^2$, the flow directions of cold and hot fluids are opposite, try to calculate the mean temperature difference of the oil cooler under this working condition.

例 10.3 对一台冷油器进行传热试验，测得下列参数：进口油温 $t'_1 = 49.9$℃，进口水温 $t'_2 = 21.4$℃。出口油温 $t''_1 = 44.6$℃，出口水温 $t''_2 = 24$℃；水的质量流量 $q_m = 21.5 \times 10^3$kg/h，传热面积 $A = 2.85\text{m}^2$，冷、热流体的流动方向相反，试计算该工况下冷油器的平均温差。

Solution: It is considered that all the assumptions for deriving the mean temperature difference are true in the following calculations. According to the meaning of the question, the solid lines in Fig. 10. 22 show the temperature change of oil and water along the path. Now then:

解： 在以下计算中，认为推导平均温差的假设均成立。根据题意，油和水的温度沿程变化如图 10. 22 中的实线所示，此时有：

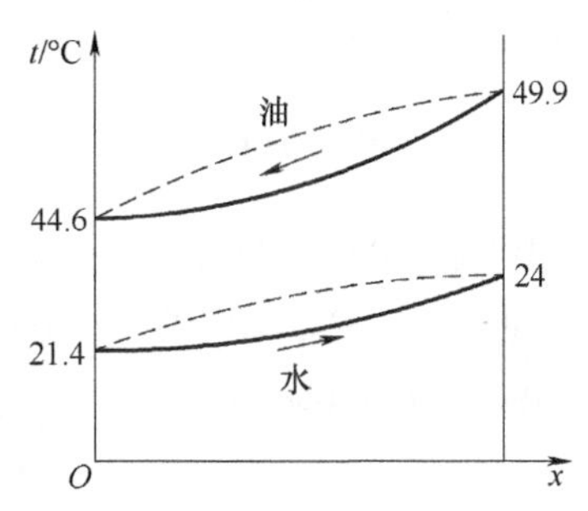

Fig. 10. 22 Illustration and analysis of Example 10. 3

图 10. 22 例 10. 3 图示和分析

$$\Delta t_{max} = 49.9℃ - 24℃ = 25.9℃$$

$$\Delta t_{min} = 44.6℃ - 21.4℃ = 23.2℃$$

$$\frac{\Delta t_{max}}{\Delta t_{min}} = \frac{25.9℃}{23.2℃} = 1.116 < 2$$

Therefore, the arithmetic mean temperature difference can be used to calculate:

所以，可以采用算术平均温差计算：

$$\Delta t_m = \frac{\Delta t_{max} + \Delta t_{min}}{2} = \frac{25.9℃ + 23.2℃}{2} = 24.6℃$$

Discussion: Fig. 10. 22 shows the temperature change curves of hot and cold fluids with solid lines. Can the temperature curves be drawn as the dotted line in the figure?

讨论： 图 10. 22 中用实线表示出了冷、热流体的温度变化曲线，试问温度曲线可否画成图中虚线所示呢？

Example 10. 4 Look at Fig. 10. 11, in a spiral plate heat exchanger, the hot water flow rate $q_{m1} = 2000\text{kg/h}$, the cold water flow rate $q_{m2} = 3000\text{kg/h}$, the inlet temperature of hot water $t_1' = 80℃$, the inlet temperature of cold water $t_2' = 10℃$, if cold water is required to be heated to $t_2'' = 30℃$, try to find the mean temperature difference in parallel-flow and counter-flow, separately.

例 10. 4 请看图 10. 11，在一台螺旋板式换热器中，热水流量 $q_{m1} = 2000\text{kg/h}$，冷水流量 $q_{m2} = 3000\text{kg/h}$，热水进口温度 $t_1' = 80℃$，冷水进口温度 $t_2' = 10℃$。如果要求将冷水加热到 $t_2'' = 30℃$，试求顺流和逆流时的平均温差。

Solution: First, the thermal equilibrium formula should be listed to calculate the outlet temperature of the hot fluid:

解： 首先应列出热平衡关系式，计算出热流体的出口温度为：

$$q_{m1}c_1(t_1'-t_1'')=q_{m2}c_2(t_2''-t_2')$$

The specific heat capacity of water $c_1=c_2=4200\text{J}/(\text{kg}\cdot\text{K})$ within the given temperature range. Substituting it into the above equation to obtain:

在本题给定的温度范围内，水的比热容 $c_1=c_2=4200\text{J}/(\text{kg}\cdot\text{K})$。代入上式即可求出：

$$(2000/3600)\text{kg/s}\times(80℃-t_1'')=(3000/3600)\text{kg/s}\times(30℃-10℃)$$

$$t_1''=50℃$$

(1) In parallel-flow, $\Delta t_{\max}=80℃-10℃=70℃$, $\Delta t_{\min}=50℃-30℃=20℃$, substitute into the calculation Eq. (10.20) of the logarithmic mean temperature difference to obtain:

(1) 顺流时，$\Delta t_{\max}=80℃-10℃=70℃$，$\Delta t_{\min}=50℃-30℃=20℃$，代入对数平均温差计算式（10.20），可得：

$$\Delta t_{\mathrm{m}}=\frac{70℃-20℃}{\ln\dfrac{70℃}{20℃}}=39.9℃$$

(2) In counter-flow, $\Delta t_{\max}=80℃-30℃=50℃$, $\Delta t_{\min}=50℃-10℃=40℃$, substitute into Eq. (10.20) to obtain:

(2) 逆流时，$\Delta t_{\max}=80℃-30℃=50℃$，$\Delta t_{\min}=50℃-10℃=40℃$，代入式（10.20），可得：

$$\Delta t_{\mathrm{m}}=\frac{50℃-40℃}{\ln\dfrac{50℃}{40℃}}=44.8℃$$

Discussion: The logarithmic mean temperature difference in counter-flow is 12.3% larger than that in parallel-flow. In other words, under the same heat transfer quantity and heat transfer coefficient, heat transfer area can be reduced by 12.3% as long as the parallel-flow system is changed to the counter-flow system.

讨论： 逆流布置时，对数平均温差比顺流时大了12.3%。也就是说，在同样的传热量和同样的传热系数下，只要将顺流系统改成逆流系统，就可以减少12.3%的换热面积。

Example 10.5 In the above example, if the 1-2 type shell and tube heat exchanger is used instead, cold water flows in the shell side, and hot water flows in the tube side. Find the mean temperature difference.

例10.5 上例中如改用1-2型管壳式换热器，冷水走壳程，热水走管程，求平均温差。

Analysis: The 1-2 type shell and tube heat exchanger has one shell side and two tube sides, therefore, the fluids in the shell side and tube side have both parallel-flow and counter-flow. Then, the mean temperature difference needs to be first calculated using counter-flow, afterwards the

分析： 1-2型管壳式换热器也就是单壳程双管程换热器，因此，壳程与管程流体既有顺流，又有逆流。那么需要先采用逆流计算出平均温差，然后再进行温差修正。

temperature difference is corrected again.

Solution: Where the subscript 1 and 2 in the calculation of parameters P and R are regarded as the shell side and the tube side, respectively.

解：这里把参数 P、R 计算中的下标 1、2，分别看成是壳侧和管侧。

$$P=\frac{t_2''-t_2'}{t_1'-t_2'}=\frac{50℃-80℃}{10℃-80℃}=0.428$$

$$R=\frac{t_1'-t_1''}{t_2''-t_2'}=\frac{10℃-30℃}{50℃-80℃}=0.667$$

Fig. 10. 18 shows that the corrected coefficient of temperature difference $\Psi=0.95$. In the above example, the mean temperature difference in counter-flow is 44. 8℃, then the mean temperature difference in the 1-2 type shell and tube heat exchanger is:

由图 10. 18 查得温差修正系数 $\Psi=0.95$。上例中已求得逆流时的平均温差为 44. 8℃。于是 1-2 型管壳式换热器中的平均温差为：

$$\Delta t_{\mathrm{m}}=0.95\times44.8℃=42.6℃$$

Discussion: If the cold water flows in the tube side and hot water flows in the shell side, then:

讨论：如果让冷水走管程，热水走壳程，则有：

$$P=\frac{t_2''-t_2'}{t_1'-t_2'}=\frac{30℃-10℃}{80℃-10℃}=0.286$$

$$R=\frac{t_1'-t_1''}{t_2''-t_2'}=\frac{80℃-50℃}{30℃-10℃}=1.50$$

Fig. 10. 18 shows that the corrected coefficient of temperature difference $\Psi=0.95$, and you can find that the P value here is PR in the above calculation, and the R here is $1/R$ in the above calculation. It can be seen that in the definition formula of P and R, the subscript 1 and 2 only represents two fluids. It is not necessary to always correspond subscript 1 and 2 to shell side and tube side or hot fluid and cold fluid for shell and tube heat exchangers. As well for cross flow heat exchangers, there is no need to correspond 1 and 2 to hot and cold or mixing and unmixing of fluids.

由图 10. 18 仍然可查得温差修正系数 $\Psi=0.95$，而且还可以发现，这里的 P 值即上述计算中的 PR，而此处的 R 则为上述计算中的 $1/R$。可见，在 P、R 的定义式中，下标 1、2 仅代表两种流体，对于管壳式换热器，没有必要一定把下标 1、2 与壳侧、管侧，或热流体、冷流体对应起来。同样，对于交叉流换热器，也没有必要一定把 1、2 与流体的冷、热或者混合、不混合联系起来。

Summary This section illustrates the calculation and correction methods of the mean temperature difference for heat exchanger with complex arrangement.

小结 本节阐述了复杂布置换热器平均温差的计算和修正方法。

Question Why is the temperature distribution curve of counter-flow in Fig. 10.22 drawn in a downward concave form?

思考题 为什么图 10.22 中逆流的温度分布曲线画成向下凹的形式？

Self-tests

自测题

1. How to calculate the mean temperature difference of heat exchanger with complex arrangement?

1. 复杂布置时换热器的平均温差如何计算？

2. How to correct the mean temperature difference of heat exchanger with complex arrangement?

2. 复杂布置时怎样对换热器的平均温差进行修正？

授课视频

10.6 The Thermal Calculation of Dividing Wall Type Heat Exchanger—The Mean Temperature Difference Method

10.6 间壁式换热器热计算——平均温差法

There are two kinds of thermal calculation of dividing wall type heat exchangers: ① Design calculation; ② Check calculation. The design calculation is to design a new heat exchanger and to determine the required heat transfer area. Check calculation is for the heat exchanger with existing or determined heat transfer area, calculating whether it can be qualified for the new working conditions under non-design conditions.

间壁式换热器的热计算有两种情况：① 设计计算；② 校核计算。设计计算就是设计一台新的换热器，以确定所需的换热面积。校核计算就是对已有或已选定换热面积的换热器，在非设计工况下，核算其能否胜任规定的新工况。

No matter design calculation or check calculation, there are two basic equations for the thermal calculation of heat exchangers:

不论是设计计算还是校核计算，换热器热计算的基本方程式都是如下两个：

(1) The total heat transfer equation:

(1) 总传热方程式：

$$\Phi = kA\Delta t_{\mathrm{m}} \tag{10.23}$$

(2) The thermal balance equation:

(2) 热平衡方程式：

$$\Phi = q_{mh}c_{\mathrm{h}}(t_{\mathrm{h}}'-t_{\mathrm{h}}'') = q_{mc}c_{\mathrm{c}}(t_{\mathrm{c}}''-t_{\mathrm{c}}') \tag{10.24}$$

Where the logarithmic mean temperature difference Δt_m is not an independent variable that depends on four temperatures, which are the inlet and outlet temperature of hot fluid and the cold fluid, and the arrangement of the heat exchanger.

式中，对数平均温差 Δt_m 不是独立变量，它取决于四个温度，也就是热流体的进/出口温度和冷流体的进/出口温度，以及换热器的布置方式。

According to the heat balance equation, as long as we know $q_m c$ of hot fluid and cold fluid, and three of the four temperatures, we can calculate the other temperature. So, there are eight quantities in the above two equations, they are the overall heat transfer coefficient k, the overall heat transfer area A, the heat capacity of hot fluid $q_{mh}c_h$, the heat capacity of cold fluid $q_{mc}c_c$ and four temperatures t'_h, t''_h, t''_c, t'_c.

由热平衡方程式可知，只要已知热流体和冷流体的 $q_m c$，以及四个温度中的3个，即可计算出另外一个温度。因此，上面两个方程式中共有8个量，也就是传热系数 k、面积 A、热流体的热容量 $q_{mh}c_h$、冷流体的热容量 $q_{mc}c_c$，以及四个温度 t'_h、t''_h、t''_c、t'_c。

Where the mass flow rate q_m and specific heat capacity c of hot and cold fluids always appear simultaneously in the calculation of heat exchangers, so, they are combined as a quantity and are usually the given quantity. Five of the eight quantities need to be given to figure out the other three variables.

其中，冷热流体的质量流量 q_m 和比热容 c 在换热器计算中总是同时出现，所以将它们合并看作一个量，通常都是已知量。8个量中需要给定其中5个量，才能计算出另外3个变量。

For design calculation, the mass flow rate and the specific heat capacity of hot and cold fluids, and three of the inlet and outlet temperatures are given, finally to find the overall heat transfer coefficient k and heat transfer area A. For check calculation, the heat transfer area A, the mass flow rate and the specific heat capacity of hot and cold fluids, and two inlet temperatures are given to calculate the two outlet temperatures t''_h, t''_c.

对于设计计算，给定的是冷热流体的质量流量和比热容，以及进出口温度中的三个，最终求总传热系数 k 和传热面积 A。对于校核计算，给定的是传热面积 A、冷热流体的质量流量和比热容，以及两个进口温度，待求的是两个出口温度 t''_h、t''_c。

There are two methods for thermal calculation of heat exchangers: ① The mean temperature difference method; ② The effectiveness-heat transfer unit number method. The mean temperature difference method is often used in engineering practice, that is, the total heat transfer equation and the heat balance equation are directly used to thermal calculation.

换热器的热计算有两种方法：① 平均温差法；② 效能-传热单元数法。在工程实际中多采用平均温差法，也就是直接应用总传热方程式和热平衡方程式进行热计算。

The specific steps of design calculation are as follows: $q_m c$ and three of inlet and outlet temperatures of hot and cold fluids are known to calculate k and A. These two parameters are related to each other, so only the iteration method can be used.

设计计算的具体步骤如下：已知冷热流体的 $q_m c$ 及进出口温度中 3 个，求 k 和 A。这两个参数相互关联，所以只能用迭代法。

(1) According to the known conditions, first, calculate the overall heat transfer quantity and the unknown outlet temperature by heat balance equation.

(1) 根据已知条件，由热平衡式先求出总传热量以及待求的出口温度。

(2) From four inlet and outlet temperatures of hot and cold fluids, logarithmic mean temperature difference can be calculated.

(2) 由冷热流体的 4 个进出口温度，可计算出对数平均温差。

(3) Assume an overall heat transfer coefficient k, then calculate the corresponding heat exchange area A according to the total heat transfer equation.

(3) 假设一个总传热系数 k，然后根据总传热方程式计算出对应的换热面积 A。

(4) Preliminarily arrange heat exchange surface, calculate the convective heat transfer coefficient on both sides of the wall and the overall heat transfer coefficient, and check heat exchange area A and fluid flow resistance.

(4) 初步布置换热面，计算壁面两侧的表面传热系数及总传热系数，并核算换热面积 A 和流体流动阻力。

(5) The hypothetical area should be about 20% larger than the checked area, and flow resistance should be less than the allowable value, otherwise, the design scheme needs to be changed and to be recalculated.

(5) “假设面积”要大于“校核面积”20%左右，流动阻力要小于允许值，否则需改变方案，进行重新计算。

The specific steps of check calculation are as follows: Knowing heat transfer area A and $q_m c$ of hot and cold fluids and two inlet temperatures, calculate two outlet temperatures.

校核计算的具体步骤如下：已知传热面积 A 和冷热流体的 $q_m c$，以及两个进口温度，求两个出口温度。

(1) Assume an outlet temperature, the other outlet temperature can be calculated using the heat balance equation.

(1) 假设一个出口温度，用热平衡方程式可算出另一个出口温度。

(2) The logarithmic mean temperature difference can be calculated according to four inlet and outlet temperatures.

(2) 根据 4 个进出口温度可求出对数平均温差 Δt_m。

(3) According to the structural arrangement of the heat exchanger, the overall heat transfer coefficient k under corresponding operating conditions can be calculated.

(3) 根据换热器的结构布置，可算出相应工况下的总传热系数 k。

(4) Knowing kA and Δt_m, a heat transfer quantity Φ_1 can be calculated using the total heat transfer equation.

(4) 已知 kA 和 Δt_m，用总传热方程式可计算出一个传热量 Φ_1。

(5) According to four inlet and outlet temperatures, another heat transfer quantity Φ_2 can be calculated using the heat balance equation.

(5) 根据4个进出口温度，用热平衡式可计算出另一个传热量 Φ_2。

(6) Neither Φ_1 nor Φ_2 is the actual heat transfer quantity, they are obtained at assumed outlet temperature. Therefore, the values of two Φ are compared, which shows that the assumptions are correct if the error meets accuracy requirement; Otherwise, re-assume outlet temperature and repeat (1)-(6) steps until the accuracy requirement is met.

(6) Φ_1 和 Φ_2 都不是真实的传热量，都是在假设出口温度下得到的。因此，比较两个 Φ 值，如果误差满足精度要求，则说明假设正确；否则重新假设出口温度，重复步骤（1）~步骤（6），直至满足精度要求。

Example 10.6 Turbine oil No. 30 with a flow rate of $39m^3/h$ is cooled from $t_1' = 56.9℃$ to $t_1'' = 45℃$ in the oil cooler. The oil cooler adopts a 1-2 type shell-and-tube structure, the tube is copper with 15mm outer diameter and 1mm wall thickness. 47.7t per hour of river water as cooling water flows through the tube side, the inlet temperature $t_2' = 33℃$. Oil is arranged on the shell side. The convective heat transfer coefficient h_o in the oil side is $450W/(m^2 \cdot K)$, and the convective heat transfer coefficient h_i in water side is $5850W/(m^2 \cdot K)$. The physical properties of turbine oil No. 30 are known to be $\rho_1 = 879kg/m^3$ and $c_1 = 1.95kJ/(kg \cdot K)$ at operating temperature, try to find the required heat transfer area.

例 10.6 流量为 $39m^3/h$ 的30号透平油，在冷油器中从 $t_1' = 56.9℃$ 冷却到 $t_1'' = 45℃$。冷油器采用1-2型管壳式结构，管子为铜管，外径为15mm，壁厚为1mm。每小时47.7t的河水作为冷却水在管侧流过，进口温度 $t_2' = 33℃$。油安排在壳侧。油侧的表面传热系数 h_o 为 $450W/(m^2 \cdot K)$，水侧的表面传热系数 h_i 为 $5850W/(m^2 \cdot K)$。已知30号透平油在运行温度下的物性参数 $\rho_1 = 879kg/m^3$ 和 $c_1 = 1.95kJ/(kg \cdot K)$，试求所需传热面积。

Analysis: Heat release Φ can be calculated from the known parameters of oil, the outlet temperature of the water can be calculated according to the heat balance equation, then the logarithmic mean temperature difference can be obtained. Physical properties of water are not given in the title, but they are easy to be consulted and are also considered as known parameters.

分析：根据油的已知参数可计算出放热量 Φ，根据热平衡方程式可计算出水的出口温度，进而可求出对数平均温差。水的物性参数虽然题目中没给，但是很容易查出，同样看作已知参数。

When calculating the overall heat transfer coefficient, it should be clear which side of the area is the bench mark.

在计算总传热系数时，要明确是以哪一侧面积为基准。如果以外侧为

If the outside is taken as the benchmark and the fouling resistance is considered, the following formula can be used.

基准，并考虑污垢热阻，可采用下式。

$$k=\frac{1}{\left(\frac{1}{h_{o}}+R_{o}\right)+R_{w}+\left(R_{i}+\frac{1}{h_{i}}\right)\frac{A_{o}}{A_{i}}}$$

Calculate the mass flow rate of oil first:

先求出油的质量流量为：

$$q_{m1}=\rho q_{V1}=879\text{kg/m}^3\times\frac{39\text{m}^3}{3600s}=9.52\text{kg/s}$$

The heat release from the oil is:

油的放热量为：

$$\Phi=q_{m1}c_1(t_1'-t_1'')=9.52\text{kg/s}\times1.95\text{kJ/(kg}\cdot\text{K)}\times(56.9℃-45℃)=2.21\times10^5\text{W}$$

The mass flow rate of water is:

水的质量流量为：

$$q_{m2}=\frac{47.7\times10^3\text{kg}}{3600\text{s}}=13.25\text{kg/s}$$

The temperature rise of cooling water can be calculated according to the heat balance equation:

根据热平衡方程式可求出冷却水的温升：

$$t_2''-t_2'=\frac{\Phi}{q_{m2}c_2}=\frac{2.21\times10^5\text{W}}{13.25\text{kg/s}\times4.19\times10^3\text{J/(kg}\cdot\text{K)}}=4℃$$

So, the outlet temperature of the cooling water is:

于是，冷却水的出口温度为：

$$t_2''=33℃+4℃=37℃$$

Considering 1-2 type shell-and-tube heat exchanger for oil cooler, the two fluids are complex arrangements, therefore, the temperature difference needs to be corrected. First calculate the parameters P and R:

考虑到冷油器采用 1-2 型管壳式换热器，两种流体为复杂布置，因此，温差需要修正。先计算出参量 P 和 R：

$$P=\frac{t_2''-t_2'}{t_1'-t_2'}=\frac{37℃-33℃}{56.9℃-33℃}=0.17$$

$$R=\frac{t_1'-t_1''}{t_2''-t_2'}=\frac{56.9℃-45℃}{37℃-33℃}=3$$

Looking at Fig. 10.18 the corrected coefficient of temperature difference is equal to 0.97, then the logarithmic mean temperature difference is:

查图 10.18 可得温差修正系数 Ψ 等于 0.97，那么对数平均温差为：

$$\Delta t_{m}=0.97\times\frac{(56.9℃-37℃)-(45℃-33℃)}{\ln\frac{56.9℃-37℃}{45℃-33℃}}=15.1℃$$

According to Table 10.1 the fouling thermal resistance inside and outside the tube are respectively: $R_i = 0.0005\text{m}^2 \cdot \text{K/W}$, $R_o = 0.0002\text{m}^2 \cdot \text{K/W}$. And omit the thermal conduction resistance of tube wall, then the overall heat transfer coefficient based on the outer side area is:

按表10.1取管内外污垢热阻分别为：$R_i = 0.0005\text{m}^2 \cdot \text{K/W}$，$R_o = 0.0002\text{m}^2 \cdot \text{K/W}$，并略去管壁导热热阻，于是以外侧面积为基准的总传热系数为：

Table 10.1 Fouling thermal resistance of water

表10.1 水的污垢热阻　　（单位：$10^{-4}\text{m}^2 \cdot \text{K/W}$）

加热介质的温度	<115℃		115~205℃	
水的温度	≤52℃		>52℃	
水的类型	水的流速/(m/s)		水的流速/(m/s)	
	≤1	>1	≤1	>1
海水	0.88	0.88	1.76	1.76
含盐的水	3.52	1.76	5.28	3.52
冷却塔和人造喷水池净化水	1.76	1.76	3.52	3.52
冷却塔和人造喷水池未净化水	5.28	5.28	8.8	7.04
自来水或井水	1.76	1.76	3.52	3.52
河水最小值	3.52	1.76	5.28	3.52
河水平均值	5.28	3.52	7.04	5.28
混浊或带有泥质的水	5.28	3.52	7.04	5.28
硬水（>256.8mg/L）	5.28	5.28	8.8	8.8
发动机冷却套用水	1.76	1.76	1.76	1.76
蒸馏水或封闭循环	0.88	0.88	0.88	0.88
冷凝液	1.76	0.88	1.76	1.76
净化的锅炉给水	1.76	0.88	1.76	1.76
锅炉排水	3.52	3.52	3.52	3.52

$$k = \frac{1}{\left(\frac{1}{h_o} + R_o\right) + \left(R_i + \frac{1}{h_i}\right)\frac{d_o}{d_i}}$$

$$= 1\Big/\left[\frac{1}{450\text{W}/(\text{m}^2 \cdot \text{K})} + 0.0002\text{m}^2 \cdot \text{K/W} + \left(\frac{1}{5850\text{W}/(\text{m}^2 \cdot \text{K})} + 0.0005\text{m}^2 \cdot \text{K/W}\right) \times \frac{15}{13}\right]$$

$$= 313\text{W}/(\text{m}^2 \cdot \text{K})$$

According to the overall heat transfer equation, the calculated area of the oil cooler is obtained:

根据总传热方程式，可得冷油器的“计算面积”为：

$$A = \frac{\Phi}{k\Delta t_m} = \frac{2.21 \times 10^5\text{W}}{313\text{W}/(\text{m}^2 \cdot \text{K}) \times 15.1℃} = 46.8\text{m}^2$$

However, 10% margin is required for the actual design area: $46.8\times1.1\text{m}^2=51.5\text{m}^2$.

而实际“设计面积”要留 10%的裕度，取为 $46.8\times1.1\text{m}^2=51.5\text{m}^2$。

Discussion: The fouling resistance has been taken into account in the heat exchanger calculation, which accounts for more than 1/4 of the total thermal resistance. But the actual heat transfer area is increased by another 10% of the redundant area to take account of some factors not taken into account, for example, possible errors in obtaining the overall heat transfer coefficient.

讨论： 虽然在换热器热计算中已经考虑了污垢热阻，占总热阻的 1/4 以上。但是，实际传热面积又增加了 10%的冗余面积，以照顾到某些未考虑到的因素，比如获得总传热系数时可能的误差等。

Summary This section illustrates thermal calculation of heat exchangers, namely design calculation and check calculation, and the mean temperature difference method and effectiveness-heat transfer unit number method, and focuses on learning the mean temperature difference method.

小结 本节阐述了换热器热计算，即设计计算和校核计算，以及平均温差法和效能-传热单元数法，并重点学习了平均温差法。

Question What is the basic idea of using mean temperature difference method to design and calculatea a heat exchanger?

思考题 用平均温差法进行换热器设计计算的基本思路是什么？

Self-tests

自测题

1. Briefly describe the basic steps of thermal calculation of heat exchanger.

1. 简述换热器热计算的基本步骤。

2. What are the two basic equations for thermal calculation of heat exchanger?

2. 换热器热计算的两个基本方程式是什么？

授课视频

10.7 The Thermal Calculation of Dividing Wall Type Heat Exchanger—The Effectiveness-Heat Transfer Unit Number Method

10.7 间壁式换热器热计算——效能-传热单元数法

The effectiveness of heat exchanger is defined based on the following ideas: Suppose that the heat exchanger is infinitely long, for a counter-flow type heat exchanger, the following situations will occur:

换热器效能的定义基于以下思想：假设换热器无限长，对于一个逆流换热器，会发生以下情况：

(1) When the heat capacity of hot fluid is much smaller than that of cold fluid, the outlet temperature of hot fluid is equal to the inlet temperature of cold fluid. Then, the possible maximum heat transfer quantity is equal to the heat capacity of hot fluid multiplied by the difference between hot fluid inlet temperature and cold fluid outlet temperature, which is the possible maximum temperature difference.

(1) 当热流体的热容量远小于冷流体的热容量时，热流体的出口温度就等于冷流体的进口温度。那么，可能的最大传热量可用热流体的热容量乘以热流体的进口温度与冷流体的进口温度之差，即可能的最大温差。

$$\Phi_{max}=q_{mh}c_h(t_h'-t_c') \tag{10.25}$$

(2) When the heat capacity of hot fluid is much greater than that of cold fluid, the outlet temperature of cold fluid is equal to the inlet temperature of hot fluid. Then, the possible maximum heat transfer quantity is equal to the heat capacity of cold fluid multiplied by the possible maximum temperature difference.

(2) 当热流体的热容量远大于冷流体的热容量时，冷流体的出口温度就等于热流体的进口温度。那么，可能的最大传热量可用冷流体的热容量乘以可能的最大温差。

$$\Phi_{max}=q_{mc}c_c(t_h'-t_c') \tag{10.26}$$

So, the two working conditions can be written uniformly as: The possible maximum heat transfer quantity is equal to the smaller heat capacity of two fluids multiplied by the possible maximum temperature difference. Where the smaller heat capacity can be recorded as C_{min}.

于是，两种工况可统一写为：可能的最大传热量等于两种流体中较小的热容量乘以可能的最大温差。这里较小的热容量可记作 C_{min}。

$$\Phi_{max}=(q_m c)_{min}(t_h'-t_c')=C_{min}(t_h'-t_c') \tag{10.27}$$

However, the actual heat transfer quantity Φ is always less than the possible maximum heat transfer quantity Φ_{max}. The ratio of the actual heat transfer quantity to the maximum heat transfer quantity can be defined as the effectiveness of the heat exchanger, and expressed by ε. Where the actual heat transfer quantity Φ can be expressed by the heat release quantity of hot fluid, or the heat absorption quantity of cold fluid.

然而，实际传热量 Φ 总是小于可能的最大传热量 Φ_{max}。将实际传热量比上最大传热量，可定义为换热器的效能，并用 ε 表示。这里，实际传热量 Φ 可用热流体的放热量表示，也可用冷流体的吸热量表示。

$$\varepsilon=\frac{\Phi}{\Phi_{max}}=\frac{C_h(t_h'-t_h'')}{C_{min}(t_h'-t_c')}=\frac{C_c(t_c''-t_c')}{C_{min}(t_h'-t_c')} \tag{10.28}$$

For an off-the-shelf heat exchanger, if its effectiveness ε

对于一台现成的换热器，如果已知

and the inlet temperature difference between hot and cold fluids are known, then the actual heat transfer quantity Φ can be easily found. It is more convenient for check calculation, and it is unnecessary to figure out the outlet temperature.

其效能 ε 和冷热流体的进口温差，那么实际传热量 Φ 可以很方便地求出。用于校核计算时会更加方便，不必求出口温度。

$$\Phi=\varepsilon\Phi_{\max}=\varepsilon C_{\min}(T'_{h}-T'_{c}) \tag{10.29}$$

Then, how to calculate the effectiveness ε? After derivation the following results are obtained, for parallel-flow:

那么，效能 ε 如何计算呢？经过推导得出以下结果，对于顺流：

$$\varepsilon_{顺}=\frac{1-\exp\left[-\frac{kA}{C_{\min}}(1+C_{r})\right]}{1+C_{r}} \tag{10.30}$$

Where C_r is equal to the ratio of the low heat capacity to the high heat capacity in two fluids, called the heat capacity ratio, namely $C_r=(q_m c)_{\min}/(q_m c)_{\max}$. kA in the effectiveness formula of heat exchanger depends on the design of heat exchanger, the low heat capacity $C_{\min}$ depends on the operating conditions of the heat exchanger. Therefore, $kA/C_{\min}$ characterizes the comprehensive technical and economic performance of a heat exchanger to a certain extent. The ratio is traditionally defined as the heat transfer unit number, which is a dimensionless number, and denoted as NTU.

式中，C_r 等于两种流体中的小热容量与大热容量之比，称为热容比，即 $C_r=(q_m c)_{\min}/(q_m c)_{\max}$。换热器效能公式中的 kA 依赖于换热器的设计，小热容量 $C_{\min}$则依赖于换热器的运行条件。因此，$kA/C_{\min}$ 在一定程度上表征了换热器的综合技术经济性能。习惯上将这个比值定义为传热单元数，是一个无量纲数，记作 NTU。

$$\mathrm{NTU}=\frac{kA}{C_{\min}} \tag{10.31}$$

The relationship between the effectiveness ε and the heat transfer unit number NTU is as follows:

效能 ε 与传热单元数 NTU 之间的关系式如下：

$$\varepsilon_{顺}=\frac{1-\exp[-\mathrm{NTU}(1+C_{r})]}{1+C_{r}} \tag{10.32}$$

$$\varepsilon_{逆}=\frac{1-\exp[-\mathrm{NTU}(1-C_{r})]}{1-C_{r}\exp[-\mathrm{NTU}(1-C_{r})]} \tag{10.33}$$

The following is an example for the check calculation of the heat exchanger.

下面是一个换热器校核计算的例题。

Example 10.7 In Example 10.6, given the mass flow rate and inlet temperature of oil in the oil cooler, the

例 10.7 在例题 10.6 中，已知冷油器中油的质量流量和进口温度、

mass flow rate and the inlet temperature of the water, and the physical parameters of oil and water, etc. If the inlet oil temperature of the oil cooler becomes 58.7℃ and the other parameters remain unchanged, find the outlet oil temperature and outlet water temperature.

水的质量流量和进口温度，以及油和水的物性参数等。如果冷油器的进口油温变为58.7℃，而其他参数均不变，求出口油温和出口水温。

Analysis: The operation conditions often change in the operation of the heat exchanger, which is called the variable condition. The change of inlet oil temperature is a possible variable condition. If the temperature of oil and water rises a lot, then the influence of physical property change on k should be considered. Now the temperature change is very small, it can be considered that the heat transfer coefficient is still 313W/(m^2 · K). This question can be calculated by the effectiveness-heat transfer unit number method.

分析： 换热器运行时经常会发生运行条件的变化，称为变工况。进口油温的变化是一种可能的变工况。如果油和水的温度升高很多，那么需考虑物性变化对 k 的影响。现在温度变化很小，可认为传热系数仍为313W/(m^2 · K)。此题可采用效能-传热单元数法计算。

Solution: The heat capacity of oil:

解： 油的热容量：

$$q_{m1}c_1 = 1.083\times10^{-2}\text{m}^3/\text{s}\times879\text{kg/m}^3\times1.95\times10^3\text{J/(kg}\cdot\text{K)} = 1.856\times10^4\text{W/K}$$

The heat capacity of water:

水的热容量：

$$q_{m2}c_2 = 13.25\text{kg/s}\times4.19\times10^3\text{J/(kg}\cdot\text{K)} = 5.552\times10^4\text{W/K}$$

The heat capacity ratio:

热容比：

$$C_r = \frac{(q_m c)_{\min}}{(q_m c)_{\max}} = \frac{1.856\times10^4\text{W/K}}{5.552\times10^4\text{W/K}} = 0.334$$

Heat transfer unit number:

传热单元数：

$$\text{NTU} = \frac{kA}{(q_m c)_{\min}} = \frac{313\times51.5\text{W/K}}{1.856\times10^4\text{W/K}} = 0.87$$

Consulting Fig. 10.23 to get the effectiveness $\varepsilon=0.54$.

查图10.23可得，效能 $\varepsilon=0.54$。

Heat transfer rate can be calculated according to Eq. (10.29):

热流量可按式（10.29）计算：

$$\begin{aligned}\Phi &= \varepsilon(q_m c)_{\min}(t_1'-t_2') \\ &= 0.54\times1.856\times10^4\text{W/K}\times(58.7℃-33℃) \\ &= 2.58\times10^5\text{W}\end{aligned}$$

The outlet temperatures of oil and water can be obtained by the heat balance equation:

由热平衡式可求得油和水的出口温度：

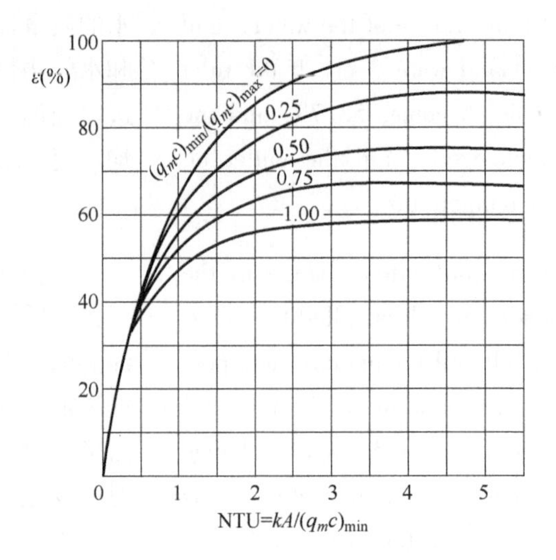

Fig. 10. 23 ε-NTU relationship for single shell side, 2, 4, 6 etc. tube side heat exchangers

图 10.23 单壳程，2、4、6 等管程换热器的 ε-NTU 关系图

$$t_1'' = t_1' - \frac{\Phi}{q_{m1} c_1} = 44.8℃$$

$$t_2'' = t_2' - \frac{\Phi}{q_{m2} c_2} = 37.6℃$$

Discussion: This question is a typical example of the heat transfer unit number method being superior to the mean temperature difference method. If the mean temperature difference method is used, then the calculation process will be more complicated.

讨论： 该题是采用传热单元数法优于平均温差法的典型例题。如果采用平均温差法，计算过程会更加复杂。

Summary This section illustrates a thermal calculation method of the heat exchanger, namely effectiveness-heat transfer unit number method.

小结 本节阐述了一种换热器热计算方法，即效能-传热单元数法。

Question In the thermal calculation of the heat exchanger, what are the characteristics of the mean temperature difference method and the effectiveness-heat transfer unit number method, respectively?

思考题 在换热器热计算中，平均温差法和效能传热单元数法各有什么特点？

Self-tests

自测题

1. What is the effectiveness and the heat transfer unit number of the heat exchanger?

1. 什么是换热器的效能和传热单元数？

2. In the check calculation of heat exchanger, why the heat transfer unit number method is more convenient?

2. 在换热器的校核计算时，为什么传热单元数法较为方便？

授课视频

10.8 The Application Examples for the Thermal Calculation of Heat Exchangers

10.8 换热器热计算应用实例

The mean temperature difference method is commonly used in the thermal calculation of heat exchangers. Whether design calculation or check calculation, two basic equations must be used for the thermal calculation of heat exchangers:

换热器热计算常用平均温差法。不论是设计计算还是校核计算，换热器热计算都必须采用两个基本方程式：

(1) The total heat transfer equation:

(1) 总传热方程式：

$$\Phi = kA\Delta t_{\mathrm{m}}$$

(2) The heat balance equation:

(2) 热平衡方程式：

$$\Phi = q_{mh}c_{\mathrm{h}}(t'_{\mathrm{h}} - t''_{\mathrm{h}}) = q_{mc}c_{\mathrm{c}}(t''_{\mathrm{c}} - t'_{\mathrm{c}}) = q_{m}c\Delta t$$

Heat transfer rate Φ is equal to the heat release quantity of hot fluid, and is also equal to the heat absorption quantity of cold fluid.

传热量 Φ 等于热流体的放热量，也等于冷流体的吸热量。

The following examples illustrate the application of thermal calculation method of heat exchanger, and focus on the analysis of solution idea rather than specific number calculation.

下面举例说明换热器热计算方法的应用，重点分析求解思路，不做具体的数值计算。

Example 10.8 A heat exchanger with two shell sides and eight tube sides for waste heat recovery, the total heat transfer area is $925\mathrm{m}^2$, the water with a mass flow rate of 45500kg/h is heated from 80℃ to 150℃. The temperature of exhaust gas entering the heat exchanger is 350℃, and the outlet temperature is 175℃, the specific heat capacity of water is 4236J/(kg · K), and the specific heat capacity of the exhaust gas is 1050J/(kg · K), the temperature difference correction factor $\Psi = 0.97$. Try to

例 10.8 一台回收余热的双壳程八管程换热器，总传热面积为 $925\mathrm{m}^2$，将质量流量为 45500kg/h 的水由 80℃ 加热至 150℃。废气进入换热器的温度为 350℃，出口温度为 175℃，水的比热容为 4236J/(kg · K)，废气的比热容为 1050J/(kg · K)，温差修正系数 $\Psi = 0.97$。试计算废气的质量流量和换热器的传热系数。

calculate the mass flow rate of the exhaust gas and the overall heat transfer coefficient of the heat exchanger.

Solution: Given mass flow rate, specific heat capacity and inlet and outlet temperatures of water, and the specific heat capacity and inlet and outlet temperatures of exhaust gas. Then, it is easy to figure out the total heat transfer quantity Φ and the mass flow rate of exhaust gas from the heat balance equation.

解：已知水的质量流量、比热容和进出口温度，以及废气的比热容和进出口温度。那么，由热平衡方程式很容易求出总传热量 Φ，以及废气的质量流量。

$$\Phi = q_{m1}c_{p1}(t_1'-t_1'') = q_{m2}c_{p2}(t_2''-t_2')$$

$$q_{m1} = \Phi/[c_{p1}(t_1'-t_1'')]$$

From four inlet and outlet temperatures of two fluids, the logarithmic mean temperature difference in counter-flow can be calculated.

由两种流体的四个进出口温度，可求出逆流时对数平均温差。

$$\Delta t_{m,c} = (\Delta t_{max} - \Delta t_{min})/\ln(\Delta t_{max}/\Delta t_{min})$$

Then, multiply by the temperature difference correction factor Ψ, the actual logarithmic mean temperature difference can be obtained.

再乘以温差修正系数 Ψ，可得到实际的对数平均温差。

$$\Delta t_m = \Psi \Delta t_{m,c}$$

The overall heat transfer coefficient k of the heat exchanger can be obtained from the total heat transfer equation.

换热器的传热系数 k 可由总传热方程式求出。

$$k = \Phi/(A\Delta t_m)$$

Example 10.9 A thin-walled concentric double-pipe heat exchanger is used to cool engine oil from 160℃ to 60℃, and the water with the inlet temperature of 25℃ is used as the coolant, the mass flow rates of engine oil and water are 2kg/s, the specific heat capacity of engine oil is 2260J/(kg · K), and the specific heat capacity of water is 4179J/(kg · K). It is known that the outer diameter of the inner tube of the heat exchanger is 0.5m, the corresponding overall heat transfer coefficient is 250W/(m^2 · K), the shell of the heat exchanger is well insulated. (1) When the required cooling temperature is reached, what is the tube length required for the heat exchanger? (2) Qualitatively draw the temperature profiles of oil and water along the way.

例 10.9 一台薄壁同心套管式换热器，用于将初温为 160℃ 的机油冷却至 60℃，用进口温度为 25℃ 的水作为冷却剂，机油和水的质量流量皆为 2kg/s，机油的比热容为 2260J/(kg · K)，水的比热容为 4179J/(kg · K)。已知换热器内管外径为 0.5m，相应的总传热系数为 250W/(m^2 · K)，换热器外壳绝热良好。（1）达到所要求的冷却温度时，换热器所需的管长是多少？（2）定性画出油和水的温度沿程变化曲线。

Solution: The mass flow rate, specific heat capacity, and inlet and outlet temperatures of engine oil, and the mass flow rate, the specific heat capacity, and the inlet temperatures of water are known. Then, it is easy to figure out the heat release quantity of engine oil, and the outlet temperature of cooling water from the heat balance equation:

解： 已知机油的质量流量、比热容和进出口温度，以及水的质量流量、比热容和进口温度。那么，由热平衡方程式很容易求出机油的放热量，以及冷却水的出口温度：

$$\Phi = q_{m1} c_1 (t_1' - t_1''), \qquad t_2'' = t_2' + \frac{\Phi}{q_{m_2} c_2}$$

The outlet temperature of cooling water is higher than that of engine oil, so it must be arranged in a counter-flow way, the logarithmic mean temperature difference can be calculated according to counter-flow:

由于冷却水的出口温度高于机油的出口温度，所以必然是逆流方式布置，可按逆流求出对数平均温差：

$$\Delta t_{\mathrm{m,c}} = \frac{(t_1' - t_2'') - (t_1'' - t_2')}{\ln[(t_1' - t_2'')/(t_1'' - t_2')]}$$

The required heat transfer area and the required tube length of the heat exchanger can be obtained from the total heat transfer equation:

由总传热方程式可求出所需传热面积及换热器所需管长：

$$A = \frac{\Phi}{k \Delta t_{\mathrm{m}}}, \qquad l = \frac{A}{\pi d_{\mathrm{o}}}$$

Qualitatively draw the temperature profiles of oil and water along the way, as shown in Fig. 10.24, the characteristic of these curves is concave downwards.

定性画出油和水的温度沿程变化曲线，如图 10.24 所示，曲线的特点是向下凹。

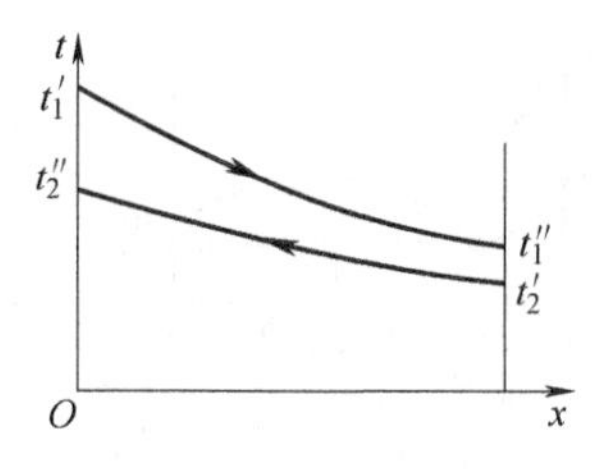

Fig. 10.24 The temperature profiles of oil and water along the way

图 10.24 油和水的温度沿程变化曲线

Example 10.10 The freon superheated vapor with the inlet temperature of 90℃ is supercooled to 30℃ liquid in the counter-flow type heat exchanger. The saturation temperature of freon is 40℃, the latent heat of vaporization is 133.33kJ/kg, the specific heat capacity in the gaseous

例 10.10 进口温度为 90℃的氟利昂过热蒸气，在逆流换热器中过冷为 30℃的液体。氟利昂的饱和温度为 40℃，汽化潜热为 133.33kJ/kg，气态时比热容为 0.68kJ/(kg·K)，液态

state is 0.68kJ/(kg · K), and the specific heat capacity in the liquid state is 1kJ/(kg · K), and the inlet and outlet temperatures of the cooling water in the heat exchanger is 20℃ and 25℃, respectively. The specific heat capacity of water is 4.183kJ/(kg · K), the heat transfer quantity of the heat exchanger is 69000kJ/h, and the overall heat transfer coefficient in the heat exchanger is 1200W/(m^2 · K). Try to calculate the heat transfer area.

时比热容为1kJ/(kg · K)，冷却水进出换热器的温度分别为20℃和25℃。水的比热容为4.183kJ/(kg · K)，换热器的传热量为69000kJ/h，在整个换热器中传热系数均为1200W/(m^2 · K)。试计算换热面积。

Analysis: From the known conditions, the subject gives the heat transfer quantity, overall heat transfer coefficient, and the inlet and outlet temperatures of two fluids in the heat exchanger, it seems that the heat transfer area can be obtained by directly substituting them into the total heat transfer equation. The problem seems to be very simple, but this calculation method is wrong.

分析： 从已知条件来看，题目给出了换热器的传热量、传热系数及两种流体的进出口温度，好像直接代入总传热方程式即可求出换热面积。看起来这个问题很简单，但是这种计算方法是错误的。

Freon vapor at 90℃ is superheated, it needs to drop to the saturation temperature of 40℃ before condensing. After the vapor condenses and continues to cool, freon liquid can reach supercooled state of 30℃. Therefore, during the entire exothermic process of freon, there are three stages: superheating, phase transition, and supercooling, then the heat release quantity shall be calculated in three stages, as shown in Fig. 10.25.

氟利昂蒸气90℃时为过热状态，需要降到饱和温度40℃才能发生冷凝。蒸气冷凝后又继续冷却，氟利昂液体才能达到30℃的过冷状态。因此，氟利昂的整个放热过程中，有过热、相变和过冷三个阶段，所以放热量需要分三段进行计算，如图10.25所示。

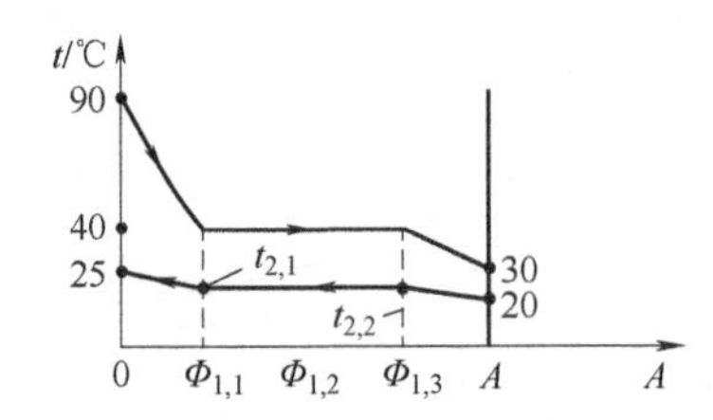

Fig. 10.25 Temperature changes of two fluids

图10.25 两种流体的温度变化

Solution: (1) Overall heat release quantity of freon is equal to the sensible heat of superheated vapor plus the latent heat of phase transition, and plus the sensible heat of the supercooled liquid again.

解： (1) 氟利昂的总放热量等于过热蒸气的显热+相变潜热+过冷液体的显热。

$$\Phi=\Phi_{1,1}+\Phi_{1,2}+\Phi_{1,3}=c_{p,g}q_{m1}(t_1'-t_s)+q_{m1}\gamma+c_{p,l}q_{m1}(t_s-t_1'')$$

Substituting the known parameters to obtain the mass flow rate of freon q_{m1}. Then, according to the above equation, the heat transfer quantity of the three stages can be calculated separately: $\Phi_{1,1}$、$\Phi_{1,2}$、$\Phi_{1,3}$

代入已知参数可求出氟利昂的质量流量 q_{m1}。然后，根据上式可分别求出三段的传热量：$\Phi_{1,1}$、$\Phi_{1,2}$、$\Phi_{1,3}$。

(2) By heat balance equation, the mass flow rate of water q_{m2} can be found:

(2) 由热平衡式可求出水的质量流量 q_{m2}：

$$\Phi = q_{m2} c_{p,\mathrm{w}} (t_2'' - t_2')$$

(3) The water temperature corresponding to each stage can be obtained from the heat balance equation: $t_{2,1}$, $t_{2,2}$.

(3) 由热平衡式可求出对应于每一段的水温：$t_{2,1}$、$t_{2,2}$。

$$\Phi_{1,1} = q_{m2} c_{p,\mathrm{w}} (t_2'' - t_{2,1}), \quad \Phi_{1,2} = q_{m2} c_{p,\mathrm{w}} (t_{2,1} - t_{2,2})$$

(4) Calculate the logarithmic mean temperature difference of each stage separately, here is the calculation formula of Δt_{m1}:

(4) 分段求出每一段的对数平均温差，这里给出了 Δt_{m1} 的计算式：

$$\Delta t_{\mathrm{m1}} = \frac{(t_1' - t_2'') - (t_\mathrm{s} - t_{2,1})}{\ln \dfrac{t_1' - t_2''}{t_\mathrm{s} - t_{2,1}}}$$

(5) The heat transfer area of the three stages can be calculated separately: A_1, A_2, A_3.

(5) 可分段求出三段的换热面积：A_1、A_2、A_3。

$$A_1 = \frac{\Phi_{1,1}}{k \Delta t_{\mathrm{m1}}}, \quad A_2 = \frac{\Phi_{1,2}}{k \Delta t_{\mathrm{m2}}}, \quad A_3 = \frac{\Phi_{1,3}}{k \Delta t_{\mathrm{m3}}}$$

Then, the overall heat transfer area is equal to the sum of the three areas.

那么，总传热面积等于三者之和。

Example 10.11 A coiled tube heat exchanger is shown in Fig. 10.26, the water vapor with the pressure of 0.361MPa corresponding to the saturation temperature of 140℃ condenses outside the tube, the convective heat transfer coefficient is 9500W/(m² · K). Cooling water flows in the coiled tube, at a velocity of 0.8m/s. The inlet temperature of cooling water is 25℃, and it is heated to 95℃. The outer diameter of brass tube is 18mm, wall thickness is 1.5mm, and the thermal conductivity is 132W/(m · K). The bending radius of the coiled tube is

例 10.11 盘管式换热器如图 10.26 所示，压力为 0.361MPa 对应饱和温度为 140℃的水蒸气在管外凝结，表面传热系数为 9500W/(m² · K)。冷却水在盘管内流动，流速为 0.8m/s，冷却水进口温度为 25℃，被加热到 95℃。黄铜管外径为 18mm，壁厚为 1.5mm，导热系数为 132W/(m · K)，盘管的弯曲半径为 90mm，不考虑管内入口效应

90mm, no consideration of the entrance effects inside the tube and the temperature difference correction. Try to find the required heat transfer area and the coiled tube length.

及温差修正。试求所需的换热面积和盘管长度。

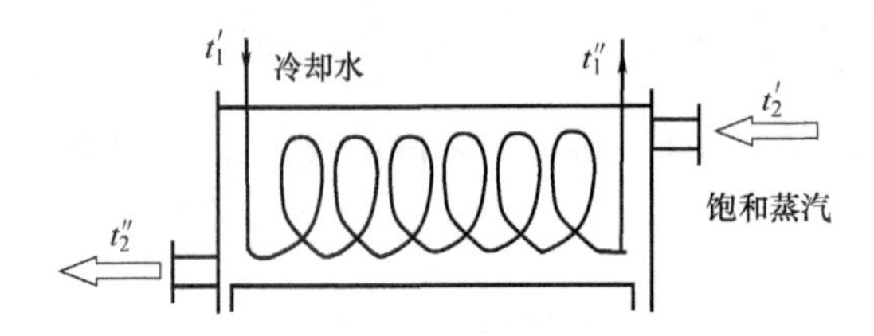

Fig. 10. 26 A coiled tube heat exchanger

图 10. 26 盘管式换热器

Solution: The reference temperature of the water is 60℃, the physical parameters of water can be identified:

解: 水的定性温度为 60℃, 可查出水的物性参数:

$$\rho=983.1\text{kg/m}^3, c_p=4.179\text{kJ/(kg}\cdot\text{K)}, \lambda=65.9\times10^{-2}\text{W/(m}\cdot\text{K)}, \nu=0.478\times10^{-6}\text{m}^2\text{/s}, Pr=2.99$$

(1) The experimental correlation of turbulent forced convection heat transfer in the tube is the Ditus-Belt formula, by multiplying the correction coefficient of the elbow, $c_r=1+10.3(d/R)^3$, the convective heat transfer coefficient h_i in the tube can be obtained:

(1) 管内湍流强制对流换热实验关联式, 即迪图斯-贝尔特公式, 乘以弯管修正系数, $c_r=1+10.3(d/R)^3$, 即可求出管内表面传热系数 h_i:

$$Nu=0.023Re_f^{0.8}Pr_f^{0.4}c_r$$

(2) The heat transfer quantity can be obtained from the heat balance equation:

(2) 由热平衡式可求出传热量:

$$\Phi=\rho u\frac{\pi}{4}d_i^2c_p(t_2''-t_2')$$

(3) The logarithmic mean temperature difference can be obtained from the four temperatures:

(3) 由两种流体的进出口温度, 可求出对数平均温差:

$$\Delta t_m=\frac{(t_1'-t_2'')-(t_1''-t_2')}{\ln\dfrac{t_1'-t_2''}{t_1''-t_2'}}$$

(4) The overall heat transfer coefficient is calculated based on the outer diameter of the tube:

(4) 以管外径为基准计算传热系数为:

$$k_o=\frac{1}{\dfrac{1}{h_i}\dfrac{d_o}{d_i}+\dfrac{d_o}{2\lambda}\ln\dfrac{d_o}{d_i}+\dfrac{1}{h_o}}$$

(5) From the total heat transfer equation, find the heat transfer area and the tube length:

(5) 由总传热方程式，求换热面积及管长：

$$A=\frac{\Phi}{k_0\Delta t_m},\quad l=\frac{A}{\pi d_o}$$

Summary This section illustrates four application examples for the thermal calculation of heat exchanger.

小结 本节阐述了换热器热计算的四个应用实例。

Question What is the connection between the two basic equations in the thermal calculation of heat exchanger?

思考题 换热器热计算必须用到的两个基本方程式之间有什么联系？

授课视频

10.9 Enhanced Heat Transfer Technologies of Heat Exchangers

10.9 换热器的强化传热技术

According to the requirement of different occasions, the actual heat transfer process sometimes needs to be enhanced and sometimes needs to be weakened. Enhanced heat transfer is mainly aimed at convection heat transfer, while weakened heat transfer is mainly aimed at heat conduction and radiation heat transfer. The purpose and means of enhancing and weakening heat transfer are different with different heat transfer modes.

根据不同场合的需求，实际传热过程有时需要强化，有时则需要弱化。强化传热主要针对对流传热，而弱化传热主要针对导热过程和辐射传热。传热方式不同，强化和弱化传热的目的及手段也不同。

Enhanced heat transfer is to increase the heat transfer quantity, or improve the heat transfer coefficient in the heat transfer process. The purpose of heat transfer enhancement is to improve heat transfer efficiency, reduce equipment and weight. The purpose of weakening heat transfer is to reduce heat loss and save energy.

强化传热就是增大传热过程的传热量，或提高传热系数。强化传热的目的是提高传热效率、缩小设备、减轻重量。弱化传热的目的是减少热量损失和节约能源。

Enhanced heat transfer technologies refer to various techniques adopted to increase heat transfer quantity or improve heat transfer coefficient while keeping the main structure of heat exchange equipment basically unchanged. Such as process technology, heat exchange surface technology,

强化传热技术，是在保持换热设备主体结构基本不变的情况下，为增大传热量或提高传热系数而采取的各种技术，如工艺技术、换热表面技术、外部技术等。需要注意，靠增加换热

external technology, etc. It should be noted that simply increasing the heat transfer area by increasing the number of heat transfer tubes does not belong to enhanced heat transfer.

管数量简单地增大换热面积不属于强化传热。

According to the overall heat transfer equation $\Phi = kA\Delta t_m$, there are three ways to enhance heat transfer:

由总传热方程式 $\Phi = kA\Delta t_m$ 可知，强化传热的途径有以下三种：

(1) Increasing the heat transfer temperature difference. For example, change the inlet and outlet temperatures of hot and cold fluids, or change the arrangement of fluid flow, such as parallel-flow into counter-flow.

(1) 增大传热温差。如改变冷热流体的进出口温度，或改变流动的布置方式，如顺流变为逆流。

(2) Increasing the heat transfer area. For example, the finned tube is an effective method to enhance heat transfer.

(2) 增大传热面积。如采用翅片管是强化传热的一种有效方法。

(3) Improving the heat transfer coefficient. This is the most commonly used way to enhance heat transfer, such as spiral groove tube, corrugated tube, etc., which can change the flow state of fluid, so as to improve the heat transfer coefficient.

(3) 提高传热系数。这是使用最多的强化传热途径，如螺旋槽管、波节管等，可改变流体的流动状态，从而提高传热系数。

According to the calculation formula of heat transfer coefficient, the heat transfer process is generally composed of three links. The principle of enhancing heat transfer is: Which link has a large thermall resistance is large, take heat transfer enhancement measures for that link, the effect will be more significant.

由传热系数计算式可知，传热过程一般由三个环节构成。强化换热的原则是：哪个环节的分热阻大，对那个环节采取强化传热措施，效果会更显著。

The influencing factors of heat transfer enhancement are discussed below. Taking fully developed turbulent heat transfer in a circular tube as an example, the experimental correlation equation, i. e. the characteristic number equation, is:

下面探讨强化传热的影响因素。以圆管内充分发展湍流换热为例，实验关联式即特征数方程为：

$$Nu = 0.023 Re_f^{0.8} Pr_f^{0.4}$$

Each characteristic number is represented by the corresponding physical quantities, after sorting out to get:

把各个特征数用对应的物理量表示，整理后得：

$$h = \frac{0.023 c_p^{0.4} \lambda^{0.6} \rho^{0.8} u^{0.8}}{\eta^{0.4} d^{0.2}} \tag{10.34}$$

From the expression of the convective heat transfer coefficient h, it can be seen that its influencing factors include geometric factors d, process factor u, and physical property factors ρ, η, λ, and c_p.

从表面传热系数 h 的表达式可以看出，其影响因素包括几何因素 d，工艺因素 u，物性因素 ρ、η、λ 和 c_p。

According to the classification method of international heat transfer expert Pro. Bergles: Enhanced heat transfer technologies can be divided into enhanced technologies with power and enhanced technologies without power.

按照国际传热专家 Bergles 的分类方式，强化传热技术分为无源强化技术和有源强化技术两类。

Enhanced technologies without power do not require additional power, also called passive enhanced technologies. The main technologies include coating surface, rough surface, extended surface, spoiler element, vortex generator, spiral tube, additives such as nanoparticles, etc., as shown in Fig. 10.27–Fig. 10.36.

无源强化技术不需要外加动力，又称为被动强化技术。其主要技术有涂层表面、粗糙表面、扩展表面、扰流元件、涡流发生器、螺旋管、添加物（如纳米粒子）等，如图 10.27~图 10.36 所示。

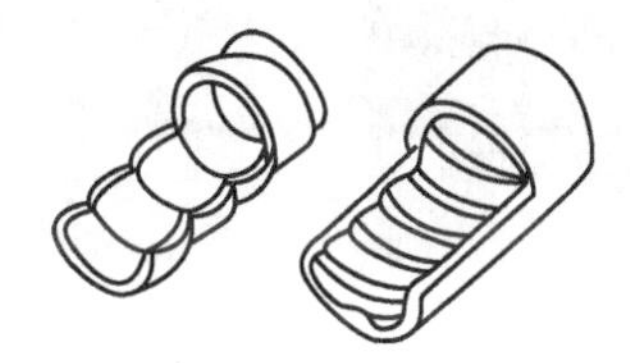

Fig. 10.27 Tube with rough surface

图 10.27 粗糙表面管

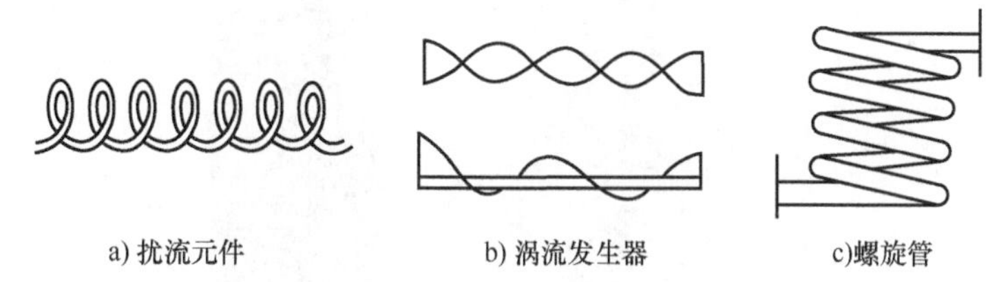

Fig. 10.28 Methods for enhancing disturbance and mixing in fluid

图 10.28 增强流体中扰动与混合的方法

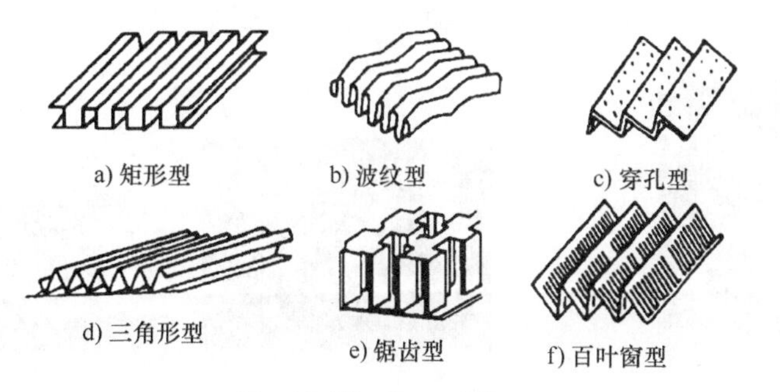

Fig. 10.29 Integral fins

图 10.29 整体式翅片管

The following figures are the achievement of passive enhanced technologies, such as transverse groove tubes, converging diverging tubes and spiral groove tubes, etc., as shown in Fig. 10.30 – Fig. 10.32. And various finned tubes such as transverse fin tube, longitudinal fin tube, serrated fin tube, needle fin tube, etc., as shown in Fig. 10.33-Fig. 10.36.

下面这些图片就是无源强化技术的成果，如横纹管、缩放管、螺旋槽管等，如图 10.30~图 10.32 所示，以及各种翅片管，如横向翅片管、纵向翅片管、锯齿状翅片管、针状翅片管等，如图 10.33 ~ 图 10.36 所示。

Fig. 10.30 Transverse groove tube

图 10.30 横纹管

Fig. 10.31 Converging diverging tube

图 10.31 缩放管

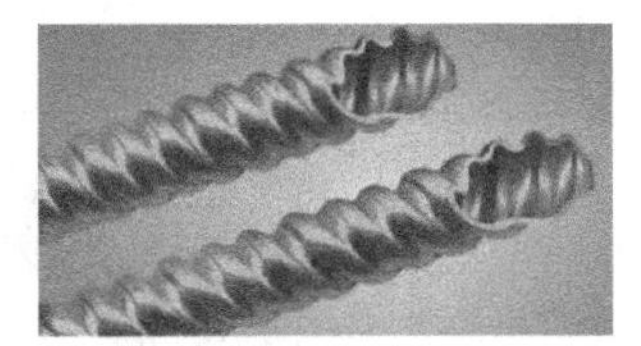

Fig. 10.32 Spiral groove tube

图 10.32 螺旋槽管

Fig. 10.33 Transverse fin tube

图 10.33 横向翅片管

Fig. 10.34 Longitudinal fin tube

图 10.34 纵向翅片管

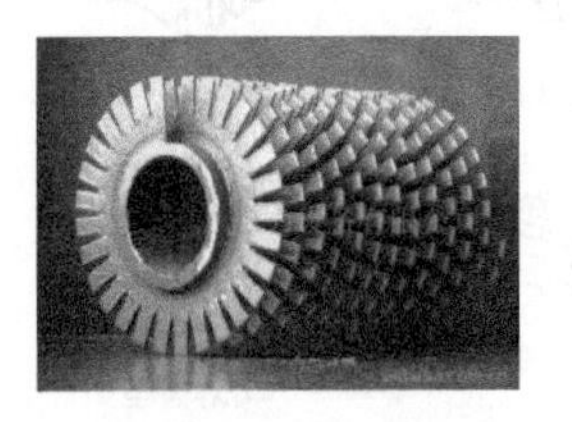

Fig. 10.35 Serrated fin tube

图 10.35 锯齿状翅片管

Fig. 10.36 Needle fin tube

图 10.36 针状翅片管

Enhanced technologies with power require additional power, also known as active enhanced technologies. The main technologies include mechanical stirring of the heat exchange medium, vibration of the heat exchange surface, vibration of the heat exchange fluid, and electromagnetic field acting on the fluid, so as to promote mixing of the fluid near the heat exchange surface.

有源强化技术需要外加动力，又称为主动强化技术。其主要技术有对换热介质做机械搅拌、使换热表面振动、使换热流体振动、将电磁场作用于流体等，促使换热表面附近流体混合。

To sum up, there are three mechanisms of convection heat transfer enhancement without phase transition: ① Thinning the boundary layer; ② Increasing temperature gradient; ③ Increasing fluid turbulence to promote mixing of all parts of fluid, particularly disturbance and mixing of fluid near the heat transfer wall.

归纳起来，无相变对流传热强化机理有三个：① 减薄边界层；② 增大温度梯度；③ 增加流体的湍动，促使流体中各部分混合，特别是换热壁面附近流体的扰动与混合。

For nucleate boiling, the key to heat transfer enhancement is to increase vaporization core. For film condensation, the key to heat transfer enhancement is to thin liquid film and accelerate the discharge of liquid film.

对于核态沸腾，强化传热的关键在于增加汽化核心。对于膜状凝结，强化传热的关键是减薄液膜，及加速液膜的排出。

We need to note that any method that can enhance convection heat transfer of a single-phase fluid will inevitably cause the increase of flow resistance. Therefore, heat transfer effect, flow resistance, manufacturing cost, operating cost and other factors should be comprehensively considered in evaluating an enhanced heat transfer technology.

需要注意：凡是能强化单相流体对流传热的方法，都不可避免地会引起流动阻力的增大。因此，评价一种强化传热技术，应综合考虑其传热效果、流动阻力、制造成本及运行费用等因素。

Summary This section illustrates the ways, principles, and influencing factors of heat transfer enhancement, as well as the classification of enhanced heat transfer technologies and enhanced heat transfer mechanisms.

小结 本节阐述了强化传热的途径、原则和影响因素，以及强化传热技术的分类和强化传热机理。

Question What are the enhanced heat transfer mechanisms of heat exchangers?

思考题 换热器强化传热的机理有哪些？

Self-tests

自测题

1. What is the principle of enhancing heat transfer?

1. 强化传热的原则是什么？

2. What are enhanced heat transfer technologies with power and enhanced heat transfer technologies without power?

2. 什么是有源强化传热技术和无源强化传热技术？

10.10 The Thermal Resistance Separation Method in the Heat Transfer Process

10.10 传热过程的热阻分离法

After the heat exchanger runs for a period of time, the heat transfer surface will often accumulate scale, sludge, oil contamination, soot and other covering scale layer, as shown in Fig. 10.37. The fluid corrodes the heat transfer surface and forms a scalc layer, and the thermal conductivity of the scale layer is generally very small, which greatly increases the thermal resistance in the heat transfer process.

换热器运行一段时间以后，换热面上常常会积聚水垢、污泥、油污、烟灰之类的覆盖物垢层，如图10.37所示。流体腐蚀换热面也会形成垢层，垢层的导热系数一般很小，大大增加了传热过程的热阻。

Fig. 10.37 Dirt on heat transfer surface

图10.37 换热表面上的污垢

With the improvement of the enhancement measures of the heat exchanger, fouling resistance sometimes becomes the main thermal resistance in the heat transfer process. Therefore, reasonable fouling resistance data should be provided in the design of the heat exchanger, which are usually determined by experiment. The calculation formula of fouling resistance is:

随着换热器强化措施的完善，污垢热阻有时会成为传热过程的主要热阻。因此换热器设计时，需要提供合理的污垢热阻数据，这些数据通常由实验来测定。污垢热阻的计算式为：

$$R_f = \frac{1}{k_o} - \frac{1}{k} \tag{10.35}$$

Where k_o is the overall heat transfer coefficient of the fouling heat exchanger, and k is the overall heat transfer coefficient of the clean heat exchanger. The overall heat transfer coefficient k is measured experimentally, how to break it down into the partial thermal resistance of each link?

式中，k_o 是结垢换热器的传热系数；k 是洁净换热器的传热系数。通过实验测出来传热系数 k，如何分解成各个环节的分热阻呢？

The most commonly used method to determine the partial thermal resistance of the heat transfer process is the Wilson graphical method. First, the overall heat transfer coefficient is required, and the steps are as follows:

确定传热过程的分热阻最常用的方法是威尔逊图解法。首先要求出总传热系数，步骤如下：

① Measure the inlet and outlet temperatures of the fluid to obtain the mean temperature difference; ② Measure the flow rates of two kinds of fluid by the flowmeter, and calculate the heat transfer rate by the heat balance equation; ③ The heat transfer area can be obtained according to the design situation; ④ The overall heat transfer coefficient k can be calculated by the overall heat transfer equation.

① 测定流体的进出口温度，从而获得平均温差；② 用流量计测得两种流体的流量，并用热平衡式计算热流量；③ 根据设计情况可获得换热面积；④ 通过传热方程式可计算出总传热系数 k。

This is the basis of the Wilson graphical method. Let's take the shell-and-tube heat exchanger as an example to present the steps of obtaining the partial thermal resistance by the Wilson graphical method:

这是威尔逊图解法的基础。下面以管壳式换热器为例，介绍威尔逊图解法获得分热阻的步骤：

(1) List expression of the overall heat transfer coefficient, which is based on the outer diameter:

(1) 列出总传热系数表达式，这里以外径为基准：

$$\frac{1}{k_o}=\frac{1}{h_o}+R_w+R_f+\frac{1}{h_i}\frac{d_o}{d_i} \tag{10.36}$$

Where R_w is the thermal conduction resistance of the tube wall, and R_f is the fouling resistance. In industrial heat exchangers, the fluid flow in the tube is generally in a fully turbulent state, the convective heat transfer coefficient h_i is directly proportional to the 0.8 power of the flow velocity u, so it can be written as:

式中，R_w 是管壁的导热热阻；R_f 是污垢热阻。在工业换热器中，管内流体的流动一般处于旺盛湍流状态，表面传热系数 h_i 和流速 u 的 0.8 次方成正比，因此可以写成：

$$h_i=c_iu_i^{0.8}$$

Substitute into the expression of the overall heat transfer coefficient to get:

代入总传热系数的表达式，可得：

$$\frac{1}{k_o}=\frac{1}{h_o}+R_w+R_f+\frac{1}{c_iu_i^{0.8}}\frac{d_o}{d_i}$$

(2) Set prerequisites: Keep h_o constant, the thermal conductivity resistance R_w of the tube wall remains unchanged, R_f is unchanged in a short time, therefore, the first three terms on the right side of the above equation can be regarded as a constant, denoted by b:

(2) 设定前提条件：保持 h_o 不变，管壁的导热热阻 R_w 不变，在短时间内 R_f 基本不变，因此，上式中右边前三项可认为是常数，用 b 表示：

$$b=\frac{1}{h_o}+R_w+R_f \tag{10.37}$$

In the case of constant physical properties, $\frac{1}{c_i}\frac{d_o}{d_i}$ is also considered a constant, denoted by m. So the expression of the overall heat transfer coefficient can be written as:

在物性不变的情况下，$\frac{1}{c_i}\frac{d_o}{d_i}$也可认为是常数，用 m 表示。于是，总传热系数的表达式可写为：

$$\frac{1}{k_o}=b+m\frac{1}{u_i^{0.8}}$$

This expression can be regarded as a linear function:

该表达式可看作一个线性函数：

$$y=b+ax$$

(3) Change the flow velocity u_i in the tube, a series of overall heat transfer coefficients k_o can be measured, and then drawn into a picture which is a straight line, as shown in Fig. 10. 38.

(3) 改变管内流速 u_i，可测得一系列总传热系数 k_o，然后绘制成图，是一条直线，如图 10. 38 所示。

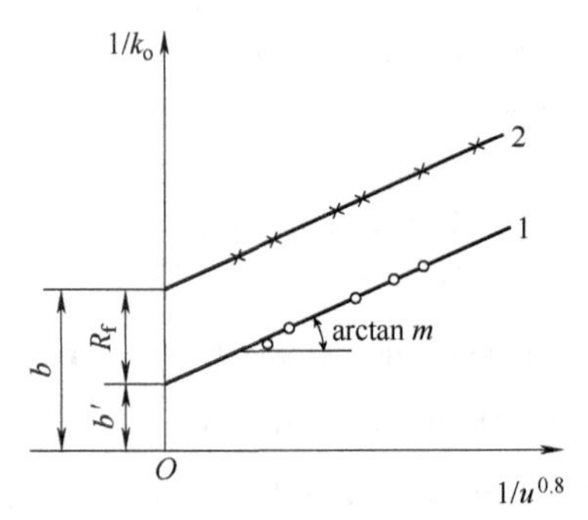

Fig. 10. 38 Wilson graphical method

图 10. 38 威尔逊图解法

The values of b and m can be obtained from the figure, the intercept of the straight line on the ordinate is the value of b, and the slope of the straight line is the value of m.

从图中可获得 b 和 m 的值，直线在纵坐标上的截距即 b 值，直线斜率即 m 值。

$$m=\frac{1}{c_i}\frac{d_o}{d_i}=\tan\theta \tag{10.38}$$

So that can be obtained:

从而可求出：

$$c_i=\frac{1}{m}\frac{d_o}{d_i},\quad h_i=c_iu_i^{0.8} \tag{10.39}$$

The convective heat transfer coefficient h_i in the tube can be separated from the overall heat transfer coefficient.

管内的表面传热系数 h_i 就从总传热系数中分离出来。

(4) After the heat exchanger runs for a period of time, another straight line can be obtained by measuring the same process, as shown in Fig. 10. 38. The difference between the intercepts of the two straight lines is the fouling resistance R_f, thus the fouling resistance is separated out.

(4) 当换热器运行一段时间后，再进行同样过程的测量，可以获得另外一条直线，如图 10. 38 所示。两条直线的截距之差就是污垢热阻 R_f，这样又把污垢热阻分离出来。

It's worth noting: The premise of Wilson graphical method is that the heat transfer resistance of one side basically remains unchanged, sometimes it is difficult to meet the condition.

值得注意：威尔逊图解法的前提是，有一侧的换热热阻基本保持不变，有时此条件很难满足。

Example 10. 12 There is an air cooler, the air flows vertically across the tube bundle outside the tube, the convective heat transfer coefficient outside the tube h_o = 90W/(m^2 · K). The cooling water flows in the tube, the convective heat transfer coefficient in the tube h_i = 6000W/(m^2 · K), the heat transfer tube is a brass tube with outer diameter of 16mm and wall thickness of 1. 5mm. Find: (1) The overall heat transfer coefficient of the air cooler; (2) If the convective heat transfer coefficient h_o outside the tube doubles, how does the overall heat transfer coefficient change? (3) If the convective heat transfer coefficient h_i in the tube also doubles, how does the overall heat transfer coefficient change?

例 10. 12 有一台空气冷却器，空气在管外垂直流过管束，管外表面传热系数 h_o = 90W/(m^2 · K)；冷却水在管内流动，管内表面传热系数 h_i = 6000W/(m^2 · K)；换热管为外径 16mm、厚度 1. 5mm 的黄铜管。求解：(1) 空气冷却器的传热系数；(2) 如果管外表面传热系数 h_o 增加 1 倍，传热系数如何变化？(3) 如果管内表面传热系数 h_i 也增加 1 倍，传热系数如何变化？

Analysis: It can be found that the thermal conductivity of brass is 111W/(m · K). Air flowing outside the tubes should be the main thermal resistance, so the outer surface is used as the basis for calculating k.

分析：可查得黄铜的导热系数为 111W/(m · K)。空气在管外流动，应该是主要热阻，所以以外表面作为 k 的计算基准。

Solution: (1) From the overall heat transfer coefficient formula:

解：(1) 由传热系数计算式：

$$k=\frac{1}{\dfrac{d_o}{h_i d_i}+\dfrac{d_o}{2\lambda}\ln\left(\dfrac{d_o}{d_i}\right)+\dfrac{1}{h_o}}$$

Substituting the given parameters to get:

代入已知参数可求得：

$$k=\frac{1}{\dfrac{1}{6000\text{W}/(\text{m}^2\cdot\text{K})}\times\dfrac{16\text{mm}}{13\text{mm}}+\dfrac{0.016}{2\times111\text{W}/(\text{m}\cdot\text{K})}\ln\left(\dfrac{16}{13}\right)+\dfrac{1}{90\text{W}/(\text{m}^2\cdot\text{K})}}$$

$$=\frac{1}{0.000205+0.0000149+0.0111}\text{W/(m}^2\cdot\text{K)}$$

$$=88.3\text{W/(m}^2\cdot\text{K)}$$

From the above calculation, it can be seen that the heat conduction resistance of the tube wall is over two orders of magnitude smaller than the convection thermal resistance of the fluid inside and outside the tube, so it can be ignored.

由上述计算可见，管壁的导热热阻比管内、外流体对流传热热阻小两个数量级以上，因此可以忽略不计。

(2) If the convective heat transfer coefficient h_o outside the tube doubles, k can be figured out:

(2) 如果管外表面传热系数 h_o 增加1倍，可求出：

$$k=\frac{1}{0.000205+0.00555}\text{W/(m}^2\cdot\text{K)}=174\text{W/(m}^2\cdot\text{K)}$$

The overall heat transfer coefficient has increased by 96%.

传热系数增加了96%。

(3) If the convective heat transfer coefficient h_i in the tube doubles, k can be figured out:

(3) 如果管内表面传热系数 h_i 增加1倍，可求出：

$$k=\frac{1}{\frac{1}{12000}\times\frac{16}{13}+0.0111}\text{W/(m}^2\cdot\text{K)}=89.3\text{W/(m}^2\cdot\text{K)}$$

The overall heat transfer coefficient has increased by less than 1% . The calculation results show that when the thermal resistance of each link is different in a heat transfer process, it is necessary to find the link with the largest thermal resistance and enhance it, then the heat transfer can be enhanced significantly.

传热系数增加了不到1%。计算结果表明，当一个传热过程的各个环节热阻不同时，需要找出分热阻最大的环节进行强化，才能收到显著的强化传热效果。

Summary This section illustrates the thermal resistance separation method in the heat transfer process, that is Wilson graphical method.

小结 本节阐述了换热器传热过程的热阻分离法，即威尔逊图解法。

Question What is the basic idea of solving the partial thermal resistance of the heat transfer process by the Wilson graphical method?

思考题 用威尔逊图解法求解传热过程分热阻的基本思路是什么？

Self-tests

自测题

1. In order to enhance the heat transfer of an oil cooler, some people use the method of increasing the cooling water flow rate, but it is found that the effect is not significant, try to analyze the reasons.

1. 为强化一台冷油器的传热，有人用提高冷却水流速的办法，但发现效果并不显著，试分析原因。

2. There is a steel tube heat exchanger, the hot water flows in the tube, and the air is deflected several times in the tube bundles and scours the tube bundle horizontally to cool the hot water in the tube. It has been proposed that in order to improve the cooling effect, fins are added outside the tube and the steel tube is replaced by the copper tube. Please evaluate the rationality of this plan.

2. 有一台钢管换热器，热水在管内流动，空气在管束间进行多次折流横向冲刷管束，以冷却管内热水。有人提出，为提高冷却效果，采用管外加装肋片并将钢管换成铜管。请你评价这一方案的合理性。

Exercises

习题

10.1 A horizontal condenser is made of heat exchange surface with brass tube with outer diameter of 25mm and wall thickness of 1.5mm. Known: The average surface heat transfer coefficient on the condensate side outside the tube $h_o = 5700W/(m^2 \cdot K)$, and the average surface heat transfer coefficient on the water side outside the tube $h_i = 4300W/(m^2 \cdot K)$. Try to calculate the overall heat transfer coefficient of the condenser in the following two cases according to the outer surface area of the tube. (1) The inner and outer surfaces of the tube are clean; (2) Seawater flows in the tube with flow rate greater than 1m/s, will scale, and its average temperature is less than 50℃. Steam contains oil.

10.1 一台卧式冷凝器采用外径为25mm、壁厚为1.5mm的黄铜管做成换热表面。已知：管外冷凝侧的平均表面传热系数 $h_o = 5700W/(m^2 \cdot K)$，管内水侧的平均表面传热系数 $h_i = 4300W/(m^2 \cdot K)$。试按管子外表面面积计算下列两种情况下冷凝器的总传热系数。（1）管子内外表面均是洁净的；（2）管内为海水，流速大于1m/s，会结水垢，平均温度小于50℃；蒸汽内含油。

10.2 It is known that $t_1' = 300℃$, $t_1'' = 210℃$, $t_2' = 100℃$, $t_2'' = 200℃$, try to calculate the logarithmic mean temperature difference of heat exchanger in the following flow arrangement: (1) Counter flow arrangement; (2) A cross flow in which neither fluid is mixed; (3) Type 1-2 shell and tube with thermal fluid on the shell side; (4) Type 2-4 shell and tube with thermal fluid on the shell side; (5) Parallel flow arrangement.

10.2 已知 $t_1' = 300℃$，$t_1'' = 210℃$，$t_2' = 100℃$，$t_2'' = 200℃$，试计算下列流动布置时换热器的对数平均温差：（1）逆流布置；（2）一次交叉，两种流体均不混合；（3）1-2型管壳式，热流体在壳侧；（4）2-4型管壳式，热流体在壳侧；（5）顺流布置。

10.3 In order to use the waste gas to heat the domestic water, a factory made a simple shell tube heat exchanger. The flue gas flows in the steel tube with an inner diameter of 30mm, the flow rate is 30m/s, the inlet temperature is 200℃, and the outlet temperature is 100℃. The cold water flows backwards with the flue gas in the space

10.3 某工厂为了利用废气来加热生活用水，自制了一台简易的管壳式换热器。烟气在内径为30mm的钢管内流动，流速为30m/s，入口温度为200℃，出口温度取为100℃。冷水在管束的空间内与烟气逆向流

of the tube bundle, which is required to be heated to 50℃ from 20℃ at thc cntrancc. Estimate the required straight tube length. The physical properties of the flue gas can be checked according to the physical properties of the standard flue gas. The surface heat transfer coefficient of the water side is much greater than that of the flue gas side, and the radiation heat transfer of the flue gas can be slightly less than.

动，从入口处 20℃ 要求加热到 50℃，估算所需的直管长度。烟气物性可按标准烟气的物性查取。水侧的表面传热系数远大于烟气侧的表面传热系数，烟气的辐射换热可忽略。

10.4 In a counter flow water-water heat exchanger, hot water is in the tube, inlet temperature $t_1' = 100℃$, and outlet temperature $t_1'' = 80℃$. Cold water is outside the tube, inlet temperature $t_2' = 20℃$, and outlet temperature $t_2'' = 70℃$. Total heat transfer rate $\Phi = 350\text{kW}$, there are a total of 53 tubes with an inner diameter of 16mm and wall thickness of 1mm. The thermal conductivity of the tube wall $\lambda = 40\text{W/(m·K)}$, the surface heat transfer coefficient of the fluid outside the tube $h_o = 1500\text{W/(m}^2\text{·K)}$, and the fluid inside the tube is a flow process. The inner and outer surfaces of the tube are clean. Try to determine the required tube length.

10.4 在一逆流式水-水换热器中，管内为热水，进口温度 $t_1' = 100℃$，出口温度 $t_1'' = 80℃$；管外流过冷水，进口温度 $t_2' = 20℃$，出口温度 $t_2'' = 70℃$。总换热量 $\Phi = 350\text{kW}$，共有 53 根内径为 16mm、壁厚为 1mm 的管子。管壁导热系数 $\lambda = 40\text{W/(m·K)}$，管外流体的表面传热系数 $h_o = 1500\text{W/(m}^2\text{·K)}$，管内流体为一个流程。管子内、外表面都是洁净的。试确定所需的管子长度。

10.5 A 1-2 type shell and tube heat exchanger uses water at 30℃ to cool hot oil at 120℃ ($c_p = 2100\text{J/(kg·K)}$), the cooling water flow rate is 1.2kg/s, and the oil flow rate is 2kg/s. The overall heat transfer coefficient $k = 275\text{W/(m}^2\text{·K)}$, and the heat transfer area $A = 20\text{m}^2$. Try to determine the respective outlet temperature of water and oil.

10.5 一台 1-2 型管壳式换热器用 30℃ 的水来冷却 120℃ 的热油 [$c_p = 2100\text{J/(kg·K)}$]，冷却水流量为 1.2kg/s，油流量为 2kg/s。设总传热系数 $k = 275\text{W/(m}^2\text{·K)}$，传热面积 $A = 20\text{m}^2$。试确定水与油各自的出口温度。

10.6 In order to use the exhaust gas of the gas turbine to heat the high-pressure water, a 1-2 type fin tube heat exchanger is used. The mass flow rate of exhaust gas obtained in one measurement is 2kg/s, the inlet temperature $t_1' = 325℃$. The mass flow rate of cooling water is 0.5kg/s, $t_2' = 25℃$, $t_2'' = 150℃$. The heat transfer area calculated according to the diameter of the base pipe on the gas side is 3.8m^2. The physical properties of exhaust

10.6 为利用燃气轮机的排气来加热高压水，采用 1-2 型肋片管换热器。在一次测定中得到排气的质量流量为 2kg/s，进口温度 $t_1' = 325℃$；冷却水质量流量为 0.5kg/s，$t_2' = 25℃$，$t_2' = 150℃$。按气体侧基管直径计算的换热面积为 3.8m^2，排气物性可近似按标准烟气的值查取。

gas can be approximated by the standard flue gas values. Try to calculate the overall heat transfer coefficient under the condition.

试计算该条件下的总传热系数。

10.7 There is a liquid-liquid heat exchanger, and two kinds of media, A and B, are used for forced convection heat exchange inside and outside the tube respectively. The variation of the heat transfer coefficient with the flow rate of the two liquids measured by the test is shown in the attached figure. Try to analyze which side of the heat exchanger is the main thermal resistance?

10.7 有一台液-液换热器，A、B两种介质分别在管内、外进行强制对流换热。实验测得的传热系数与两种液体流速的变化情况如图10.39所示。试分析该换热器的主要热阻在哪一侧？

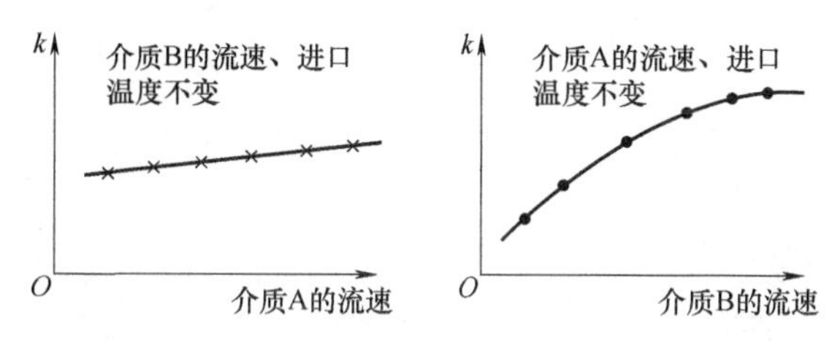

Fig. 10.39 Figure for exercise 10.7

图10.39 习题10.7图

Appendix (QR Code)

附　录（二维码）

Appendix A　Heat Transfer Common Information (Units, Physical Properties, Functions)

附录 A　传热学常用资料（单位、物性、函数）

表 A-1　常用单位换算表

表 A-2　部分金属材料的密度、比热容和导热系数

表 A-3　保温、建筑及其他材料的密度和导热系数

表 A-4　几种保温、耐火材料的导热系数与温度的关系

表 A-5　大气压力（$p=1.01325\times10^5$ Pa）下干空气的热物理性质

表 A-6　大气压力（$p=1.01325\times10^5$ Pa）下标准烟气的热物理性质

表 A-7　大气压力（$p=1.01325\times10^5$ Pa）下过热水蒸气的热物理性质

表 A-8　生物材料的热物理性质

表 A-9　饱和水的热物理性质

表 A-10　干饱和水蒸气的热物理性质

表 A-11　几种饱和液体的热物理性质

表 A-12　几种液体的体胀系数

表 A-13　液态金属的热物理性质

表 A-14　第一类贝塞尔（Bessel）函数选择

表 A-15　误差函数选摘

表 A-16　常用材料的表面发射率

表 A-17　常用换热器传热系数的大致范围

表 A-18　黑体辐射函数表

Appendix B　English-Chinese Comparison of Heat Transfer Termi-nology

附录 B　传热学专业词汇英汉对照（按中文拼音排序）

Appendix C English-Chinese Comparison of the Physical Meanings of the Main Symbols in This Book

附录C 本书主要符号物理意义英汉对照

参考文献

References

［1］ 杨世铭，陶文铨．传热学［M］．4版．北京：高等教育出版社，2006.

［2］ 霍尔曼．传热学：原书第10版［M］．北京：机械工业出版社，2011.

［3］ 邓元望，唐爱坤．传热学［M］．北京：机械工业出版社，2021.

［4］ 苏亚欣．传热学［M］．北京：机械工业出版社，2023.

［5］ 刘彦丰，高正阳，梁秀俊．传热学［M］．2版．北京：中国电力出版社，2021.

［6］ 邬田华，王晓墨，许国良．工程传热学［M］．2版．武汉：华中科技大学出版社，2020.

［7］ 王秋旺，曾敏．传热学要点与题解［M］．西安：西安交通大学出版社，2006.

［8］ 周根明．传热学学习指导及典型习题分析［M］．北京：中国电力出版社，2004.

［9］ 吴金星，韩东方，曹海亮．高效换热器及其节能应用［M］．北京：化学工业出版社，2009.

［10］ 钱颂文．换热器设计手册［M］．北京：化学工业出版社，2006.

［11］ 林宗虎，汪军，李瑞阳，等．强化传热技术［M］．北京：化学工业出版社，2007.

［12］ 尾花英朗．热交换器设计手册［M］．徐中权，译．北京：烃加工出版社，1987.

［13］ 张利，李友荣．换热器原理与计算［M］．北京：中国电力出版社，2017.

［14］ 吴金星，刘敏珊，董其伍，等．换热器管束支撑结构对壳程性能的影响［J］．化工机械，2002，29（2）：108-112.

［15］ 吴金星，魏新利，朱登亮．流体旋转流动强化对流传热场协同分析［J］．冶金能源，2006，25（1）：18-20.

［16］ 吴金星，董其伍，刘敏珊，等．折流杆换热器壳程湍流和传热的数值模拟［J］．高校化学工程学报，2006，20（2）：213-216.

［17］ 吴金星，朱登亮，魏新利，等．螺旋肋片自支撑换热器强化换热试验研究［J］．热能动力工程，2008，23（2）：157-160.

［18］ 吴金星，魏新利，董其伍，等．花瓣孔板纵流式换热器的研发及试验研究［J］．高校化学工程学报，2008，22（2）：205-209.

［19］ WU J X，LIU S L，WANG M Q. Process calculation method and optimization of the spiral-wound heat exchanger with bilateral phase chang［J］. Applied Thermal Engineering，2018，134：360-368.

［20］ WU J X，SUN X Z，LI X，et al. Research on film condensation heat transfer at the inside of spiral coil tube［J］. International Journal Heat and Mass Transfer，2020，147：1-7.